An Introduction to Spinors and Geometry with Applications in Physics

An Introduction to Spinors and Geometry with Applications in Physics

I M Benn

Faculty of Science,
University College of The Northern Territory, Australia

R W Tucker

Department of Physics,
University of Lancaster, UK

Adam Hilger, Bristol and New York

British Library Cataloguing in Publication Data
Benn, I. M.
An introduction to spinors and geometry
with applications in physics.
1. Spinor analysis
I. Title II. Tucker, R. W.
512'.57 QA433

ISBN 0-85274-169-3
ISBN 0-85274-261-4 (pbk)

Library of Congress Cataloging-in-Publication Data
Benn, I. M. (Ian M.)
An introduction to spinors and geometry with applications in physics
Bibliography: p.
Includes index.
1. Spinor analysis. 2. Geometry, Differential
I. Tucker, R. W. (Robin W.) II. Title.
QC20.7.S65B46 1988 515'.63 87-21117

ISBN 0-85274-169-3
ISBN 0-85274-261-4 (pbk)

Consultant Editor: **Professor R F Streater**
King's College, London

First published 1987
Paperback edition 1989

Published under the Adam Hilger imprint by IOP Publishing Ltd
Techno House, Redcliffe Way, Bristol BS1 6NX, England
335 East 45th Street, New York, NY 10017–3483, USA

Typeset by KEYTEC, Bridport, Dorset
Printed and bound in Great Britain by
Butler & Tanner Ltd, Frome and London

To Shian
To Daniel and Edmund

Contents

Preface		**ix**
1	**Tensor Algebra**	**1**
1.1	The tensor algebra	2
1.2	The exterior algebra of antisymmetric tensors	4
1.3	The exterior algebra as a quotient of the tensor algebra	10
1.4	The Hodge map	13
1.5	The mixed tensor algebra	16
	Bibliography	20
2	**Clifford Algebras and Spinors**	**21**
2.1	The Clifford algebra	23
2.2	The structure of the real Clifford algebras	28
2.3	The even subalgebra	39
2.4	The Clifford group	42
2.5	Spinors	54
2.6	Spin-invariant inner products	62
2.7	The complexified Clifford algebras	80
2.8	The confusion of tongues	85
	Bibliography	105
3	**Pure Spinors and Triality**	**106**
3.1	Pure spinors	106
3.2	Triality	117
	Bibliography	122
4	**Manifolds**	**123**
4.1	Topological manifolds	124
4.2	Derivatives of functions $\mathbb{R}^m \to \mathbb{R}^n$	127

4.3 Differentiable manifolds 129
4.4 Parametrised curves 134
4.5 Tangent vectors 136
4.6 Vector fields 141
4.7 The tangent bundle 143
4.8 Differential 1-forms 146
4.9 Tensor fields 150
4.10 Exterior derivatives 154
4.11 One-parameter diffeomorphisms and integral curves 156
4.12 Lie derivatives 161
4.13 Integration on manifolds 167
4.14 Metric tensor fields 171
Bibliography 174

5 Applications in Physics **176**
5.1 Galilean spacetimes 176
5.2 Maxwell's equations and Minkowski spacetime 178
5.3 Observer curves 183
5.4 Electromagnetism 188
Bibliography 197

6 Connections **199**
6.1 Linear connections 200
6.2 Examples and Newtonian force 204
6.3 Covariant differentiation of tensors 206
6.4 Curvature and torsion tensors of ∇ 208
6.5 Bianchi identities 212
6.6 Metric-compatible connections 214
6.7 The covariant exterior derivative 216
6.8 The curvature scalar and Einstein tensor 219
6.9 The pseudo-Riemannian connection 221
6.10 Sectional curvature 223
6.11 The conformal tensor 225
6.12 Some curvature relations in low dimensions 227
6.13 Killing's equation 229
Bibliography 231

7 Gravitation **232**
7.1 Lorentzian connections 232
7.2 Fermi–Walker transport 234
7.3 The Einstein field equations 234
7.4 Conservation laws 237
7.5 Some matter fields 239
7.6 The Reissner–Nordström solution 243
7.7 Gravitation with torsion 249
Bibliography 250

8 Clifford Calculus on Manifolds **251**
8.1 Covariant differentiation of Clifford products 252
8.2 The operator $\not{d}$ 254
8.3 The Kahler equation 256
8.4 The Duffin–Kemmer–Petiau equations 260
Bibliography 260

9 Spinor Fields **261**
9.1 Spinor bundles 261
9.2 Inner products on spinor fields 264
9.3 Covariant differentiation of spinor fields 267
9.4 Lie derivatives of spinor fields 271
9.5 Representing spinor fields with differential forms 275
Bibliography 277

10 Spinor Field Equations **278**
10.1 The Dirac operator 278
10.2 Covariances of the Dirac equation and conserved currents 280
10.3 The Dirac equation in spacetime 282
10.4 The stress tensor 290
10.5 Tensor spinors 294
10.6 The Lichnerowicz theorem 299
10.7 Killing spinors 300
10.8 Parallel spinors 303

Appendix A: Algebra **307**
Bibliography 341

Appendix B: Vector Calculus on $\mathbb{R}^3$ **342**

References **349**

Index **351**

Preface

A student of theoretical physics who wishes to follow recent trends in current research is liable to be confronted with a bewildering amalgam of ideas from physics and mathematics. In particular, much of the terminology permeating developments in the theories of matter and gravitation is borrowed from classical differential geometry. In many of these theories spinors play a prominent role. A further notable development is the introduction of spaces with 'exotic' topologies and geometries in formulating the basic laws of Nature. Consequently the student finds it necessary to possess a broad knowledge of mathematical techniques that encompasses such generalities as well as the computational skills necessary to use this information.

In this book we have attempted to provide a concise but self-contained introduction to the basic properties of differential geometry and spinors accommodating some of the needs mentioned above. We feel that physicists learn most rapidly by seeing new concepts spelled out in some detail. We have attempted a blend of mathematics and theoretical physics which we hope will assist in the assimilation of new ideas and give readers a feeling that they are closer to the 'nuts and bolts' of the subject material. In writing any introduction to a subject as broad as this we have had to face the problem of what prerequisites we expect our readers to possess. Fundamental to any appreciation of tensor methods is a firm familiarity with linear algebra. Thus our book begins with algebraic notions. We have tried to encapsulate the necessary concepts used in Chapters 1 and 2 into Appendix A. This should provide a reservoir of compact information for those who may find some foreign vocabulary in these early chapters. Our emphasis here is on real vector spaces and their complexifications. We feel that this approach makes closest contact with what most physicists actually use when working with the complexified Clifford algebra of spacetime. We introduce a spinor as an element carrying an irreducible representation of some Clifford algebra. This emphasis on the Clifford algebras rather than the spin groups is slightly different from that commonly adopted by most working physicists. However, the spin groups are most easily defined as sitting in the Clifford algebra, and thus we may induce representations of these groups from those of the algebras. No doubt some readers will be surprised at the classical tone that dominates our description of spinors. We offer little apology. As a mathematical entity the notion of a spinor requires no quantum theoretical overtones. Although we would have liked to develop further the basic role played by spinors in quantum field theory we feel that their role in physical models need not intrude into their basic relation to geometry. Moreover, a proper appreciation of this relation is essential in relativistic quantum field theory.

The introduction to differential manifolds (Chapter 4) is fairly elementary and presupposes only a basic knowledge of the calculus of many variables. We have interrupted its development with a chapter on physical applications before formally introducing the idea of a linear connection. This chapter illustrates the importance of Lorentzian geometry in relativistic physics, and is motivated by a discussion of electromagnetism. Chapter 7 is devoted to the field theory of gravitation and its sources in which many of the mathematical tools introduced earlier are put to use. The two main themes of Clifford algebras and differentiable manifolds are drawn together in the final chapters on Clifford forms and spinor fields. Here readers will find physical applications involving spinors on manifolds and are introduced to some recent developments that relate geometrical properties of a space to the existence of spinor fields with particular properties. Earnest readers are invited to test their expertise by working out some of the illustrative examples that have been inserted at strategic points in the text.

In the course of writing this book we have benefited from dialogues with many colleagues. In particular, we wish to thank Graeme Segal, R Al-Saad, J Brooke, C T J Dodson, E Kahler, K McCrimmond, and D Towers for helpful comments on various aspects of our enterprise. We are also grateful for correspondence with A Crumeyrolle, K McKenzie and D Plyman on aspects of Clifford algebras. The production of our manuscript was greatly assisted with the aid of TEXnical facilities generously provided by A B Clegg and P M Lee. We also thank G Hughes for all the time and effort he spent teaching us to drive the Vax-editor and its peripherals. Finally, we are happy to acknowledge the support provided by the University of Lancaster Research Fund.

I M Benn
R W Tucker

1

Tensor Algebra

This first chapter will provide a foundation for the two initially separate directions the book will take; algebra and geometry. In Appendix A we have gathered together a number of ideas relating to the study of vector spaces and algebras. These notions will be used freely within the first two chapters. The reader who is initially confronted with foreign vocabulary or new concepts should consult this Appendix for definitions where a concise development of rudimentary ideas is also to be found.

The first section of this chapter introduces the tensor algebra of an arbitrary vector space. In Chapter 2 this will be the starting point for our construction of the Clifford algebra, which will be defined as a quotient of the tensor algebra. In Chapter 4 and subsequent chapters when beginning geometry we will be interested in the tangent space (and the cotangent space) of a manifold. We will then be able to apply the material of this chapter immediately to that vector space. In fact it will be the cotangent space that is taken for the arbitrary vector space V. Anticipating this we have (identifying the second dual space of V with itself) written elements of V as acting on V^*, rather than the other way around.

Particularly important on manifolds are the totally antisymmetric tensor fields; the differential forms. In §1.2 we introduce the exterior forms on an arbitrary vector space. These will also play a prominent role in our treatment of the Clifford algebra. To facilitate a comparison with the Clifford algebra we re-introduce the exterior algebra in §1.3 as a quotient of the tensor algebra.

Only in §1.4 does a metric enter. (Our meaning of a metric is given in Appendix A.) This allows us to introduce the Hodge map which is a key ingredient of the calculus of differential forms on (pseudo-) Riemannian manifolds.

We have delayed introducing the mixed tensor algebra until §1.5. Here contact is made with the classical definition of a tensor in terms of transformation properties of components. Index conventions will be established that allow the traditional 'raising and lowering' of indices.

1.1 The Tensor Algebra

If V is any vector space over some field F then the set of F-valued linear maps on V forms a vector space; the dual space, V^*. If, as we now assume, V is finite dimensional then there is a natural way to regard elements of V as linear maps on V^*. That is, if $x \in V$ and $X \in V^*$ such that X acts on x to produce the scalar $X(x)$ then we can equivalently think of this as defining an action of x on X, $x(X) = X(x)$. In the following it will be convenient to adopt this seemingly perverse view of regarding V as the space of linear mappings on V^*. Just as the F-valued linear maps on V^* form a vector space so do the multilinear maps on ordered sets of elements from V^*. The F-valued multilinear maps on $V^* \times V^* \times \ldots \times V^*$ (r times) are called *tensors* of degree r. The notion of multilinearity is an obvious extension of the notion of a linear map; for any fixed choice of $r-1$ elements of V^* the map is linear in the remaining variable. Multilinearity ensures that a tensor of degree r is completely specified by its action on all ordered sets of basis vectors for V^*, thus if V (and hence V^*) is n-dimensional then the tensors of degree† r form an n^r-dimensional vector space, $T_r(V)$.

We may associate a set of r elements from V with a tensor of degree r. For $x^i \in V$, $i = 1, \ldots, r$ and $X_i \in V^*$, $i = 1, \ldots, r$ we define

$$(x^1 \otimes x^2 \otimes \ldots \otimes x^r)(X_1, X_2, \ldots, X_r) = x^1(X_1)x^2(X_2) \ldots x^r(X_r).$$

In particular, if $\{e^i\}$ is a basis for V then the set of all n^r elements $\{e^{i_1} \otimes e^{i_2} \otimes \ldots \otimes e^{i_r}\}$, where the indices take all values from 1 to n, forms a basis for $T_r(V)$. The vector space $T_r(V)$ is called the tensor product of V r times

$$T_r(V) = V \otimes V \ldots \otimes V \equiv \otimes^r V.$$

More generally, the tensor product defines a mapping between tensors of different degrees

$$\otimes : T_r(V) \times T_s(V) \longrightarrow T_{r+s}(V) \qquad (1.1.1)$$
$$a, b \longmapsto a \otimes b$$

where

† Formerly called rank.

$$(a \otimes b)(X_1, X_2, \ldots, X_r, X_{r+1}, \ldots, X_{r+s})$$
$$= a(X_1 \ldots, X_r)b(X_{r+1}, \ldots, X_{r+s}).$$

We may take the (external) direct sum of the vector spaces $T_r(V)$ for all r to form an infinite-dimensional vector space. The direct sum of such a vector space with a one-dimensional space spanned by an identity element forms an associative (but not commutative) algebra under the tensor product, the *tensor algebra* $T(V)$. The subspace spanned by the identity is written as $T_0(V)$, and since this is just another copy of the field F with an identical rule for multiplication on tensors we shall not distinguish between these two spaces. The tensor algebra is generated by V and the identity element; any element can be written as a sum of tensor products of elements from V and the identity. Those tensors that are simply a product of vectors from V are called *decomposable*. By construction we have the direct sum vector space decomposition

$$T(V) = \sum_{p=0}^{\infty} \oplus T_p(V).$$

The tensor product is such that the tensor algebra is a $\mathbb{Z}$-graded algebra; elements in $T(V)$ that are sums of products of p elements from V being homogeneous of degree p. The zero element (which is homogeneous for every degree) is the only term that is homogeneous for negative degree. The grading naturally gives rise to an involutary *automorphism* η defined on homogeneous elements by†

$$\eta a = (-1)^{\deg a} a. \tag{1.1.2}$$

This is certainly an automorphism since if a and b are homogeneous

$$\begin{aligned}\eta(a \otimes b) &= (-1)^{\deg a \otimes b} a \otimes b \\ &= (-1)^{\deg a + \deg b} a \otimes b \text{ (since the algebra is graded)} \\ &= (-1)^{\deg a}(-1)^{\deg b} a \otimes b\end{aligned}$$

and so

$$\eta(a \otimes b) = \eta a \otimes \eta b. \tag{1.1.3}$$

To say that η is involutary means that $\eta^2 = 1$, which indeed follows from (1.1.2). The homomorphism $\mathbb{Z} \to \mathbb{Z}_2$ induces a coarser $\mathbb{Z}_2$-gradation in $T(V)$. The $\mathbb{Z}_2$-homogeneous subspaces consist of the sum of all $\mathbb{Z}$-homogeneous subspaces of even (odd) degree. Thus the $\mathbb{Z}_2$-homogeneous subspaces are eigenspaces for the automorphism η with eigenvalues plus (minus) one. Elements of these spaces will be called even or odd, respectively.

† The notation a^{η} is also employed.

The tensor algebra is isomorphic to its opposite algebra† and admits an involutary anti-automorphism, or simply an *involution*, ξ defined on homogeneous elements by

$$(x^1 \otimes x^2 \ldots \otimes x^p)^\xi = x^p \otimes \ldots \otimes x^2 \otimes x^1. \tag{1.1.4}$$

It is straightforward to see that this really is an anti-automorphism, namely $(a \otimes b)^\xi = b^\xi \otimes a^\xi$ such that $\xi^2 = 1$.

If X is in V^* then the *interior derivative* with respect to X is denoted i_X. It is defined to be a linear transformation that is an anti-derivation with respect to the automorphism η, that is

$$\mathrm{i}_X(a \otimes b) = \mathrm{i}_X a \otimes b + \eta a \otimes \mathrm{i}_X b. \tag{1.1.5}$$

If $x \in V$ then $\mathrm{i}_X x \equiv X(x)$, whilst for λ in the subspace spanned by the identity $\mathrm{i}_X \lambda \equiv 0$, and so the interior derivative is a homogeneous linear mapping on $T(V)$ (with respect to the $\mathbb{Z}$-gradation) of degree -1. These properties completely characterise the interior derivative. Since i_X is an anti-derivative with respect to the involution η, with $\mathrm{i}_X \eta = -\eta \mathrm{i}_X$, it follows that $\mathrm{i}_X \mathrm{i}_Y + \mathrm{i}_Y \mathrm{i}_X$ is a derivation on $T(V)$. For $x \in V$ or the subspace spanned by the identity $(\mathrm{i}_X \mathrm{i}_Y + \mathrm{i}_Y \mathrm{i}_X)x = 0$, and since $T(V)$ is generated by this space

$$(\mathrm{i}_X \mathrm{i}_Y + \mathrm{i}_Y \mathrm{i}_X)a = 0 \quad \text{for all } a \in T(V). \tag{1.1.6}$$

In particular $\mathrm{i}_X \mathrm{i}_X = 0$.

1.2 The Exterior Algebra of Antisymmetric Tensors

A tensor is a multilinear mapping on an ordered set of vectors, the ordering being in general important. Many important tensors have symmetries, however, the result of the evaluation on a set of vectors being invariant under the interchange of certain pairs of vectors. To formalise this we introduce the interchange permutation π_{jk}, which rearranges the set of numbers $\{1, 2, \ldots, p\}$ such that $\pi_{jk}(i) = i$ if $i \neq j$ or k, $\pi_{jk}(j) = k$ and $\pi_{jk}(k) = j$. Then a degree-p tensor T is symmetric (antisymmetric) in the j,k entries if

$$T(X_{\pi_{jk}(1)}, X_{\pi_{jk}(2)}, \ldots, X_{\pi_{jk}(p)}) = +(-)T(X_1, X_2, \ldots, X_p).$$

A tensor that is symmetric (antisymmetric) under all such interchanges is called totally symmetric (totally antisymmetric). The totally antisymmetric tensors are particularly important. The subspace of totally

† See Appendix A.

antisymmetric tensors in $T_p(V)$ is denoted by $\Lambda_p(V)$, the elements of this space being called *exterior p-forms*, or simply p-forms. The total antisymmetry ensures that a p-form is determined by its evaluation on all distinct combinations of p vectors from a basis for V^*. So if V is n-dimensional and

$$\binom{n}{p}$$

denotes the number of distinct combinations of p objects chosen from n then

$$\dim \Lambda_p = \binom{n}{p}.$$

In particular, the only p-forms for $p>n$ are zero, and $\dim \Lambda_n = 1$. In analogy with the case of the tensor algebra it will be convenient to identify the field F with a space $\Lambda_0(V)$.

Given an arbitrary element $a \in T_p(V)$, we define a new tensor $\mathcal{ALT}\, a \in T_p(V)$ by

$$\mathcal{ALT}\, a(X_1, X_2, \ldots, X_p) = \frac{1}{p!}\sum_\sigma \varepsilon(\sigma) a(X_{\sigma(1)}, X_{\sigma(2)}, \ldots, X_{\sigma(p)}) \qquad \forall X_i \in V^* \quad (1.2.1)$$

where the sum is over all permutations σ, $\varepsilon(\sigma)$ being $+1$ if this permutation is even (i.e. an even number of pair interchanges rearranges the elements $1, 2, \ldots, p$ into the order $\sigma(1), \sigma(2), \ldots, \sigma(p)$) or -1 if the permutation is odd (an odd number of such interchanges). From the definition of $\mathcal{ALT} a$ we see that it is totally antisymmetric and that $\mathcal{ALT}(\mathcal{ALT} a) = \mathcal{ALT} a$. Hence $\mathcal{ALT}$ is a projection operator, $\mathcal{ALT}: T_p(V) \rightarrow \Lambda_p(V)$. Although $\otimes: T_s(V) \times T_t(V) \rightarrow T_{s+t}(V)$, the map $\otimes$ will not map $\Lambda_s(V) \times \Lambda_t(V)$ into $\Lambda_{s+t}(V)$. Thus we devise a new composition map in terms of $\otimes$ and $\mathcal{ALT}$ that does have this property. It is called the *exterior product*† and is denoted by a $\wedge$ placed between the elements of $\Lambda_s(V)$ and $\Lambda_t(V)$

$$\wedge : \Lambda_s(V) \times \Lambda_t(V) \longrightarrow \Lambda_{s+t}(V)$$
$$a, b \longmapsto a \wedge b = \mathcal{ALT}(a \otimes b). \quad (1.2.2)$$

† The reader is cautioned that there are other conventions for the definition of the exterior product. Other conventions involve a numerical factor which depends on the degrees of a and b. The reader should convince himself that such numerical factors cannot be arbitrarily inserted with impunity! (Why not?) The convention we have adopted is convenient for regarding the exterior algebra as a quotient of the tensor algebra modulo the kernel of $\mathcal{ALT}$, as we shall do in the next section.

It follows from this definition that

$$(a \wedge b)(X_1, X_2, \ldots X_s, X_{s+1}, \ldots, X_{s+t})$$

$$= \frac{1}{(s+t)!} \sum \varepsilon(\sigma)(a \otimes b)(X_{\sigma(1)}, \ldots, X_{\sigma(s+t)})$$

$$= \frac{1}{(s+t)!} \sum \varepsilon(s, t) a(X_{j_1}, \ldots, X_{j_s}) b(X_{k_1}, \ldots, X_{k_t})$$

where the sum is over all partitions of $(1, 2, \ldots, s+t)$ into $(j_1, j_2, \ldots, j_s)$ and $(k_1, k_2, \ldots, k_t)$, and $\varepsilon(s, t)$ is the sign of the permutation

$$(1, 2, \ldots, s+t) \longmapsto (j_1, j_2, \ldots, j_s, k_1, k_2, \ldots, k_t).$$

The exterior product has a well defined symmetry under the interchange of factors such as a and b above. To see this we introduce a permutation v,

$$(1, 2, \ldots, s, s+1, \ldots, s+t) \overset{v}{\longmapsto} (t+1, \ldots, t+s, 1, 2, \ldots, t).$$

We write any permutation σ as $\sigma = \tau v$, giving $\varepsilon(\sigma) = \varepsilon(v)\varepsilon(\tau)$. Inserting this in the above gives

$$(a \wedge b)(X_1, X_2, \ldots, X_s, X_{s+1}, \ldots, X_{s+t})$$

$$= \frac{1}{(s+t)!} \sum \varepsilon(\sigma) a(X_{\sigma(1)}, \ldots, X_{\sigma(s)}) b(X_{\sigma(s+1)}, \ldots, X_{\sigma(s+t)})$$

$$= \frac{\varepsilon(v)}{(s+t)!} \sum_{\tau} \varepsilon(\tau) a(X_{\tau(t+1)}, \ldots, X_{\tau(t+s)}) b(X_{\tau(1)}, \ldots, X_{\tau(t)})$$

$$= \varepsilon(v)(b \wedge a)(X_1, \ldots, X_{s+t}).$$

A trivial combinatorial calculation gives $\varepsilon(v) = (-1)^{st}$, and so we have for any s-form a and t-form b

$$a \wedge b = (-1)^{st} b \wedge a. \tag{1.2.3}$$

The *exterior algebra* $\Lambda(V)$ is formed by the direct vector space sum of all the spaces of p-forms

$$\Lambda(V) = \sum_{p=0}^{n} \oplus \Lambda_p(V)$$

with multiplication given by the exterior product. The exterior product is defined on non-homogeneous elements by extending $\mathscr{ALT}$ to be distributive over addition, ensuring that the exterior product is. Unlike the tensor algebra this algebra is finite dimensional: we have

$$\dim \Lambda(V) = \sum_{p=0}^{n} \binom{n}{p} = 2^n. \tag{1.2.4}$$

The exterior algebra is, in fact, associative. This will be seen to follow from the observation that if $\mathscr{ALT}m = 0$ then $\mathscr{ALT}(a \otimes m) = \mathscr{ALT}(m \otimes a) = 0$, $\forall\ a \in T(V)$, as will now be established.

Let C_k be the group of all permutations of k objects. Then the subgroup of C_{s+t} that only permutes the first s objects is obviously isomorphic to C_s, and we shall identify it as such. Let H be a set that contains one and only one element from each left coset of C_{s+t} relative to C_s. So for each $\sigma \in C_{s+t}$ $\sigma = h\tau$, $\tau \in C_s$ and $h \in \mathrm{H}$, with $\varepsilon(\sigma) = \varepsilon(\tau)\varepsilon(h)$, and then

$$\mathscr{ALT}(m \otimes a)(X_1, \ldots, X_{s+t}) = \frac{1}{(s+t)!} \sum_{h \in \mathrm{H}} \varepsilon(h) \sum_{\tau \in C_s} \varepsilon(\tau)(m \otimes a)(X_{\sigma(1)}, \ldots, X_{\sigma(s+t)}).$$

For some fixed h let $X_{h(i)} = Y_i$, then $X_{\sigma(i)} = X_{h\tau(i)} = Y_{\tau(i)}$, and thus

$$\begin{aligned}\sum_{\tau \in C_s} \varepsilon(\tau)(m \otimes a)(X_{\sigma(1)}, \ldots, X_{\sigma(s+t)}) &= \sum_{\tau \in C_s} \varepsilon(\tau)(m \otimes a)(Y_{\tau(1)}, \ldots, Y_{\tau(s+t)}) \\ &= \sum_{\tau \in C_s} \varepsilon(\tau) m(Y_{\tau(1)}, \ldots, Y_{\tau(s)}) a(Y_{s+1}, \ldots, Y_{s+t}). \\ &= \mathscr{ALT}m(Y_1, \ldots, Y_s) a(Y_{s+1}, \ldots, Y_{s+t}).\end{aligned}$$

So indeed $\mathscr{ALT}(m \otimes a) = 0$ if $\mathscr{ALT}m = 0$. Similarly, it follows that $\mathscr{ALT}(a \otimes m) = 0$.

Since, as we have remarked, $\mathscr{ALT}$ is a projection operator, if we set $(1 - \mathscr{ALT})(a \otimes b) = m$ then $a \otimes b = \mathscr{ALT}(a \otimes b) + m$, with $\mathscr{ALT}m = 0$. From the definition of the exterior product we have

$$\begin{aligned}(a \wedge b) \wedge c &= \mathscr{ALT}\ (\mathscr{ALT}\ (a \otimes b) \otimes c) \\ &= \mathscr{ALT}\ (a \otimes b \otimes c - m \otimes c)\end{aligned}$$

since $\otimes$ is associative

$$= \mathscr{ALT}\ (a \otimes b \otimes c)$$

from the above result, which may be used once more to give

$$\begin{aligned}(a \wedge b) \wedge c &= \mathscr{ALT}\ (a \otimes \mathscr{ALT}\ (b \otimes c)) \\ &= a \wedge (b \wedge c).\end{aligned}$$

The exterior algebra inherits a $\mathbb{Z}$-gradation from the tensor algebra. The zero element is the only homogeneous element of degree greater than n in the exterior algebra. Since $\mathscr{ALT}$ is a homogeneous mapping of degree zero on the tensor algebra it follows that η is also an automorphism of the exterior algebra, that is

$$\eta(a \wedge b) = \eta a \wedge \eta b. \tag{1.2.5}$$

Similarly exterior forms are called even or odd according to their $\mathbb{Z}_2$-gradation in the tensor algebra. The involution ξ commutes with $\mathscr{ALT}$ and so it is also an involution of $\Lambda(V)$. Taking the definition of $\mathscr{ALT}$ and rearranging the permutations gives the following simple expression for ξ acting on a p-form ω,

$$\omega^{\xi} = (-1)^{[p/2]}\omega. \tag{1.2.6}$$

where [ ] denotes the integer part.

The interior derivative i_X has already been defined on tensors, and so it is defined the same way on exterior forms. In fact this is where it will mainly be utilised. We need to show that the result of i_X on an exterior form is another exterior form, of one lower degree, and that the anti-derivation property (1.1.5) goes over to the exterior algebra with $\otimes$ replaced by $\wedge$. It will be sufficient to consider decomposable tensors. If

$$T = x^1 \otimes x^2 \otimes \ldots \otimes x^p$$

then

$$\begin{aligned}\mathrm{i}_{X_1} T = {} & x^1(X_1)x^2 \otimes \ldots \otimes x^p - x^2(X_1)x^1 \otimes x^3 \ldots \otimes x^p \\ & + x^3(X_1)x^1 \otimes x^2 \otimes x^4 \ldots \otimes x^p \\ & + \ldots + (-1)^{p-1}x^p(X_1)x^1 \otimes \ldots \otimes x^{p-1}\end{aligned}$$

that is

$$(\mathrm{i}_{X_1} T)(X_2, \ldots, X_p) = \sum_{\nu} \varepsilon(\nu) T(X_{\nu(1)}, \ldots, X_{\nu(p)}) \tag{1.2.7}$$

where ν is any of the p permutations such that

$$(1, 2, \ldots, r, \ldots, p) \longrightarrow (2, 3, \ldots, r-1, 1, r, \ldots, p).$$

Substituting $\mathscr{ALT}\,T$ into (1.2.7) gives

$$(\mathrm{i}_{X_1}\mathscr{ALT}\,T)(X_2, \ldots, X_p) = p\mathscr{ALT}\,T(X_1, \ldots, X_p). \tag{1.2.8}$$

From the definition we have

$$\begin{aligned}&(\mathscr{ALT}\mathrm{i}_{X_1} T)(X_2, \ldots, X_p) \\ &= \frac{x^1(X_1)}{(p-1)!} \sum_{\tau} \varepsilon(\tau)(x^2 \otimes \ldots \otimes x^p)(X_{\tau(2)}, \ldots, X_{\tau(p)}) \\ &\quad - \frac{x^2(X_1)}{(p-1)!} \sum_{\tau} \varepsilon(\tau)(x^1 \otimes x^3 \otimes \ldots \otimes x^p)(X_{\tau(2)}, \ldots, X_{\tau(p)}) \\ &\quad + \ldots + \frac{(-1)^{p-1}x^p(X_1)}{(p-1)!} \sum_{\tau} \varepsilon(\tau)(x^1 \otimes \ldots \otimes x^{p-1})(X_{\tau(2)}, \\ &\qquad \ldots, X_{\tau(p)}) \qquad \text{where } \tau \in C_{p-1}\end{aligned}$$

$$= \frac{1}{(p-1)!} \sum_\sigma \varepsilon(\sigma)(x^1 \otimes \ldots \otimes x^p)(X_{\sigma(1)}, \ldots, X_{\sigma(p)}) \qquad \text{for } \sigma \in C_p$$

$$= \frac{p!}{(p-1)!} (\mathscr{ALT}\, T)(X_1, \ldots, X_p)$$

and thus

$$(\mathscr{ALT} \mathrm{i}_{X_1} T)(X_2, \ldots, X_p) = (p\mathscr{ALT}\, T)(X_1, \ldots, X_p). \qquad (1.2.9)$$

So (2.8) and (2.9) give $\mathrm{i}_X \mathscr{ALT} = \mathscr{ALT} \mathrm{i}_X$. Thus $\mathrm{i}_X : \Lambda_p \to \Lambda_{p-1}$, and $\mathrm{i}_X(a \wedge b) = \mathrm{i}_X \mathscr{ALT}(a \otimes b) = \mathscr{ALT}(\mathrm{i}_X a \otimes b + \eta a \otimes \mathrm{i}_X b)$ and hence

$$\mathrm{i}_X(a \wedge b) = \mathrm{i}_X a \wedge b + \eta a \wedge \mathrm{i}_X b. \qquad (1.2.10)$$

If $\omega \in \Lambda_p(V)$ then $\mathscr{ALT}\omega = \omega$ and $\mathscr{ALT}\mathrm{i}_X \omega = \mathrm{i}_X \omega$, so (1.2.9) reduces to

$$(\mathrm{i}_{X_1}\omega)(X_2, \ldots, X_p) = p\omega(X_1, \ldots X_p). \qquad (1.2.11)$$

Just as the space formed by V together with the identity generates $T(V)$ under the product $\otimes$, it generates $\Lambda(V)$ with the product $\wedge$. Thus any element of $\Lambda(V)$ can be written as a sum of decomposable forms, these being the ones consisting of products of elements from V. If $\{e^i\}$ is any basis for the n-dimensional V then the $\binom{n}{p}$ p-forms $e^{i_1} \wedge e^{i_2} \wedge \ldots \wedge e^{i_p}$ for $i_1 < i_2 < \ldots < i_p$ $(p \geq 1)$ form a basis for $\Lambda_p(V)$. It is often convenient to label such p-forms by an ordered multi-index,

$$I = (i_1, i_2, \ldots, i_p) \text{ with } i_1 < i_2 < \ldots < i_p$$

with each index i_j varying from 1 to n. So if ω is an arbitrary p-form

$$\omega = \sum_{i_1 < i_2 < \ldots < i_p} \omega_{i_1 i_2 \ldots i_p} e^{i_1} \wedge e^{i_2} \wedge \ldots \wedge e^{i_p}$$

$$= \sum \omega_I e^I$$

where $\omega_I \equiv \omega_{i_1 i_2 \ldots i_p} \in F$ are the components of ω in this basis. Care must be exercised when using the summation convention (see Appendix A) with ordered multi-indices. Since this convention operates with unconstrained summations one may equivalently write

$$\omega = \frac{1}{p!} \omega_{i_1 i_2 \ldots i_p} e^{i_1} \wedge e^{i_2} \wedge \ldots \wedge e^{i_p}$$

it being understood that the components are totally antisymmetric in the indices.

If $\{f^i\}$ is a new basis for V related to $\{e^i\}$ by $f^i = M^i{}_j e^j$, $\{M^i{}_j\} \in \mathrm{Gl}(n, F)$†, then we can induce a corresponding change in the

† The group of $n \times n$ invertible matrices with elements from F, see Appendix A.

components of a p-form. Since the space of n-forms is one-dimensional the n-forms formed by the products of the two bases must be related by a multiple of F. In fact it follows from the antisymmetry that

$$f^1 \wedge f^2 \wedge \ldots \wedge f^n = \det M e^1 \wedge e^2 \wedge \ldots \wedge e^n \qquad (1.2.12)$$

where $\det M$ is the determinant of the matrix $\{M^i{}_j\}$ that relates the bases. Any n-form Ω can be used to classify frames $\{X_i\}$ for V^*. These frames fall into two classes according to the sign of $\Omega(X_1, X_2, \ldots, X_n)$. Frames in different classes are said to be of opposite orientation. The $\mathrm{Gl}(n, F)$ related frames $\{e^i\}$ and $\{f^i\}$ are of the same orientation if and only if $\det M$ is positive. This is consistent since the determinant of a product of two matrices is positive if the determinant of each factor is positive.

1.3 The Exterior Algebra as a Quotient of the Tensor Algebra

We have introduced the exterior algebra as the set of totally antisymmetric tensors with the product $\wedge$ constructed out of $\otimes$ and $\mathcal{ALT}$. This algebra is isomorphic to a quotient of the tensor algebra; indeed the definition in terms of the quotient offers certain advantages. In the next chapter we will define the Clifford algebra as a quotient of the tensor algebra, and it is useful to see the exterior algebra introduced in a parallel way. We will use bold-face type to denote the quotient algebra and its product, the use of the same symbols anticipating its isomorphism with the exterior algebra of antisymmetric tensors already defined.

Let I be the ideal in $T(V)$ consisting of sums of terms of the form $a \otimes x \otimes x \otimes b$ where $x \in V$ and a, b are arbitrary elements of $T(V)$. Then we define the exterior algebra $\mathbf{\Lambda}(V)$ by

$$\mathbf{\Lambda}(V) = T(V)/\mathrm{I}. \qquad (1.3.1)$$

Elements in $\mathbf{\Lambda}(V)$ are equivalence classes of elements in $T(V)$, where the equivalence relation is defined by $a \sim b$ if $a = b + c$ for some $c \in \mathrm{I}$. The equivalence class that contains a is denoted $[a]$. The vector space structure of $\mathbf{\Lambda}(V)$ is defined by

$$[a] + \lambda[b] = [a + \lambda b] \qquad a, b \in T(V), \lambda \in F \qquad (1.3.2)$$

and the multiplication which is denoted by $\boldsymbol{\wedge}$ is given by

$$[a] \boldsymbol{\wedge} [b] = [a \otimes b]. \qquad (1.3.3)$$

The ideal I is a $\mathbb{Z}$-graded† subspace of $T(V)$ and so $\mathbf{\Lambda}(V)$ inherits a natural $\mathbb{Z}$-gradation given by $\deg[a] = \deg a$. The automorphism η and the involution ξ preserve the ideal I and they thus extend in an obvious way to $\mathbf{\Lambda}(V)$ by

$$\begin{aligned} \eta[a] &= [\eta a] \\ [a]^{\xi} &= [a^{\xi}]. \end{aligned} \tag{1.3.4}$$

Similarly interior multiplication preserves I and so we may define

$$\mathrm{i}_X\,[a] = [\mathrm{i}_X a]. \tag{1.3.5}$$

If $x, y \in V$ then

$$2x\otimes y = (x\otimes y - y\otimes x) + (x + y)\otimes(x + y) - x\otimes x - y\otimes y$$

hence

$$x\otimes y = x \wedge y + \tfrac{1}{2}\{(x + y)\otimes(x + y) - x\otimes x - y\otimes y\}. \tag{1.3.6}$$

The $\wedge$ denotes the antisymmetrised tensor product as defined in (1.2.2). The term in brackets is in I and so $x\otimes y \sim x \wedge y$. That is, $[x]\wedge[y] = [x\otimes y] = [x \wedge y]$.

More generally, it follows that the ideal I is just the kernel of $\mathscr{ALT}$, and so $[a] = [\mathscr{ALT} a]$. We have already seen, in proving that $\wedge$ is associative, that this kernel is an ideal. To see that it is in fact I we will prove that

$$x\otimes\omega \sim x \wedge \omega \qquad \text{for } x \in V,\ \omega \in \Lambda(V). \tag{1.3.7}$$

The recursive application of this result gives

$$x^1\otimes x^2\otimes \ldots \otimes x^p \sim \mathscr{ALT}\,(x^1\otimes x^2\otimes\ldots\otimes x^p) = x^1 \wedge x^2 \wedge \ldots \wedge x^p.$$

We will prove (1.3.7) by induction on the degree of ω. It is certainly true when ω is a 1-form; we assume it is true for ω of degree less than p. It is sufficient to consider the case of ω decomposable. The definition of $\wedge$ involves the permutation of the arguments in the evaluation, but this is obviously equivalent to permuting the factors in the product. Thus from the definition of $\wedge$ we have

$$y^0 \wedge y^1 \ldots \wedge y^p = \frac{1}{(p+1)!}\sum_{\sigma} \varepsilon(\sigma) y^{\sigma(0)}\otimes y^{\sigma(1)}\otimes \ldots \otimes y^{\sigma(p)}$$

where σ permutes the set $(0, 1, 2, \ldots, p)$. We will characterise each permutation according to the first number in the reordered set. With one interchange we swap the elements 0 and r, and with $r - 1$ further

† Grading is discussed in Appendix A.

interchanges bring the 0 to the second position. So if ν_r is the permutation such that

$$(0, 1, \ldots, r, \ldots, p) \xrightarrow{\nu_r} (r, 0, 1, \ldots, \hat{r}, \ldots, p)$$

where $\hat{r}$ denotes that r is missing from this sequence, then $\varepsilon(\nu_r) = (-1)^r$. We can now write any permutation σ as $\sigma = \tau_r \nu_r$ for some r, where τ_r permutes the set with r removed, then

$$\begin{aligned} & y^0 \wedge y^1 \ldots \wedge y^p \\ & = \frac{1}{(p+1)!} \sum_{\tau_0} \varepsilon(\tau_0) y^0 \otimes y^{\tau_0(1)} \otimes \ldots \otimes y^{\tau_0(p)} \\ & \quad + \frac{1}{(p+1)!} \sum_{r=1}^{p} (-1)^r y^r \otimes \sum_{\tau_r} \varepsilon(\tau_r) y^{\tau_r(0)} \otimes \ldots \otimes y^{\tau_r(r-1)} \otimes y^{\tau_r(r+1)} \\ & \qquad \ldots \otimes y^{\tau_r(p)} \\ & = \frac{1}{(p+1)} \Big(y^0 \otimes (y^1 \wedge \ldots \wedge y^p) \\ & \quad + \sum_{r=1}^{p} (-1)^r y^r \otimes (y^0 \wedge \ldots \wedge y^{\hat{r}} \wedge \ldots \wedge y^p) \Big). \end{aligned}$$

Substituting x for y^0 gives

$$x \wedge y^{12 \ldots p} = \frac{1}{(p+1)} \left(x \otimes y^{12 \ldots p} + \sum_{r=1}^{p} (-1)^r y^r \otimes (x \wedge y^{1 \ldots \hat{r} \ldots p}) \right)$$

where $y^{12 \ldots p} \equiv y^1 \wedge y^2 \wedge \ldots \wedge y^p$, and again the hat means that a term is missing.

Now $y^r \otimes (x \wedge y^{1 \ldots \hat{r} \ldots p}) \sim y^r \otimes x \otimes y^{1 \ldots \hat{r} \ldots p}$ since (1.3.7) is assumed true for $(p-1)$-forms

$$\begin{aligned} & \sim -x \otimes y^r \otimes y^{1 \ldots \hat{r} \ldots p} && \text{since } x \otimes y + y \otimes x \sim 0 \\ & \sim -x \otimes (y^r \wedge y^{1 \ldots \hat{r} \ldots p}) && \text{from (1.3.7) again,} \\ & \sim (-1)^r x \otimes y^{12 \ldots p} \end{aligned}$$

where the sign comes from moving y^r through $r-1$ terms.

So $x \wedge y^1 \wedge \ldots \wedge y^p \sim x \otimes y^{12 \ldots p}$. Thus if (1.3.7) holds for ω of degree less than p it is also true when ω is a p-form. This completes the proof.

Thus every equivalence class of $\mathbf{\Lambda}(V)$ is represented by an element of $\Lambda(V)$, and the product of the classes under $\wedge$ is the class of the product of the representatives under $\wedge$. Thus $\mathbf{\Lambda}(V)$ is indeed isomorphic to $\Lambda(V)$. In practice it is more convenient to work with representatives, the antisymmetric tensors, rather than with their equivalence classes.

1.4 The Hodge Map

When the vector space V has a (non-degenerate) metric g then the Hodge dual, or $*$ map, may be defined on exterior forms. Since

$$\binom{n}{p} = \binom{n}{n-p}$$

we have $\dim \Lambda_p(V) = \dim \Lambda_{n-p}(V)$, and thus these two vector spaces are isomorphic. We may use the metric g to set up a standard isomorphism between these spaces: the Hodge map, denoted by $*$. (Although one can define a Hodge map for a non-symmetric non-degenerate metric, we shall take g to be symmetric as well as non-degenerate.)

If V has a metric then one can use a g-orthonormal frame $\{e^i\}$ to construct a standard n-form ω,

$$\omega = e^1 \wedge e^2 \wedge \ldots \wedge e^n. \tag{1.4.1}$$

Since the determinant of the matrix relating orthonormal frames is plus or minus one, depending on the relative orientations, we see from (1.2.12) that there are two possibilities for ω, differing by a sign. The members of a g-orthonormal frame for V are sometimes called n-beins in the physics literature, generalising the familiar triad of orthonormal vectors in Euclidean three space. Some authors, however, associate this term with the r^2 elements $\{N^i{}_j\} \in \mathrm{Gl}(n, F)$ that relate an orthonormal frame to an arbitrary one $\{f^i\}$,

$$e^i = N^i{}_j f^j.$$

If the components of g in the frame $\{e^i\}$ are η^{ij}, where $\eta^{ij} = 0$ if $i \neq j$ and for each value of i, $\eta^{ii} = \pm 1$, and the components in the frame $\{f^i\}$ are $g^{ij(f)}$, then

$$\eta^{ij} = N^i{}_k N^j{}_l g^{kl(f)}.$$

Hence $\det(\eta^{ij}) = \det(g^{ij(f)})(\det N)^2$.

The components of the metric on the dual space form the inverse matrices, $\eta_{ij}\eta^{jk} = \delta_i^k$ and $g_{ij}^{(f)} g^{jk(f)} = \delta_i^k$. (For further details see Appendix A.) So if $t = \det(\eta_{ij}) = \pm 1$ then, since $\det(m^{-1}) = (\det m)^{-1}$ for all matrices m,

$$\det(g_{ij}^{(f)}) = t(\det N)^2.$$

But $\omega = (\det N) f^1 \wedge f^2 \wedge \ldots \wedge f^n$, so if we write the sign of $\det N$ as

$$\mu_N = \frac{\det N}{|\det N|}$$

then

$$\omega = \mu_N\{t\det(g_{ij}^{(f)})\}^{1/2} f^1 \wedge f^2 \wedge \ldots \wedge f^n. \tag{1.4.2}$$

If the frames $\{e^i\}$ and $\{f^i\}$ are related by a $\mathrm{Gl}(n, F)$ transformation that preserves the orientation, then $\mu_N = 1$.

A metric on V naturally gives rise to a metric on $\Lambda(V)$. We start by defining a metric g_p on the space of p-forms, $\Lambda_p(V)$, for any $p > 1$. Since g_p is defined to be bilinear it is sufficient to specify its action on decomposable p-forms. If $A = \alpha^1 \wedge \alpha^2 \wedge \ldots \wedge \alpha^p$ and $B = \beta^1 \wedge \beta^2 \wedge \ldots \wedge \beta^p$ then

$$g_p(A, B) = \det\{g(\alpha^i, \beta^j)\}. \tag{1.4.3}$$

It is convenient to define g_0 to simply multiply the two 0-forms. Having defined a metric on the homogeneous subspaces we define a metric G on $\Lambda(V)$ by requiring it to be diagonal in the homogeneous subspaces. That is, if $\Phi, \Psi \in \Lambda(V)$ with, for example, Φ_p denoting the projection of Φ into the subspace of degree p, then

$$G(\Phi, \Psi) = \sum_{p=0}^{n} g_p(\Phi_p, \Psi_p). \tag{1.4.4}$$

As we have remarked the spaces of p-forms and $(n-p)$-forms are of the same dimension, and we are now in a position to establish a standard isomorphism between them. The Hodge map, $*$, is a linear map from the space of p-forms to the space of $(n-p)$-forms:

$$*: \Lambda_p(V) \longrightarrow \Lambda_{n-p}(V)$$
$$a \longmapsto *a$$

where $*a$ is given implicitly by

$$b \wedge *a = g_p(b, a)\omega \qquad \forall\, b \in \Lambda_p(V). \tag{1.4.5}$$

The standard n-form ω is defined as in (1.4.1). The definition may be completed by defining the map on a 0-form, $*1 = \omega$. This is called the *volume n-form*. Linearity extends the definition to inhomogeneous elements of the exterior algebra. Thus the definition of the Hodge map depends not only on the metric but on a choice of orientation. The non-degeneracy of g (and hence of g_p) ensures that such a definition does indeed determine the $*$ map. It immediately follows from the symmetry of g (and hence of g_p) that

$$a \wedge *b = b \wedge *a \qquad \forall\, a, b \in \Lambda_p(V). \tag{1.4.6}$$

A useful calculus can be set up relating the $*$ map to the interior product. We may use the metric g to establish an isomorphism (denoted by a tilde) between V and V^*. If $x \in V$ then the metric dual, $\tilde{x}$, is in V^*;

given by

$$y(\tilde{x}) = g(x, y) \qquad \forall y \in V.$$

It then follows from the definition of $*$ that

$$*(\Phi \wedge x) = \mathrm{i}_{\tilde{x}} * \Phi \qquad x \in V,\ \Phi \in \Lambda(V). \tag{1.4.7}$$

This formula can be applied recursively to a decomposable p-form to produce

$$*(x^1 \wedge x^2 \wedge \ldots \wedge x^p) = \mathrm{i}_{\tilde{x}^p} \mathrm{i}_{\tilde{x}^{p-1}} \ldots \mathrm{i}_{\tilde{x}^1} *1. \tag{1.4.8}$$

It is convenient to display the action of $*$ on exterior products of basis vectors. Suppose that $\{e^i\}$ and $\{X_i\}$ are dual bases, with $e^i(X_j) = \delta^i{}_j$. We will often use the shorthand

$$\mathrm{i}_{X_j} \equiv i_j.$$

The metric dual, $\tilde{e}^a$, of e^a is $g^{ab}X_b \equiv X^a$ and we write $\mathrm{i}_{X^a} \equiv i^a$.

Equation (1.4.8) takes the following simple form for the product of p basis vectors

$$*(e^1 \wedge e^2 \wedge \ldots \wedge e^p) = i^p i^{p-1} \ldots i^1 *1.$$

From this it can be seen that the dual of a product of p orthonormal 1-forms is the product of their complement in the basis. Duals of the orthonormal basis forms can be expressed in terms of the Levi–Civita antisymmetric ε-symbol. This is defined such that

$$\varepsilon_{i_1 i_2 \ldots i_n} = \begin{cases} 0 \text{ if } i_j = i_k. \\ +1\ (-1) \text{ if } (i_1, i_2, \ldots, i_n) \text{ is an even (odd) permutation} \\ \text{of the standard sequence } (1, 2, 3, \ldots, n). \end{cases} \tag{1.4.9}$$

With the summation convention the volume n-form can be written in the orthonormal frame $\{e^i\}$ as

$$*1 = \frac{1}{n!} \varepsilon_{i_1 i_2 \ldots i_n} e^{i_1} \wedge e^{i_2} \wedge \ldots \wedge e^{i_n}. \tag{1.4.10}$$

If the components of the metric in this orthonormal frame are η^{ij} we have

$$*(e^{i_1} \wedge e^{i_2} \wedge \ldots \wedge e^{i_p}) = \frac{1}{(n-p)!} \varepsilon^{i_1 i_2 \ldots i_p}{}_{i_{p+1} \ldots i_n} e^{i_{p+1}} \wedge \ldots \wedge e^{i_n}$$

where

$$\varepsilon^{i_1 i_2 \ldots i_p}{}_{i_{p+1} \ldots i_n} = \eta^{i_1 j_1} \eta^{i_2 j_2} \ldots \eta^{i_p j_p} \varepsilon_{j_1 j_2 \ldots j_p i_{p+1} \ldots i_n}.$$

It is sometimes necessary to rearrange expressions such as

$$e^a \wedge *(e^{b_1} \wedge e^{b_2} \wedge \ldots \wedge e^{b_p}).$$

This may be accomplished by using (1.4.8), for example

$$\begin{aligned} e^a \wedge *(e^b \wedge e^c) &= e^a \wedge i^c * e^b = -i^c(e^a \wedge * e^b) + g^{ca} * e^b \\ &= -g^{ab} i^c * 1 + g^{ca} * e^b \qquad (\text{since } e^a \wedge * e^b = g^{ab} * 1) \\ &= -g^{ab} * e^c + g^{ca} * e^b. \end{aligned}$$

and similarly

$$e^a \wedge *(e^b \wedge e^c \wedge e^d) = g^{ab} *(e^c \wedge e^d) - g^{ac} *(e^b \wedge e^d) + g^{ad} *(e^b \wedge e^c).$$

1.5 The Mixed Tensor Algebra

Just as the tensor product $\otimes^r V$ is the space of multilinear mappings on $V^* \times V^* \times \ldots \times V^*$ (r times), the tensor product of V^* with itself, $\otimes^r V^*$, is the space of multilinear mappings on $V \times V \times \ldots \times V$ (r times). More generally we have the vector space of multilinear mappings on

$$\underbrace{V^* \times V^* \times \ldots \times V^*}_{r \text{ times}} \quad \times \quad \underbrace{V \times V \times \ldots \times V}_{s \text{ times}},$$

the space $\otimes^r V \otimes^s V^*$. This space is called the space of mixed tensors of covariant degree r and contravariant degree s, $T_r{}^s(V)$. (The assignment of the terms covariant and contravariant is a matter of convention. The way we have indexed our bases accords with the classical component conventions.) Tensors in $T_r{}^s(V)$ will be referred to as being of type (r, s). It will be seen that we have defined tensors to be multilinear maps on sets of vectors ordered such that those from V^* occur first; that is, our space of tensors is formed by tensor products of V with itself followed by products with V^*. One might envisage a more general definition that formed the tensor product of the spaces V and V^* in no definite order. However, such tensor product spaces are naturally isomorphic to the canonically ordered product. For example, the ordered pairs $V \times V^*$ are certainly distinct from $V^* \times V$, the bilinear mappings on these spaces being $V^* \otimes V$ and $V \otimes V^*$ respectively. However, we may define a map φ by

$$\varphi : V^* \otimes V \longmapsto V \otimes V^*$$
$$T \longmapsto \varphi T$$

where

$$(\varphi T)(X, \omega) = T(\omega, X) \qquad \forall X \in V^*, \omega \in V.$$

It is easy to see that φ defines an isomorphism between $V^*\otimes V$ and $V\otimes V^*$. Further it is natural (or canonical), depending on no choice of bases for these spaces. Similarly, any tensor product containing V r times and V^* s times is naturally isomorphic to the canonically ordered $\otimes^r V \otimes^s V^*$. We shall not distinguish between these naturally isomorphic spaces, and shall always form tensor products with the factors from V collected at the left. Thus we adopt the convention that tensors will be evaluated on a set ordered with elements from V^* occurring first.

If $\{e^i\}$ is a basis for V, with $\{X_i\}$ a dual basis for V^*, such that $e^i(X_j) = \delta^i{}_j$, then a basis for $T_r{}^s(V)$ is provided by the $n^{(r+s)}$ elements

$$\{e^{i_1}\otimes e^{i_2}\otimes \ldots \otimes e^{i_r}\otimes X_{j_1}\otimes X_{j_2}\otimes \ldots \otimes X_{j_s}\}.$$

If T is any element of $T_r{}^s(V)$ then

$$T = T^{j_1 j_2 \ldots j_s}_{i_1 i_2 \ldots i_r}\, e^{i_1}\otimes e^{i_2}\otimes \ldots \otimes e^{i_r}\otimes X_{j_1}\otimes \ldots \otimes X_{j_s}$$

where the summation convention is employed. If $\{e'^i\}$ is a different basis for V, with dual basis $\{X'_i\}$, then if $e'^i = M^i{}_j e^j$ and $X'_i = N_i{}^j X_j$ it follows from $e'^i(X'_j) = \delta^i{}_j$ that

$$M^i{}_k N_j{}^k = \delta^i{}_j.$$

So if the transformation coefficients are arranged into matrices M and N, the transpose of N is the inverse of M. If the components of T in the basis labelled with a prime are

$$T'^{j_1 \ldots j_s}_{i_1 \ldots i_r}$$

then

$$\begin{aligned} T'^{j_1 \ldots j_s}_{i_1 \ldots i_r} &= T(X'_{i_1}, X'_{i_2}, \ldots, X'_{i_r}, e'^{j_1}, \ldots, e'^{j_s}) \\ &= M^{j_1}{}_{q_1} \ldots M^{j_s}{}_{q_s}\, N_{i_1}{}^{p_1} \ldots N_{i_r}{}^{p_r}\, T^{q_1 \ldots q_s}_{p_1 \ldots p_r}. \end{aligned}$$

This is the classical expression for the change in the components of a tensor induced by a change of basis. The contravariant components, placed as superscripts, transform contragradiently to the covariant components, placed as subscripts.

We may classify the symmetry of a mixed tensor according to the behaviour under permutations of the vectors from V, and those from V^*: of course it makes no sense to talk of a symmetry that mixes these spaces.

Since dual bases transform contragradiently we can define a contraction map that reduces both the contravariant and the covariant degrees by one:

$$\begin{aligned} C^l_k : T^s_r(V) &\longrightarrow T^{s-1}_{r-1}(V) \\ T &\longmapsto C^l_k T \end{aligned}$$

$$C^l_k T(\,,\,,\ldots,\,;\,,\ldots,) = T(\underbrace{,\,,\ldots, X_i}_{k\text{th entry}}, \ldots,\overbrace{;\,,\,,\ldots, e^i}^{l\text{th entry}},\,,\ldots,) \tag{1.5.1}$$

where $\{e^i\}$ is dual to $\{X_i\}$.

Since the dual frames transform contragradiently the linearity of T ensures that the definition of C^l_k is basis independent. If, in some basis, T has the components

$$T^{j_1 \ldots j_s}_{i_1 \ldots i_r}$$

then the components of $C^l_k T$ are

$$T^{j_1 \ldots j_{l-1} m\, j_{l+1} \ldots j_s}_{i_1 \ldots i_{k-1} m\, i_{k+1} \ldots i_r}$$

where the 'dummy' index m is summed over. For the special case of $T \in T^1_1(V)$ the contraction C^1_1 maps T to the field F. In this case the contraction map is sometimes called the trace of T, Tr T.

When V has a metric there is a canonical isomorphism $\tilde{}$, between V and V^*. Similarly we can use a metric on V to define a mapping between tensors of different contravariant and covariant degrees. For example, given a tensor $T \in T^s_r(V)$ we can define an $S \in T^{s+1}_{r-1}(V)$ as follows:

$$\begin{aligned} &S(X_1, \ldots X_{r-1}; e^1, \ldots, e^s, e^{s+1}) \\ &\quad = T(X_1, \ldots, X_{k-1}, \tilde{e}^j, X_k, \ldots, X_{r-1}; e^1, \ldots, e^{j-1}, e^{j+1}, \ldots, e^{s+1}). \end{aligned}$$

In a similar way we could associate with T a tensor in $T^{s-1}_{r+1}(V)$ or more generally a tensor in $T^p_q(V)$ with $p + q = r + s$. We give an example. Given $T \in T^1_2(V)$ we define $S \in T^1_2(V)$ by

$$S(W, Y, \omega) = T(W, \tilde{\omega}, \tilde{Y}) \qquad \forall\, W, Y \in V^*, \omega \in V. \tag{1.5.2}$$

If $\{e^i\}$ and $\{x_i\}$ are dual bases for V and V^* respectively such that

$$T = T^k_{ij} e^i \otimes e^j \otimes X_k \qquad S = S^k_{ij} e^i \otimes e^j \otimes X_k$$

then writing W, Y and ω in this basis gives

$$W^i Y^j \omega_k S^k_{ij} = W^i Y^j \omega_k g^{qk} g_{pj} T^p_{iq}.$$

Since this must hold for all W, Y and ω

$$S^k_{ij} = g^{qk} g_{pj} T^p_{iq}. \tag{1.5.3}$$

Such expressions can be simplified by adopting a convention for raising and lowering indices with the components of the metric, similar to the case for vectors. However, such a procedure would be ambiguous with the tensor components arranged in the way we have them: it not being clear, for example, where the upper index should be lowered to. To enable a raising and lowering convention to be employed, from now on we will order the upper indices relative to the lower ones. We can always specify a tensor with the indices in a canonical order; the lower indices occurring first. Components can then be raised and lowered with the components of the metric, maintaining the ordering. Thus one obtains an array of components not in canonical order, some superscripts occurring before subscripts. If we return to the example we were considering, only this time stagger the components in the canonical order,

$$T = T_{ij}{}^{k} e^i \otimes e^j \otimes X_k \qquad S = S_{ij}{}^{k} e^i \otimes e^j \otimes X_k$$

then the relationship (1.5.2) between S and T relates the components by

$$S_{ij}{}^{k} = g^{qk} g_{pj} T_{iq}{}^{p}.$$

This can now be compactly written as

$$S_{ij}{}^{k} = T_{i}{}^{k}{}_{j}. \tag{1.5.4}$$

There are a couple of points relating to this index convention that are worth emphasising. The first is that a raising and lowering convention need not be adopted at all: in which case there is no need to order the upper indices relative to the lower ones. No inconsistencies would arise, only relationships between tensors such as (1.5.2) would have the untidy component form of (1.5.3). The second point concerns the ordering of the basis. We have decided to work always with tensors formed with products from V to the left. Nevertheless relationships such as (1.5.4) involve components that are not indexed in the canonical order. As we earlier remarked one could work with the larger class of tensors in which the factors from V and V^* occur in no definite order. In this case one might adopt the convention that the basis is attached in the order in which the components occur; an element from V going with a subscript for example. Such a tensor would, however, as we have pointed out, be naturally isomorphic to a tensor with the same components but with a canonically ordered basis. Thus the adopted ordering of the basis is in no real sense a restriction, and in particular we have the freedom to employ the raising and lowering conventions that introduce the non-canonically ordered components.

Sometimes we may speak, for example, of a degree two tensor being symmetric and trace free. Such imprecise statements should be understood to mean that T is a symmetric tensor in $T^0_2(V)$, and that $S \in T^1_1(V)$ is traceless, where

$$S(X, \omega) = T(X, \tilde{\omega}) \qquad \forall\, X \in V^*, \omega \in V.$$

Equivalently, $T(X^i, X_i) = 0$, where $X^i = g^{ij} X_j$.

Bibliography

Abrahams R, Marsden J E and Ratiu T 1983 *Manifolds, Tensor Analysis and Applications* (New York: Addison-Wesley)

Dodson C T J and Poston T 1977 *Tensor Geometry* (London: Pitman)

Greub W 1978 *Multilinear Algebra* 2nd edn (Heidelberg: Springer)

Schutz B F 1980 *Geometrical Methods of Mathematical Physics* (Cambridge: Cambridge University Press)

2

Clifford Algebras and Spinors

In this chapter we present an account of Clifford algebras and spinors. Taken with Appendix A it is fairly self-contained. Whereas in some places we have explicitly referred to Appendix A we have often tacitly assumed knowledge of something that is to be found there. Thus a reader confronted with concepts or terminology that are unfamiliar should consult Appendix A where (we hope) further details may be found.

The Clifford algebra is constructed so as to facilitate a study of orthogonal transformations. It leads to a systematic way of introducing the spin groups (the covering groups of the orthogonal groups and various subgroups) for arbitrary dimensions and signature. The irreducible representations of the Clifford algebra give rise to irreducible representations of the spin groups: spinors. If the real vector space V with bilinear form g is an orthogonal space then we wish to imbed V and a copy of the real numbers as vector subspaces in the real associative algebra $\mathbf{C}(V, g)$ in such a way that $x^2 = g(x, x)$, $\forall x \in V$. The square of x denotes its product with itself in this algebra, and the right-hand side is a real number which lies in the vector subspace of the algebra spanned by the identity. If S is any invertible element of the algebra and $x' = SxS^{-1}$ then obviously $x'^2 = g(x, x)$. So if x' is in V we have an orthogonal transformation. Those elements S such that x' is in V form a group, the Clifford group. Obviously elements of the Clifford group which differ by a multiple of the centre will produce the same orthogonal transformation, so that the mapping from the Clifford group to the orthogonal group is many-to-one. By suitably normalising elements of the Clifford group we obtain a subgroup such that the mapping into the orthogonal group is two-to-one, and we have a double covering of the orthogonal group. Being able to write an orthogonal transformation in terms of simultaneous multiplication from both sides by an element of the Clifford group we are led to consider those transformations obtained by multiplying from one side only; the spin transformations.

The Clifford algebra can be constructed as a quotient of the tensor algebra. This is in close parallel with §1.3, where we considered the exterior algebra as a quotient of the tensor algebra. Rather than regarding elements of the Clifford algebra as equivalence classes in the tensor algebra it is more convenient to work with representatives of these classes. We show how we can choose these representatives to be the exterior forms, the Clifford product being given in terms of the exterior and interior products. In §2.2 we determine the structure of the real Clifford algebras. These algebras are $\mathbb{Z}_2$-graded†, and we give the structure of the even subalgebra in §2.3. In §2.4 we introduce the Clifford group and show the relation of it and its subgroups to the orthogonal group and its subgroups. After examining the irreducible representations of the Clifford algebra and group, spinors, we move on to spin-invariant products. At this point some readers will probably feel the furthest removed from what they feel they want to know, and from relevance to physics. However, such readers should be assured that this section will enable them to determine all the spin-invariant products in whichever dimension is currently in fashion, and, for example, whether the charge conjugation matrix (defined in either of two ways) is symmetric or antisymmetric. The reader with a trusting disposition may be content to learn how to interpret the tables that summarise the results. In §2.7 we consider the complexified Clifford algebras. Anyone familiar with the γ-matrices, which are usually assumed to be complex, may wonder why we have postponed the complex case for so long. However, although the γ-matrices are usually assumed to be complex, conjugate–linear operations, such as the Dirac adjoint, are considered as well as complex–linear ones. Thus an underlying real structure is singled out and so one way or another we need the results of the real case. The account we have given is logically complete at the end of §2.7. It makes no reference, however, to such things as Dirac spinors and charge conjugation with which most physicists are familiar. Whilst not being intended as a dictionary, §2.8 makes contact with the γ-matrices and physics vocabulary. We also mention the 'two-component spinor formalism' for Lorentzian spinors.

Having outlined what we shall do, it is in order to state what is omitted. There are two main restrictions we have imposed: we only consider algebras over the real or complex field and we assume the bilinear form is non-degenerate. The important topic of pure spinors has been given a chapter of its own.

† Grading is discussed in Appendix A.

2.1 The Clifford Algebra

We assume now that the vector space V has an F-valued non-degenerate symmetric bilinear form, or metric, g. Let J be the ideal of $T(V)$ consisting of sums of terms of the form $a\otimes\{x\otimes x - g(x, x)\}\otimes b$, $a, b \in T(V)$, $x \in V$. Then the Clifford algebra associated with V is $\underline{\mathbf{C}}(V, g)$ defined by

$$\underline{\mathbf{C}}(V, g) = T(V)/J. \tag{2.1.1}$$

The product will be denoted $\vee$, satisfying $[a]\vee[b] = [a\otimes b]$. The ideal J is not a $\mathbb{Z}$-graded subspace and so $\underline{\mathbf{C}}(V, g)$ does not inherit a $\mathbb{Z}$-gradation. However, $x\otimes x - g(x, x)$ is homogeneous with respect to the induced $\mathbb{Z}_2$-gradation of $T(V)$ making J a $\mathbb{Z}_2$-graded subspace. Thus $\underline{\mathbf{C}}(V, g)$ inherits a $\mathbb{Z}_2$-gradation. The ideal J is preserved by η, ξ and i_X and so all of these naturally induce operations (denoted by the same symbol) in $\underline{\mathbf{C}}(V, g)$. If $x, y \in V$ then

$$x\otimes y = x\wedge y + g(x, y) + \tfrac{1}{2}\{(x + y)\otimes(x + y) - g(x + y, x + y) - x\otimes x + g(x, x) - y\otimes y + g(y, y)\}.$$

The term in brackets is in J and so

$$x\otimes y \sim x\wedge y + g(x, y). \tag{2.1.2}$$

More generally for ω a p-form and $x \in V$ we have

$$x\otimes\omega \sim x\wedge\omega + \mathrm{i}_{\tilde{x}}\omega. \tag{2.1.3}$$

Here $\tilde{x} \in V^*$ is the metric dual of x, defined by $\tilde{x}(y) = g(x, y)$, $\forall y \in V$. For ω a 1-form (2.1.3) reduces to (2.1.2). We may prove its general validity by induction. This will be closely analogous to the proof of (1.3.7). Suppose that (2.1.3) is true for ω of degree less than or equal to $p - 1$, then it will be true for all p-forms if it holds for ω the product of p orthogonal 1-forms. As we showed in the proof of (1.3.7) it follows from the definition of the exterior product that if x, y^i, $i = 1, \ldots, p$ are in V then

$$x\wedge y^1\wedge\ldots\wedge y^p = 1/(p+1)\left(x\otimes y^{1\ldots p} + \sum_{r=1}^{p}(-1)^r y^r\otimes(x\wedge y^{1\ldots\hat{r}\ldots p})\right) \tag{2.1.4}$$

where $y^{1\ldots\hat{r}\ldots p} = y^1\wedge y^2\wedge\ldots\wedge y^{r-1}\wedge y^{r+1}\wedge\ldots\wedge y^p$. Since (2.1.3) is assumed true for ω of degree $p-1$ or less

$$y^r\otimes(x\wedge y^{1\ldots\hat{r}\ldots p}) \sim y^r\otimes(x\otimes y^{1\ldots\hat{r}\ldots p} - \mathrm{i}_{\tilde{x}}y^{1\ldots\hat{r}\ldots p}).$$

Use of (2.1.2) gives

$$y^r \otimes (x \wedge y^{1 \dots \hat{r} \dots p})$$
$$\sim 2g(y^r, x) y^{1 \dots \hat{r} \dots p} - x \otimes y^r \otimes y^{1 \dots \hat{r} \dots p} - y^r \otimes \mathrm{i}_{\tilde{x}} y^{1 \dots \hat{r} \dots p}.$$

Since the y^i are assumed orthogonal we may use (2.1.3) for ω a $(p-1)$ or $(p-2)$-form to show that

$$y^r \otimes (x \wedge y^{1 \dots \hat{r} \dots p})$$
$$\sim 2g(y^r, x) y^{1 \dots \hat{r} \dots p} - x \otimes (y^r \wedge y^{1 \dots \hat{r} \dots p}) - y^r \wedge \mathrm{i}_{\tilde{x}} y^{1 \dots \hat{r} \dots p}.$$

We may pull the interior derivative to the front of the last term and use $y^r \wedge y^{1 \dots \hat{r} \dots p} = (-1)^{r-1} y^{1 \dots p}$ to produce

$$y^r \otimes (x \wedge y^{1 \dots \hat{r} \dots p})$$
$$\sim g(y^r, x) y^{1 \dots \hat{r} \dots p} + (-1)^r x \otimes y^{1 \dots p} - (-1)^r \mathrm{i}_{\tilde{x}} y^{1 \dots p}$$

so

$$\sum_{r=1}^{p} (-1)^r y^r \otimes (x \wedge y^{1 \dots \hat{r} \dots p})$$
$$\sim \sum_{r=1}^{p} (-1)^r g(y^r, x) y^{1 \dots \hat{r} \dots p} + p x \otimes y^{1 \dots p} - p \mathrm{i}_{\tilde{x}} y^{1 \dots p}$$
$$\sim p x \otimes y^{1 \dots p} - (1+p) \mathrm{i}_{\tilde{x}} y^{1 \dots p}.$$

Returning to (2.1.4) shows that if (2.1.3) is true for ω a q-form with $q \leqslant p-1$ then it is true for ω a p-form. Thus (2.1.2) shows that indeed (2.1.3) holds for all p-forms. Repeated use of (2.1.3) shows that an arbitrary tensor product is equivalent to a sum of exterior forms, for example

$$x^1 \otimes x^2 \otimes x^3 \sim x^1 \otimes \{x^2 \wedge x^3 + g(x^2, x^3)\}$$
$$\sim x^1 \wedge x^2 \wedge x^3 + g(x^1, x^2) x^3 - g(x^1, x^3) x^2 + g(x^2, x^3) x^1.$$

In principle we could write down an explicit formula for the relation between the class of a homogeneous tensor and classes of exterior forms. However, it is generally sufficient to know that (2.1.3) determines such a relation and for practical purposes we shall be content with (2.1.3) and the following other special case. If ω is an arbitrary p-form and x a 1-form then

$$\omega \otimes x \sim x \wedge \eta\omega - \mathrm{i}_{\tilde{x}} \eta\omega. \tag{2.1.5}$$

For ω a 1-form this is certainly true since it reduces to (2.1.1). Again we prove its general validity by induction. Suppose that (2.1.5) holds for ω of degree less than or equal to p, then

$$(y \wedge \omega) \otimes x \sim (y \otimes \omega - \mathrm{i}_{\tilde{y}} \omega) \otimes x \qquad \text{by (2.1.3)}$$
$$\sim y \otimes (x \wedge \eta\omega - \mathrm{i}_{\tilde{x}} \eta\omega) - x \wedge \eta \mathrm{i}_{\tilde{x}} \omega + \mathrm{i}_{\tilde{x}} \eta \mathrm{i}_{\tilde{y}} \omega \qquad \text{by (2.1.5).}$$

A second application of (2.1.3) gives

$$(y \wedge \omega)\otimes x \sim y \wedge (x \wedge \eta\omega - \mathrm{i}_{\tilde{x}}\eta\omega) + \mathrm{i}_{\tilde{y}}x\eta\omega - x \wedge \mathrm{i}_{\tilde{y}}\eta\omega - \mathrm{i}_{\tilde{y}}\mathrm{i}_{\tilde{x}}\eta\omega$$
$$+ x \wedge \mathrm{i}_{\tilde{y}}\eta\omega - \mathrm{i}_{\tilde{x}}\mathrm{i}_{\tilde{y}}\eta\omega$$

where we have used $\eta\mathrm{i}_X = -\mathrm{i}_X\eta$. Dropping the terms that cancel and a little rearranging gives

$$(y \wedge \omega)\otimes x \sim x \wedge \eta(y \wedge \omega) - \mathrm{i}_{\tilde{x}}\eta(y \wedge \omega)$$

and so if (2.1.5) holds for all ω of degree less than or equal to p it also holds for all $(p+1)$-forms. This completes the inductive proof of the general validity of (2.1.5).

We have shown that the classes of a basis for the space of all exterior forms provide a basis for $\underset{\sim}{\mathbf{C}}(V, g)$. There is thus a natural way of introducing a product, $\vee$, on the space of exterior forms that turns this vector space into an algebra, $\mathbf{C}(V, g)$ say, where $\mathbf{C}(V, g) = \underset{\sim}{\mathbf{C}}(V, g)$. If α and ω are exterior forms then the exterior form $\alpha \vee \omega$ is defined by

$$[\alpha] \vee [\omega] = [\alpha \vee \omega]. \tag{2.1.6}$$

Since $[\alpha] \vee [\omega] = [\alpha\otimes\omega]$ the equivalence in (2.1.3) gives for x a 1-form

$$x \vee \omega = x \wedge \omega + \mathrm{i}_{\tilde{x}}\omega. \tag{2.1.7}$$

Similarly (2.1.5) gives

$$\omega \vee x = x \wedge \eta\omega - \mathrm{i}_{\tilde{x}}\eta\omega. \tag{2.1.8}$$

As we noted earlier, the associativity of the product together with (2.1.7) completely determines $\vee$ on arbitrary forms. Thus the vector space of exterior forms together with the antisymmetrised tensor product $\wedge$ is an exterior algebra, whereas the product $\vee$ turns the same vector space into a Clifford algebra. The products are related as in (2.1.7).

By quotienting the tensor algebra in a particular way we have been led to an algebra $\mathbf{C}(V, g)$ which satisfies the familiar relations

$$x \vee y + y \vee x = 2g(x, y) \qquad \forall x, y \in V. \tag{2.1.9}$$

It is because of this relation that the Clifford algebra is adapted to the study of orthogonal transformations of V. We would like to know if there are any other associative algebras, apart from the one we have constructed, whose product satisfies the relation (2.1.9). Suppose that $\mathbf{C}'(V, g)$ is an associative algebra with product $\triangle$ and that φ is a linear mapping of V into a subspace of $\mathbf{C}'(V, g)$, V', which generates the algebra, and that

$$\varphi(x)\triangle\varphi(y) + \varphi(y)\triangle\varphi(x) = 2g(x, y) \qquad \forall x, y \in V. \tag{2.1.10}$$

The right-hand side is understood to contain the identity in $\mathbf{C}'(V, g)$. The mapping φ can be extended to a homomorphism Φ from $T(V)$ to $\mathbf{C}'(V, g)$:

$$\Phi: T(V) \longrightarrow \mathbf{C}'(V, g)$$
$$\Phi(x \otimes y) = \varphi(x) \vartriangle \varphi(y). \tag{2.1.11}$$

Since V' generates $\mathbf{C}'(V, g)$, $\Phi[T(V)] = \mathbf{C}'(V, g)$. It follows from (2.1.10) and (2.1.11) that $\Phi\{x \otimes x - g(x, x)\} = 0$, and so $\Phi(J) = 0$ where J is the ideal used to construct $\underline{\mathbf{C}}(V, g)$. Thus if π is the mapping of $T(V)$ onto $\underline{\mathbf{C}}(V, g)$ defined by $\pi a = [a]$ then $\Phi = \psi \circ \pi$, where ψ is some homomorphism from $\underline{\mathbf{C}}(V, g)$ to $\mathbf{C}'(V, g)$. So the dimension of $\mathbf{C}'(V, g)$ certainly cannot be greater than that of $\underline{\mathbf{C}}(V, g)$, and if the dimensions are the same then the algebras are isomorphic. Since the kernel of ψ is an ideal of $\underline{\mathbf{C}}(V, g)$ if the dimension of $\mathbf{C}'(V, g)$ is less than that of $\underline{\mathbf{C}}(V, g)$ it must be a (non-trivial) quotient of that algebra. So the only possibility of a $\mathbf{C}'(V, g)$ which is not isomorphic to $\underline{\mathbf{C}}(V, g)$ arises if $\underline{\mathbf{C}}(V, g)$ is not simple. Conversely, it readily follows that if $\underline{\mathbf{C}}(V, g)$ is not simple then any quotient satisfies the conditions assumed for $\mathbf{C}'(V, g)$. Sometimes any algebra like $\mathbf{C}'(V, g)$ is called a Clifford algebra, the algebra $\underline{\mathbf{C}}(V, g)$ being termed the universal Clifford algebra.

From now on, unless indicated otherwise, by Clifford algebra we shall mean the algebra of the vector space of exterior forms with the product given in (2.1.7), and shall reserve the notation $\mathbf{C}(V, g)$ for this algebra. We shall also henceforth omit the symbol $\vee$, it being understood that juxtapositioning of exterior forms denotes this product. Although the Clifford algebra is not a $\mathbb{Z}$-graded algebra the vector space of exterior forms is a $\mathbb{Z}$-graded vector space and it will be convenient to use the decomposition into $\mathbb{Z}$-homogeneous subspaces:

$$\mathbf{C}(V, g) = \sum_{p=0}^{n} \mathcal{S}_p\,(\mathbf{C}(V, g)) \tag{2.1.12}$$

where n is the dimension of V and the projection operators $\mathcal{S}_p$ project out the homogeneous subspaces of p-forms. If A and B are homogeneous of degree p and q respectively then their Clifford product will not in general be homogeneous; rather

$$AB = \mathcal{S}_{p+q}(AB) + \mathcal{S}_{p+q-2}(AB) + \ldots + \mathcal{S}_{|p-q|}(AB). \tag{2.1.13}$$

This follows directly from (2.1.7) and (2.1.8). If φ and ψ are arbitrary elements of the algebra then

$$\mathcal{S}_0(\varphi\psi) = \sum_p \mathcal{S}_0(\varphi_p \psi_p) \tag{2.1.14}$$

where $\varphi_p \equiv \mathcal{S}_p\varphi$ and (2.1.13) has been used. If g_p denotes the metric on p-forms induced from g, as introduced in the previous chapter, then we may introduce a metric on inhomogeneous forms, G, by defining

$$G(\varphi, \psi) = \sum_p g_p(\varphi_p, \psi_p) \tag{2.1.15}$$

that is, G is diagonal in the homogeneous subspaces. This metric on forms can be related to Clifford multiplication

$$G(\varphi, \psi) = \mathcal{S}_0(\varphi^\xi \psi). \tag{2.1.16}$$

From (2.1.14) the right-hand side is seen to be diagonal in the homogeneous components of φ and ψ and so to verify (2.1.16) all we need to check is that $g_p(\varphi_p, \psi_p) = \mathcal{S}_0(\varphi_p^\xi \psi_p)$. Since both sides are linear in φ_p and ψ_p it suffices to consider the case of φ_p and ψ_p products of orthonormal 1-forms. If $\varphi_p = a^1 a^2 \ldots a^p$ and $\psi_p = b^1 b^2 \ldots b^p$ then from (2.1.7)

$$\mathcal{S}_0(\varphi_p^\xi \psi_p) = \mathrm{i}_{\tilde{a}_p} \ldots \mathrm{i}_{\tilde{a}_i} (b^1 b^2 \ldots b^p).$$

If the $\{a^i\}$ and $\{b^i\}$ are subsets of an orthonormal basis then the right-hand side is zero unless these sets are the same up to a relabelling. Since

$$\begin{aligned} \mathrm{i}_{\tilde{a}_p} \ldots \mathrm{i}_{\tilde{a}_i} (a^1 a^2 \ldots a^p) &= g(a^1, a^1) g(a^2, a^2) \ldots g(a^p, a^p) \\ &= g_p(a^1 a^2 \ldots a^p, a^1 a^2 \ldots a^p) \end{aligned}$$

we have verified (2.1.16).

One trivial result that is important for calculations is

$$\mathcal{S}_0(\varphi\psi) = \mathcal{S}_0(\psi\varphi) \tag{2.1.17}$$

as

$$\mathcal{S}_0(\varphi\psi) = \sum_p \mathcal{S}_0(\varphi_p \psi_p) = \sum_p (-1)^{[p/2]} g_p(\varphi_p, \psi_p)$$

where $[p/2]$ denotes the integer part of $p/2$, and the result follows from the symmetry of g_p.

It will sometimes be useful to expand an arbitrary element of the Clifford algebra in a G-orthonormal basis. If $\{e^a\}$ is a g-orthonormal basis then $\{e^A\}$ is a G-orthonormal basis where the multi-index A takes on all naturally ordered sequences of distinct indices. We use the notation

$$e^{12\ldots p} \equiv e^1 \wedge e^2 \wedge \ldots \wedge e^p = e^1 e^2 \ldots e^p.$$

If $g(e^a, e^b) = \eta^{ab}$ and η_{ab} denotes the inverse matrix then we set $e_a = \eta_{ab} e^b$, giving e_A an obvious meaning. Then $\mathcal{S}_0(e_A^\xi e^B) = \delta_A{}^B$ where

$\delta_A{}^B$ denotes the Krönecker function that takes the value zero, unless the sequences A and B are the same in which case its value is one. If a is any element of the Clifford algebra then we can expand in this basis

$$a = \sum_A \mathcal{S}_0(ae_A{}^{\xi})e^A. \tag{2.1.18}$$

The Hodge dual of a form may also be related to Clifford multiplication. The definition of the Hodge dual, (1.4.5), of ψ_p, $*\psi_p$, is given by $\varphi_p \wedge *\psi_p = g_p(\varphi_p, \psi_p)*1$ for all p-forms φ_p. Setting $z \equiv *1$ (2.1.16) enables this to be rewritten as $\varphi_p \wedge *\psi_p = \mathcal{S}_0(\varphi_p^{\xi}\psi_p)z = \mathcal{S}_0(\varphi_p\psi_p^{\xi})z$. It immediately follows from (2.1.7) that $\mathcal{S}_n(\varphi_p*\psi_p) = \varphi_p \wedge *\psi_p$ and from (2.1.13) that $\mathcal{S}_0(\varphi_p\psi_p^{\xi})z = S_n(\varphi_p\psi_p^{\xi})$. Thus $\mathcal{S}_n(\varphi_p*\psi_p) = \mathcal{S}_n(\varphi_p\psi_p^{\xi}z)$ giving

$$*\psi = \psi^{\xi}z. \tag{2.1.19}$$

Exercise 2.1
If $\{e^a\}$, $\{X_b\}$ are any dual bases, $e^a(X_b) = \delta^a_b$, and α, β are any exterior forms, derive the relations

$$\alpha \vee \beta = \sum_{p=0}^{n} \frac{(-1)^{[p/2]}}{p!}(\eta^p \mathrm{i}_{X_{a_1}} \ldots \mathrm{i}_{X_{a_p}}\alpha) \wedge (\mathrm{i}_{\widetilde{e^{a_1}}} \ldots \mathrm{i}_{\widetilde{e^{a_p}}}\beta)$$

$$\alpha \wedge \beta = \sum_{p=0}^{n} \frac{(-1)^{[p/2]}}{p!}(\mathrm{i}_{X_{a_1}} \ldots \mathrm{i}_{X_{a_p}}\eta^p\alpha) \vee (\mathrm{i}_{\widetilde{e^{a_1}}} \ldots \mathrm{i}_{\widetilde{e^{a_p}}}\beta).$$

2.2 The Structure of the Real Clifford Algebras

In this section we take the field F to be the real numbers $\mathbb{R}$. We shall determine the structure of $\mathbf{C}(V, g)$ for all real symmetric non-degenerate g. If g has a signature with p plus and q minus signs, then the structure of the Clifford algebra can only depend on p and q. We shall anticipate this by setting $\mathbf{C}(V, g) \equiv \mathbf{C}_{p,q}(\mathbb{R})$.

One thing we know about the Clifford algebras is their dimension. Since we have identified the underlying vector space with the space of exterior forms the dimension of $\mathbf{C}_{p,q}(\mathbb{R})$ is 2^n where $p + q = n$. Given a basis for V we can repeatedly use (2.1.7) to construct a multiplication table for the Clifford algebras, and in this sense we know its structure completely. What we would like to do is to relate the Clifford algebra to other 'standard' algebras. In particular we have already seen that if $\mathbf{C}_{p,q}(\mathbb{R})$ is not simple then we can construct a smaller algebra that satisfies the relation (2.1.9). Some low-dimensional examples will clarify how (2.1.7) is used in practice. It will also transpire that we can relate any Clifford algebra to a number of low-dimensional Clifford algebras.

We will denote an orthonormal basis for V by $\{e^i, f^j\}$ for $i = 1, \ldots, p$, $j = 1, \ldots, q$ where $g(e^i, e^i) = -g(f^j, f^j) = 1$. It will be convenient to set $z = e^1 \wedge e^2 \wedge \ldots e^p \wedge f^1 \wedge \ldots \wedge f^q$.

The two-dimensional algebra $\mathbf{C}_{0,\,1}(\mathbb{R})$ has as basis $\{1, f\}$ where $f^2 = -1$. It is thus isomorphic to the algebra of complex numbers,

$$\mathbf{C}_{0,\,1}(\mathbb{R}) \simeq \mathbb{C}(\mathbb{R}). \tag{2.2.1}$$

A basis for $\mathbf{C}_{1,\,0}(\mathbb{R})$ is $\{1, e\}$, and this algebra might not be so immediately recognisable. If $P_1 = \frac{1}{2}(1+e)$ and $P_2 = \frac{1}{2}(1-e)$ then $\{P_1, P_2\}$ is obviously a new basis. The multiplication table is given in table 2.1. Thus P_1 and P_2 each span mutually orthogonal one-dimensional subalgebras, each of which is isomorphic to the field $\mathbb{R}$, so that

$$\mathbf{C}_{1,\,0}(\mathbb{R}) \simeq \mathbb{R} \oplus \mathbb{R}. \tag{2.2.2}$$

Table 2.1

	P_1	P_2
P_1	P_1	0
P_2	0	P_2

Rather than simply determine the structure of $\mathbf{C}_{1,\,1}(\mathbb{R})$ we shall take this opportunity to demonstrate some general features of associative algebras. A basis is $\{1, e, f, z\}$ where $z \equiv e \wedge f = ef$ since e and f are orthogonal. The multiplication table is readily completed (see table 2.2). (For example, $ez = eef = f$ since e is of unit norm.)

Table 2.2

	1	e	f	z
1	1	e	f	z
e	e	1	z	f
f	f	$-z$	-1	e
z	z	$-f$	$-e$	1

It is straightforward to see that the identity spans the centre. An immediate consequence of this is that $\mathbf{C}_{1,\,1}(\mathbb{R})$ is not reducible. More generally, all $\mathbf{C}_{p,q}(\mathbb{R})$ have an identity. If the algebra were reducible then the identity would be the sum of the identities in the component algebras. The identities of the component algebras must all lie in the centre, so if an algebra with a unit element is reducible then the identity can be written as a sum of pairwise orthogonal central idempotents. Conversely if the centre of an algebra contains a set of mutually orthogonal idempotents then the algebra is reducible. Thus either $\mathbf{C}_{1,1}(\mathbb{R})$ has a radical or it is simple. The multiplication table enables the two-dimensional Clifford algebras we have already encountered to be recognised as subalgebras. Both $\{1, e\}$ and $\{1, z\}$ span subalgebras isomorphic to $\mathbb{R}\oplus\mathbb{R}$, whereas the algebra spanned by $\{1, f\}$ is isomorphic to $\mathbb{C}(\mathbb{R})$. We can use the pair of orthogonal idempotents in one of the $\mathbb{R}\oplus\mathbb{R}$ subalgebras to write $\mathbf{C}_{1,1}(\mathbb{R})$ as a sum of two left ideals. For example, if $P_1 = \frac{1}{2}(1 + z)$, $P_2 = \frac{1}{2}(1 - z)$ then $\mathbf{C}_{1,1}(\mathbb{R}) = \mathbf{C}_{1,1}(\mathbb{R})P_1 + \mathbf{C}_{1,1}(\mathbb{R})P_2$. Since $fP_1 = eP_1$ and $zP_1 = P_1$ a basis for the left ideal $\mathbf{C}_{1,1}(\mathbb{R})P_1$ is $\{P_1, eP_1\}$. Similarly a basis for $\mathbf{C}_{1,1}(\mathbb{R})P_2$ is $\{P_2, eP_2\}$. It is instructive to look at the multiplication table for the algebra in this basis (see table 2.3).

Table 2.3

	P_1	eP_1	P_2	eP_2
P_1	P_1	0	0	eP_2
eP_1	eP_1	0	0	P_2
P_2	0	eP_1	P_2	0
eP_2	0	P_1	eP_2	0

The left ideals $\mathbf{C}_{1,1}(\mathbb{R})P_1$ and $\mathbf{C}_{1,1}(\mathbb{R})P_2$ are both minimal; they contain no smaller left ideals. So P_1 and P_2 are primitive† idempotents, for if $P_1 = P + Q$ where P and Q are orthogonal idempotents then $\mathbf{C}_{1,1}(\mathbb{R})P_1 = \mathbf{C}_{1,1}(\mathbb{R})P + \mathbf{C}_{1,1}(\mathbb{R})Q$. The sum must be a direct vector space sum. For suppose that $bP = cQ$ for some b and c. Then since P is idempotent $bP = bPP$, but $bPP = cQP = 0$ since Q and P are orthogonal. Thus $b = c = 0$. So if P_1 were not primitive $\mathbf{C}_{1,1}(\mathbb{R})P_1$ would be a sum of two smaller left ideals. Could $\mathbf{C}_{1,1}(\mathbb{R})$ contain any two-sided ideals? Suppose I is a two-sided ideal and that $a \in \mathrm{I}$. We can write $a = a_1 + a_2$ where $a_1 \in \mathbf{C}_{1,1}(\mathbb{R})\ P_1$, $a_2 \in \mathbf{C}_{1,1}(\mathbb{R})P_2$. Now $\mathbf{C}_{1,1}(\mathbb{R})a_1$ is a left ideal which is contained in the left ideal $\mathbf{C}_{1,1}(\mathbb{R})P_1$ since a_1 is. But this left ideal is minimal and so $\mathbf{C}_{1,1}(\mathbb{R})a_1 = \mathbf{C}_{1,1}P_1$. Thus if $a_1 \neq 0$ there is a b such that $ba_1 = P_1$ and so $ba = P_1 + ba_2$

† The notion of 'primitive idempotents' is discussed in (A11)–(A19) of Appendix A.

and $baP_1 = P_1$, which shows that P_1 must be in I since a is. Similarly, there exists a c such that $caP_1 = eP_1$, which must be in I. But from the multiplication table we see that right multiplying P_1 and eP_1 by eP_2 generates the remainder of the basis for the whole algebra.

The situation is the same if we assume that $a_2 \neq 0$. Thus the only ideals are the zero ideal and the algebra itself which is thus simple. Wedderburn's structure theorem, together with Frobenius's theorem on real division algebras, shows that the only simple four-dimensional associative algebras over the reals are the total matrix algebra $\mathcal{M}_2(\mathbb{R})$ and the quaternions, $H(\mathbb{R})$. The quaternion algebra is a division algebra whose only idempotent is the identity and so we must have

$$\mathbf{C}_{1,1}(\mathbb{R}) \simeq \mathcal{M}_2(\mathbb{R}). \tag{2.2.3}$$

Of course we could have obtained this result directly, for if $\{\mathbf{e}_{ij}\}$, i, $j = 1, 2$ is an ordinary matrix basis for $\mathcal{M}_2(\mathbb{R})$ then a set of generators is $\{e, f\}$ where $e = \mathbf{e}_{12} + \mathbf{e}_{21}$, $f = \mathbf{e}_{12} - \mathbf{e}_{21}$. These generators anticommute and satisfy $e^2 = -f^2 = 1$.

A basis for $\mathbf{C}_{0,2}(\mathbb{R})$ is $\{1, f^1, f^2, z\}$ and the multiplication table is given in table 2.4. This may be recognised as the multiplication table of the standard basis for the quaternion algebra by relabelling $f^1 = i$, $f^2 = j$, $z = k$:

$$\mathbf{C}_{0,2}(\mathbb{R}) \simeq H(\mathbb{R}). \tag{2.2.4}$$

Table 2.4

	1	f^1	f^2	z
1	1	f^1	f^2	z
f^1	f^1	-1	z	$-f^2$
f^2	f^2	$-z$	-1	f^1
z	z	f^2	$-f^1$	-1

$\mathbf{C}_{0,3}(\mathbb{R})$ is generated by an orthonormal basis for V, $\{f^1, f^2, f^3\}$. Since $z = f^1f^2f^3$ it will commute with these generators, and hence must lie in the centre. Furthermore, $z^2 = 1$ and so $P_1 = \frac{1}{2}(1 + z)$, $P_2 = \frac{1}{2}(1 - z)$ are a pair of orthogonal idempotents in the centre. Thus $\mathbf{C}_{0,3}(\mathbb{R})$ is reducible, $\mathbf{C}_{0,3}(\mathbb{R}) = \mathbf{C}_{0,3}(\mathbb{R})P_1 \oplus \mathbf{C}_{0,3}(\mathbb{R})P_2$. A basis for $\mathbf{C}_{0,3}(\mathbb{R})$ is $\{1, f^1, f^2, f^3, f^1f^2, f^2f^3, f^3f^1, z\}$ and since $zP_1 = P_1$, $f^1f^2P_1 = -f^3P_1$, $f^2f^3P_1 = -f^1P_1$, $f^3f^1P_1 = -f^2P_1$ a basis for $\mathbf{C}_{0,3}(\mathbb{R})P_1$ is $\{P_1, f^1P_1, f^2P_1, f^3P_1\}$. The resulting multiplication table is given in table 2.5. The identity in this algebra is P_1. Again we have the quaternion algebra with a standard basis $\{P_1, f^1P_1, f^2P_1, -f^3P_1\}$. The mapping η is an automorphism of $\mathbf{C}_{0,3}(\mathbb{R})$, but maps one component algebra into the other since $\eta z = -z$. It thus establishes an isomorphism between these component algebras and so

$$\mathbf{C}_{0,3}(\mathbb{R}) \simeq H(\mathbb{R}) \oplus H(\mathbb{R}). \tag{2.2.5}$$

Table 2.5

	P_1	f^1P_1	f^2P_1	f^3P_1
P_1	P_1	f^1P_1	f^2P_1	f^3P_1
f^1P_1	f^1P_1	$-P_1$	$-f^3P_1$	f^2P_1
f^2P_1	f^2P_1	f^3P_1	$-P_1$	$-f^1P_1$
f^3P_1	f^3P_1	$-f^2P_1$	f^1P_1	$-P_1$

It is unlikely that we will recognise the sixteen-dimensional algebra $\mathbf{C}_{0,4}(\mathbb{R})$ by writing out the multiplication table. An orthonormal basis for V $\{f^1, f^2, f^3, f^4\}$ generates the algebra. These generators mutually anticommute and square to minus one. If we can find a new set of generators that splits into two mutually commuting subsets then these subsets will generate mutually commuting subalgebras. If the product of the dimensions of these subalgebras is the dimension of $\mathbf{C}_{0,4}(\mathbb{R})$ then we can express that algebra as the tensor product of these subalgebras. Such a set is provided by $\{f^1, z, f^2f^3, f^3f^4\}$. The first two elements certainly commute with the last two but we need to verify that they do indeed generate the algebra. We do this by checking that we can recover the original generators by forming sums of products of this new set. In fact, $f^1zf^2f^3 = f^4$, $f^1zf^3f^4 = f^2$ and so $f^1zf^3f^4f^2f^3 = -f^3$ and, indeed, we have a new set of generators. The generators $\{f^1, z\}$ mutually anticommute satisfying $z^2 = -(f^1)^2 = 1$. They therefore generate an algebra isomorphic to $\mathbf{C}_{1,1}(\mathbb{R})$, that is $\mathcal{M}_2(\mathbb{R})$. The anticommuting pair $\{f^2f^3; f^3f^4\}$ both square to minus one, and so they generate the quaternion algebra. (In the standard basis we may choose $\{i, j\}$ as generators.) Both $\mathcal{M}_2(\mathbb{R})$ and $H(\mathbb{R})$ are four dimensional and so we have

$$\mathbf{C}_{0,4}(\mathbb{R}) \simeq H(\mathbb{R}) \otimes \mathcal{M}_2(\mathbb{R}). \tag{2.2.6}$$

Of course, in a similar way, we could have quickly identified the structure of the algebras previously considered.

It has been anticipated that a knowledge of some low-dimensional Clifford algebras will enable the structure of an arbitrary Clifford algebra to be determined. In fact, given that we know the structure of $\mathbf{C}_{1,1}(\mathbb{R})$, $\mathbf{C}_{1,0}(\mathbb{R})$ and $\mathbf{C}_{0,q}(\mathbb{R})$ for $q = 1, 2, 3, 4$ the following determine the structure of all the real Clifford algebras:

$$\mathbf{C}_{p+1,q}(\mathbb{R}) \simeq \mathbf{C}_{q+1,p}(\mathbb{R}) \tag{2.2.7}$$

$$\mathbf{C}_{p+1,q+1}(\mathbb{R}) \simeq \mathbf{C}_{p,q}(\mathbb{R}) \otimes \mathbf{C}_{1,1}(\mathbb{R}) \tag{2.2.8}$$

$$\mathbf{C}_{p,\,q+4}(\mathbb{R}) \simeq \mathbf{C}_{p,q}(\mathbb{R}) \otimes \mathbf{C}_{0,4}(\mathbb{R}). \tag{2.2.9}$$

Before demonstrating the truth of the above assertion we have to prove these relations. This will be done by choosing suitable generators. A set of generators for $\mathbf{C}_{p+1,q}(\mathbb{R})$ is provided by an orthonormal basis for V, $\{e^i, f^j\}$ for $i = 1, \ldots, p+1$, $j = 1, \ldots, q$. Alternatively, we could generate the algebra with $\{e^{p+1}, e^{p+1}e^i, e^{p+1}f^j\}$, $i = 1, \ldots, p$, $j = 1, \ldots, q$. This follows since we can easily recover the original generators from products of this set. The new generators are mutually anticommuting and for $i = 1, \ldots, p$, $(e^{p+1}e^i)^2 = e^{p+1}e^ie^{p+1}e^i = -(e^{p+1})^2(e^i)^2 = -1$; similarly $(e^{p+1}f^j)^2 = 1$. So we have a set of mutually anticommuting generators, $q+1$ of which square to plus one and p of which square to minus one and so (2.2.7) indeed holds.

$\mathbf{C}_{p+1,q+1}(\mathbb{R})$ is generated by $\{e^{p+1}, e^i, f^{q+1}, f^j\}$ for $i = 1, \ldots, p$, $j = 1, \ldots, q$. A new set of generators are $\{e^{p+1}, f^{q+1}, e^{p+1}f^{q+1}e^i, e^{p+1}f^{q+1}f^j\}$ with $i = 1, \ldots, p$, $j = 1, \ldots, q$. (Although the notation assumes $p \geqslant 1$ and $q \geqslant 1$ the argument obviously goes through with $p = 0$ or $q = 0$.) We have only to verify that the original generators are recovered by products of the new set to be sure that they are indeed generators. The first pair of mutually anticommuting generators commute with the second mutually anticommuting pair. For $i = 1, \ldots, p$

$$(e^{p+1}f^{q+1}e^i)^2 = e^{p+1}f^{q+1}e^ie^{p+1}f^{q+1}e^i = (e^{p+1})^2f^{q+1}e^if^{q+1}e^i$$
$$= -(e^{p+1})^2(f^{q+1})^2(e^i)^2 = (e^i)^2 = 1.$$

Similarly $(e^{p+1}f^{q+1}f^j)^2 = -1$. Thus the second pair of the set generate $\mathbf{C}_{p,\,q}(\mathbb{R})$, whereas the first pair obviously generate $\mathbf{C}_{1,\,1}(\mathbb{R})$. The product of the dimensions of these mutually commuting subalgebras is indeed the dimension of $\mathbf{C}_{p+1,\,q+1}(\mathbb{R})$ and we have proved (2.2.8).

The proof of (2.2.9) proceeds in the same spirit. An orthonormal basis for V provides a set of mutually anticommuting generators for $\mathbf{C}_{p,\,q+4}(\mathbb{R})$. We partition the generators into two subsets, and form new generators out of the first subset and the elements of the second subset multiplied by the product of all the elements in the first set. If the first set is of even dimension, we will then have two mutually commuting subsets of generators. That is, we replace the generators

$$\{e^i, f^j, f^{q+1}, f^{q+2}, f^{q+3}, f^{q+4}\} \qquad i = 1, \ldots, p;\ j = 1, \ldots, q$$

with the set

$$\{\hat{z}e^i, \hat{z}f^j, f^{q+1}, f^{q+2}, f^{q+3}, f^{q+4}\} \qquad i = 1, \ldots, p;\ j = 1, \ldots, q$$

where $\hat{z} = f^{q+1}f^{q+2}f^{q+3}f^{q+4}$. Then $\hat{z}f^{q+1} = -f^{q+1}\hat{z}$, for example, and the last four generators commute with the first $p+q$. Since $\hat{z}e^i = e^i\hat{z}$,

$\hat{z}f^j = f^j\hat{z}$ for $i = 1, \ldots, p$, $j = 1, \ldots, q$ and $\hat{z}^2 = 1$ we have $\mathbf{C}_{p,q}(\mathbb{R})$ and $\mathbf{C}_{0,4}(\mathbb{R})$ as mutually commuting subalgebras. The dimensions of the algebras are such that we have proved (2.2.9). Of course we could equally well have shown that $\mathbf{C}_{p+4,q}(\mathbb{R}) \simeq \mathbf{C}_{p,q}(\mathbb{R}) \otimes \mathbf{C}_{4,0}(\mathbb{R})$.

The low-dimensional examples and periodicity relations we have given have been judiciously chosen to enable the structure of an arbitrary Clifford algebra to be determined. We show first how the structure of $\mathbf{C}_{p,q}(\mathbb{R})$ can be determined assuming $q > p$. Repeated use of (2.2.8) gives

$$\mathbf{C}_{p,q}(\mathbb{R}) \simeq \mathbf{C}_{0,q-p}(\mathbb{R}) \otimes \underbrace{\mathbf{C}_{1,1}(\mathbb{R}) \otimes \ldots \otimes \mathbf{C}_{1,1}(\mathbb{R})}_{p \text{ terms}}.$$

If we set $q - p = 4\lambda + \mu$ with $\mu < 4$ then use of (2.2.9) shows that

$$\mathbf{C}_{p,q}(\mathbb{R}) \simeq \mathbf{C}_{0,\mu}(\mathbb{R}) \otimes \underbrace{\mathbf{C}_{0,4}(\mathbb{R}) \otimes \ldots \mathbf{C}_{0,4}(\mathbb{R})}_{\lambda \text{ terms}} \otimes \underbrace{\mathbf{C}_{1,1}(\mathbb{R}) \otimes \ldots \otimes \mathbf{C}_{1,1}(\mathbb{R})}_{p \text{ terms}}.$$

Since we know the structure of all the $\mathbf{C}_{0,\mu}(\mathbb{R})$ for $\mu < 4$, we have expressed $\mathbf{C}_{p,q}(\mathbb{R})$ as a tensor product of factors of known structure. Now we do the same thing assuming that $p < q$; by (2.2.8)

$$\mathbf{C}_{p,q}(\mathbb{R}) \simeq \mathbf{C}_{p-q,0}(\mathbb{R}) \otimes \underbrace{\mathbf{C}_{1,1}(\mathbb{R}) \otimes \ldots \otimes \mathbf{C}_{1,1}(\mathbb{R})}_{q \text{ terms}}.$$

Now we use (2.2.7) for the first time:

$$\mathbf{C}_{p,q}(\mathbb{R}) \simeq \mathbf{C}_{1,p-q-1}(\mathbb{R}) \otimes \underbrace{\mathbf{C}_{1,1}(\mathbb{R}) \ldots \mathbf{C}_{1,1}(\mathbb{R})}_{q \text{ terms}}.$$

If $p - q = 1$ or 2 then there is nothing left to do, and in the former case we will need our knowledge of the structure of $\mathbf{C}_{1,0}(\mathbb{R})$. If not then one more application of (2.2.8) gives

$$\mathbf{C}_{p,q}(\mathbb{R}) \simeq \mathbf{C}_{0,p-q-2}(\mathbb{R}) \otimes \underbrace{\mathbf{C}_{1,1}(\mathbb{R}) \otimes \ldots \otimes \mathbf{C}_{1,1}(\mathbb{R})}_{q+1 \text{ terms}}.$$

If we set $p - q - 2 = 4\alpha + \beta$, with $\beta < 4$ then (2.2.9) produces

$$\mathbf{C}_{p,q}(\mathbb{R}) \simeq \mathbf{C}_{0,\beta}(\mathbb{R}) \otimes \underbrace{\mathbf{C}_{0,4}(\mathbb{R}) \otimes \ldots \otimes \mathbf{C}_{0,4}(\mathbb{R})}_{\alpha \text{ terms}} \otimes \underbrace{\mathbf{C}_{1,1}(\mathbb{R}) \otimes \ldots \otimes \mathbf{C}_{1,1}(\mathbb{R})}_{q+1 \text{ terms}}$$

Again we have expressed the algebra in terms of products of algebras whose structures are known. So what are the possibilities for $\mathbf{C}_{p,q}(\mathbb{R})$? Since $\mathbf{C}_{1,1}(\mathbb{R}) \simeq \mathcal{M}_2(\mathbb{R})$ and $\mathcal{M}_m(\mathbb{R}) \otimes \mathcal{M}_n(\mathbb{R}) \simeq \mathcal{M}_{mn}(\mathbb{R})$, repeated tensor products of $\mathbf{C}_{1,\,1}(\mathbb{R})$ are isomorphic to a total matrix algebra. We have seen that $\mathbf{C}_{0,\,4}(\mathbb{R}) \simeq H(\mathbb{R})\otimes\mathcal{M}_2(\mathbb{R})$, and since $H(\mathbb{R})\otimes H(\mathbb{R}) \simeq \mathcal{M}_4(\mathbb{R})$ the tensor product of $C_{0,\,4}(\mathbb{R})$ an even number of times is isomorphic to a total matrix algebra, whereas an odd number of products produces the product of the quaternions and a total matrix algebra. So any Clifford algebra is either isomorphic to a total matrix algebra or isomorphic to the tensor product of $\mathbf{C}_{0,\,\beta}(\mathbb{R})$, $\beta < 4$, with either a total matrix algebra, or the tensor product of the quaternions and a total matrix algebra. In the former case equations (2.2.1), (2.2.4) and (2.2.5) show that

$$\mathbf{C}_{p,\,q}(\mathbb{R}) = \mathscr{A}(\mathbb{R})\otimes\mathcal{M}_r(\mathbb{R}) \tag{2.2.10}$$

where $\mathscr{A} \simeq \mathbb{C}$, H or $H\oplus H$, and $r^2 \dim\mathscr{A} = 2^{p+q}$. Since $\mathbb{C}(\mathbb{R})\otimes H(\mathbb{R}) \simeq \mathbb{C}(\mathbb{R})\otimes\mathcal{M}_2(\mathbb{R})$, and as we have already noted $H(\mathbb{R})\otimes H(\mathbb{R}) \simeq \mathcal{M}_4(\mathbb{R})$ the second case would lead to (2.2.10) with $\mathscr{A} = \mathbb{C}$, $\mathbb{R}$ or $\mathbb{R} + \mathbb{R}$. So any real Clifford algebra can be expressed as in (2.2.10) with $\mathscr{A} = \mathbb{R}$, $\mathbb{C}$, H, $\mathbb{R}\oplus\mathbb{R}$ or $H\oplus H$. Since we know the dimension of the real Clifford algebras their structure is characterised by the algebra $\mathscr{A}$. The possibilities for $\mathscr{A}$ show that the real Clifford algebras are either simple or semi-simple, in the latter case being the direct sum of two isomorphic simple components. Obviously the values of p and q determine $\mathscr{A}$, in fact from (2.2.8) it can be seen that $\mathscr{A}$ is determined by $p - q$. Two applications of (2.2.9) give

$$\begin{aligned}\mathbf{C}_{p,\,q+8}(\mathbb{R}) &\simeq \mathbf{C}_{p,\,q+4}(\mathbb{R})\otimes\mathbf{C}_{0,\,4}(\mathbb{R}) \simeq \mathbf{C}_{p,\,q}(\mathbb{R})\otimes\mathbf{C}_{0,\,4}(\mathbb{R})\otimes\mathbf{C}_{0,\,4}(\mathbb{R})\\ &\simeq \mathbf{C}_{p,\,q}(\mathbb{R})\otimes H(\mathbb{R})\otimes\mathcal{M}_2(\mathbb{R})\otimes H(\mathbb{R})\otimes\mathcal{M}_2(\mathbb{R}) \quad \text{(by (2.2.6))}\end{aligned}$$

thus $\mathbf{C}_{p,\,q+8}(\mathbb{R}) \simeq \mathcal{M}_{16}(\mathbb{R})\otimes\mathbf{C}_{p,q}(\mathbb{R})$. So in fact $\mathscr{A}$ is determined by $p - q \bmod 8$. The low-dimensional algebras given in equations (2.2.1) to (2.2.6) provide examples of $p - q \bmod 8$ being 7, 1, 0, 6, 5 and 4. So all that is missing is $p - q \bmod 8$ equal to 2 and 3. From (2.2.7) we have $\mathbf{C}_{2,\,0}(\mathbb{R}) \simeq \mathbf{C}_{1,\,1}(\mathbb{R}) \simeq \mathcal{M}_2(\mathbb{R})$ and $\mathbf{C}_{3,\,0}(\mathbb{R}) \simeq \mathbf{C}_{1,\,2}(\mathbb{R})$, and so by (2.2.8), $\mathbf{C}_{3,\,0}(\mathbb{R}) \simeq \mathbf{C}_{1,\,1}(\mathbb{R})\otimes\mathbf{C}_{0,\,1}(\mathbb{R}) \simeq \mathcal{M}_2(\mathbb{R})\otimes\mathbb{C}(\mathbb{R})$. We now have the structure of all the Clifford algebras, namely $\mathbf{C}_{p,\,q}(\mathbb{R}) \simeq \mathscr{A}\otimes\mathcal{M}$ where $\mathscr{A}$ is given in table 2.6. Some of this table is easy to understand and remember. If $p + q$ is even, then $\mathbf{C}_{p,\,q}(\mathbb{R})$ is central simple, whereas for $p + q$ odd the centre is spanned by the identity and z. If $z^2 = -1$ then the centre must be $\mathbb{C}$, and this will be the case if $p - q \bmod 8$ is 3 or 7. If $z^2 = 1$ then the centre is isomorphic to $\mathbb{R}\oplus\mathbb{R}$ and the algebra is reducible. It can be checked that $z^2 = 1$ for $p - q \bmod 8$ equal to 1 or 5. The involution ξ will induce an involution on the components of one of the reducible algebras if and only if $z^\xi = z$. The only reducible Clifford algebras occur when V has odd dimension and in that case $z^{\xi\eta} = -z^\xi$ and so either ξ or $\xi\eta$ induce an involution on the simple components.

Table 2.6

$p - q \bmod 8$	$\mathscr{A}$
0 , 2	$\mathbb{R}$
3 , 7	$\mathbb{C}$
4 , 6	H
1	$\mathbb{R} \oplus \mathbb{R}$
5	$H \oplus H$

Of paramount physical importance is the algebra $\mathbf{C}_{3,1}(\mathbb{R})$. From table 2.6 we see that $\mathbf{C}_{3,1}(\mathbb{R}) \simeq \mathscr{M}_4(\mathbb{R})$ and so the algebra admits an ordinary matrix basis $\{\mathbf{e}_{ij}\}$ with $i, j = 1, \ldots, 4$. It is instructive to construct such a basis. This construction provides a concrete example of Wedderburn's structure theorem for simple algebras. The identity is of rank four and first we seek a set of four pairwise orthogonal primitive idempotents. We seek an a and b which commute and square to one, for then taking all sign choices the set $\{\frac{1}{2}(1 \pm a)\frac{1}{2}(1 \pm b)\}$ consists of pairwise orthogonal idempotents. For example, if $\{e^a\}$, $a = 0, 1, 2, 3$ is an orthonormal coframe with $(e^0)^2 = -1$ we choose $a = e^1$, $b = e^{02}$ and set

$$\begin{aligned} P_1 &= \tfrac{1}{4}(1 + e^1)(1 + e^{02}) \\ P_2 &= \tfrac{1}{4}(1 + e^1)(1 - e^{02}) \\ P_3 &= \tfrac{1}{4}(1 - e^1)(1 + e^{02}) \\ P_4 &= \tfrac{1}{4}(1 - e^1)(1 - e^{02}) \end{aligned} \tag{2.2.11}$$

where $e^{02} = e^0 \wedge e^2$. These four primitives are all similar, for example

$$\begin{aligned} e^3 P_1 (e^3)^{-1} &= P_3 \\ e^0 P_1 (e^0)^{-1} &= P_4 \\ e^{03} P_1 (e^{03})^{-1} &= P_2. \end{aligned} \tag{2.2.12}$$

Thus, $e^{03}P_1 \subset P_2\mathbf{C}_{3,1}(\mathbb{R})P_1$, $e^3P_1 \subset P_3\mathbf{C}_{3,1}(\mathbb{R})P_1$ and $e^0P_1 \subset P_4\mathbf{C}_{3,1}(\mathbb{R})P_1$ and we set

$$\begin{aligned} \mathbf{e}_{11} &= P_1 \\ \mathbf{e}_{21} &= e^{03} P_1 \\ \mathbf{e}_{31} &= e^3 P_1 \\ \mathbf{e}_{41} &= e^0 P_1. \end{aligned} \tag{2.2.13}$$

If the $\{\mathbf{e}_{1j}\}$ for $j = 1, \ldots, 4$ are given by

$$\begin{aligned} \mathbf{e}_{11} &= P_1 \\ \mathbf{e}_{12} &= (e^{03})^{-1} P_2 \\ \mathbf{e}_{13} &= (e^3)^{-1} P_3 \\ \mathbf{e}_{14} &= (e^0)^{-1} P_4 \end{aligned} \tag{2.2.14}$$

then $\mathbf{e}_{1j} \subset P_1 \mathbf{C}_{3,1}(\mathbb{R}) P_j$ and $\mathbf{e}_{1j}\mathbf{e}_{j1} = \mathrm{P}_1$. If now $\mathbf{e}_{ij} = \mathbf{e}_{i1}\mathbf{e}_{1j}$ then the $\mathbf{e}_{ij}$ do indeed form an ordinary matrix basis. The resulting $\mathbf{e}_{ij}$ are tabulated in table 2.7.

Table 2.7

$\mathbf{e}_{ij}$ $\mapsto$ ↓			
P_1	$e^{03}P_2$	e^3P_3	$-e^0P_4$
$e^{03}P_1$	P_2	e^0P_3	$-e^3P_4$
e^3P_1	$-e^0P_2$	P_3	$e^{03}P_4$
e^0P_1	$-e^3P_2$	$e^{03}P_3$	P_4

Any element of $\mathbf{C}_{3,1}(\mathbb{R})$ can be expanded in this basis. In particular, the orthonormal 1-forms can be written as

$$e^a = \sum_{i,j} \gamma^a{}_{ij}\, \mathbf{e}_{ij} \tag{2.2.15}$$

where the arrays of components form a real representation (or Majorana representation) of the familiar Dirac γ-matrices. In principle we can determine these components from the formula

$$\gamma^a_{ij} = \sum_k \mathbf{e}_{ki} e^a \mathbf{e}_{jk}$$

but it is here easier to proceed by inspection. From (2.2.11)

$$e^1 = P_1 + P_2 - P_3 - P_4$$

so if the components γ^1_{ij} are arranged as a matrix:

$$(\gamma^1_{ij}) = \begin{pmatrix} 1 & 0 & 0 & 0 \\ 0 & 1 & 0 & 0 \\ 0 & 0 & -1 & 0 \\ 0 & 0 & 0 & -1 \end{pmatrix}$$

and again from (2.2.11)

$$e^{02} = P_1 + P_3 - P_2 - P_4$$

so

$$\begin{aligned} e^2 &= -e^0P_1 - e^0P_3 + e^0P_2 + e^0P_4 \\ &= -\mathbf{e}_{41} - \mathbf{e}_{23} - \mathbf{e}_{32} - \mathbf{e}_{14} \end{aligned}$$

similarly

$$\begin{aligned} e^0 &= -e^2P_1 - e^2P_3 + e^2P_2 + e^2P_4 \\ &= -e^2e^{02}P_1 - e^2e^{02}P_3 - e^2e^{02}P_2 - e^2e^{02}P_4 \\ &= e^0P_1 + e^0P_3 + e^0P_2 + e^0P_4 \\ &= \mathbf{e}_{41} + \mathbf{e}_{23} - \mathbf{e}_{32} - \mathbf{e}_{14}. \end{aligned}$$

Thus we have

$$(\gamma^2_{ij}) = \begin{pmatrix} 0 & 0 & 0 & -1 \\ 0 & 0 & -1 & 0 \\ 0 & -1 & 0 & 0 \\ -1 & 0 & 0 & 0 \end{pmatrix}$$

$$(\gamma^0_{ij}) = \begin{pmatrix} 0 & 0 & 0 & -1 \\ 0 & 0 & 1 & 0 \\ 0 & -1 & 0 & 0 \\ 1 & 0 & 0 & 0 \end{pmatrix}.$$

Writing $e^3 = e^3(P_1 + P_2 + P_3 + P_4)$ gives

$$e^3 = \mathbf{e}_{31} - \mathbf{e}_{42} + \mathbf{e}_{13} - \mathbf{e}_{24}$$

and hence

$$(\gamma^3_{ij}) = \begin{pmatrix} 0 & 0 & 1 & 0 \\ 0 & 0 & 0 & -1 \\ 1 & 0 & 0 & 0 \\ 0 & -1 & 0 & 0 \end{pmatrix}.$$

Since the algebra $\mathbf{C}_{3,\,1}(\mathbb{R})$ is central simple, the transposition can be related to the involution ξ by an inner automorphism, namely

$$a^{\mathrm{T}} = \mathbf{C}^{-1}a^{\xi}\mathbf{C} \qquad \forall\, a \in \mathbf{C}_{3,\,1}(\mathbb{R}) \tag{2.2.16}$$

where $\mathbf{C}$ can be chosen such that $\mathbf{C}^{\xi} = \pm\mathbf{C}$. The choice of a $\mathbf{C}$ in (2.2.16) is determined up to a multiple of the centre, and so we have no choice in the symmetry of $\mathbf{C}$ under ξ. For the basis given in table 2.7 we may take $\mathbf{C} = e^1e^2e^3$, and have $\mathbf{C}^{\xi} = -\mathbf{C}$. Since e^1, e^2 and e^3 commute with $\mathbf{C}$ their components will form symmetric matrices (as we have already seen). The components of $\mathbf{C}$ are related to the charge conjugation matrix: exactly how will be seen in §2.8.

In the above example of $\mathbf{C}_{3,\,1}(\mathbb{R})$ the Clifford algebra was isomorphic to a total matrix algebra, generated by a real set of Dirac γ-matrices. As a less familiar example we now consider $\mathbf{C}_{4,0}(\mathbb{R}) \simeq H \otimes \mathcal{M}_2(\mathbb{R})$. (Although many physicists will be used to working with γ-matrices that satisfy the anticommutation relations with a positive-definite metric such matrices are always complex, generating the complexified Clifford algebra. This complexified algebra will be discussed in §2.7.) As usual z denotes the volume 4-form with here $z^2 = 1$. Thus a pair of orthogonal primitive idempotents is given by $P_1 = \frac{1}{2}(1 + z)$, $P_2 = \frac{1}{2}(1 - z)$. Since $P_2 = e^1 P_1 e^1$ we may choose a basis for $\mathcal{M}_2(\mathbb{R})$ as follows:

$$\mathbf{e}_{ij} \to \atop \downarrow \quad \begin{matrix} P_1 & e^1 P_2 \\ e^1 P_1 & P_2 \end{matrix}$$

$\{P_1, e^{23}P_1, e^{34}P_1, e^{24}P_1\}$ is a basis for $P_1\mathbf{C}_{4,0}(\mathbb{R})P_1$. This is a canonical basis for the quaternion algebra. Replacing P_1 with P_2 gives a basis for $P_2\mathbf{C}_{4,0}(\mathbb{R})P_2$. Thus, a quaternion subalgebra of $C_{4,0}(\mathbb{R})$ that commutes with all the $\mathbf{e}_{ij}$ is spanned by $\{1, e^{23}, e^{34}, e^{24}\}$.

2.3 The Even Subalgebra

The $\mathbb{Z}_2$-gradation of the Clifford algebra ensures that elements of even degree form a subalgebra, $\mathbf{C}^+_{p,q}(\mathbb{R})$. That is, $a \in \mathbf{C}^+_{p,q}(\mathbb{R})$ if and only if $\eta a = a$. Since V generates the Clifford algebra the 2-forms must generate the even subalgebra. However, a basis for 2-forms provides a set of generators with redundant elements, that is, a subset will generate the even subalgebra. If $\{e^i, f^j\}$ for $i = 1, \ldots, p+1$, $j = 1, \ldots, q$ are an orthonormal basis with $(e^i)^2 = -(f^j)^2 = 1$ then a set of generators, with no redundant members, for $\mathbf{C}^+_{p+1,q}(\mathbb{R})$ is $\{e^{p+1}e^i, e^{p+1}f^j\}$ for $i = 1, \ldots, p$, $j = 1, \ldots, q$. Since, for example, $e^{p+1}e^ie^{p+1}e^j = -e^ie^j$ we see that products of this set produce a basis for 2-forms and so the set generates $\mathbf{C}^+_{p+1,q}(\mathbb{R})$. It is not hard to see that there are no redundant generators. These generators are mutually anticommuting with $(e^{p+1}e^i)^2 = -1$ and $(e^{p+1}f^j)^2 = 1$ thus

$$\mathbf{C}^+_{p+1,q}(\mathbb{R}) \simeq \mathbf{C}_{q,p}(\mathbb{R}). \tag{2.3.1}$$

So if $\mathbf{C}_{p,q}(\mathbb{R}) \simeq \mathcal{A} \otimes \mathcal{M}_r$ and $\mathbf{C}^+_{p,q}(\mathbb{R}) \simeq \mathcal{B} \otimes \mathcal{M}_{r'}$ the algebra $\mathcal{B}$ is obtained by relabelling table 2.6. Since $\dim \mathbf{C}^+_{p,q}(\mathbb{R}) = \frac{1}{2} \dim \mathbf{C}_{p,q}(\mathbb{R})$ it follows that $r'^2 \dim \mathcal{B} = 2^{n-1}$ (see table 2.8). Whereas more than one value of $p - q \bmod 8$ can give rise to the same $\mathcal{A}$ or $\mathcal{B}$ no combination of $\mathcal{A}$ and $\mathcal{B}$ is repeated in table 2.8.

An important example of the even subalgebra is provided by $\mathbf{C}^+_{3,1}(\mathbb{R})$. From table 2.8 we see that this algebra is isomorphic to the algebra of

Table 2.8

$p-q$ mod 8	$\mathscr{A}$	$\mathscr{B}$
0	$\mathbb{R}$	$\mathbb{R}\oplus\mathbb{R}$
1	$\mathbb{R}\oplus\mathbb{R}$	$\mathbb{R}$
2	$\mathbb{R}$	$\mathbb{C}$
3	$\mathbb{C}$	H
4	H	$H\oplus H$
5	$H\oplus H$	H
6	H	$\mathbb{C}$
7	$\mathbb{C}$	$\mathbb{R}$

complex matrices of order two. The centre of the algebra, which is isomorphic to $\mathbb{C}$, is spanned by $\{1, z\}$ where, as usual, $z = e^1e^2e^3e^0$. The involution ξ leaves z invariant and so induces an involution on $\mathbf{C}^+_{3,1}(\mathbb{R})$ which is similar to transposition. That is, if $\{\boldsymbol{\varepsilon}_{\alpha\beta}\}$, α, $\beta = 1, 2$ is an ordinary matrix basis and the involution over $\mathbb{C}$, t, is defined by $\boldsymbol{\varepsilon}_{\alpha\beta}{}^t = \boldsymbol{\varepsilon}_{\beta\alpha}$ then there is a $c \in \mathbf{C}^+_{3,1}(\mathbb{R})$ such that

$$a^t = c^{-1}a^{\xi}c \qquad \forall\, a \in \mathbf{C}^+_{3,1}(\mathbb{R}) \tag{2.3.2}$$

with $c^{\xi} = \pm c$. The element c is determined up to a multiple of the centre and so we can have only one of these signs. In fact it must be the minus sign since elements are invariant under ξ if and only if they are in the centre, so c must be a 2-form. Thus, although similar, t and ξ cannot be equivalent since $c^{\xi} = -c$. Equation (2.3.2) may be naturally extended to define t on the whole of $\mathbf{C}_{3,1}(\mathbb{R})$.

If j is any odd regular element of $\mathbf{C}_{3,1}(\mathbb{R})$ then the involution $\mathscr{J}$ defined by

$$a^{\mathscr{J}} = ja^{\xi}j^{-1} \qquad \forall\, a \in \mathbf{C}_{3,1}(\mathbb{R}) \tag{2.3.3}$$

will induce an involution in $\mathbf{C}^+_{3,1}(\mathbb{R})$. Since $z^{\mathscr{J}} = -z$ this involution must be similar to Hermitian conjugation in $\mathbf{C}^+_{3,1}(\mathbb{R})$. That is, if † is the involution over $\mathbb{R}$ in $\mathbf{C}^+_{3,1}(\mathbb{R})$ defined by $\boldsymbol{\varepsilon}_{\alpha\beta}{}^{\dagger} = \boldsymbol{\varepsilon}_{\beta\alpha}$ then there is a $b \in \mathbf{C}^+_{3,1}(\mathbb{R})$ such that

$$a^{\mathscr{J}} = b^{-1}a^{\dagger}b \qquad \forall\, a \in \mathbf{C}^+_{3,1}(\mathbb{R}) \tag{2.3.4}$$

where $b^{\dagger} = \pm b$. Since b is only determined up to an element of the centre, which is $\mathbb{C}$, we can have either sign. This equation is naturally extended to define † on $\mathbf{C}_{3,1}(\mathbb{R})$. Equations (2.3.2) to (2.3.4) show that transposition and Hermitian conjugation in $\mathbf{C}^+_{3,1}(\mathbb{R})$ differ by an inner automorphism of $\mathbf{C}_{3,1}(\mathbb{R})$. This automorphism is not an inner automorphism of $\mathbf{C}^+_{3,1}(\mathbb{R})$. We have

$$a^{\dagger} = va^tv^{-1} \tag{2.3.5}$$

where $v = bjc$. The inner automorphism of $\mathbf{C}_{3,1}(\mathbb{R})$, $a \to vav^{-1}$, induces the involutary outer automorphism $\#$ on $\mathbf{C}^+_{3,1}(\mathbb{R})$, where $\#$ complex conjugates the matrix components in the basis $\{\boldsymbol{\varepsilon}_{\alpha\beta}\}$. It in fact follows that we can find a unit-norm 1-form x such that

$$a^\# = xax \qquad \forall\, a \in \mathbf{C}^+_{3,1}(\mathbb{R}) \text{ and } xc = cx \tag{2.3.6}$$

for an appropriate choice of c in (2.3.2). For we know that $a^\# = vav^{-1}$ for some odd v, and since $\#^2 = 1$, v^2 lies in the centre of $\mathbf{C}^+_{3,1}(\mathbb{R})$. Suppose that $v = y + wz$ for the 1-forms y and w. Then $v^2 = y^2 + w^2 + (yw - wy)z = y^2 + w^2 + 2(y \wedge w)z$. The first two terms are 0-forms, whilst the last is a 2-form, and so for $w \neq 0$ we must have $y = \lambda w$, $\lambda \in \mathbb{R}$. Thus $v = (\lambda - z)w$, and since $\lambda - z$ is in the centre of the even subalgebra $a^\# = waw^{-1}$ for all even a. Now $v^2 \neq 0$ and so $w^2 \neq 0$, so we have $a^\# = xax^{-1}$ where $x = w/(|w^2|)^{1/2}$, giving $x^2 = \pm 1$. Since $\boldsymbol{\varepsilon}_{\alpha\beta}{}^\# = \boldsymbol{\varepsilon}_{\alpha\beta}$, x must commute with the matrix basis, giving $\mathrm{i}_{\tilde{x}}\boldsymbol{\varepsilon}_{\alpha\beta} = 0$. Thus the $\boldsymbol{\varepsilon}_{\alpha\beta}$ must lie in the even subalgebra of the orthogonal complement to x, whereas $\mathbf{C}^+_{2,1}(\mathbb{R}) \simeq \mathcal{M}_2(\mathbb{R})$, $\mathbf{C}^+_{3,0}(\mathbb{R}) \simeq H$ and so we must have $x^2 = 1$. We can choose the c of (2.3.2) to lie in the subalgebra $\mathbf{C}^+_{2,1}(\mathbb{R})$ and then $xc = cx$. We give an explicit example.

A basis for $\mathbf{C}^+_{3,1}(\mathbb{R})$ is $\{1, e^{01}, e^{02}, e^{03}, e^{12}, e^{23}, e^{31}, z\}$, where we use the previously introduced notation. In exactly the same way as we constructed a matrix basis for $\mathbf{C}_{3,1}(\mathbb{R})$, we can construct the matrix basis given in table 2.9 for $\mathbf{C}^+_{3,1}(\mathbb{R})$ where $P^+_1 = \frac{1}{2}(1 + e^{02})$ and $P^+_2 = \frac{1}{2}(1 - e^{02})$. This matrix basis spans the even subalgebra associated with the vector space spanned by $\{e^0, e^2, e^3\}$. We may choose the c of equation (2.3.2) to be e^{23}. The 1-form e^1 commutes with the matrix basis and squares to one, and we may choose it to be the x of equation (2.3.6). This element can be used together with the primitives in the even subalgebra to form primitives in the full algebra. For example, if $P_1 = \frac{1}{2}(1 + x)P^+_1$, $P_2 = \frac{1}{2}(1 + x)P^+_2$, $P_3 = \frac{1}{2}(1 - x)P^+_1$ and $P_4 = \frac{1}{2}(1 - x)P^+_2$ then we have a set of pairwise orthogonal primitives of $\mathbf{C}_{3,1}(\mathbb{R})$. These are the primitives used to construct the matrix basis given in table 2.7. Notice that the involution that corresponded to transposition in that matrix basis induces Hermitian conjugation in the basis for the even subalgebra given here.

Table 2.9

$\boldsymbol{\varepsilon}_{\alpha\beta} \rightarrow$ $\downarrow$		
	P^+_1	$e^{03}P^+_2$
	$e^{03}P^+_1$	P^+_2

2.4 The Clifford Group

Those regular (that is, invertible) elements, s, such that

$$sxs^{-1} \in V \qquad \forall x \in V \tag{2.4.1}$$

form the *Clifford group*, Γ. It is straightforward to see that they do indeed form a group. The *vector representation* of Γ, χ, maps Γ into the group of automorphisms of the Clifford algebra:

$$\begin{aligned} \chi : \Gamma &\longrightarrow \text{Aut}\, \mathbf{C}_{p,q}(\mathbb{R}) \\ s &\longmapsto \chi(s) \qquad \text{where } \chi(s)x = sxs^{-1}. \end{aligned} \tag{2.4.2}$$

Since

$$2g(\chi(s)x, \chi(s)y) = sxs^{-1}sys^{-1} + sys^{-1}sxs^{-1} = 2g(x, y)$$

χ clearly maps the Clifford group into the orthogonal group. If n is the dimension of V then the range of χ depends on n. If n is even then

$$\chi(\Gamma) = \mathrm{O}(p, q) \tag{2.4.3a}$$

whereas for n odd

$$\chi(\Gamma) = \mathrm{SO}(p, q). \tag{2.4.3b}$$

Let σ be any orthogonal transformation on V. Then since V generates the Clifford algebra, σ extends uniquely to an automorphism of the algebra, that is, we define $\sigma(x_1x_2 \ldots x_p) = \sigma x_1 \sigma x_2 \ldots \sigma x_p$. If n is even then the Clifford algebra is central simple and all automorphisms are inner, so in this case $\chi(\Gamma) = \mathrm{O}(p, q)$. If n is odd then the centre is spanned by $\{1, z\}$, the identity and the volume n-form. Clearly, any orthogonal automorphism that does not leave the volume n-form invariant cannot be inner. However, any automorphism that does leave the centre invariant is inner. For if $C_{p,q}(\mathbb{R})$ is simple all automorphisms over the centre are inner. If $C_{p,q}(\mathbb{R})$ is not simple then it is the sum of two central simple components

$$C_{p,q}(\mathbb{R}) = C_{p,q}(\mathbb{R})P_1 \oplus C_{p,q}(\mathbb{R})P_2$$

where $\{P_1, P_2\}$ are orthogonal idempotents that span the centre. If σ is an orthogonal automorphism that leaves the centre invariant then it induces an automorphism on the simple components, and this must be an inner automorphism of the component algebras. That is, for any a, $\sigma(aP_i) = S_i(aP_i)S_i^{-1}$ where $S_iS_i^{-1} = P_i$, the identity in $\mathbf{C}_{p,q}(\mathbb{R})P_i$, $i = 1, 2$. If $S = S_1 + S_2$ then $S^{-1} = S_1^{-1} + S_2^{-1}$, for $SS^{-1} = S_1S_1^{-1} + S_2S_2^{-1}$ since $S_1S_2^{-1} = S_2S_1^{-1} = 0$ and $P_1 + P_2 = 1$. Now

$$\begin{aligned} \sigma a &= \sigma(aP_1) + \sigma(aP_2) = S_1aP_1S_1^{-1} + S_2aP_2S_2^{-1} \\ &= (S_1 + S_2)(aP_1 + aP_2)(S_1 + S_2)^{-1} = SaS^{-1}. \end{aligned}$$

We have shown that for n odd any orthogonal automorphism that leaves the volume n-form invariant is inner, that is, $\chi(\Gamma) = \mathrm{SO}(p, q)$.

Obviously the Clifford algebra does not transform irreducibly under the vector representation of Γ, the $\mathbb{Z}$-homogeneous subspaces being preserved. In fact these spaces of p-forms carry irreducible representations.

It will be convenient to be able to express any element of the Clifford group in a standard form. To do this we firstly show how any element of the orthogonal group can be written in a standard form, as the product of reflections. Let y be a non-null (non-isotropic) vector with $g(y, y) = a$, $a \neq 0$. Then the reflection of x in the plane orthogonal to y is given by

$$S_y x = x - 2a^{-1} g(x, y) y \qquad \forall x \in V. \tag{2.4.4}$$

If we write

$$x = \frac{g(x, y)}{a} y + r$$

where r is orthogonal to y then

$$S_y x = r - \frac{g(x, y)}{a} y$$

so S_y indeed corresponds to the usual notion of a reflection. It is readily verified that reflections are orthogonal transformations, for

$$\begin{aligned} g(S_y x, S_y x) &= g(x, x) + 4a^{-2} g(x, y)^2 g(y, y) - 4a^{-1} g(x, y) g(x, y) \\ &= g(x, x). \end{aligned}$$

The following theorem has already been anticipated.

Any orthogonal transformation of a finite-dimensional vector space with non-degenerate bilinear form is expressible as the product of a finite number of reflections. (2.4.5)

The truth of this statement will be proved by induction on the dimension of the vector space V. Note firstly that any two vectors of the same non-zero length can be related by at most two reflections. For if $g(x, x) = g(y, y) \neq 0$ and $x - y$ is not null then

$$\begin{aligned} S_{x-y} x &= x - \frac{2g(x, x - y)}{g(x - y, x - y)} (x - y) \\ &= x - \frac{2[g(x, x) - g(x, y)]}{[g(x, x) + g(y, y) - 2g(x, y)]} (x - y) \\ &= x - (x - y) \qquad \text{if } g(x, x) = g(y, y) \\ &= y. \end{aligned}$$

If $x - y$ is null then $x + y$ cannot be since x and y are not. Then

$$S_{x+y}x = x - \frac{2g(x, x + y)}{g(x + y, x + y)}(x + y) = -y$$

and so $S_y S_{x+y} x = -S_y y = y$. Suppose now that (2.4.5) is true for n-dimensional orthogonal spaces and that V is of dimension $n + 1$. If y is any non-null vector then its conjugate space (the space of all vectors orthogonal to y) is an n-dimensional orthogonal space (since g is non-degenerate). Furthermore, since y is non-null the restriction of the non-degenerate g to its conjugate is also non-degenerate. If σ is any orthogonal transformation of V then, since it has the same length as y, σy can be transformed into y by the product of at most two reflections. That is, there exists a u which is a product of reflections such that $u\sigma y = y$. Since $u\sigma$ leaves y invariant it must transform the conjugate space into itself, that is it is an orthogonal transformation on this n-dimensional orthogonal space. By hypothesis then $u\sigma = v$, where v is a product of reflections and so $\sigma = u^{-1}v$ which is also a product of reflections. For $n = 1$ relation (2.4.5) is obviously true and so we have proved its general validity.

As a step towards writing an arbitrary element of the Clifford group in a standard form we observe the following.

If $x \in V$ and $g(x, x) \neq 0$ then $x \in \Gamma$ and $\chi(x) = \eta S_x$. (2.4.6)

It is sufficient to show that $\chi(x)y = -S_x y$ for $y \in V$ since V generates the algebra. We have

$$\chi(x)y = xyx^{-1} = \{2g(x, y) - yx\}x^{-1} = -y + 2g(x, y)x^{-1}$$

and since $x^2 = g(x, x) \neq 0$ then

$$x^{-1} = \frac{x}{g(x, x)} \qquad \text{and } xyx^{-1} = -y + \frac{2g(x, y)}{g(x, x)}x = -S_x y.$$

Together (2.4.5) and (2.4.6) give a canonical form for any element of the Clifford group.

If $s \in \Gamma$ then $s = \lambda x^i \ldots x^h$ where λ is in the centre and the x^i are non-isotropic vectors in V. (2.4.7)

Suppose firstly that n is odd, and so if $s \in \Gamma$, $\chi(s) \in \mathrm{SO}(p, q)$. Since $\det S_x = -1$ (as is readily seen in a basis consisting of x and vectors from its orthogonal complement) it follows that $\chi(x)$ can be written as an even number of reflections. If then $\chi(s) = S_{x^1} \ldots S_{x^h}$ with h even, then $\chi(s) = \chi(x^1 \ldots x^h)$. The kernel of the vector representation is obviously the centre and so (2.4.7) follows. If n is even then $\eta = \chi(z)$ where z is the volume n-form and $S_x = \chi(zx)$. Since zx is a product of

$n-1$ non-isotropic vectors it follows that for any $s \in \Gamma$, $\chi(s) = \chi(x^1 \ldots x^h)$, where h need not now be even, and so (2.4.7) again follows.

If n is even then the Clifford algebra is central simple and so in this case elements of the Clifford group are even or odd. If $\Gamma^{\pm}$ is the subgroup of Γ consisting of all elements that are either even or odd, then for n odd $\Gamma^{\pm}$ is a non-trivial subgroup. When n is odd the vector representation maps the Clifford group onto the special orthogonal group and not the whole orthogonal group. The *twisted vector representation* is introduced to map $\Gamma^{\pm}$ onto $O(p, q)$ for n odd as well as even:

$$\varphi : \Gamma^{\pm} \longrightarrow \operatorname{Aut} \mathbf{C}_{p,q}(\mathbb{R})$$

$$s \longmapsto \varphi(s) \text{ where } \varphi(s)x = s^{\eta} x s^{-1} \qquad \text{for } x \in V. \tag{2.4.8}$$

Notice that (2.4.8) gives the action of $\varphi(s)$ on elements of V by Clifford multiplication, and since V generates the algebra the action on the whole algebra is defined:

$$\varphi(\Gamma^{\pm}) = O(p, q). \tag{2.4.9}$$

If x is a regular element of V then $x \in \Gamma^{\pm}$ and for $y \in V$ $\varphi(x)y = -\chi(x)y = S_x y$. Thus (2.4.9) follows from (2.4.5).

If n is even then $\Gamma^{\pm} = \Gamma$ and if $s^{\eta} = s$ then $\varphi(s) = \chi(s)$. If $s^{\eta} = -s$ then $\varphi(s)x = -sxs^{-1} = szxz^{-1}s^{-1} = \chi(sz)x$. The kernel of φ is the multiplicative group of non-zero real numbers, $\mathbb{R}^*$. For if $s^{\eta}xs^{-1} = x$ $\forall x \in V$ and s is written in terms of even and odd parts as $s = s_+ + s_-$ we have $s_+ x = x s_+$ and $x s_- + s_- x = 0$ $\forall x \in V$. The condition on the odd part of s is $i_{\tilde{x}} s_- = 0$ for all x and so $s_- = 0$. Thus s is in the even part of the centre which is $\mathbb{R}^*$. (Sometimes the Clifford group is defined differently. It is defined to be the group G consisting of all regular s such that $s^{\eta}xs^{-1} \in V$, $\forall x \in V$. It follows that $G = \Gamma^{\pm}$.)

The even elements in the Clifford group form a subgroup Γ^+. In this case the 'twisted' representation and the vector representation coincide and we have

$$\chi(\Gamma^+) = SO(p, q). \tag{2.4.10}$$

It follows from (2.4.7) that if n is even and $s \in \Gamma^+$ then $s = \lambda x^1 \ldots x^h$ where $\lambda \in \mathbb{R}$ and h is even. From (2.4.6) then $\chi(s) = (-1)^h S_{x^1} \ldots S_{x^h}$ which, since h is even, is an even number of reflections. Hence in this case $\chi(\Gamma^+) = SO(p, q)$. If n is odd then $\chi(\Gamma^+) \subset SO(p, q)$. It again follows from (2.4.7) that if $s \in \Gamma$ then $\chi(s) = \chi(x^1 \ldots x^h)$ for some x^i. If h were odd then $\chi(x^1 \ldots x^h) = \chi(zx^1 \ldots x^h)$ where z is the volume n-form which, for n odd, lies in the centre. If h is odd then $zx^1 \ldots x^h$ is even and in Γ^+ so $\chi(\Gamma^+) = \chi(\Gamma) = SO(p, q)$

If $s \in \Gamma^+$ and n is even then s is a product of an even number of non-singular 1-forms whereas if n is odd, s can be written as a product of non-singular $(n-1)$-forms. (2.4.11)

The case of n even is taken care of by (2.4.7). For n odd we can write $s = \lambda x^1 \ldots x^h$ with λ in the centre. Since s is even if h is even then $\lambda \in \mathbb{R}$ and $s = \pm\lambda(x^1 z) \ldots (x^h z)$. By redefining x^1 the factor of $\pm\lambda$ can be absorbed. If h were odd then λ would be proportional to the volume form, say $s = \mu z x^1 \ldots x^h$ with $\mu \in \mathbb{R}$. Once more, $s = \pm\mu(zx^1) \ldots (zx^h)$ and we have proved (2.4.11).

The kernel of the 'twisted' representation (and the vector representation for n even) is $\mathbb{R}^*$. By suitably 'normalising' elements of $\Gamma^\pm$ we obtain a subgroup whose image under these representations is the same as that of $\Gamma^\pm$, whereas the kernel is smaller. The norm homomorphism λ is a group homomorphism:

$$\begin{aligned} \lambda : \Gamma^\pm &\longrightarrow \mathbb{R}^* \\ s &\longmapsto \lambda(s) = s^\xi s. \end{aligned} \tag{2.4.12}$$

If s is invertible then so is s^ξ with $(s^\xi)^{-1} = (s^{-1})^\xi$. If $s \in \Gamma$ then $(sxs^{-1})^\xi = sxs^{-1}\ \forall x \in V$ so $(s^{-1})^\xi x s^\xi = sxs^{-1}$ or $s^\xi s x = x s^\xi s$. Since V generates the algebra $s^\xi s$ lies in the centre. If s is even or odd then $s^\xi s$ is even, and so λ does map $\Gamma^\pm$ into $\mathbb{R}^*$. It is straightforward to see that $\lambda(s_1 s_2) = \lambda(s_1)\lambda(s_2)$.

We denote the subgroup of $\Gamma^\pm$ which consists of those elements whose norm is plus or minus one by ${}_\pm\Gamma^\pm$; the subgroup of unit norm elements ${}_+\Gamma^\pm$. We define ${}_\pm\Gamma^+$ and ${}_+\Gamma^+$ similarly. The group ${}_\pm\Gamma^\pm$ is sometimes called PIN(p, q), ${}_\pm\Gamma^+$ called SPIN (p, q) and ${}_+\Gamma^+$ called SPIN$^+(p, q)$. If $s \in \Gamma^\pm$ then $s/(|\lambda(s)|)^{1/2} \in {}_\pm\Gamma^\pm$ and $\varphi\{s/(|\lambda(s)|)^{1/2}\} = \varphi(s)$ and so indeed the image of ${}_\pm\Gamma^\pm$ under φ is $O(p, q)$ and the kernel consists of the multiplicative group formed by plus and minus one, which is isomorphic to $\mathbb{Z}_2$. Similarly $\chi({}_\pm\Gamma^+) = SO(p, q)$ with kernel $\mathbb{Z}_2$.

We can introduce a slightly different norm, μ:

$$\begin{aligned} \mu : \Gamma^\pm &\longrightarrow \mathbb{R}^* \\ s &\longmapsto \mu(s) = s^{\xi\eta} s. \end{aligned} \tag{2.4.13}$$

Obviously $\mu(s) = \pm\lambda(s)$ depending on whether s is even or odd and so the only new subgroup is the group of those s with $\mu(s) = 1$, ${}^+\Gamma^\pm$.

The various subgroups of $\Gamma^\pm$ that have been introduced can be arranged as follows:

$${}_+\Gamma^+ \lhd {}_\pm\Gamma^\pm \begin{array}{l} \vdots \longrightarrow {}_+\Gamma^\pm \vdots \\ \vdots \longrightarrow {}^+\Gamma^\pm \vdots \\ \vdots \longrightarrow {}_\pm\Gamma^+ \vdots \end{array} \rhd {}_+\Gamma^+. \tag{2.4.14}$$

In this last diagram (2.4.14) the appropriate mathematical symbol here for $\longleftarrow$ is ◢ and for $\longrightarrow$ is ◣. Here ${}_{+}\Gamma^{+}$ ◢ ${}_{\pm}\Gamma^{\pm}$ denotes that ${}_{+}\Gamma^{+}$ is a normal subgroup of ${}_{\pm}\Gamma^{\pm}$. This is certainly the case, for if $\sigma \in {}_{+}\Gamma^{+}$ and $s \in {}_{\pm}\Gamma^{\pm}$ then $(s\sigma s^{-1})^{\eta} = s\sigma s^{-1}$ since $s^{\eta} = \pm s$ and $\lambda(s\sigma s^{-1}) = 1$ since $\lambda(s) = \pm 1$. If we look at all (four) quotients modulo ${}_{+}\Gamma^{+}$ this gives all (seven) quotients obtainable from this diagram. For example,

$$ {}_{\pm}\Gamma^{\pm}/{}_{+}\Gamma^{\pm} \simeq \frac{{}_{\pm}\Gamma^{\pm}/{}_{+}\Gamma^{+}}{{}_{+}\Gamma^{\pm}/{}_{+}\Gamma^{+}}. $$

Firstly consider ${}_{+}\Gamma^{\pm}/{}_{+}\Gamma^{+}$. If there are no odd elements of unit norm then obviously ${}_{+}\Gamma^{\pm} \simeq {}_{+}\Gamma^{+}$, so assume that σ is odd with $\lambda(\sigma) = 1$. If s_{-} is any odd element in ${}_{+}\Gamma^{\pm}$ then $s_{-} = (s_{-}\sigma^{-1})\sigma$, where $s_{-}\sigma^{-1}$ is even with norm plus one so that $s_{-} \sim \sigma$. Similarly if s_{+} is any even element $s_{+} \sim 1$ and so ${}_{+}\Gamma^{\pm}/{}_{+}\Gamma^{+}$ is the multiplicative group of plus and minus one, isomorphic to $\mathbb{Z}_2$. The argument above applies in exactly the same way to ${}^{+}\Gamma^{\pm}/{}_{+}\Gamma^{+}$ and ${}_{\pm}\Gamma^{+}/{}_{+}\Gamma^{+}$.

In the general case ${}_{\pm}\Gamma^{\pm}$ will contain even elements with norms plus and minus one, ${}_{+}\gamma^{+}$ and ${}_{-}\gamma^{+}$, and odd elements with both norms, ${}_{+}\gamma^{-}$ and ${}_{-}\gamma^{-}$. It readily follows that ${}_{\pm}\Gamma^{\pm}/{}_{+}\Gamma^{+}$ has four elements $[{}_{+}\gamma^{+}]$, $[{}_{-}\gamma^{+}]$, $[{}_{+}\gamma^{-}]$ and $[{}_{-}\gamma^{-}]$. Each element is labelled by an ordered pair of indices which take the values plus or minus one. The multiplication rule is defined by multiplying the values of these indices pairwise, and so ${}_{\pm}\Gamma^{\pm}/{}_{+}\Gamma^{+} \simeq \mathbb{Z}_2 \times \mathbb{Z}_2$. In various special cases this quotient group can have less than four elements as will be made clear in the following.

The kernel of φ from $\Gamma^{\pm}$ to $\mathrm{O}(p, q)$ is the group of plus and minus one, $\mathbb{Z}_2$, which is contained in all the subgroups in (2.6.14), and so the kernel of φ restricted to these subgroups is the same. Thus, for example

$$ \frac{\varphi(\Gamma^{\pm})}{\varphi({}_{+}\Gamma^{\pm})} \simeq \frac{\Gamma^{\pm}/\mathbb{Z}_2}{{}_{+}\Gamma^{\pm}/\mathbb{Z}_2} \simeq \frac{\Gamma^{\pm}}{{}_{+}\Gamma^{\pm}}. $$

We have already determined the images of ${}_{\pm}\Gamma^{\pm}$ and ${}_{\pm}\Gamma^{+}$ under φ, and now turn to the unit-norm subgroups.

If x is a non-singular element of V then $\varphi(x) = S_x$ and $\lambda(x) = g(x, x)$. So the image of unit-norm elements of $\Gamma^{\pm}$ under φ contains an even number of reflections in planes orthogonal to negative length, 'timelike', vectors. Such orthogonal transformations are said to be 'orthochronous'; the subgroup of orthochronous transformations being denoted $\mathrm{O}^{\uparrow}(p, q)$. For $x \in V$, $\mu(x) = -g(x, x)$ and so the unit μ-norm elements have images in the orthogonal group containing an even number of reflections in planes orthogonal to positive length, spacelike, vectors. Such orthogonal transformations will be called 'parity preserving' and the subgroup denoted $\mathrm{O}^{+}(p, q)$. If elements of $\mathrm{SO}(p, q)$ are orthochronous then they must also be parity preserving and so the notation $\mathrm{SO}^{+}(p, q)$ is unambiguous. The following summarises the images of the various subgroups under φ·

$$
\begin{aligned}
{}_{\pm}\Gamma^{\pm} &\xrightarrow{\varphi} \mathrm{O}(p,q)\\
{}_{+}\Gamma^{\pm} &\xrightarrow{\varphi} \mathrm{O}^{\uparrow}(p,q)\\
{}^{+}\Gamma^{\pm} &\xrightarrow{\varphi} \mathrm{O}^{+}(p,q)\\
{}_{\pm}\Gamma^{+} &\xrightarrow{\varphi} \mathrm{SO}(p,q)\\
{}_{+}\Gamma^{+} &\xrightarrow{\varphi} \mathrm{SO}^{+}(p,q).
\end{aligned}
\tag{2.4.15}
$$

If the dimension of V is even then the image of the Clifford group under χ is the same as under φ. If q is even then the volume form is of unit norm, $\lambda(z)=\mu(z)=1$. As has already been noted if s is an even element of Γ then $\chi(s)=\varphi(s)$, whereas if s is odd $\chi(s)=\varphi(sz)$. Since, for q even, $\lambda(sz)=\lambda(s)$ and $\mu(sz)=\mu(s)$ the images of the subgroups under χ are the same as under φ. If, however, q is odd then $\lambda(sz)=-\lambda(s)$ and $\mu(sz)=-\mu(s)$ and thus for s odd $\lambda(sz)=\mu(s)$ and $\mu(sz)=\lambda(s)$. So in this case $\chi({}_{+}\Gamma^{\pm})=\mathrm{O}^{+}(p,q)$ and $\chi({}^{+}\Gamma^{\pm})=\mathrm{O}^{\uparrow}(p,q)$.

The groups $\mathrm{O}^{\uparrow}(p,q)$ and $\mathrm{O}^{+}(p,q)$ have been identified with subgroups whose elements contain an even number of reflections in timelike and spacelike planes respectively. (A timelike (spacelike) plane is the conjugate of a timelike (spacelike) vector.) The nomenclature reflects the fact that these groups preserve the timelike and spacelike orientations of V in a way that will now be defined. Let V be written as a direct sum of a p-dimensional positive-definite orthogonal space and a q-dimensional negative-definite conjugate space, $V=P\oplus Q$. If $\sigma\in\mathrm{O}(p,q)$ then we define a linear mapping on P:

$$
\begin{aligned}
m(\sigma):P &\longrightarrow P\\
x &\longmapsto m(\sigma)x=\mathcal{P}(\sigma x)
\end{aligned}
$$

where $\mathcal{P}$, $\mathcal{Q}$ denote the projections onto the subspaces P and Q. This mapping must be one-to-one, for if $m(\sigma)x=0$ then $\sigma x\in Q$ and since σ is an orthogonal transformation x must be zero. Thus $\det m(\sigma)\neq 0$. If $\det m(\sigma)>0$ then σ will be said to preserve the spatial orientation of V. Of course for this definition to make sense it is necessary to verify that this criterion does not depend on the particular orthogonal decomposition of V chosen. If $x_1, x_2\in P$ then

$$
\begin{aligned}
g(x_1, m(\sigma)x_2) &= g(x_1,\mathcal{P}(\sigma x_2)) = g(x_1,\sigma x_2) = g(\sigma\sigma^{-1}x_1,\sigma x_2)\\
&= g(\sigma^{-1}x_1, x_2) = g(\mathcal{P}(\sigma^{-1}x_1), x_2) = g(m(\sigma^{-1})x_1, x_2).
\end{aligned}
$$

So if $m(\sigma)^t$ denotes the adjoint map, with respect to the induced positive-definite orthogonal metric on P, we have $m(\sigma)^t = m(\sigma^{-1})$. Since reflections are involutary the linear transformation associated with a reflection is symmetric. It is thus diagonalisable with determinant the product of the eigenvalues. If y is non-singular then $y = u + v$ where u, v are in P and Q respectively and

$$S_y x = x - \frac{2g(x, y)}{g(y, y)} y = x - \frac{2g(x, u)}{g(y, y)} y \qquad \text{for } x \in P$$

$$m(S_y)x = x - \frac{2g(x, u)}{g(y, y)} u.$$

There are $p - 1$ linearly independent vectors in P orthogonal to u and these are obviously eigenvectors of $m(S_y)$ with eigenvalues one. A basis of eigenvectors is completed by u, with

$$m(S_y)u = \left(1 - \frac{2g(u, u)}{g(y, y)}\right)u = \left(\frac{g(v, v) - g(u, u)}{g(y, y)}\right)u$$

thus

$$\det m(S_y) = \frac{g(v, v) - g(u, u)}{g(y, y)}.$$

The numerator is negative-definite and so reflections in timelike planes preserve spatial orientation. Any orthogonal transformation is a product of reflections and it will preserve a spatial orientation if it contains an even number of reflections in spacelike planes. This criterion obviously does not depend on any particular orthogonal decomposition of V. In exactly the same way any orthogonal transformation induces a linear transformation on the negative-definite space Q. If the determinant is positive then the orthogonal transformation is called time-orientation preserving, or orthochronous. Such transformations contain an even number of reflections in timelike planes.

The orthogonal group has (in general) four disconnected pieces containing 1, P, T and PT respectively. Here $P(T)$ denote transformations which change the spacelike (timelike) orientation whilst perserving the timelike (spacelike) orientation. The component containing the identity is a subgroup as is the sum of that component with any other component.

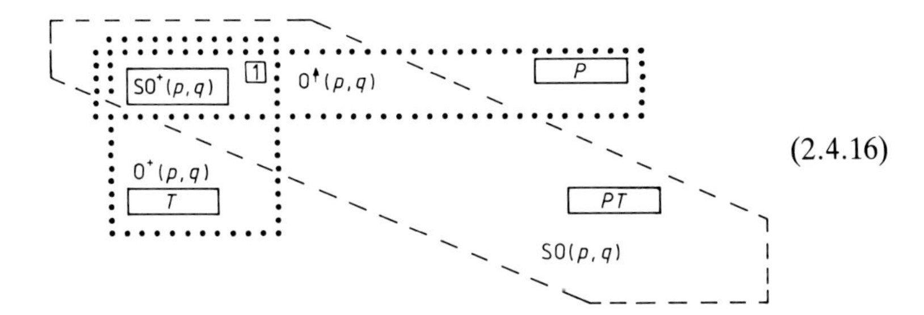

(2.4.16)

The Clifford group is in fact a Lie group, and its Lie algebra can be identified with a subspace of the Clifford algebra, the Lie bracket being the Clifford commutator. The regular representation maps the Clifford algebra into a total matrix algebra, thus the group of all regular elements, $\mathbf{C}^*_{p,q}(\mathbb{R})$, and hence the Clifford group and its subgroups are all subgroups of some general linear group. The general linear group is certainly a Lie group and the charts of this group induce charts on $\mathbf{C}^*_{p,q}(\mathbb{R})$ and Γ which give them a manifold structure. The exponential map is defined on the Clifford algebra in the obvious way

$$\exp a = \sum_{n=0}^{\infty} \frac{a^n}{n!} \qquad a \in \mathbf{C}_{p,q}(\mathbb{R}). \tag{2.4.17}$$

Since the Clifford algebra is isomorphic to a subalgebra of a total matrix algebra where the exponential map can be defined, the limit implicit in this definition does indeed exist. Since $\exp(-a) = (\exp a)^{-1}$ the exponential maps the Clifford algebra into the group of all invertible elements, $\mathbf{C}^*_{p,q}(\mathbb{R})$. Thus the vector space of the Clifford algebra with the product of Clifford commutation can be identified with the Lie algebra of $\mathbf{C}^*_{p,q}(\mathbb{R})$. With this identification the vector representation of $\mathbf{C}^*_{p,q}(\mathbb{R})$, χ, is seen to map the group into the automorphism group of the Lie algebra; this corresponds to the adjoint representation of $\mathbf{C}^*_{p,q}(\mathbb{R})$, Ad. Similarly if we define

$$\begin{aligned} \mathrm{ad}: \mathbf{C}_{p,q}(\mathbb{R}) &\longrightarrow \mathrm{End}\,\mathbf{C}_{p,q}(\mathbb{R}) \\ a &\longmapsto \mathrm{ad}\; a \end{aligned} \tag{2.4.18}$$

where $(\mathrm{ad}\, a)b = [a,\ b]$ with the bracket denoting a Clifford commutator, $[a,\ b] = ab - ba$, then ad is the adjoint representation of the Lie algebra of $\mathbf{C}^*_{p,q}(\mathbb{R})$.

The Clifford group is a Lie subgroup of the group of all invertible elements and its Lie algebra must be a vector subspace of the Clifford algebra. Suppose that m is in the Lie algebra of Γ, then

$$\exp(\lambda m)x\exp(-\lambda m) \subset V \qquad \forall x \in V,\ \forall \lambda \in \mathbb{R}. \tag{2.4.19}$$

The standard group theory result, $\mathrm{Ad}\exp(\lambda m) = \exp(\mathrm{ad}\,\lambda m)$, shows that this can hold for all λ if and only if the Clifford commutator of m with x is in V. This can be seen directly by defining, for fixed m and x, the Clifford-algebra-valued function

$$f(\lambda) = \exp(\lambda m)x\exp(-\lambda m).$$

We then have

$$\mathrm{d}f(\lambda)/\mathrm{d}\lambda = \exp(\lambda m)[m,\ x]\exp(-\lambda m)$$

and more generally

$$\mathrm{d}^n f(\lambda)/\mathrm{d}\lambda^n = \exp(\lambda m)\ (\mathrm{ad}\ m)^n x\exp(-\lambda m).$$

By expanding $f(\lambda)$ in a Taylor series about $\lambda = 0$ it can easily be seen that $f(\lambda) \in V \; \forall \lambda$ if and only if

$$[m, x] \in V.$$

If m is written in terms of even and odd parts, $m = m_+ + m_-$, then, from (2.3.7), for m to be in the Lie algebra of Γ we must have

$$\begin{aligned} x \wedge m_- &= 0 \\ \mathrm{i}_{\tilde{x}} m_+ &\in V \qquad \forall x \in V. \end{aligned} \tag{2.4.20}$$

If the dimension of V is even then the odd part of m must be zero, whilst if the dimension is odd m_- can be an n-form, which is then in the centre. The even part of m has to be a sum of 0-forms and 2-forms.

The exponential of an even element will be even whilst the one-parameter subgroup generated by the volume form for n odd will consist of elements that are in general neither even nor odd. Thus the Lie algebra of $\Gamma^{\pm}$ consists of the $\frac{1}{2}n(n-1)$ 2-forms and the identity. Since $(\exp \lambda m)^{\xi} = \exp \lambda m^{\xi}$ we must have $m^{\xi} = -m$ if m is in the Lie algebra of ${}_{+}\Gamma^{\pm}$, similarly $m^{\xi\eta} = -m$ if m is in the Lie algebra of ${}^{+}\Gamma^{\pm}$. Thus the Lie algebra of these groups is the commutator algebra of the 2-forms.

The exponential map sends the Lie algebra into that component of the group which is connected to the identity. This connected component is a subgroup, so products of exponentials are also connected to the identity. Conversely, every element of that component of the group which is connected to the identity can be written as a finite product of exponentials. Since 2-forms are even under η and odd under ξ the exponential maps the Lie algebra of ${}_{+}\Gamma^{\pm}$ into ${}_{+}\Gamma^{+}$, and so this must contain the component of ${}_{+}\Gamma^{\pm}$ connected to the identity. We will now demonstrate that, except for one exceptional case, ${}_{+}\Gamma^{+}$ is a connected group.

If $s \in {}_{+}\Gamma^{+}$ then $s = \alpha x^1 x^2 \ldots x^{2h}$, with $\alpha \in \mathbb{R}^*$ and the x^i non-singular elements of V. By suitably scaling α we can obviously arrange that $x^{i2} = \varepsilon^i = \pm 1$. Then the norm of s is given by $\lambda(s) = \lambda(\alpha)\varepsilon^1 \ldots \varepsilon^{2h}$, and so if $s \in {}_{+}\Gamma^{+}$ we must have an even number of negative-norm vectors and $\alpha = \pm 1$. The negative-norm elements can be collected at the left-hand side, for if $\varepsilon^i = 1$ and $\varepsilon^{i+1} = -1$ then we write $x^i x^{i+1} = (x^i x^{i+1} x^i) x^i \equiv x'^i x'^{i+1}$ where $(x'^i)^2 = x^i x^{i+1} x^i x^i x^{i+1} x^i = -1$. The overall factor of plus or minus one can be absorbed by redefining x^1 and thus if $s \in {}_{+}\Gamma^{+}$ $s = \sigma^1 \sigma^2 \ldots \sigma^h$ where each σ can be written

$$\sigma = xy \qquad x, y \in V \text{ with } x^2 = y^2 = \pm 1. \tag{2.4.21}$$

Thus every element of ${}_{+}\Gamma^{+}$ will be connected to the identity if and only if all such products of vectors are. If $y = \pm x$ then $\sigma = \pm x^2$ and so for ${}_{+}\Gamma^{+}$ to be connected -1 must be connected to $+1$. For an indefinite metric in two dimensions the Lie algebra of ${}_+\Gamma^+$ is spanned by the volume 2-form z, where $z^2 = 1$. In this case $\exp(\alpha z)\exp(\beta z) = \exp[(\alpha+\beta)z]\ \forall\, \alpha, \beta \in \mathbb{R}$, so if -1 were connected to $+1$ we would in fact be able to write it as an exponential. However, $\exp(\theta z) = \cosh\theta + z\sinh\theta$, and so $\exp(\theta z) \neq -1$ for any θ. Thus in this case ${}_+\Gamma^+$ is not connected. Ruling out this exceptional case we always have a pair of orthogonal vectors a, b with $a^2 = b^2 = \pm 1$. So $(ab)^2 = -1$ and since $\exp(\pi ab) = -1$ the identity is connected to minus one by a one-parameter subgroup. We still need to show that a general σ is connected to the identity. We consider three cases.

Suppose firstly that x and y are linearly independent, spanning an orthogonal plane with positive- or negative-definite metric. Then we have an orthonormal basis $\{x, u\}$ where $x^2 = u^2 = \varepsilon$, $\varepsilon = \pm 1$. Since $y^2 = x^2$ we can write $y = \cos\theta\, x + \sin\theta\, u$, and $xy = \varepsilon(\cos\theta + \sin\theta\,\varepsilon xu) = \varepsilon\exp(\varepsilon\theta xu)$. We have already shown that -1 is connected to $+1$ and so xy is also connected to $+1$.

If x and y span a non-degenerate orthogonal plane with orthonormal basis $\{x, u\}$ with $x^2 = -u^2 = \varepsilon$ then $(xu)^2 = 1$. Now we must have

$$y = \cosh\theta\, x + \sinh\theta\, u$$

and

$$xy = \varepsilon(\cosh\theta + \sinh\theta\,\varepsilon xu) = \varepsilon\exp(\varepsilon\theta xu).$$

Again the fact that -1 is connected to the identity ensures that all such products xy are.

If x and y span an isotropic plane then we let $\{x, u\}$ denote a basis in which u is an isotropic vector orthogonal to x. Then $y = \pm(x + \theta u)$ and $xy = \pm\varepsilon(1 + \theta\varepsilon xu)$. Since xu is nilpotent we have $xy = \pm\varepsilon\exp(\theta\varepsilon xu)$ and, since -1 is connected to $+1$, we have demonstrated that xy is connected to the identity.

We have shown that ${}_+\Gamma^+$ is a connected Lie group unless V is two-dimensional with indefinite metric. Thus save for this exceptional case ${}_+\Gamma^+$ is a connected double covering of that component of the orthogonal group which is connected to the identity, and it follows from the topology of the orthogonal group that ${}_+\Gamma^+$ is simply connected.

In suitably low dimensions it is particularly easy to identify the spin groups, due to the following:

If $\dim V \leqslant 5$ then if $s^\eta = \pm s$ and $s^\xi = \pm s^{-1}$ then $s \in {}_\pm\Gamma^\pm$. (2.4.22)

All we need to check is that if $x \in V$ then $sxs^{-1} \subset V$ for such an s. If we set $x' = sxs^{-1}$ then $x'^{\eta} = -x'$ and $x'^{\xi} = x'$ if $x \in V$ and s is even or odd under both η and ξ. In five or fewer dimensions the only elements that are both odd under η and even under ξ are linear combinations of

1-forms and 5-forms. So if $n < 5$ the result follows immediately. If $n = 5$ then the 5-form is in the centre of the algebra, so if $x' = a + b$ where a is a 1-form and b a 5-form then $x'^2 = a^2 + b^2 + 2ab$. Now a^2 and b^2 are both 0-forms whereas ab is a 4-form. But since x^2 is a 0-form and $x' = sxs^{-1}$ then x'^2 is a 0-form and thus $ab = 0$, that is either $a = 0$ or $b = 0$. However, a cannot be zero since an inner automorphism cannot take an element that is not in the centre into the centre, and so $b = 0$ and (2.4.22) follows.

Needless to say (2.4.22) does not go through in six dimensions. For example, if $\{e^i\}$ $i = 1, \ldots, 6$ is an orthonormal basis for a positive-definite orthogonal space and $s = (1/\sqrt{2})(e^{12} + e^{3456})$, where $e^{12} = e^1e^2$ etc, then $s^\eta = s$ and $s^\xi = s^{-1}$. However, $se^1s^{-1} = -e^{23456}$ and so $s \notin \Gamma$.

The results of this section will now be illustrated by considering the algebra $\mathbf{C}_{3,1}(\mathbb{R})$. In this case $\Gamma^\pm = \Gamma$. An orthonormal basis for V is $\{e^a\}$ $a = 0, 1, 2, 3$ where $-(e^0)^2 = (e^i)^2 = 1$. Let $P = e^0$, $T = e^{123}$ then

$$\begin{aligned} Te^0T^{-1} &= -e^0 \quad Pe^0P^{-1} = e^0 \\ Te^iT^{-1} &= e^i \qquad Pe^iP^{-1} = -e^i \quad i = 1, 2, 3. \end{aligned} \tag{2.4.23}$$

The norms of these elements are easily seen to be $\lambda(P) = -1$, $\mu(P) = 1$, $\lambda(T) = 1$ and $\mu(T) = -1$. So $P \in {}^+\Gamma$ whereas $T \in {}_+\Gamma$. Suppose now that $s_1 \in {}_\pm\Gamma$ such that $s_1^\eta = s_1$, $\lambda(s_1) = -1$. Then $s_1 = (s_1PT)(PT)^{-1}$ and $(s_1PT)^\eta = s_1PT$, $\lambda(s_1PT) = 1$ thus $s_1 = \sigma_1(PT)$ where $\sigma_1 \in {}_+\Gamma^+$. Similarly if $s_2 \in {}_\pm\Gamma$ such that $s_2^\eta = -s_2$, $\lambda(s_2) = 1$, then $s_2 = \sigma_2 T$; and if $s_3 \in {}_\pm\Gamma$ such that $s_3^\eta = -s_3$, $\lambda(s_3) = -1$, then $s_3 = \sigma_3 P$, $\sigma_2, \sigma_3 \in {}_+\Gamma^+$.

We know that the six 2-forms generate ${}_+\Gamma^+$, the Lie bracket being a Clifford commutator. If we take a product of two spacelike 1-forms, for example e^{12}, then $(e^{12})^2 = -1$ and $\exp(\theta e^{12}) = \cos\theta + \sin\theta\, e^{12}$. Such elements thus generate rotations, and the Clifford commutators are seen to give the familiar Lie algebra of the rotation group, $[e^{12}, e^{23}] = 2e^{13}$ etc. Elements such as e^{01} generate 'boosts', with $(e^{01})^2 = 1$ giving $\exp(\theta e^{01}) = \cosh\theta + \sinh\theta e^{01}$. The commutator of two boosts gives a rotation, for example $[e^{01}, e^{02}] = 2e^{12}$. The remaining structure constants are determined by looking at the commutator of a boost with a rotation, for example $[e^{01}, e^{12}] = 2e^{02}$. The group ${}_+\Gamma^+$ can be recognised as a matrix group by using (2.4.22). This result shows that ${}_+\Gamma^+$ is the group of unit-norm regular elements of $\mathbf{C}_{3,1}^+(\mathbb{R})$. In §2.3 it was shown that the even subalgebra was isomorphic to the algebra of all complex two by two matrices, and so ${}_+\Gamma^+$ must be the subgroup of Gl(2, $\mathbb{C}$) consisting of unit-norm elements. Since we have already explicitly constructed a matrix basis for $\mathbf{C}_{3,1}^+(\mathbb{R})$ it can be directly verified that the norm corresponds to the determinant. If $\{\boldsymbol{\varepsilon}_{\alpha\beta}\}$ is the basis given in table 2.9 then $\boldsymbol{\varepsilon}_{11}{}^{\xi} = \boldsymbol{\varepsilon}_{22}$, $\boldsymbol{\varepsilon}_{12}{}^{\xi} = -\boldsymbol{\varepsilon}_{12}$, $\boldsymbol{\varepsilon}_{21}{}^{\xi} = -\boldsymbol{\varepsilon}_{21}$ and $\boldsymbol{\varepsilon}_{22}{}^{\xi} = \boldsymbol{\varepsilon}_{11}$. So if

$$s = \sum_{\alpha,\beta=1}^{2} s_{\alpha\beta}\, \boldsymbol{\varepsilon}_{\alpha\beta}$$

then

$$ss^{\xi} = (s_{11}s_{22} - s_{12}s_{21})(\boldsymbol{\varepsilon}_{11} + \boldsymbol{\varepsilon}_{22}) = \det(s)1.$$

Thus in this four-dimensional Lorentzian case, ${}_{+}\Gamma^{+} \simeq \mathrm{Sl}(2, \mathbb{C})$, the group of complex matrices of order two with unit determinant. The group of matrices with determinants of plus or minus one is obviously isomorphic to ${}_{\pm}\Gamma^{+}$, ${}_{\pm}\Gamma^{+} \simeq {}_{\pm}\mathrm{Sl}(2, \mathbb{C})$. The PIN group, ${}_{\pm}\Gamma$, is obviously a subgroup of $\mathrm{Gl}(4, \mathbb{R})$. However, since we have identified ${}_{\pm}\Gamma^{+}$ as a matrix group it is convenient to identify ${}_{\pm}\Gamma$ as a product of this matrix group with a discrete subgroup. We have already seen how any element of ${}_{\pm}\Gamma$ can be written as a product of an element of ${}_{+}\Gamma^{+}$ with either 1, P, T or PT. Since $P^2 = T^2 = -1$ these elements do not form a subgroup and so it is convenient to introduce a unit-norm 1-form, x say, so that $\{1, x\}$ form a subgroup, Ω, isomorphic to $\mathbb{Z}_2$. If s is any element of ${}_{\pm}\Gamma$ then it can be uniquely written as $s = \sigma t$, $\sigma \in {}_{\pm}\Gamma^{+}$ and $t \in \Omega$. The multiplication of two elements s_1 and s_2 is given by

$$s_1 s_2 = \sigma_1 t_1 \sigma_2 t_2 = \sigma_1 t_1 \sigma_2 t_1{}^{-1} t_1 t_2 = \sigma_1 \{\chi(t_1)\sigma_2\} t_1 t_2.$$

Now $\chi(t_1)$ acts on ${}_{\pm}\Gamma^{+}$ as an outer automorphism and $\chi(\Omega) \simeq \Omega$. We can equivalently write elements of ${}_{\pm}\Gamma$ as an ordered pair of an element of ${}_{\pm}\Gamma^{+}$ and an element of Ω with the multiplication defined by $(\sigma_1, t_1)(\sigma_2, t_2) = (\sigma_1\chi(t_1)\sigma_2, t_1 t_2)$. In this form ${}_{\pm}\Gamma$ is recognised as a semidirect product of ${}_{\pm}\Gamma^{+}$ and a $\mathbb{Z}_2$ group of automorphisms, ${}_{\pm}\Gamma \simeq {}_{\pm}\Gamma^{+} \odot \mathbb{Z}_2$. We have shown that ${}_{\pm}\Gamma^{+} \simeq {}_{\pm}\mathrm{Sl}(2, \mathbb{C})$ and the generator of the automorphism group, x, sends σ to $x\sigma x$. As was discussed in §2.1 we can always choose such an x, which complex conjugates the matrix components and thus ${}_{\pm}\Gamma \simeq {}_{\pm}\mathrm{Sl}(2, \mathbb{C}) \odot \mathbb{Z}_2$, where the automorphism group is generated by complex conjugation.

2.5 Spinors

From the irreducible representations of the Clifford algebra and its even subalgebra we obtain irreducible representations of the Clifford group: the spinor representations. It should be noted that minor variations exist in the literature as to the precise nomenclature for these representations.

The regular representation maps the Clifford algebra into its endomorphism algebra; that is, into the algebra of linear transformations on the vector space structure of the Clifford algebra. This representation will not be irreducible; certain vector subspaces will be preserved under multiplication from the left, namely the left ideals. It is a truism to say that the minimal left ideals transform irreducibly under the regular representation. If the Clifford algebra is simple then the regular representation induces a faithful representation on any minimal left ideal. The mapping into the endomorphism algebra of any minimal left ideal induced by the regular representation is called the *spinor representation* of the simple Clifford algebra and the minimal left ideal is called the space of spinors. The choice of a different minimal left ideal gives another equivalent representation. When the Clifford algebra is not simple it is the sum of two simple component algebras, and any minimal left ideal must lie in one of these simple components. The regular representation of a non-simple Clifford algebra induces a faithful representation on the left ideal which is the sum of two minimal left ideals, one lying in each simple component. The mapping into such an endomorphism algebra induced by the regular representation will be called the *spinor representation* of the non-simple Clifford algebra, and such an ideal will be termed the spinor space. The minimal left ideals will be termed *semi-spinor spaces* and the mapping that the regular representation induces on a minimal left ideal will be called the *semi-spinor representation* of the Clifford algebra. The kernel of such a representation is obviously the simple component algebra that does not contain the semi-spinor space. Thus the spinor representation of a non-simple Clifford algebra is reducible, being the sum of two inequivalent semi-spinor representations. The spinor representation of the Clifford algebra induces a representation of any subset by restricting to left multiplication on the ideal by elements of that set. In particular it induces a representation of the Clifford group.

Irreducible representations of the Clifford algebra induce irreducible representations of the Clifford group. (2.5.1)

That is, the spinor representation of a simple Clifford algebra, or the semi-spinor representation of a non-simple one, induces an irreducible representation of the Clifford group. This will also be called the spinor or semi-spinor representation. The proof of the statement follows immediately from the observation that non-singular vectors generate the Clifford group and the Clifford algebra. In fact the Clifford group could be replaced with the subgroup ${}_{\pm}\Gamma^{\pm}$ and the statement would obviously still be true.

If an irreducible representation of the Clifford algebra induces a reducible representation of the even subalgebra then that induced representation is the sum of two irreducible ones. For suppose that I is a minimal left ideal of the Clifford algebra that splits into invariant subspaces under left multiplication by the even subalgebra. Let W be such an invariant subspace of smallest dimension. Then if x is any odd regular element let $xW = X$, giving $\dim X = \dim W$. If $S = W + X$, where the sum is not necessarily direct, then S is preserved under multiplication by the Clifford algebra. For

$$\mathbf{C}_{p,q}(\mathbb{R}) = \mathbf{C}^+_{p,q}(\mathbb{R}) + \mathbf{C}^+_{p,q}(\mathbb{R})x$$

so

$$\mathbf{C}_{p,q}(\mathbb{R})W = \mathbf{C}^+_{p,q}(\mathbb{R})W + x\mathbf{C}^+_{p,q}(\mathbb{R})W \subset W + xW$$

and

$$\mathbf{C}_{p,q}(\mathbb{R})xW = \mathbf{C}_{p,q}(\mathbb{R})W.$$

Since $S \subset \mathrm{I}$ and I is a minimal left ideal we must have $S = \mathrm{I}$. If $W \cap X = Y$ then $\mathbf{C}^+_{p,q}(\mathbb{R})Y \subset Y$ since W and hence X are preserved under left multiplication by $\mathbf{C}^+_{p,q}(\mathbb{R})$. But W is an invariant subspace of minimal dimension and so either $Y = 0$, and I is the sum of two invariant subspaces, or $Y = W = X = \mathrm{I}$ and I transforms irreducibly.

Having shown that irreducible representations of the Clifford algebra induce a representation of the even subalgebra that is either irreducible or the sum of two irreducible representations, we would like to know in which cases each possibility occurs. Suppose firstly that the even subalgebra is reducible; this can only occur in even dimensions in which case the Clifford algebra is simple. Then the spinor representation of the Clifford algebra induces a faithful representation of the even subalgebra, that is, the kernel is zero. This must therefore be a reducible representation of the reducible subalgebra, being the sum of the two inequivalent irreducible representations whose kernels are the different simple ideals. The irreducible representations of a non-simple even subalgebra will again be called semi-spinor representations of that algebra.

Suppose now that the Clifford algebra is reducible; this can only occur in odd dimensions in which case the even subalgebra is simple. In this case the semi-spinor representations induce irreducible representations of the even subalgebra. For let I be a minimal left ideal (the semi-spinor space) and z denote the volume form. Then if, for example, the kernel of the semi-spinor representation is the simple ideal $\mathbf{C}_{p,q}(\mathbb{R})(1 + z)$ the semi-spinor space is an eigenspace of the volume form, $z\varphi = -\varphi \; \forall \varphi \in \mathrm{I}$. Since z is odd and regular we have $\mathbf{C}_{p,q}(\mathbb{R}) = \mathbf{C}^+_{p,q}(\mathbb{R}) + \mathbf{C}^+_{p,q}(\mathbb{R})z$ and $\mathbf{C}_{p,q}(\mathbb{R})\mathrm{I} = \mathbf{C}^+_{p,q}(\mathbb{R})\mathrm{I}$. So I can have no invariant subspaces under multiplication by $\mathbf{C}^+_{p,q}(\mathbb{R})$ since it is a minimal left ideal of $\mathbf{C}_{p,q}(\mathbb{R})$.

The irreducible representations of the Clifford algebra can induce a reducible representation on the even subalgebra even when that algebra is simple. The general criterion is given by the following.

Irreducible representations of the Clifford algebra induce reducible representations of the even subalgebra if and only if primitives in the subalgebra are primitive in the full algebra. (2.5.2)

What we need to show is that the minimal left ideals of the full algebra have twice the dimension of the minimal left ideals of the even subalgebra if and only if primitives in the subalgebra are primitive in the full algebra. Let P^+ be a primitive idempotent of $\mathbf{C}^+_{p,q}(\mathbb{R})$. Then $\mathbf{C}_{p,q}(\mathbb{R})P^+$ is a left ideal and $\mathbf{C}_{p,q}(\mathbb{R})P^+ = \mathbf{C}^+_{p,q}(\mathbb{R})P^+ + x\mathbf{C}^+_{p,q}(\mathbb{R})P^+$ for any odd regular x. Since P^+ is primitive in $\mathbf{C}^+_{p,q}(\mathbb{R})$ then $\mathbf{C}^+_{p,q}(\mathbb{R})P^+$ is a minimal left ideal of the even subalgebra and so the dimension of $\mathbf{C}_{p,q}(\mathbb{R})P^+$ is twice that of the minimal left ideals of $\mathbf{C}^+_{p,q}(\mathbb{R})$. So the minimal left ideals of the full algebra are twice the dimension of those of the subalgebra if and only if $\mathbf{C}_{p,q}(\mathbb{R})P^+$ is a minimal left ideal, that is, if and only if P^+ is primitive in $\mathbf{C}_{p,q}(\mathbb{R})$. If a minimal left ideal of the full algebra is projected out by a primitive of the subalgebra then the $\mathbf{C}^+_{p,q}(\mathbb{R})$-irreducible subspaces are obviously the even and odd subspaces.

Just as the irreducible representations of the Clifford algebra gave representations of the Clifford group the irreducible representations of the even subalgebra induce representations of the even Clifford group and in particular:

Irreducible representations of the even subalgebra induce irreducible representations of ${}_+\Gamma^+$. (2.5.3)

Again this follows from the fact that ${}_+\Gamma^+$ generates $\mathbf{C}^+_{p,q}(\mathbb{R})$. First we note that the Clifford algebra is generated by non-singular vectors of the same norm. For if $\{e^i, f^j\}$ with $i = 1, \ldots, p$, $j = 1, \ldots, q$ is an orthonormal basis, and if $p \neq 0$, then a new basis of unit-norm vectors is $\{e^i, \sqrt{2}e^1 + f^j\}$. Thus $\mathbf{C}^+_{p,q}(\mathbb{R})$ is generated by products of unit-norm vectors, and such products are in ${}_+\Gamma^+$.

The relationship between the irreducible representations of the Clifford algebra and its even subalgebra is summarised in table 2.10. The structure of $\mathbf{C}_{p,q}(\mathbb{R})$ is determined by $p - q \bmod 8$ where $p + q = n$. The eight different cases have been grouped in pairs. For the first pair the semi-spinor representation of the full algebra induces an irreducible representation of the subalgebra; whereas for the second pair the Clifford spinor representation induces an irreducible even Clifford spinor representation. For the third pair of algebras the spinor representation splits into a pair of equivalent spinor representations of the subalgebra, whereas in the final case the spinor representation is the sum of two inequivalent semi-spinor representations of the subalgebra.

In table 2.10 we give the dimensions of the irreducible representations of the Clifford algebra and its even subalgebra. We have used $C - S/S$ to denote that the irreducible representation of the Clifford algebra is a

Table 2.10 Dimensions of the irreducible representations of the Clifford algebra **C** and its even subalgebra $\mathbf{C}^+$. S denotes a spinor representation and S/S a semi-spinor representation.

	$(p-q) \bmod 8$						
Dimension	5	1	3	2	6, 7	4	0
$2(2^{[n/2]})$	$C - S/S$		$C - S$		$C - S$	$C - S$	
	$C^+ - S$		$C^+ - S$				
$2^{[n/2]}$		$C - S/S$		$C - S$	$C^+ - S$	$C^+ - S/S$	$C - S$
		$C^+ - S$		$C^+ - S$			
$\frac{1}{2}(2^{[n/2]})$							$C^+ - S/S$

semi-spinor representation, $C^+ - S$ to denote the induced spinor representation of the even subalgebra, and similarly for the other two cases. The integer part of $n/2$ is denoted by $[n/2]$. This table is an immediate consequence of table 2.8.

Since we are concerned with algebras over the real field the spinor spaces are $\mathbb{R}$-linear vector spaces, the dimensions of which are given in table 2.10. As well as obviously being left $\mathbf{C}_{p,q}(\mathbb{R})$ modules the spinor spaces are also right $\mathscr{A}$-modules where $\mathscr{A}$ is the algebra given in table 2.8. Whereas, in general, right multiplication will not preserve a left ideal it will be preserved under right multiplication by elements of $\mathscr{A}$. When the Clifford algebra is simple $\mathscr{A}$ is a division algebra, whereas when the Clifford algebra is not simple $\mathscr{A} = \mathscr{D}\oplus\mathscr{D}$ where $\mathscr{D}$ is a division algebra. In this case the semi-spinor spaces are right $\mathscr{D}$-modules. It is an immediate consequence of associativity that left multiplication induces a $\mathscr{D}$-linear transformation on the minimal left ideals. Similarly the irreducible representations of the even subalgebra may be regarded as $\mathscr{D}$-linear transformations where $\mathscr{D}$ is one of the real division algebras $\mathbb{R}$, $\mathbb{C}$ or H. The dimensions of the spinor and the semi-spinor spaces regarded as $\mathscr{D}$-linear spaces can be found from tables 2.8 and 2.10 since $d(\dim_{\mathscr{D}}) = \dim_{\mathbb{R}}$ where $\mathscr{D}$ is a d-dimensional $\mathbb{R}$-algebra.

For those Clifford algebras whose centre is $\mathbb{C}$ the spinor space may be regarded as a $\mathbb{C}$-linear space by using the complex structure of right (or left) multiplication by the volume form z. For example, we may define multiplication by the imaginary unit by $\mathrm{i}\psi = \psi z$, where ψ lies in a minimal left ideal. Alternatively, we could define $\mathrm{i}\psi = -\psi z$. Although we have already noted that all irreducible representations of a simple algebra are equivalent, when representing a simple $\mathbb{R}$-algebra on a $\mathbb{C}$-linear space the question of equivalence needs treating carefully. If ρ and ρ' are representations of any simple $\mathbb{R}$-algebra $\mathscr{A}$, where $\mathscr{A}(\mathbb{R}) \simeq \mathbb{C}(\mathbb{R})\otimes\mathscr{M}_m(\mathbb{R})$, on $\mathbb{R}$-linear spaces V and V' then there is an $\mathbb{R}$-linear transformation S from V' to V such that $\rho'(a) = S^{-1}\rho(a)S$ for all a of $\mathscr{A}$. If, however, V and V' are regarded as complex vector spaces by defining $\mathrm{i}v = \rho(z)v \;\forall\, v \in V$ (where z generates the centre) then there is a $\mathbb{C}$-linear transformation S such that $\rho'(a) = S^{-1}\rho(a)S \;\forall\, a$ if and only if $\mathrm{i}v' = \rho'(z)v'$. Thus by defining $\rho(z)v = \mathrm{i}v$ and $\rho'(z)v' = -\mathrm{i}v'$ we get two complex-inequivalent representations of a simple $\mathbb{R}$-algebra. This is easily understood in terms of the complexified algebra. Regarded as a complex vector space V carries an irreducible representation of the complexified algebra $\mathscr{A}^{\mathbb{C}} \simeq \mathscr{A}\otimes\mathbb{C}$. The representation ρ extends by $\mathbb{C}$-linearity to $\mathscr{A}^{\mathbb{C}}$, $\rho(\mathrm{i}a)v = \mathrm{i}\rho(a)v$. Since $\mathbb{C}\otimes\mathbb{C} \simeq \mathbb{C}\oplus\mathbb{C}$, $\mathscr{A}^{\mathbb{C}}$ is reducible and its irreducible representations have as kernel one of the simple ideals, and the irreducible representations are equivalent if and only if the kernels are the same. If $\rho(z)v = \mathrm{i}v$ then $\rho(1 + \mathrm{i}z) = 0$ and the kernel of ρ is projected by the central idempotent $\frac{1}{2}(1 + \mathrm{i}z)$. If, however, $\rho'(z)v' = -\mathrm{i}v'$ then $\frac{1}{2}(1 - \mathrm{i}z)$ is in the kernel, thus ρ' and ρ are inequivalent representations of $\mathscr{A}^{\mathbb{C}}$.

When the spinor space is a right H-module† then it can be regarded as a complex vector space by choosing as complex structure any complex subalgebra of the quaternions. If $q \in H$ such that $q^2 = -1$ then we may define multiplication by complex numbers on spinors by $\mathrm{i}\psi = \psi q$. Again, regarded as complex vector spaces, these minimal left ideals carry irreducible representations of the complexified algebra by extending the spinor representation by $\mathbb{C}$-linearity. Since $H \otimes \mathbb{C} \simeq \mathbb{C} \otimes \mathscr{M}_2$ the complexified algebra is simple and hence all irreducible representations are equivalent. Thus in this case all irreducible representations of the simple $\mathbb{R}$-algebra on complex vector spaces are complex-equivalent. We shall return to a discussion of the complexified Clifford algebras later.

The first example we give is of $\mathbf{C}_{0,2}(\mathbb{R}) \simeq H(\mathbb{R})$. Here the spinor space is the algebra itself. If $\{f^1, f^2\}$ is an orthonormal basis then $\{1, f^1, f^2, f^1f^2 = z\}$ is a standard basis for the quaternions. We may choose as complex structure right multiplication by z and define $\mathrm{i}a = az$ for $a \in \mathbf{C}_{0,2}(\mathbb{R})$. Then $\{1, f^1\}$ is a basis for the corresponding complex vector space. If ρ denotes the spinor representation then with respect to this basis and choice of complex structure we have the matrices of the transformations, $\rho(a)$, as follows:

$$\boldsymbol{\rho}(1) = \begin{pmatrix} 1 & 0 \\ 0 & 1 \end{pmatrix} \qquad \boldsymbol{\rho}(f^1) = \begin{pmatrix} 0 & -1 \\ 1 & 0 \end{pmatrix}$$

$$\boldsymbol{\rho}(f^2) = \begin{pmatrix} 0 & -\mathrm{i} \\ -\mathrm{i} & 0 \end{pmatrix} \qquad \boldsymbol{\rho}(z) = \begin{pmatrix} \mathrm{i} & 0 \\ 0 & -\mathrm{i} \end{pmatrix}.$$

Had we instead chosen $\mathrm{i}a = -az$ then we would have the complex conjugate matrices. These give a complex-equivalent representation; we have $\boldsymbol{\rho}(a)^* = \boldsymbol{\rho}(f^1 a (f^1)^{-1})$.

Regarded as a complex vector space, H carries an irreducible representation of the complexified algebra $H \otimes \mathbb{C}$. If $P_\pm = \frac{1}{2}(1 \pm \mathrm{i}z)$ then $P_\pm$ are primitive idempotents in $H \otimes \mathbb{C}$ and $(H \otimes \mathbb{C})P_\pm$ are minimal left ideals such that $u^\pm z = \mp \mathrm{i} u^\pm$ for all $u^\pm \in (H \otimes \mathbb{C})P_\pm$. Since $P_- = (f^1)^{-1} P_+ f^1$ then right multiplication by f^1 is a $\mathbb{C}$-linear transformation between the two left ideals which obviously commutes with left multiplication and hence establishes the equivalence of these complex representations.

The even subalgebra is isomorphic to $\mathbb{C}(\mathbb{R})$ and the spinor representation of $\mathbf{C}_{0,2}(\mathbb{R})$ induces a reducible representation of $\mathbf{C}^+_{0,2}(\mathbb{R})$, the even and odd quaternions transforming irreducibly. These irreducible repre-

† The notion of an 'H-module' is to be found at the end of Appendix A where the quaternion algebra, H, is also introduced.

sentations of the simple algebra are equivalent: right multiplication by any odd quaternion interchanges the even and odd subspaces and commutes with left multiplication. However, right multiplication by z induces a complex structure on the even and odd subspaces that enables them to be regarded as complex one-dimensional vector spaces. These are complex-inequivalent, right multiplying by any odd element not being $\mathbb{C}$-linear.

We next consider $\mathbf{C}_{3,1}(\mathbb{R}) \simeq \mathcal{M}_4(\mathbb{R})$. Here the four-dimensional spinor representation induces an irreducible representation of the even subalgebra $\mathbf{C}^+_{3,1}(\mathbb{R}) \simeq \mathbb{C}(\mathbb{R}) \otimes \mathcal{M}_2(\mathbb{R})$. We may choose as spinor space the minimal left ideal whose basis is the first column in table 2.7. By defining $\mathrm{i}\psi = z\psi$ for all spinors ψ the spinor space may be regarded as a complex vector space with left multiplication by the even subalgebra a $\mathbb{C}$-linear transformation. A basis for this complex vector space is $\{P_1, e^0 P_1\}$. With this basis and choice of complex structure the matrices of these transformations for a basis for the even subalgebra are as follows:

$$\boldsymbol{\rho}(1) = \begin{pmatrix} 1 & 0 \\ 0 & 1 \end{pmatrix} \qquad \boldsymbol{\rho}(z) = \mathrm{i}\boldsymbol{\rho}(1)$$

$$\boldsymbol{\rho}(\mathrm{e}^{12}) = \begin{pmatrix} 0 & -1 \\ 1 & 0 \end{pmatrix} \qquad \boldsymbol{\rho}(e^{03}) = \mathrm{i}\boldsymbol{\rho}(e^{12})$$

$$\boldsymbol{\rho}(e^{23}) = \begin{pmatrix} 0 & -\mathrm{i} \\ -\mathrm{i} & 0 \end{pmatrix} \qquad \boldsymbol{\rho}(e^{01}) = \mathrm{i}\boldsymbol{\rho}(e^{23})$$

$$\boldsymbol{\rho}(e^{31}) = \begin{pmatrix} -\mathrm{i} & 0 \\ 0 & \mathrm{i} \end{pmatrix} \qquad \boldsymbol{\rho}(e^{02}) = \mathrm{i}\boldsymbol{\rho}(e^{31}).$$

The matrix representations of the generators of the rotation group will be recognised as the Pauli matrices (up to conventional factors of i). Defining $\mathrm{i}\psi = -z\psi$ gives the complex conjugate representation which is complex-inequivalent.

In this section we have naturally represented the Clifford algebra, and hence the Clifford group, on its left ideals. We can also represent the algebra on its right ideals. Associating each element of the algebra with the linear transformation obtained by multiplying with that element from the right gives a mapping into the endomorphism algebra, namely

$$R\colon \mathbf{C}_{p,q}(\mathbb{R}) \longrightarrow \operatorname{End} \mathbf{C}_{p,q}(\mathbb{R})$$
$$a \longmapsto R(a),\ R(a)b = ba.$$

Since $\{R(a)R(b)\}c \equiv R(a)\{R(b)c\} = cba = R(ba)c$ this correspondence is not an algebraic isomorphism. Given an involution $\mathcal{J}$ of the Clifford algebra we can use this correspondence to define a representation $\tilde{\rho}$:

$$\widetilde{\rho}\colon \mathbf{C}_{p,q}(\mathbb{R}) \longrightarrow \operatorname{End}\mathbf{C}_{p,q}(\mathbb{R})$$
$$a \longmapsto \widetilde{\rho}(a) = R(a^{\mathcal{J}}). \tag{2.5.4}$$

Indeed we have a representation since $\widetilde{\rho}(a)\widetilde{\rho}(b) = R(a^{\mathcal{J}})R(b^{\mathcal{J}}) = R(b^{\mathcal{J}}a^{\mathcal{J}}) = R([ab]^{\mathcal{J}}) = \widetilde{\rho}(ab)$. Obviously the minimal right ideals transform irreducibly under this representation. Just as the minimal left ideals may be regarded as right $\mathcal{D}$-modules these minimal right ideals can be regarded as left $\mathcal{D}$-modules.

A minimal right ideal is naturally identified with the space of $\mathcal{D}$-valued $\mathcal{D}$-linear mappings on a minimal left ideal. For if $\psi \in \mathbf{C}_{p,q}(\mathbb{R})P$ and $\Phi \in P\mathbf{C}_{p,q}(\mathbb{R})$ with P primitive then we may write

$$\Phi\colon \psi \longrightarrow \Phi(\psi) = \Phi\psi$$

with $\Phi(\psi) \in P\mathbf{C}_{p,q}(\mathbb{R})P \simeq \mathcal{D}$. Obviously $\Phi(\psi q) = \Phi(\psi)q$ for $q \in \mathcal{D}$. Similarly the Clifford algebra itself (or a simple component thereof) may be identified with the space of $\mathcal{D}$-valued linear transformations on the Cartesian product of a minimal left ideal and a minimal right ideal. For if $\Phi \in P\mathbf{C}_{p,q}(\mathbb{R})$ and $\psi \in \mathbf{C}_{p,q}(\mathbb{R})P$ then for any $a \in \mathbf{C}_{p,q}(\mathbb{R})$ we may write

$$a(\Phi, \psi) = \Phi a \psi$$

giving

$$a(q\Phi, \psi) = qa(\Phi, \psi),\ a(\Phi, \psi q) = a(\Phi, \psi)q \text{ for } q \in \mathcal{D}.$$

If the minimal left ideal carries the spinor representation ρ and the minimal right ideal carries the representation $\widetilde{\rho}$ then we may induce a representation τ on the Clifford algebra (or a simple component) by defining

$$\tau(s)(\psi\Phi) = [\rho(s)\psi][\widetilde{\rho}(s)\Phi] = s\psi\Phi s^{\mathcal{J}}.$$

If we choose $\mathcal{J} = \xi$ then $s^{\mathcal{J}} = s^{-1}$ for $s \in {}_{+}\Gamma^{+}$ and the representation τ and the vector representation χ coincide on ${}_{+}\Gamma^{\pm}$. In this case the representations ρ and $\widetilde{\rho}$ induce contragradient representations of ${}_{+}\Gamma^{\pm}$, and since we have seen how the minimal right ideal can be identified with the dual space of the left ideal we can construct a $\mathcal{D}$-valued ${}_{+}\Gamma^{\pm}$-invariant product. This will be discussed in the following section.

2.6 Spin-Invariant Inner Products

Having identified the elements of certain minimal left ideals as spinors we now examine spin-invariant products of two such elements. Since Clifford multiplication from the left induces a linear transformation on the spinor space we may use a product on the spinor space to define an involution on the Clifford algebra by sending every element to that which induces the adjoint linear transformation. Such an involution will be termed the adjoint involution. We shall construct a product of spinors φ and ψ which is the same as that of $s\varphi$ and $s\psi$ when $s \in {}_{+}\Gamma^{+}$. The adjoint involution of such a product will be either ξ or $\xi\eta$. Conversely, any product on the minimal left ideal with ξ or $\xi\eta$ as adjoint involution will be invariant under (at least) ${}_{+}\Gamma^{+}$. We shall first consider an arbitrary simple $\mathbb{R}$-algebra and show how any involution is the adjoint of some product on the minimal left ideals. These products fall into a finite number of distinct classes, and any two involutions are equivalent (as defined in Appendix A) if and only if the associated products are in the same class. The case of the direct sum of two isomorphic simple algebras is treated similarly. Returning to the Clifford algebras we shall determine into which class the products associated with ξ and $\xi\eta$ fall. Similarly, we can classify the products on the minimal left ideals of the even subalgebra. As a corollary in up to five dimensions we can use (2.4.22) to express ${}_{+}\Gamma^{+}$ as the invariance group of some product.

Let $\mathscr{A}$ be simple over $\mathbb{R}$ and $\mathscr{J}$ be some involution.

If P is any primitive idempotent then $P^{\mathscr{J}} = JPJ^{-1}$ for some element J with $J^{\mathscr{J}} = \varepsilon J$, $\varepsilon = \pm 1$. (2.6.1)

For if $\mathscr{J}$ leaves elements of the centre invariant and $\mathscr{T}$ denotes transposition in a matrix basis in which P is diagonal then we are assured (by (A23) of Appendix A) of a J with $J^{\mathscr{J}} = \pm J$ such that $a^{\mathscr{T}} = J^{-1}a^{\mathscr{J}}J \; \forall \, a \in \mathscr{A}$; in particular, $P^{\mathscr{T}} = P = J^{-1}P^{\mathscr{J}}P$. In the same way if the centre is $\mathbb{C}$ with $\mathscr{J}$ inducing complex conjugation the argument can be repeated with Hermitian conjugation replacing transposition. If then $\varphi \in \mathscr{A}P$ then $J^{-1}\varphi^{\mathscr{J}} \in P\mathscr{A}$ and we define

$$(\, , \,) : \mathscr{A}P \times \mathscr{A}P \longrightarrow P\mathscr{A}P \equiv \mathscr{D}$$

$$\varphi, \psi \longmapsto (\varphi, \psi) = J^{-1}\varphi^{\mathscr{J}}\psi. \tag{2.6.2}$$

If a is any element of $\mathscr{A}$ then

$$(\varphi, a\psi) = (a^{\mathscr{J}}\varphi, \psi) \tag{2.6.3}$$

and $\mathscr{J}$ is the adjoint involution of this product. The minimal left ideal $\mathscr{A}P$ is a right $\mathscr{D}$-module. If $q \in \mathscr{D}$ then

$$(\varphi, \psi q) = (\varphi, \psi)q \tag{2.6.4}$$

and the product is $\mathscr{D}$-linear in the second entry. If we define $q^{j} = J^{-1}q^{\mathscr{J}}J$ for $q \in \mathscr{D}$ then j is readily seen to be an involution of $\mathscr{D}$ such that

$$(\varphi q, \psi) = q^j(\varphi, \psi). \qquad (2.6.5)$$

The involution j will reverse the order of terms in a product; in fact

$$\begin{aligned}(\varphi, \psi)^j &= J^{-1}(\varphi, \psi)^{\mathcal{J}} J = J^{-1}(J^{-1}\varphi^{\mathcal{J}}\psi)^{\mathcal{J}} J = J^{-1}\psi^{\mathcal{J}}\varphi J^{-1\mathcal{J}} J \\ &= \varepsilon J^{-1}\psi^{\mathcal{J}}\varphi\end{aligned}$$

and thus

$$(\psi, \varphi) = \varepsilon(\varphi, \psi)^j. \qquad (2.6.6)$$

Such a product will be called $\mathcal{D}^j$-symmetric or $\mathcal{D}^j$-skew as ε is plus or minus one. We may use this product to define a mapping from the minimal left ideal to its dual space. If $\mathrm{L}(\mathcal{A}P, \mathcal{D})$ is the space of $\mathcal{D}$-linear maps from $\mathcal{A}P$ to $\mathcal{D}$ then we define

$$\begin{aligned}\sim : \mathcal{A}P &\longrightarrow \mathrm{L}(\mathcal{A}P, \mathcal{D}) \\ \varphi &\longmapsto \tilde{\varphi} \qquad \text{where } \tilde{\varphi}(\psi) = (\varphi, \psi). \qquad (2.6.7)\end{aligned}$$

We shall refer to $\tilde{\varphi}$ as the adjoint of φ with respect to (,). We remarked in the previous section that $\mathrm{L}(\mathcal{A}P, \mathcal{D})$ is naturally identified with a minimal right ideal; elements acting on the minimal left ideal by the algebraic product. With this identification we have

$$\tilde{\varphi} = J^{-1}\varphi^{\mathcal{J}}. \qquad (2.6.8)$$

Having chosen some arbitrary minimal left ideal on which to define a product we can obtain a product on any other minimal left ideal. If P and P' are primitives then the simplicity of $\mathcal{A}$ ensures an element S such that $P' = SPS^{-1}$. Given the product of (2.6.2) we define

$$\begin{aligned}\{\ ,\ \} : \mathcal{A}P' \times \mathcal{A}P' &\longrightarrow P'\mathcal{A}P' \equiv \mathcal{D}' \\ \alpha, \beta &\longmapsto \{\alpha, \beta\} = S(\alpha S, \beta S)S^{-1}. \qquad (2.6.9)\end{aligned}$$

We can write this as $\{\alpha, \beta\} = J'^{-1}\alpha^{\mathcal{J}}\beta$ where $J'^{-1} = SJ^{-1}S^{\mathcal{J}}$ and which satisfies $P'^{\mathcal{J}} = J'P'J'^{-1}$. An involution on $\mathcal{D}'$ equivalent to the involution j on $\mathcal{D}$ is defined by $p^{j'} = S(S^{-1}pS)^jS^{-1}$ for $p \in \mathcal{D}'$. It then follows that $\{\beta, \alpha\} = \varepsilon\{\alpha, \beta\}^{j'}$ and $\{\alpha p, \beta\} = p^{j'}\{\alpha, \beta\}$.

The product we have constructed in (2.6.2) involves not only the involution $\mathcal{J}$ but also the element J as defined in (2.6.1). Obviously such an element cannot be unique. Suppose that $P^{\mathcal{J}} = J'PJ'^{-1}$ with $J'^{\mathcal{J}} = \varepsilon' J'$. Then $J'^{-1}JP = PJ'^{-1}J$ and so $J'^{-1}JP = PJ'^{-1}J = \lambda$ say, where $\lambda \in \mathcal{D}$. Since

$$\lambda^j = J^{-1}\lambda^{\mathcal{J}}J = J^{-1}J^{\mathcal{J}}J'^{-1\mathcal{J}}P^{\mathcal{J}}J = \varepsilon\varepsilon' J'^{-1}JP$$

then

$$\lambda^j = \varepsilon\varepsilon'\lambda. \qquad (2.6.10)$$

If $(\varphi, \psi)' = J'^{-1}\varphi^{\mathcal{J}}\psi$ then $(\varphi, \psi)' = J'^{-1}JJ^{-1}\varphi^{\mathcal{J}}\psi$ and since $(\varphi, \psi) \in \mathcal{D}$ we have

$$(\varphi, \psi)' = \lambda(\varphi, \psi). \tag{2.6.11}$$

When $\mathcal{D} \simeq \mathbb{R}$, then j must be the identity and so we must have $\varepsilon = \varepsilon'$ and the products are related by a real multiple. The complex numbers have two distinct involutions, the identity and complex conjugation. When j is the former then the products are related by an arbitrary complex multiple. When j is complex conjugation then $\mathcal{J}$ is the adjoint of a (pseudo-) Hermitian-symmetric product, determined up to a real multiple, or equivalently the Hermitian-skew product which differs from it by a multiple of the imaginary unit. The quaternions have two inequivalent involutions, conjugation and reversion. Quaternion conjugation, denoted by a bar, is the only involution in its equivalence class. In contrast there are distinct involutions equivalent to some 'standard' representative called reversion and denoted ^. Suppose that $(\psi, \varphi) = \varepsilon(\varphi, \psi)^\wedge$. Then if $(\varphi, \psi)' = \lambda(\varphi, \psi)$ with $\lambda = \lambda^\wedge$ then

$$(\psi, \varphi)' = \varepsilon\lambda(\varphi, \psi)^\wedge = \varepsilon\lambda\{\lambda(\varphi, \psi)\}^\wedge\lambda^{-1} = \varepsilon\lambda\{(\varphi, \psi)'\}^\wedge\lambda^{-1}.$$

Since $\lambda = \lambda^\wedge$ then (as demonstrated in Appendix A) we can set $\lambda = \mu\mu^\wedge$ for some μ. Thus $\lambda q^\wedge\lambda^{-1} = \mu(\mu^{-1}q\mu)^\wedge\mu^{-1}$ and we see that if $\mathcal{J}$ is the adjoint of an $H^\wedge$-symmetric (or skew) product then it is also the adjoint of an H^j-symmetric (or skew-) product for any j equivalent to reversion. If $\mathcal{J}$ is the adjoint of a quaternion-conjugate-symmetric product then it will also be the adjoint of the reversion-skew product obtained by multiplying this product by any vector quaternion. The conjugate-skew and reversion-symmetric products are likewise related.

The above considerations show how any involution is the adjoint of some $\mathcal{D}^j$-symmetric or $\mathcal{D}^j$-skew product. Certain of these products can be further labelled by a signature. First we note that these products are non-degenerate; for if $(J^{-1}n^{\mathcal{J}})\psi = 0 \;\forall\, \psi \in \mathcal{A}P$ then $J^{-1}n^{\mathcal{J}} = 0$ since the regular representation of a simple $\mathcal{A}$ induces a faithful representation on any minimal left ideal. Consider now a non-degenerate $\mathcal{D}^j$-symmetric product on a right $\mathcal{D}$-module. Then if the mapping from $\mathcal{D}$ into the j-symmetric quantities of $\mathcal{D}$, $q \to q^jq$, is surjective then there is an orthogonal basis of unit-norm elements. If this mapping is not surjective but any j-symmetric quantity can be written as $\pm q^jq$, then there is an orthogonal basis of elements normalised to plus or minus one. This is just an obvious generalisation of the result guaranteeing an orthonormal basis for a real symmetric product and can be proved by induction on the dimension of the module. The two different cases are seen to arise when normalising a non-zero-norm quantity. Suppose that $(\psi, \psi) = \lambda$, then if the product is $\mathcal{D}^j$-symmetric $\lambda = \lambda^j$. If we can write $\lambda = q^jq$ for some $q \in \mathcal{D}$ then ψq^{-1} will have unit norm. The mapping $q \to q^jq$ is not a surjection from $\mathscr{D}$ to the j-symmetric quantities when $\mathscr{D}$ is $\mathbb{R}$, $\mathbb{C}$ or H with j the identity, complex and quaternion conjugation respectively. Thus the $\mathbb{R}$-, $\mathbb{C}^*$- and H^--symmetric products are further characterised by their signatures (the number of positive- and negative-norm elements in an orthogonal basis). The smallest of these two numbers will be called the *index* (or *Witt index*). The complex numbers have the important property that any complex number can be written as a square. Similarly any reversion-symmetric quaternion can be written as a square of a reversion-symmetric quantity (as demonstrated in Appendix A). Thus any $\mathbb{C}$-symmetric or $H^\wedge$-symmetric product has an orthogonal basis of unit-norm elements. An $\mathbb{R}$-skew or $\mathbb{C}$-skew product can be non-degenerate only if the vector space is of even dimension, $2n$ say. In this case there is a canonical basis $\{p_i, q_i\}$ for $i = 1, \ldots, n$ with $(p_i, q_j) = \delta_{ij}$.

Example 2.1
Take $\mathscr{A} = \mathbf{C}_{4,0}(\mathbb{R})$ with $\mathscr{J} \equiv \xi$. At the end of §2.2 we constructed a basis for this algebra. Let P be what was there called P_1, that is $P = \frac{1}{2}(1 + z)$. The division algebra $P\mathscr{A}P \equiv \mathscr{D}$ is isomorphic to the quaternion algebra, with standard basis $\{P, e^{23}P, e^{34}P, e^{24}P\}$. Since $P^\xi = P$, in this case the involution ξ induces an involution on $\mathscr{D}$. This is quaternion conjugation since, for example, $(e^{23}P)^\xi = -e^{23}P$. An H^--symmetric product on $\mathscr{A}P$ is given by

$$(\varphi, \psi) = \varphi^\xi \psi.$$

An H-linearly independent basis for $\mathscr{A}P$ is $\{P, e^1P\}$. We have

$$(P, P) = P$$

$$(P, e^1P) = Pe^1P = 0$$

$$(e^1P, e^1P) = Pe^1e^1P = P.$$

So this basis is in fact orthonormal, the product being of index zero.

Thus far we have shown how any involution can be put into one and only one class determined by the $\mathscr{D}^j$-symmetric or $\mathscr{D}^j$-skew product (further labelled by an index where appropriate) for which it is the adjoint involution. We may choose the representatives given in table 2.11 for the classes of product. Where we have chosen, for example, a $\mathbb{C}^*$-symmetric product we could have chosen a $\mathbb{C}^*$-skew one. For the same reason we only further classify products by the index rather than the signature. These classes of product define an equivalence relation on the associated involutions. We have already termed involutions $\mathscr{J}$ and $\mathscr{K}$ equivalent if there is an automorphism $\mathscr{S}$ such that $a^{\mathscr{K}} = ((a^{\mathscr{S}})^{\mathscr{J}})^{(\mathscr{S}^{-1})}$ for all $a \in \mathscr{A}$. In fact it follows that with this notion of equivalence:

Two involutions are equivalent if and only if they are the adjoints of equivalent products. (2.6.12)

Table 2.11

(1) $\mathbb{R}$ -symmetric, of index v
(2) $\mathbb{R}$ -skew (only in even dimensions)
(3) $\mathbb{C}$ -symmetric
(4) $\mathbb{C}$ -skew (only in even dimensions)
(5) $\mathbb{C}^*$-symmetric, of index v
(6) H^--symmetric, of index v
(7) $H^\wedge$-symmetric

Here products are equivalent if they are both of the same one of seven main types and, where appropriate, of the same index. If $\mathcal{J}$ and $\mathcal{K}$ are adjoints of equivalent products then we can introduce two $\mathcal{D}^j$-symmetric or skew products, $(\ ,\ )_J$ and $(\ ,\ )_K$ of (where appropriate) the same signature, with $\mathcal{J}$ and $\mathcal{K}$ their respective adjoints. Both products admit a canonical basis of the same type, and any change of basis may be effected by left multiplication by a regular element. Since both products are j-linear in the first variable, and linear in the second there must be a regular σ such that $(\varphi, \psi)_J = (\sigma\varphi, \sigma\psi)_K$ for all φ and ψ in the minimal left ideal, that is

$$\begin{aligned} J^{-1}\varphi^{\mathcal{J}}\psi &= K^{-1}(\sigma\varphi)^{\mathcal{K}}\sigma\psi = K^{-1}\varphi^{\mathcal{K}}\sigma^{\mathcal{K}}\sigma\psi \\ &= K^{-1}(\sigma^{\mathcal{K}}\sigma)(\sigma^{\mathcal{K}}\sigma)^{-1}\varphi^{\mathcal{K}}(\sigma^{\mathcal{K}}\sigma)\psi. \end{aligned}$$

If we introduce an involution $\mathcal{T}$ defined by

$$a^{\mathcal{T}} = (\sigma^{\mathcal{K}}\sigma)^{-1}a^{\mathcal{K}}(\sigma^{\mathcal{K}}\sigma) \qquad \forall\, a \in \mathcal{A}$$

then $P^{\mathcal{T}} = TPT^{-1}$ with $T = (\sigma^{\mathcal{K}}\sigma)^{-1}K$. We have $J^{-1}\varphi^{\mathcal{J}}\psi = T^{-1}\varphi^{\mathcal{T}}\psi$, that is

$$(\varphi, \psi)_J = (\varphi, \psi)_T \qquad \forall\, \varphi, \psi \in \mathcal{A}P.$$

But $(\varphi, a\psi)_J = (a^{\mathcal{J}}\varphi, \psi)_J$ and $(\varphi, a\psi)_T = (a^{\mathcal{T}}\varphi, \psi)_T = (a^{\mathcal{T}}\varphi, \psi)_J$ giving $((a^{\mathcal{J}} - a^{\mathcal{T}})\varphi, \psi)_J = 0$. Since this is true for all $\psi \in \mathcal{A}P$ and the product is non-degenerate $(a^{\mathcal{J}} - a^{\mathcal{T}})\varphi = 0\ \forall\, \varphi \in \mathcal{A}P$ and $a^{\mathcal{J}} = a^{\mathcal{T}}\ \forall\, a \in \mathcal{A}$. Recalling the definition of $\mathcal{T}$ we have $a^{\mathcal{J}} = \sigma^{-1}(\sigma a \sigma^{-1})^{\mathcal{K}}\sigma$, that is, $\mathcal{J}$ and $\mathcal{K}$ are equivalent. To prove the converse we suppose that $\mathcal{J} = \mathcal{S}\mathcal{K}\mathcal{S}^{-1}$ for some automorphism $\mathcal{S}$. Then if

$$(\ ,\ ) : \mathcal{A}P \times \mathcal{A}P \longrightarrow P\mathcal{A}P$$

is a product with $\mathcal{K}$ as adjoint-involution we define

$$\{\ ,\ \} : \mathcal{A}P^{\mathcal{S}^{-1}} \times \mathcal{A}P^{\mathcal{S}^{-1}} \longrightarrow P^{\mathcal{S}^{-1}}\mathcal{A}P^{\mathcal{S}^{-1}}$$

by $\{\alpha, \beta\} = (\alpha^{\mathcal{G}}, \beta^{\mathcal{G}})^{\mathcal{G}^{-1}}$, then

$$\{\alpha, m\beta\} = (\alpha^{\mathcal{G}}, m^{\mathcal{G}}\beta^{\mathcal{G}})^{\mathcal{G}^{-1}} = (m^{\mathcal{G}\mathcal{K}}\alpha^{\mathcal{G}}, \beta^{\mathcal{G}})^{\mathcal{G}^{-1}} = ([m^{\mathcal{G}\mathcal{K}\mathcal{G}^{-1}}\alpha]^{\mathcal{G}}, \beta^{\mathcal{G}})^{\mathcal{G}^{-1}}$$
$$= \{m^{\mathcal{J}}\alpha, \beta\}.$$

Similarly if

$$(\beta, \alpha) = \varepsilon(\alpha, \beta)^k \text{ then } \{\beta, \alpha\} = (\beta^{\mathcal{G}}, \alpha^{\mathcal{G}})^{\mathcal{G}^{-1}}$$
$$= \varepsilon(\alpha^{\mathcal{G}}, \beta^{\mathcal{G}})^{k\mathcal{G}^{-1}} = \varepsilon\{\alpha, \beta\}^{\mathcal{G}k\mathcal{G}^{-1}} = \varepsilon\{\alpha, \beta\}^j$$

where $j = \mathcal{G}k\mathcal{G}^{-1}$ is equivalent to k. Thus we have a product on $\mathcal{A}P^{\mathcal{G}^{-1}}$ with $\mathcal{J}$ as an adjoint-involution of the same type as the product on $\mathcal{A}P$, which has $\mathcal{K}$ as adjoint-involution. As already noted a product on any given minimal left ideal enables an equivalent product to be defined on any other minimal left ideal. Thus if $\mathcal{J} = \mathcal{G}\mathcal{K}\mathcal{G}^{-1}$ then $\mathcal{J}$ and $\mathcal{K}$ are adjoint-involutions of equivalent products on any minimal left ideal.

Although for complex matrices not all automorphisms are inner a corollary to the above is that if involutions are related by $\mathcal{J} = \mathcal{G}\mathcal{K}\mathcal{G}^{-1}$ for any automorphism $\mathcal{G}$ then in fact there is an inner automorphism Σ such that $\mathcal{J} = \Sigma\mathcal{K}\Sigma^{-1}$.

The result of (2.6.10), together with table 2.11, gives the number of inequivalent involutions for a simple $\mathbb{R}$-algebra. This is displayed in table 2.12.

Table 2.12 The number of inequivalent involutions.

	$\mathcal{M}_r$	$\mathbb{C}\otimes\mathcal{M}_r$	$H\otimes\mathcal{M}_r$
r even	$\frac{1}{2}r + 2$	$\frac{1}{2}r + 3$	$\frac{1}{2}r + 2$
r odd	$\frac{1}{2}(r + 1)$	$\frac{1}{2}(r + 3)$	$\frac{1}{2}(r + 3)$

Since not all Clifford algebras are simple we now consider involutions of semi-simple algebras that have two simple components. Let $\mathcal{A} = \mathcal{B}\oplus\mathcal{C}$ where $\mathcal{B}$ and $\mathcal{C}$ are simple with $\mathcal{B} = \mathcal{A}P$, $\mathcal{C} = \mathcal{A}Q$ for central idempotents P, Q with $PQ = QP = 0$ and $P + Q = 1$. If $\mathcal{K}$ is an involution of $\mathcal{A}$ then $P^{\mathcal{K}}$, $Q^{\mathcal{K}}$ are central idempotents with $P^{\mathcal{K}}Q^{\mathcal{K}} = Q^{\mathcal{K}}P^{\mathcal{K}} = 0$ and $P^{\mathcal{K}} + Q^{\mathcal{K}} = 1$. So $\mathcal{A} = \mathcal{A}P^{\mathcal{K}}\oplus\mathcal{A}Q^{\mathcal{K}}$ and, since the expression of a semi-simple algebra as a sum of simple ones is unique up to ordering ((A17) of Appendix A) then either $\mathcal{A}P^{\mathcal{K}} = \mathcal{B}$ and $\mathcal{A}Q^{\mathcal{K}} = \mathcal{C}$, or $\mathcal{A}Q^{\mathcal{K}} = \mathcal{B}$ and $\mathcal{A}P^{\mathcal{K}} = \mathcal{C}$. In the former case $\mathcal{K}$ induces an involution on the simple algebras $\mathcal{B}$ and $\mathcal{C}$ and may thus be classified in the manner already treated. In the second case every element of $\mathcal{B}$ is sent to $\mathcal{C}$, and this can only arise when $\mathcal{C}$ is isomorphic to the opposite algebra of $\mathscr{B}$, $\mathscr{C} \simeq \mathscr{B}^{\mathrm{op}}$. There is in fact only one such involution, up to equivalence.

If $\mathscr{A}$ is the sum of two simple algebras, with $\mathscr{K}$ and $\mathscr{J}$ involutions that do not preserve these simple components, then $\mathscr{K}$ and $\mathscr{J}$ are equivalent. (2.6.13)

If $\mathscr{K}$ and $\mathscr{J}$ are as described then $\mathscr{J}\mathscr{K}$ is an automorphism of $\mathscr{A}$ that induces automorphisms on the simple components $\mathscr{B}$ and $\mathscr{C}$. We introduce an automorphism $\mathscr{S}$ of $\mathscr{A}$ by defining

$$b^{\mathscr{S}} = b^{\mathscr{K}\mathscr{J}} \qquad \forall\, b \in \mathscr{B}$$

$$c^{\mathscr{S}} = c \qquad \forall\, c \in \mathscr{C}.$$

Then $\mathscr{S}$ is an automorphism of $\mathscr{A}$ such that for $b \in \mathscr{B}$ $b^{\mathscr{S}\mathscr{J}} = b^{\mathscr{K}}$, which is in $\mathscr{C}$, and so $b^{\mathscr{S}\mathscr{J}\mathscr{S}^{-1}} = b^{\mathscr{K}\mathscr{S}^{-1}} = b^{\mathscr{K}}$. Similarly for $c \in \mathscr{C}$ $c^{\mathscr{S}\mathscr{J}} = c^{\mathscr{J}}$ and so $c^{\mathscr{S}\mathscr{J}\mathscr{S}^{-1}} = c^{\mathscr{J}\mathscr{S}^{-1}} = c^{\mathscr{J}\mathscr{J}^{-1}\mathscr{K}^{-1}} = c^{\mathscr{K}}$ and we have established the equivalence of $\mathscr{J}$ and $\mathscr{K}$.

When $\mathscr{A} = \mathscr{D}\otimes\mathscr{M}\oplus\mathscr{D}\otimes\mathscr{M}$ where $\mathscr{M}$ is a total matrix algebra and the division algebra satisfies $\mathscr{D} \simeq \mathscr{D}^{\mathrm{op}}$ then any involution in the class not preserving the simple components will be called a $\mathscr{D}$-swap. (The real division algebras $\mathbb{R}$, $\mathbb{C}$ and H are isomorphic to their opposite algebras.) Any such involution is the adjoint of a non-degenerate product on the left ideal formed from the direct sum of two minimal left ideals from the two different simple components. For let P be some primitive idempotent (necessarily in one simple component). If $\mathscr{J}$ is some $\mathscr{D}$-swap involution then $Q = P + P^{\mathscr{J}}$ is a $\mathscr{J}$-symmetric idempotent. We may define

$$\begin{aligned} (\ ,\ )\colon \mathscr{A}Q \times \mathscr{A}Q &\longrightarrow Q\mathscr{A}Q \simeq \mathscr{D}\oplus\mathscr{D} \\ \alpha, \beta &\longmapsto (\alpha, \beta) = \alpha^{\mathscr{J}}\beta. \end{aligned} \tag{2.6.14}$$

Such a product is non-degenerate for if $\alpha^{\mathscr{J}}\beta = 0\ \forall\, \beta$ then choosing β to lie in one simple component shows that the component of α in the other must vanish, and hence $\alpha = 0$. It immediately follows from (2.6.14) that

$$(\alpha, m\beta) = (m^{\mathscr{J}}\alpha, \beta) \quad \forall\, m \in \mathscr{A}$$

$$(\beta, \alpha) = (\alpha, \beta)^{\mathscr{J}}$$

$$(\alpha q, \beta) = q^{\mathscr{J}}(\alpha, \beta) \quad \forall\, q \in \mathscr{D}\oplus\mathscr{D}.$$

We can equally well introduce a product with different symmetry. If s is any regular element lying in $\mathscr{D}$, $ss^{-1} = 1_{\mathscr{D}}$, then let $S = s - s^{\mathscr{J}}$. This $\mathscr{J}$-skew element of $\mathscr{D}\oplus\mathscr{D}^{\mathscr{J}}$ has an inverse given by

$$S^{-1} = s^{-1} - (s^{-1})^{\mathscr{J}} \qquad SS^{-1} = ss^{-1} + (s^{-1}s)^{\mathscr{J}} = 1_{\mathscr{D}} + 1_{\mathscr{D}^{\mathscr{J}}} = 1.$$

We may now define

$$\{\ ,\ \}\colon \mathscr{A}Q \times \mathscr{A}Q \longrightarrow \mathscr{D}\oplus\mathscr{D}$$

$$\alpha, \beta \longmapsto \{\alpha, \beta\} = S^{-1}\alpha^{\mathscr{J}}\beta.$$

This product satisfies

$$\{\alpha, m\beta\} = \{m^{\mathscr{J}}a, \beta\}$$

$$\{\beta, \alpha\} = -S^{-1}\{\alpha, \beta\}^{\mathscr{J}}S$$

$$\{\alpha q, \beta\} = S^{-1}q^{\mathscr{J}}S\{\alpha, \beta\}.$$

We are now ready to return to Clifford algebras. Having established how a given involution is the adjoint of a non-degenerate product on the minimal left ideals the problem of finding products invariant under ${}_{+}\Gamma^{+}$ reduces to finding involutions $\mathscr{J}$ such that $s^{\mathscr{J}} = s^{-1}\ \forall s \in {}_{+}\Gamma^{+}$. Of course ξ and $\xi\eta$ are such involutions (which may or may not be equivalent), but before classifying the associated products we confirm that these are the only such involutions (up to equivalence). It is convenient to consider the cases of even and odd dimensions separately.

Suppose firstly that $p + q = n$ is even, so that $\mathbf{C}_{p,q}(\mathbb{R})$ is central simple. Then if $\mathscr{J}$ is any involution there exists some J such that $a^{\mathscr{J}} = Ja^{\xi}J^{-1}$ with $J^{\xi} = \pm J$. If $s \in {}_{+}\Gamma^{+}$ then $s^{\mathscr{J}} = Js^{\xi}J^{-1} = Js^{-1}J^{-1}$; thus $s^{\mathscr{J}} = s^{-1}$ if and only if $Js^{-1} = s^{-1}J \quad \forall s \in {}_{+}\Gamma^{+}$. We know (from the proof of (2.5.3)) that ${}_{+}\Gamma^{+}$ generates the even subalgebra, and so $s^{\mathscr{J}} = s^{-1} \quad \forall s \in {}_{+}\Gamma^{+}$ if and only if J commutes with all elements of the even subalgebra. If J has this property then so will its even and odd parts separately. But the volume n-form Z is even and it anticommutes with all odd elements and so the odd part of J must vanish, hence J must be in the centre of the even subalgebra. This centre is spanned by $\{1, z\}$.

If $Z^{\xi} = -Z$ then, since $J^{\xi} = \pm J$, either $J \in \mathbb{R}$ giving $\mathscr{J} = \xi$ or $J = \lambda z$ for $\lambda \in \mathbb{R}$ and for any a, $a^{\mathscr{J}} = za^{\xi}z^{-1} = a^{\xi\eta}$.

If $z^{\xi} = z$ and $z^{2} = -1$ then the centre of $\mathbf{C}^{+}_{p,q}(\mathbb{R})$ is $\mathbb{C}(\mathbb{R})$. If $J \in \mathbb{C}$ then, since $\mathbb{C}$ is algebraically closed, $J = \sigma^{2} = \sigma\sigma^{\xi}$ for some $\sigma \in \mathbb{C}$ and $a^{\mathscr{J}} = \sigma\sigma^{\xi}a^{\xi}\sigma^{\xi-1}\sigma^{-1} = \sigma(\sigma^{-1}a\sigma)^{\xi}\sigma^{-1}$ showing that $\mathscr{J}$ is equivalent to ξ.

If $z^{\xi} = z$ and $z^{2} = 1$ then $\mathbf{C}^{+}_{p,q}(\mathbb{R})$ is reducible and the centre is spanned by the orthogonal idempotents $\{P_{+}, P_{-}\}$ where $P_{\pm} = \frac{1}{2}(1 \pm z)$. If J is regular then $J = \lambda P_{+} + \mu P_{-}$ with λ, μ non-zero reals. If λ and μ are both of the same sign then there is no loss of generality in assuming them positive since multiplying J by an element of the centre does not alter $\mathscr{J}$. In this case we set

$$J = \sigma^{2}P_{+} + \nu^{2}P_{-} = (\sigma P_{+} + \nu P_{-})(\sigma P_{+} + \nu P_{-})^{\xi}$$

and $\mathscr{J}$ is equivalent to ξ. Similarly if λ and μ are of opposite sign then, with no loss of generality, we assume $J = \sigma^{2}P_{+} - \nu^{2}P_{-}$. If

$k = \sigma P_+ + \nu P_-$ then

$$kzk^{\xi} = (\sigma P_+ + \nu P_-)(P_+ - P_-)(\sigma P_+ + \nu P_-)$$
$$= \sigma^2 P_+ - \nu^2 P_- = J.$$

Thus

$$a^{\mathscr{J}} = kzk^{\xi}a^{\xi}(k^{\xi})^{-1}z^{-1}k^{-1} = kz(k^{-1}ak)^{\xi}z^{-1}k^{-1} = k(k^{-1}ak)^{\xi\eta}k^{-1}$$

and $\mathscr{J}$ is equivalent to $\xi\eta$.

For the case in which $z^{\xi} = z$ we have seen that the requirement that $s^{\mathscr{J}} = s^{-1}\ \forall s \in {}_+\Gamma^+$ only requires $\mathscr{J}$ to be equivalent to either ξ or $\xi\eta$. If, however, $p \neq 0$ then ${}_+\Gamma^{\pm}$ contains an odd element and requiring $s^{\mathscr{J}} = s^{-1}\ \forall s \in {}_+\Gamma^{\pm}$ uniquely determines $\mathscr{J}$ to be ξ. For if $J = \lambda + \mu z$ then the requirement that J commute with an odd element forces μ to vanish. Similarly, if $q \neq 0$ then $\xi\eta$ is the unique involution such that $s^{\mathscr{J}} = s^{-1}\ \forall s \in {}^+\Gamma^{\pm}$.

We turn now to the case in which n is odd, with $\{1, z\}$ spanning the centre of $\mathbf{C}_{p,q}(\mathbb{R})$, and $\mathbf{C}^+_{p,q}(\mathbb{R})$ central simple. Since $z^{\eta} = -z$ one of the involutions ξ and $\xi\eta$ will leave the centre invariant, the other will not. It follows that if $\mathscr{J}$ is any involution then either $a^{\mathscr{J}} = Ja^{\xi}J^{-1}$ with $J^{\xi} = \pm J$, or $a^{\mathscr{J}} = Ja^{\xi\eta}J^{-1}$ with $J^{\xi\eta} = \pm J$.

Consider the former case. Then requiring $s^{\mathscr{J}} = s^{-1}\ \forall s \in {}_+\Gamma^+$ shows, exactly as before, that J must commute with elements of the even subalgebra. Then if J_- is the odd part of J we can write $J_- = (J_-z^{-1})z$ and, since the odd element z commutes with everything, J_- will commute with the even subalgebra if and only if J_-z^{-1} does. Since the even subalgebra is central simple J_- must be proportional to z. It then follows that J is in the centre of $\mathbf{C}_{p,q}(\mathbb{R})$ and $\mathscr{J} = \xi$.

In exactly the same way it follows that if $s^{\mathscr{J}} = Js^{\xi\eta}J^{-1}$ and $s^{\mathscr{J}} = s^{-1}\ \forall s \in {}_+\Gamma^+$ then $\mathscr{J} = \xi\eta$.

We may summarise as follows:

if ${}_+\Gamma^{\pm} \not\subset {}_+\Gamma^+$ then $s^{\mathscr{J}} = s^{-1}\ \forall s \in {}_+\Gamma^{\pm}$ iff $\mathscr{J} = \xi$

if ${}^+\Gamma^{\pm} \not\subset {}_+\Gamma^+$ then $s^{\mathscr{J}} = s^{-1}\ \forall s \in {}^+\Gamma^{\pm}$ iff $\mathscr{J} = \xi\eta$.

If $s^{\mathscr{J}} = s^{-1}\ \forall s \in {}_+\Gamma^+$ then $\mathscr{J}$ is equivalent to ξ or $\xi\eta$; if $n \neq 4$ mod 4 then either $\mathscr{J} = \xi$ or $\mathscr{J} = \xi\eta$. (2.6.15)

The involutions ξ and $\xi\eta$ induce the same involution on the even subalgebra. This is the only such involution that inverts elements of ${}_+\Gamma^+$. For if $\mathscr{J}$ is any involution of a central simple $\mathbf{C}^+_{p,q}(\mathbb{R})$ then $a^{\mathscr{J}} = Ja^{\xi}J^{-1}$ for some even J. Thus $s^{\mathscr{J}} = s^{-1}\ \forall s \in {}_+\Gamma^+$ only if $\mathscr{J}$ is in the centre, giving $\mathscr{J} = \xi$. If $\mathbf{C}^+_{p,q}(\mathbb{R})$ is not central simple then $a^{\mathscr{J}} = Ja^{\xi}J^{-1}$ with J even if $\mathscr{J}\xi$ leaves elements of the centre invariant, or odd if $\mathscr{J}\xi$ induces a non-trivial automorphism on the centre. In the former case we have seen that only if $\mathcal{J} = \xi$ is $s^{\mathcal{J}} = s^{-1}\ \forall s \in {}_{+}\Gamma^{+}$. There can be no involution with this property in the second category, for there is no odd element that commutes with the even subalgebra since this contains z which anticommutes with all odd elements.

We wish now to classify the involutions ξ and $\xi\eta$ and the involution they induce on the even subalgebra. This will be done for an arbitrary Clifford algebra using the same isomorphisms that enabled its structure to be determined. We use those isomorphisms for which ξ and $\xi\eta$ on the factors of a tensor product induce ξ or $\xi\eta$ on the product algebra. In this way knowing the class of ξ and $\xi\eta$ on the factors enables the class of the involution on the product to be determined, and ξ and $\xi\eta$ on arbitrary algebras can be classified by explicitly classifying these involutions for a few low-dimensional algebras. First we consider involutions of tensor products.

Let $\mathcal{A} \simeq \mathcal{B} \otimes \mathcal{M}_n$ where $\mathcal{B}$ is a simple algebra over $\mathbb{R}$ and $\mathcal{M}_n$ is the algebra of order n real matrices. Let $\mathcal{J}$ be an involution on $\mathcal{A}$ that induces involutions $\mathcal{T}$ on $\mathcal{M}_n$ and $\mathcal{K}$ on $\mathcal{B}$. We shall write this as $\mathcal{J} = \mathcal{K}\otimes\mathcal{T}$. If $\mathcal{B} = \mathcal{D}\otimes\mathcal{M}_m$ and $\mathcal{K}$ is $\mathcal{D}^k$-symmetric or skew then $\mathcal{J}$ is certainly either $\mathcal{D}^k$-symmetric or skew, the symmetry being determined by that of $\mathcal{K}$ and $\mathcal{T}$, in a way to be determined. If Q is primitive in $\mathcal{B}$ and R is primitive in $\mathcal{M}_n$ then $P = QR$ is primitive in $\mathcal{A}$. If $Q^{\mathcal{K}} = KQK^{-1}$ and $R^{\mathcal{T}} = TRT^{-1}$ then $P^{\mathcal{J}} = JPJ^{-1}$ with $J = KT$. Since $J^{\mathcal{J}} = K^{\mathcal{K}}T^{\mathcal{T}}$, J is symmetric if K and T are both symmetric or both skew, and skew if K and T are of different symmetry. Thus if either $\mathcal{K}$ is $\mathcal{D}^k$-symmetric with $\mathcal{T}$ $\mathbb{R}$-skew, or $\mathcal{K}$ is $\mathcal{D}^k$-skew with $\mathcal{T}$ $\mathbb{R}$-skew, then $\mathcal{J}$ is $\mathcal{D}^k$-symmetric; otherwise it is $\mathcal{D}^k$-skew. We now investigate the signature in the case in which $\mathcal{J}$ is $\mathcal{D}^k$-symmetric. If $(\ ,\ )_J$ is the product on $\mathcal{A}P$ associated with $\mathcal{J}$ then for $b_i, b_j \in \mathcal{B}$ and $m_\alpha, m_\beta \in \mathcal{M}_n$

$$\begin{aligned}(b_i m_\alpha, b_j m_\beta)_J &= J^{-1} m_\alpha{}^{\mathcal{J}} b_i{}^{\mathcal{J}} b_j m_\beta = T^{-1}K^{-1} m_\alpha{}^{\mathcal{T}} b_i{}^{\mathcal{K}} b_j m_\beta \\ &= T^{-1} m_\alpha{}^{\mathcal{T}} m_\beta K^{-1} b_i{}^{\mathcal{K}} b_j\end{aligned}$$

since $\mathcal{B}$ and $\mathcal{M}_n$ are mutually commuting subalgebras of $\mathcal{A}$. So

$$(b_i m_\alpha, b_j m_\beta)_J = (m_\alpha, m_\beta)_T (b_i, b_j)_K$$

where the products on the right-hand side are those on the subalgebras associated with the involutions indicated. Thus if, for example, $(\ ,\ )_T$ is symmetric admitting an orthonormal basis with r vectors normalised to plus one and s to minus one (of signature r, s) and $(\ ,\ )_K$ has a basis with r' normalised to plus one and s' to minus one then $(\ ,\ )_J$ has a basis of $rr' + ss'$ positive-norm and $rs' + sr'$ negative-norm vectors. Similarly it follows that if $(\ ,\ )_T$ is $\mathcal{R}$-skew and $(\ ,\ )_K$ is $\mathcal{D}^k$-skew then $(\ ,\ )_J$ admits an orthonormal basis with as many positive- as negative-norm basis vectors. We summarise the situation below. If $\mathcal{J}$ is an involution on $\mathcal{A}$, where $\mathcal{A} = \mathcal{B}\otimes\mathcal{M}_n$ with $\mathcal{J} = \mathcal{K}\otimes\mathcal{T}$ then:

(i) if either $\mathcal{K}$ is $\mathcal{D}^k$-skew with $\mathcal{T}$ $\mathbb{R}$-symmetric, or $\mathcal{K}$ is $\mathcal{D}^k$-symmetric with $\mathcal{T}$ $\mathbb{R}$-skew, then $\mathcal{J}$ is $\mathcal{D}^k$-skew;
(ii) if $\mathcal{K}$ is $\mathcal{D}^k$-skew and $\mathcal{T}$ is $\mathbb{R}$-skew then $\mathcal{J}$ is $\mathcal{D}^k$-symmetric, of maximal index (if any);
(iii) if $\mathcal{K}$ is $\mathcal{D}^k$-symmetric, with signature r, s, and $\mathcal{T}$ is $\mathbb{R}$-symmetric, of signature r', s' then $\mathcal{J}$ is $\mathcal{D}^k$-symmetric of signature $rr' + ss'$, $rs' + sr'$. (2.6.16)

Having shown how to classify the involution on the tensor product of a simple $\mathbb{R}$-algebra with a matrix algebra in terms of involutions on the factors, to classify the involutions on the product of two simple algebras it only remains to consider involutions on products of the division algebras. When one of these division algebras is $\mathbb{R}$ itself there is nothing to do. For the product of two copies of the complex numbers we have

$$\mathbb{C}(\mathbb{R}) \otimes \mathbb{C}(\mathbb{R}) \simeq \mathbb{C}(\mathbb{R}) \oplus (\mathbb{R})$$

$$\text{identity} \otimes \text{identity} \simeq \text{identity} \oplus \text{identity}$$

$$\text{identity} \otimes \text{conjugation} \simeq \mathbb{C}\text{-swap}$$

$$\text{conjugation} \otimes \text{conjugation} \simeq \text{conjugation} \oplus \text{conjugation}. \quad (2.6.17)$$

We use the obvious notation for an involution on a reducible algebra that induces involutions on the simple component algebras. The above can be verified by choosing a specific basis. If $\{1, i\}$, $\{1, j\}$ are standard bases for the factors and $P_{\pm} = \frac{1}{2}(1 \pm ij)$ then $\{P_+, \mathrm{i}P_+\}$ and $\{P_-, \mathrm{i}P_-\}$ are bases for the simple components. Similarly

$$\mathbb{C}(\mathbb{R}) \otimes H(\mathbb{R}) \simeq \mathbb{C}(\mathbb{R}) \otimes \mathcal{M}_2(\mathbb{R})$$

$$\text{identity} \otimes \text{quaternion conjugation} \simeq \mathbb{C}\text{-skew}$$

$$\text{identity} \otimes \text{reversion} \simeq \mathbb{C}\text{-symmetric}$$

$$\text{complex conjugation} \otimes \text{quaternion conjugation}$$

$$\simeq \mathbb{C}^*\text{-symmetric, zero index}$$

$$\text{complex conjugation} \otimes \text{reversion} \simeq \mathbb{C}^*\text{-symmetric, index one}. \quad (2.6.18)$$

If $\{1, z\}$ and $\{1, i, j, k\}$ are standard bases for the factors then $\{\mathbf{e}_{ij}\}$ is an ordinary matrix basis where $\mathbf{e}_{11} = \frac{1}{2}(1 + zi)$, $\mathbf{e}_{22} = \frac{1}{2}(1 - zi)$, $\mathbf{e}_{21} = j\mathbf{e}_{11} = \mathbf{e}_{22}j$ and $\mathbf{e}_{12} = -j\mathbf{e}_{22} = -\mathbf{e}_{11}j$. Finally,

$$H(\mathbb{R}) \otimes H(\mathbb{R}) \simeq \mathcal{M}_4(\mathbb{R})$$

$$\text{conjugation} \otimes \text{conjugation} \simeq \mathbb{R}\text{-symmetric, zero index}$$

$$\text{reversion} \otimes \text{reversion} \simeq \mathbb{R}\text{-symmetric, index two}$$

$$\text{conjugation} \otimes \text{reversion} \simeq \mathbb{R}\text{-skew}. \quad (2.6.19)$$

Again this can be verified by constructing a basis. It is sufficient to note that a primitive is given by $P = \frac{1}{4}(1 + iI)(1 + jJ)$ where $\{1, i, j, k\}$ and $\{1, I, J, K\}$ are standard bases for the factors. A basis for the minimal left ideal in $\mathcal{M}_4(\mathbb{R})$ is $\{P, iP, jP, kP\}$ and the index of the symmetric products can be explicitly evaluated.

We are now in a position to classify all involutions on simple $\mathbb{R}$-algebras, or sums of two such algebras, that are obtained from involutions on the factors of a tensor product. If $\mathcal{A} \simeq \mathcal{B} \otimes \mathcal{C}$ and $\mathcal{J} = \mathcal{K} \otimes \mathcal{T}$ then table 2.12 gives the class of the involution $\mathcal{J}$ in terms of the classes of $\mathcal{K}$ and $\mathcal{T}$. These classes are encoded into the types of table 2.11, with the $\oplus$ symbol denoting an involution on a direct sum of algebras inducing involutions on the components, and the types (8), (9) and (10) (see table 2.13) being $\mathcal{D}$-swaps for $\mathcal{D} = \mathbb{R}$, H and $\mathbb{C}$. The class of $\mathcal{K}$ determines the row of the table whilst $\mathcal{T}$ determines the column, the class of $\mathcal{J}$ being given in the intersection. (The symmetry of the table reflects the fact that $\mathcal{A} \otimes \mathcal{B} \simeq \mathcal{B} \otimes \mathcal{A}$.) For example, (2.6.16) is encoded into the first two rows and the first two columns, whilst the diagonal block formed by the intersection of the third and fourth rows and columns is given by (2.6.16) and (2.6.17). There are various blanks in table 2.13, corresponding to those cases when the tensor product of the factors would be a reducible algebra with more than two components.

To use the above to classify the involutions ξ and $\xi\eta$ on arbitrary Clifford algebras we need to build up the Clifford algebras from tensor products of smaller ones such that the standard involutions on the factors induce ξ and $\xi\eta$ on the product. In §2.2 the structure of an arbitrary Clifford algebra was determined using the relations (2.2.7), (2.2.8) and (2.2.9) together with a knowledge of certain low-dimensional algebras. Examining the isomorphisms that established these relations shows that the involutions ξ and $\xi\eta$ of the left-hand side of (2.2.7) do not induce either of the standard involutions on the factors. However, equations (2.2.8) and (2.2.9) give a relation between the standard involutions on the factors and the standard involutions on the product. As was noted in §2.2 there is another relation similar to (2.2.9), and in this case the standard involutions on the factors are related to those on the product. The relations are given below.

$$\begin{aligned} \mathbf{C}_{p+1,\,q+1}(\mathbb{R}) &\simeq \mathbf{C}_{p,\,q}(\mathbb{R}) \otimes \mathbf{C}_{1,\,1}(\mathbb{R}) \\ \xi &\simeq \xi\eta \otimes \xi \\ \xi\eta &\simeq \xi \otimes \xi\eta \end{aligned} \tag{2.6.20}$$

$$\begin{aligned} \mathbf{C}_{p,\,q+4}(\mathbb{R}) &\simeq \mathbf{C}_{p,\,q}(\mathbb{R}) \otimes \mathbf{C}_{0,\,4}(\mathbb{R}) \\ \xi &\simeq \xi \otimes \xi \\ \xi\eta &\simeq \xi\eta \otimes \xi\eta \end{aligned} \tag{2.6.21}$$

Table 2.13 The classification of involutions on tensor products induced from involutions of the factors. (See table 2.11 for the classification scheme.)

	Involutions on tensor products									
	1	2	3	4	5	6	7	8	9	10
1	1	2	3	4	5	6	7	8	9	10
2	2	1	4	3	5	7	6	8	9	10
3	3	4	$3 \oplus 3$	$4 \oplus 4$	10	4	3	10	10	
4	4	3	$4 \oplus 4$	$3 \oplus 3$	10	3	4	10	10	
5	5	5	10	10	$5 \oplus 5$	5	5	10	10	
6	6	7	4	3	5	1	2	9	8	10
7	7	6	3	4	5	2	1	9	8	10
8	8	8	10	10	10	9	9			
9	9	9	10	10	10	8	8			
10	10	10				10	10			

(8) $\equiv \mathbb{R}$-swap, (9) $\equiv H$-swap, (10) $\equiv \mathbf{C}$-swap.

$$\begin{aligned} \mathbf{C}_{p+4,\,q}(\mathbb{R}) &\simeq \mathbf{C}_{p,\,q}(\mathbb{R}) \otimes \mathbf{C}_{4,\,0}(\mathbb{R}) \\ \xi &\simeq \xi \otimes \xi \\ \xi\eta &\simeq \xi\eta \otimes \xi\eta. \end{aligned} \tag{2.6.22}$$

Here we have used the same symbol to denote involutions on different algebras. For example, in (2.6.21) the ξ on the left-hand side is the standard involution on $\mathbf{C}_{p,q+4}(\mathbb{R})$ whereas the same symbol on the right-hand side denotes firstly the standard involution on $\mathbf{C}_{p,q}(\mathbb{R})$, and then that on $\mathbf{C}_{0,4}(\mathbb{R})$.

Those low-dimensional algebras whose involutions must be classified by inspection are given in table 2.14. The products on the left ideals associated with the involutions are readily contructed for these algebras. Some of these products are further labelled by an index; we have only indicated the index where it is zero. This is justified by the following theorem.

When the involution ξ on $\mathbf{C}_{p,\,q}(\mathbb{R})$ is associated with a spinor product labelled by an index then if $q \neq 0$ that index is maximal, whilst for $q = 0$ the index is zero. Similarly, any index associated with $\xi\eta$ is maximal unless $p = 0$ in which case the index is zero. (2.6.23)

Table 2.14 Involution classes of some low-dimensional Clifford algebras.

	ξ	$\xi\eta$
$\mathbf{C}_{1,0}(\mathbb{R})$	$1 \oplus 1$	8
$\mathbf{C}_{2,0}(\mathbb{R})$	1(zero index)	2
$\mathbf{C}_{3,0}(\mathbb{R})$	5(zero index)	4
$\mathbf{C}_{4,0}(\mathbb{R})$	6(zero index)	6
$\mathbf{C}_{0,1}(\mathbb{R})$	3	5
$\mathbf{C}_{0,2}(\mathbb{R})$	7	6
$\mathbf{C}_{0,3}(\mathbb{R})$	9	$6 \oplus 6$
$\mathbf{C}_{0,4}(\mathbb{R})$	6	6(zero index)
$\mathbf{C}_{1,1}(\mathbb{R})$	1	2

Suppose that (,) is a product on some left ideal with ξ as adjoint involution, and that $\{\varepsilon_i\}$ is an orthonormal basis with r positive-norm and s negative-norm elements. If $q \neq 0$ then there is a vector x with $x^2 = -1$. Then $\{x\varepsilon_i\}$ is a new orthonormal basis for the left ideal and $(x\varepsilon_i,\ x\varepsilon_i) = (x^{\xi}x\varepsilon_i,\ \varepsilon_i) = (x^2\varepsilon_i,\ \varepsilon_i) = -(\varepsilon_i,\ \varepsilon_i)$. Thus this basis has r negative- and s positive-norm elements and so if the signature is well defined we must have $r = s$. So if $q \neq 0$ the index is maximal. That it is in fact zero for $q = 0$ can be verified by repeated use of (2.6.22), together with table 2.14. The involution $\xi\eta$ is treated in exactly the same way.

We can now give the class of ξ and $\xi\eta$ for all $\mathbf{C}_{p,q}(\mathbb{R})$. First we note that this class only depends on p mod 8 and q mod 8: two applications of (2.6.21) and (2.6.22) enable this to be inferred from table 2.13. We have given the classes in table 2.15. To complete this table we use (2.6.21) and (2.6.22) with table 2.14 to complete the first row and the first column. Then the classes of ξ and $\xi\eta$ are simultaneously entered in the diagonals by using (2.6.20) with the multiplication of table 2.13. The third entry for a given p and q in table 2.15 gives the class of the involution that ξ and $\xi\eta$ induce on the even subalgebra. (The case of $p = q = 0$ is a degenerate case for which the class is 1.) Reference to the derivation of (2.3.1) gives

$$\begin{aligned} \mathbf{C}^{+}_{p+1,q}(\mathbb{R}) &\simeq \mathbf{C}_{q,p}(\mathbb{R}) \\ \xi &\simeq \xi\eta. \end{aligned} \tag{2.6.24}$$

Thus the class of the involution on the subalgebra is obtained from a relabelling of the classification of $\xi\eta$.

When $p + q \leqslant 5$ we can use the classification of the involution ξ on the even subalgebra to obtain ${}_{+}\Gamma^{+}$ as the group of automorphisms of the associated product. This follows from (2.4.22). The automorphism group of an $\mathbb{R}$-skew product on an n-dimensional vector space is denoted $\mathrm{Sp}(n, \mathbb{R})$, similarly $\mathrm{Sp}(n, \mathbb{C})$ is the automorphism group of a $\mathbb{C}$-skew product. For a $\mathbb{C}^{*}$-symmetric product with signature r, s the automorphism group is $U(r, s)$, whilst we use $\mathrm{Sp}(r, s, H)$ to denote the automorphism group of an H^{-}-symmetric product with this signature. When $s = 0$ then we simply write $U(r)$ and $\mathrm{Sp}(r, H)$. The products associated with the $\mathscr{D}$-swap have the general linear groups as automorphism groups. For, taking the product of (2.6.14), a general element of the automorphism group is $S = s + s^{-1\mathscr{I}}$ with s any regular element of $\mathscr{D}\otimes\mathscr{M}$. We have arranged these spin groups in table 2.16. From this table we have, for example, for $\mathbf{C}_{3,1}(\mathbb{R})$ ${}_{+}\Gamma^{+} \simeq \mathrm{Sp}(2, \mathbb{C})$ whereas at the end of §2.4 we demonstrated that ${}_{+}\Gamma^{+} \simeq \mathrm{SL}(2, \mathbb{C})$. These groups are isomorphic; in fact we have

$$\begin{aligned} \mathrm{Sp}(1, H) &\simeq \mathrm{SU}(2) \\ \mathrm{Sp}(2, \mathbb{R}) &\simeq \mathrm{SL}(2, \mathbb{R}) \\ \mathrm{Sp}(2, \mathbb{C}) &\simeq \mathrm{SL}(2, \mathbb{C}). \end{aligned} \tag{2.6.25}$$

It can be seen that in two dimensions ${}_{+}\Gamma^{+}$ is isomorphic to the orthogonal group.

Table 2.15 Classes of involutions of the real Clifford algebras. For each p and q the classes of ξ, $\xi\eta$ and the involution they induce on the even subalgebra are given.

p \ q	0	1	2	3	4	5	6	7
	Classes of involution on the real Clifford algebras $\mathbf{C}_{p,q}(\mathbb{R})$							
0	1 1 $1 \oplus 1$	3 5 1	7 6 5	9 $6 \oplus 6$ 6	6 6 $6 \oplus 6$	4 5 6	2 1 5	8 $1 \oplus 1$ 1
1	$1 \oplus 1$ 8 1	1 2 8	5 4 2	6 6 4	$6 \oplus 6$ 9 6	6 7 9	5 3 7	1 1 3
2	1 2 5	8 $2 \oplus 2$ 2	2 2 $2 \oplus 2$	4 5 2	6 7 5	9 $7 \oplus 7$ 7	7 7 $7 \oplus 7$	3 5 7
3	5 4 6	2 2 4	$2 \oplus 2$ 8 2	2 1 8	5 3 1	7 7 3	$7 \oplus 7$ 9 7	7 6 9
4	6 6 $6 \oplus 6$	4 5 6	2 1 5	8 $1 \oplus 1$ 1	1 1 $1 \oplus 1$	3 5 1	7 6 5	9 $6 \oplus 6$ 6
5	$6 \oplus 6$ 9 6	6 7 9	5 3 7	1 1 3	$1 \oplus 1$ 8 1	1 2 8	5 4 2	6 6 4

Table 2.15 *(cont.)*

Classes of involution on the real Clifford algebras $\mathbf{C}_{p,q}(\mathbb{R})$

p \ q	0	1	2	3	4	5	6	7
6	6 7 5	9 7 ⊕ 7 7	7 7 7 ⊕ 7	3 5 7	1 2 5	8 2 ⊕ 2 2	2 2 2 ⊕ 2	4 5 2
7	5 3 1	7 7 3	7 ⊕ 7 9 7	7 6 9	5 4 6	2 2 4	2 ⊕ 2 8 2	2 1 8

(1) $\mathbb{R}$-symmetric, of index ν
(2) $\mathbb{R}$-skew (only in even dimensions)
(3) $\mathbb{C}$-symmetric
(4) $\mathbb{C}$-skew (only in even dimensions)
(5) $\mathbb{C}^*$-symmetric, of index ν
(6) H^--symmetric, of index ν
(7) $H\wedge$-symmetric
(8) $\mathbb{R}$-swap
(9) H-swap
(10) $\mathbb{C}$-swap

Table 2.16

p \ q	0	1	2	3	4	5
0	1	1	U(1)	Sp(1,H)	Sp(1,H) × Sp(1,H)	Sp(2,H)
1	1	$\mathbb{R}^*$	Sp(2,$\mathbb{R}$)	Sp(2,$\mathbb{C}$)	Sp(1,1,H)	
2	U(1)	Sp(2,$\mathbb{R}$)	Sp(2,$\mathbb{R}$) × Sp(2,$\mathbb{R}$)	Sp(4,$\mathbb{R}$)		
3	Sp(1,H)	Sp(2,$\mathbb{C}$)	Sp(4,$\mathbb{R}$)			
4	Sp(1,H) × Sp(1,H)	Sp(1,1,H)				
5	Sp(2,H)					

Sp(1,H) ≅ SU(2), Sp(2,$\mathbb{R}$) ≅ Sl(2,$\mathbb{R}$), Sp(2, $\mathbb{C}$) ≅ Sl(2,$\mathbb{C}$)
$_+\Gamma^+$ for all $\mathbf{C}_{p,q}(\mathbb{R})$ with $p + q \leqslant 5$.

2.7 The Complexified Clifford Algebras

So far we have only considered real orthogonal spaces and their associated real Clifford algebras. Much of the discussion could, however, be repeated with the real field replaced by an arbitrary field; in particular the complex field. If W is a complex vector space with h a complex valued, symmetric, non-degenerate $\mathbb{C}$-bilinear form then the Clifford algebra can be constructed as in §2.1. Since h is not characterised by any signature the structure of the Clifford algebra can only depend on n, and it will be denoted $\mathbf{C}_n(\mathbb{C})$. The structure of the algebra can be determined as in §2.2, only here the situation is even simpler. We have

$$\mathbf{C}_{n+2}(\mathbb{C}) \simeq \mathbf{C}_n(\mathbb{C}) \otimes \mathbf{C}_2(\mathbb{C}) \tag{2.7.1}$$

$$\mathbf{C}_1(\mathbb{C}) \simeq \mathbb{C} \oplus \mathbb{C} \tag{2.7.2}$$

$$\mathbf{C}_2(\mathbb{C}) \simeq \mathcal{M}_2(\mathbb{C}). \tag{2.7.3}$$

These are the analogues of (2.2.8), (2.2.2) and (2.2.3) and they may be proved in a similar way. They give the structure of all $\mathbf{C}_n(\mathbb{C})$.

If n is even then

$$\mathbf{C}_n(\mathbb{C}) \simeq \mathcal{M}_{2^{n/2}}(\mathbb{C}) \tag{2.7.4a}$$

whereas if n is odd

$$\mathbf{C}_n(\mathbb{C}) \simeq \mathcal{M}_{2^{(n-1)/2}}(\mathbb{C}). \tag{2.7.4b}$$

The structure of the even subalgebra follows from the analogue of (2.3.1), namely

$$\mathbf{C}_n^+(\mathbb{C}) \simeq \mathbf{C}_{n-1}(\mathbb{C}). \tag{2.7.5}$$

If n is even then

$$\mathbf{C}_n^+(\mathbb{C}) \simeq \mathcal{M}_{2^{n/2-1}}(\mathbb{C}) \oplus \mathcal{M}_{2^{n/2-1}}(\mathbb{C}). \tag{2.7.6a}$$

whereas if n is odd

$$\mathbf{C}_n^+(\mathbb{C}) \simeq \mathcal{M}_{2^{(n-1)/2}}(\mathbb{C}). \tag{2.7.6b}$$

Rather than proceed with the study of the Clifford groups and their relation to the complex orthogonal groups we shall show how the complex Clifford algebras may be related to real orthogonal spaces.

If V is a real n-dimensional orthogonal space with bilinear form g then $V^{\mathbb{C}}$, the complexification of V, is an n- dimensional complex vector space. The real bilinear form g may be extended by $\mathbb{C}$-linearity to a $\mathbb{C}$-bilinear form on $V^{\mathbb{C}}$, $g^{\mathbb{C}}$. If g is non-degenerate then so is $g^{\mathbb{C}}$. If $V^{\mathbb{C}}$ is regarded as a $2n$-dimensional real vector space then V is canonically identified with an n-dimensional subspace. The complex algebra

ERRATUM

Page 149

In lines 1, 2 and 3, replace first occurrence of x^j by y^j

Page 151

On line 33, replace 'section' by 'is a section'

Page 154

On line 17, $X \in \Gamma TM$

Page 185

On line 34, $\boldsymbol{P} = \mathcal{E}\boldsymbol{v}$

Page 197

Line 1 should read: 3-chain Σ and $\int_{S^2} \mathrm{i}_V \tau_V$

Page 203

After line 21 insert: describes how 'curved' a curve is!)

Page 206

On line 21, $\Phi = -GM/r$
On line 24,

$$= -(GMm/r^2)\partial_r. \tag{6.2.8}$$

Page 231

On line 8,

$$\delta \tilde{K} = -n\lambda \tag{6.13.11}$$

Page 239

On line 7, replace '(4.13.12)' by '(6.13.12)'

Page 258

On line 11,

$$= kY_k^m(\theta, \phi)\mathrm{d}r + r\mathrm{d}Y_k^m$$

Page 342

Replace last equation by: $\phi = \tan^{-1} \frac{x^2}{x^1}$

Page 346

On line 19,

$$\mathrm{div}(\boldsymbol{v} \times \boldsymbol{u}) = g(\boldsymbol{u}, \mathrm{curl}\, \boldsymbol{v}) - g(\boldsymbol{v}, \mathrm{curl}\, \boldsymbol{u})$$

Throughout the book for 'Kahler' read 'Kähler'

$\mathbf{C}(V^{\mathbb{C}}, g^{\mathbb{C}})$ may be regarded as a 2^{n+1}-dimensional real algebra. Thus regarded $\mathbf{C}(V, g)$ is a subalgebra which certainly commutes with the subalgebra generated by the identity over the complex field. So we have the following isomorphism of real algebras

$$\mathbf{C}(V^{\mathbb{C}}, g^{\mathbb{C}}) \simeq \mathbf{C}(V, g) \otimes \mathbb{C}(\mathbb{R}). \tag{2.7.7}$$

For $\mathbf{C}(V, g)\otimes\mathbb{C}(\mathbb{R})$ we shall write $\mathbf{C}^{\mathbb{C}}(V, g)$. We may define the conjugate-linear operation of complex conjugation, *, on $V^{\mathbb{C}}$: if $z \in V^{\mathbb{C}}$ then $z = x + \mathrm{i}y$ for $x, y \in V$ and $z^* = x - \mathrm{i}y$. Complex conjugation extends to an automorphism of $\mathbf{C}(V^{\mathbb{C}}, g^{\mathbb{C}})$ regarded as a real algebra (although not of course as a complex algebra). If the real subalgebra consists of all elements equal to their complex conjugates then the real subalgebra of $\mathbf{C}(V^{\mathbb{C}}, g^{\mathbb{C}})$ is of course $\mathbf{C}(V, g)$. It is worth stressing that for an arbitrary complex vector space there is no naturally defined operation of complex conjugation. It is here well defined because the complex vector space is obtained from the complexification of some underlying real vector space. We have already shown that the complex Clifford algebras are isomorphic to complex matrix algebras or sums of two such algebras. The operation of complex conjugation, *, as defined above will not, however, necessarily simply complex conjugate the components of these matrices. The situation is clarified below.

Suppose that $\mathcal{A}(\mathbb{R}) \simeq \mathbb{C}(\mathbb{R})\otimes\mathcal{M}_r(\mathbb{R})$ has some involutory automorphism, *, that induces a non-trivial automorphism on the centre. Let $\mathcal{B}$ be the real subalgebra, that is $a \in \mathcal{B}$ if and only if $a = a^*$. Since any $a \in \mathcal{A}$ can be written as a sum of real and imaginary parts it follows that $\mathcal{A} \simeq \mathbb{C}\otimes\mathcal{B}$. So we have $\mathbb{C}\otimes\mathcal{B} \simeq \mathbb{C}\otimes\mathcal{M}_r$. For this to be true $\mathcal{B}$ must certainly be simple, and since the only simple real algebras are isomorphic to $\mathcal{D}\otimes\mathcal{M}$ with $\mathcal{D} = \mathbb{R}$, $\mathbb{C}$ or H we must have either $\mathcal{B} \simeq \mathcal{M}_r$ or $\mathcal{B} \simeq H\otimes\mathcal{M}_{r/2}$. By writing $\mathcal{A} = \mathbb{C}\otimes\mathcal{M}_r$ for some particular matrix subalgebra $\mathcal{M}_r$ we can define another involutory automorphism $^\#$ that leaves elements of $\mathcal{M}_r$ invariant and conjugates elements of the centre. If $\{\mathbf{e}_{ij}\}$ is an ordinary matrix basis for $\mathcal{M}_r$ then $\{\mathbf{e}^*_{ij}\}$ is another ordinary matrix basis, for $\mathcal{M}_r'$ say. It follows from the uniqueness of the Wedderburn decomposition ((A24) of Appendix A) that $\mathbf{e}^*_{ij} = m\mathbf{e}_{ij}m^{-1}$ for some $m \in \mathcal{A}$. So if

$$a = \sum_{i,j=1}^{r} a_{ij}\mathbf{e}_{ij}$$

then

$$a^* = \sum_{i,j=1}^{r} a^*_{ij}m\mathbf{e}_{ij}m^{-1} = m\sum_{i,j=1}^{r} a^*_{ij}\mathbf{e}_{ij}m^{-1}$$

that is

$$a^* = ma^\# m^{-1}. \tag{2.7.8}$$

Since * and $^\#$ are involutory (2.7.8) gives $m^*m = \rho$ where ρ is in the centre. Now * and $^\#$ induce the same automorphism on the centre, giving $(m^*m)^* = (m^*m)^\# = m^{-1}(m^*m)m$. That is, $mm^* = m^*m$ and ρ is in fact real. The defining property of m, (2.7.8), only determines it up to a multiple of the centre and so by a suitable scaling we can arrange either $m^*m = 1$, or $m^*m = -1$. (Equivalently $m^\#m = 1$ or $m^\#m = -1$.) We summarise as follows: $\mathscr{A} = \mathscr{B}\otimes\mathbb{C}$ with * an automorphism that conjugates $\mathbb{C}$ and leaves $\mathscr{B}$ invariant, and $\mathscr{A} = \mathscr{M}_r\otimes\mathbb{C}$ with $^\#$ an automorphism that conjugates $\mathbb{C}$ and leaves $\mathscr{M}_r$ invariant. The two automorphisms are related by $a^* = ma^\#m^{-1}$. There are two possibilities for $\mathscr{B}$, either $\mathscr{B} \simeq \mathscr{M}_r$ or $\mathscr{B} \simeq H\otimes\mathscr{M}_{r/2}$; and two possibilities for m, either $mm^* = 1$ or $mm^* \equiv -1$. These possibilities are in fact related.

If $\mathscr{A} \simeq \mathbb{C}\otimes\mathscr{B} \simeq \mathbb{C}\otimes\mathscr{M}_r$ with * and $^\#$ automorphisms that conjugate the centre and leave $\mathscr{B}$ and $\mathscr{M}_r$ respectively invariant, then $a^* = ma^\#m^{-1}$ where we can choose either $mm^* = 1 \Leftrightarrow \mathscr{B} = \mathscr{M}_r$, or $mm^* = -1 \Leftrightarrow \mathscr{B} = H\otimes\mathscr{M}_{r/2}$. (2.7.9)

We now consider the proof. Since there are two and only two mutually exclusive possibilities for $\mathscr{B}$, and similarly for m, if we can prove that $mm^* = 1 \Leftrightarrow \mathscr{B} \simeq \mathscr{M}_r$ then we must have $mm^* = -1 \Leftrightarrow \mathscr{B} \simeq H\otimes\mathscr{M}_{r/2}$. Suppose firstly that $\mathscr{B} \simeq \mathscr{M}_r$. Then if $\{\mathbf{b}_{ij}\}$ and $\{\mathbf{e}_{ij}\}$ are ordinary matrix bases for $\mathscr{B}$ and $\mathscr{M}_r$ respectively then $\mathbf{e}_{ij} = s\mathbf{b}_{ij}s^{-1}$ for some $s \in \mathscr{A}$. If we write

$$a = \sum_{i,j} a_{ij}\mathbf{e}_{ij}$$

then

$$\begin{aligned} a^* &= \sum_{i,j} a_{ij}^*(s\mathbf{b}_{ij}s^{-1})^* = \sum_{i,j} a_{ij}^* s^*\mathbf{b}_{ij}s^{*-1} \\ &= \sum_{i,j} a_{ij}^* s^*s^{-1}\mathbf{e}_{ij}ss^{*-1} = s^*s^{-1}a^\#(s^*s^{-1})^{-1}. \end{aligned}$$

That is, we may choose $m = s^*s^{-1}$ giving $m^* = m^{-1}$. Now the converse: we introduce a $\mathbb{C}$-conjugate-linear transformation on $\mathscr{A}$ by defining $a^{\mathrm{c}} = a^*m = ma^\#$. Thus ^{c} preserves the columns of $\mathscr{M}_r$. If $m^* = m^{-1}$ then ^{c} is involutory and for any $a \in \mathscr{A}$ we write $a = \frac{1}{2}(a + a^{\mathrm{c}}) + \frac{1}{2}(a - a^{\mathrm{c}})$. In particular, the minimal left ideals of $\mathscr{A}$ that are the columns of $\mathscr{M}_r$ with entries in $\mathbb{C}$ can be decomposed into eigenspaces of ^{c}. Since the real dimension of a minimal left ideal of $\mathscr{A}$ is $2r$ these eigenspaces are r-dimensional. Let ψ be an element of one of these eigenspaces. Then if $a \in \mathscr{B}$ $a\psi$ is certainly in the minimal left ideal of $\mathscr{A}$ and since $(a\psi)^{\mathrm{c}} = a^*\psi^{\mathrm{c}} = a\psi^{\mathrm{c}}$ it is in fact in the eigenspace. Hence these eigenspaces carry representations of $\mathscr{B}$. That is, if $m^* = m^{-1}$ then irreducible representations of $\mathscr{A}$ induce reducible representations of $\mathscr{B}$. But either $\mathscr{B} \simeq \mathscr{M}_r$, in which case its irreducible representations are r-dimensional, or $\mathcal{B} \simeq H \otimes \mathcal{M}_{r/2}$ with $2r$-dimensional irreducible representations. Thus $m^* = m^{-1}$ implies $\mathcal{B} \simeq \mathcal{M}_r$. The argument of the proof is summarised below.

$$m = s^* s^{-1} \Rightarrow mm^* = 1$$

$$\mathcal{B} \simeq \mathcal{M}_r \quad \Rightarrow \quad \Downarrow$$

$\Leftarrow$ irreducible representations of $\mathcal{A}$ induce reducible representations of $\mathcal{B}$.

$$\text{Hence } mm^* = 1 \Leftrightarrow \mathcal{B} \simeq \mathcal{M}_r$$

$$\text{and so } mm^* = -1 \Leftrightarrow \mathcal{B} \simeq H \otimes \mathcal{M}_{r/2}.$$

The complexification of the real Clifford algebra associated with an even-dimensional orthogonal space is isomorphic to the algebra of complex matrices. Complex conjugation (that leaves the real Clifford algebra invariant) is equivalent to the automorphism that conjugates the components of these matrices when the real algebra is a total matrix algebra: in this case complex conjugation will simply conjugate the components in an appropriate basis. The algebra associated with the complexification of an odd-dimensional real orthogonal space is a direct sum of two matrix algebras. Complex conjugation is equivalent to the automorphism that conjugates the components of these matrices if and only if the real algebra is a sum of two total matrix algebras. When the real algebra is the sum of two simple algebras whose Wedderburn decomposition involves the quaternions then complex conjugation induces an automorphism on the simple components of the complexified algebra that is inequivalent to conjugating the matrix components. When the real algebra is isomorphic to the algebra of complex matrices then complex conjugation of the complexified algebra interchanges the simple components.

The irreducible representations of the complex algebras will again be called spinor representations, or semi-spinor representations when the algebra is reducible, the spinor (or semi-spinor) spaces being identified with minimal left ideals. These minimal left ideals are obviously of complex dimension $2^{[n/2]}$ where $[n/2]$ denotes the integer part of $n/2$. When n is even there is only one such representation, up to equivalence, whereas if n is odd there are two inequivalent semi-spinor representations.

Irreducible representations of $\mathbf{C}(V^{\mathbb{C}}, g^{\mathbb{C}})$ induce representations of $\mathbf{C}(V, g)$ which may or may not be reducible. The question of the reducibility of these representations has to some extent been anticipated in §2.5. It was shown there that when the division algebra occurring in the Wedderburn decomposition of the real Clifford algebra (or a simple component of that algebra) was $\mathbb{C}$ or H the complex structure of right multiplication by the generator of a complex subalgebra enabled the spinor (or semi-spinor) space to be regarded as a complex vector space. In this case irreducible representations of the real algebra can be extended by $\mathbb{C}$-linearity to representations of the complexified algebra. Thus conversely, in these cases irreducible representations of the complexified algebra induce irreducible representations of the real algebra. When the real Clifford algebra is isomorphic to the algebra of real matrices, or the sum of two such algebras, then its irreducible representations are of real dimension $2^{[n/2]}$; that is, half that of the real dimension of the irreducible representations of the complexified algebra. Thus, in these cases, irreducible representations of the complexified algebra induce reducible representations of the real algebra. The way in which this reduction can be performed was given in the proof of (2.7.9). The induced representations of the real even subalgebra may be treated in exactly the same way. The irreducible representations of the even subalgebra of the complexified algebra are of real dimension $2(2^{[(n-1)/2]})$, whilst the dimensions of those of the real even subalgebra are given in table 2.10.

We turn now to classifying involutions of the complexified algebras. The $\mathbb{C}$-linear involutions ξ and $\xi\eta$ induce the standard involutions on the real subalgebra, and these have already been classified. Thus we may classify these involutions on the complexified algebra from a knowledge of the involutions that they induce on the factors of a tensor product. The involution induced on the factor $\mathbb{C}$ is of class 3, and so if we multiply the entries in table 2.15 by 3, using the multiplication of table 2.13, then we obtain the class of ξ and $\xi\eta$ on the complexified algebra, and that of the involution they induce on its even subalgebra. (The classes are given in table 2.11.) Now these involutions are of course involutions of the Clifford algebra associated with the complex vector space $V^{\mathbb{C}}$, and so they can only depend on the dimension of V and not the signature of g. The classes depend on n mod 8, and are given in table 2.17. The involutions ξ and $\xi\eta$ commute with complex conjugation, and so they may be composed with it to form involutions ξ^* and $\xi\eta^*$ which again induce the standard involutions on the real subalgebra. These involutions are certainly not involutions of $\mathbf{C}(V^{\mathbb{C}}, g^{\mathbb{C}})$ regarded as a complex algebra, but are real algebra involutions. The classes can be obtained by multiplying the entries in table 2.15 by five using the multiplication of table 2.13. Of course on the simple algebras these involutions can only be of class 5, whilst in the reducible case they either induce involutions of class 5 on the component algebras or interchange those components. The classes depend on p mod 2 and q mod 2, and are given in table 2.18. It follows from (2.6.23) that ξ^* is the adjoint of a zero index Hermitian-symmetric product if and only if

$q = 0$; otherwise any index is maximal. Similarly $\xi\eta^*$ is the adjoint of a zero index product if and only if $p = 0$, otherwise maximal.

Table 2.17 Classification of involutions of the complexified Clifford algebras.

$p + q = n$	ξ	$\xi\eta$	ξ on $\mathbf{C}^+_{p,q}(\mathbb{R}) \otimes \mathbb{C}$
1	$3 \oplus 3$	10	3
2	3	4	10
3	10	$4 \oplus 4$	4
4	4	4	$4 \oplus 4$
5	$4 \oplus 4$	10	4
6	4	3	10
7	10	$3 \oplus 3$	3
8	3	3	$3 \oplus 3$

Table 2.18 The classes of ξ^* and $\xi\eta^*$ on $\mathbf{C}_{p,q}(\mathbb{R}) \otimes \mathbb{C}$ and ξ^* on $\mathbf{C}^+_{p,q}(\mathbb{R}) \otimes \mathbb{C}$.

p \ q	0	1
0	5 5 $5 \oplus 5$	10 $5 \oplus 5$ 5
1	$5 \oplus 5$ 10 5	5 5 10

2.8 The Confusion of Tongues

The theory of spinors was developed independently by physicists and mathematicians, and this historical apartheid has continued. Of particular physical interest is the case of a four-dimensional real vector space with a Lorentzian metric, and it was in this case that much of the terminology and notation used by physicists originated. More recently there has been much interest in physical theories set in a variety of different dimensions and the nomenclature and terminology has been extrapolated to these situations. Thus there is now a language, with many dialects, for discussing spinors in physics which makes little contact with the expositions of the theory to be found in the mathematics literature. Physicist readers may at this point vehemently declare that it also makes little contact with the exposition given here. We will now try to redress this situation.

The Dirac matrices, or γ-matrices, are usually defined to be complex square matrices of minimal order that satisfy

$$\gamma^a \gamma^b + \gamma^b \gamma^a = 2\boldsymbol{\eta}^{ab} \tag{2.8.1}$$

where $\boldsymbol{\eta}$ is diagonal with p entries of plus one and q of minus one. If $p + q = n$ then the order of these marices is $2^{[n/2]}$ with the bracket denoting the integer part. These matrices are also usually assumed to have certain Hermiticity properties, and we shall examine this shortly. Here we note that the presence of such operations that are not $\mathbb{C}$-linear is sufficient to infer that the γ-matrix algebra is not to be regarded as a complex algebra. In fact from (2.7.4) we recognise that these matrices generate an algebra isomorphic to that of the complexified Clifford algebra or, in odd dimensions, a simple component of that algebra.

For the case of n even we have $\mathbf{C}^{\mathbb{C}}_{p,q} \simeq \mathcal{M}_{2^{n/2}}(\mathbb{C})$. If $\{e^a\}$ is a basis for the real vector space that generates $\mathbf{C}_{p,q}(\mathbb{R})$ then

$$e^a = \sum_{i,\,j=1}^{n/2} \gamma^a_{ij} \mathbf{e}_{ij}$$

where $\{\mathbf{e}_{ij}\}$ is some ordinary matrix basis for the complexified algebra. The arrays of complex components, γ^a_{ij}, with the usual rules of matrix multiplication, will obviously satisfy (2.8.1). All matrix bases of the complexified algebra are related by an inner automorphism, the change of matrix basis giving a new set of matrix components for the $\{e^a\}$; an equivalent representation of the γ-matrices.

The way in which a matrix basis can be constructed and the matrix components of any element found is contained in the proof of the Wedderburn structure theorem, (A23) of Appendix A. An explicit example was given at the end of §2.2. We now further restrict ourselves to the complexification of $\mathbf{C}_{p,1}(\mathbb{R})$, for p odd, and show how a 'standard' representation of the γ-matrices can be given. (Although we shall have no need of such representations this will hopefully strengthen the link with the standard physics literature.) For p odd

$$\mathbf{C}^{\mathbb{C}}_{p,0} = \mathcal{M}_{2^{(p-1)/2}}(\mathbb{C}) \oplus \mathcal{M}_{2^{(p-1)/2}}(\mathbb{C})$$

and thus $\mathbf{C}^{\mathbb{C}}_{p,1}$ is isomorphic to a total matrix algebra with $\mathbf{C}_{p,0}(\mathbb{C})$ a subalgebra isomorphic to the direct sum of two algebras of matrices of half the order. In a suitable matrix basis, therefore, $\mathbf{C}^{\mathbb{C}}_{p,0}$ is the subalgebra of elements whose matrix components are block-diagonal; the two simple component algebras having matrix components in only the upper or lower blocks, that is

$$\gamma^i = \begin{pmatrix} \boldsymbol{\sigma}^i & 0 \\ 0 & \boldsymbol{\Sigma}^i \end{pmatrix} \qquad i = 1, \ldots, p$$

with $\boldsymbol{\sigma}^i$ and $\boldsymbol{\Sigma}^i$ matrices of order $2^{(p-1)/2}$.

From table 2.18 and the remark at the end of §2.7 it follows that the involution ξ^* on $\mathbf{C}^{\mathbb{C}}_{r,0}$ is the adjoint involution of a zero-index $\mathbb{C}^*$-symmetric product; that is, it is equivalent to Hermitian conjugation. Thus we can arrange a basis in which ξ^* on $\mathbf{C}^{\mathbb{C}}_{r,1}$ induces Hermitian conjugation on the diagonal blocks (but not, of course, on the off-diagonal blocks), and in such a basis $\boldsymbol{\sigma}^i$ and $\boldsymbol{\Sigma}^i$ are Hermitian. If $\check{z} = \lambda e^1 \ldots e^p$ with $\lambda = 1$ or i such that $\check{z}^2 = 1$, and $P_{\pm} = \frac{1}{2}(1 \pm \check{z})$, then P_+ and P_- are the identities in the simple components of $\mathbf{C}^{\mathbb{C}}_{p,0}$. Since $\check{z} = P_+ - P_-$ then if $\check{\gamma} = \lambda\gamma^1 \ldots \gamma^p$ then

$$\check{\gamma} = \begin{pmatrix} \mathbf{I} & 0 \\ 0 & -\mathbf{I} \end{pmatrix}.$$

But e^0 anticommutes with $\check{z}$ and so γ^0 can only have off-diagonal components. Since also $(e^0)^2 = -1$ we must have

$$\gamma^0 = \begin{pmatrix} 0 & \mathbf{T} \\ -\mathbf{T}^{-1} & 0 \end{pmatrix}$$

for some non-singular matrix $\mathbf{T}$. Since e^0 anticommutes with all e^i we must in fact have $\boldsymbol{\Sigma}^i = - -\boldsymbol{T}^{-1}\boldsymbol{\sigma}\boldsymbol{T}$. If we now change basis so that the components transform $\gamma^a \rightarrow \mathbf{S}\gamma^a\mathbf{S}^{-1}$ with

$$\mathbf{S} = \frac{1}{\sqrt{2}}\begin{pmatrix} \mathbf{I} & -\mathrm{i}\mathbf{T} \\ \mathbf{I} & \mathrm{i}\mathbf{T} \end{pmatrix} \qquad \mathbf{S}^{-1} = \frac{1}{\sqrt{2}}\begin{pmatrix} \mathbf{I} & \mathbf{I} \\ \mathrm{i}\mathbf{T}^{-1} & -\mathrm{i}\mathbf{T}^{-1} \end{pmatrix}$$

then we arrive at the following 'standard' representation of the γ-matrices:

$$\gamma^0 = \mathrm{i}\begin{pmatrix} \mathbf{I} & 0 \\ 0 & -\mathbf{I} \end{pmatrix} \qquad \gamma^i = \begin{pmatrix} 0 & \boldsymbol{\sigma}^i \\ \boldsymbol{\sigma}^i & 0 \end{pmatrix}. \qquad (2.8.2)$$

Here $\boldsymbol{\sigma}^i$, and hence γ^i, are Hermitian whilst γ^0 is manifestly anti-Hermitian. The case of $\mathbf{C}^{\mathbb{C}}_{1,q}$ may be treated similarly. Since $\xi\eta^*$ is equivalent to Hermitian conjugation in $\mathbf{C}^{\mathbb{C}}_{0,q}$ we are lead to a 'standard' representation as above, but with the $\boldsymbol{\sigma}^i$ anti-Hermitian and the i removed from γ^0. (In this case e^0 denoting the one positive-norm vector.)

To illustrate further the relation between the γ-matrices and the more abstract approach to Clifford algebras that we have pursued, we examine $\mathbf{C}^{\mathbb{C}}_{3,1}$ in more detail. First we shall choose a matrix basis for a simple component of $\mathbf{C}^{\mathbb{C}}_{3,0}$ in which ξ^* coincides with Hermitian conjugation giving the Hermitian $\{\boldsymbol{\sigma}^i\}$. We then have from (2.8.2) a standard representation of the γ-matrices and shall reverse the argument to construct the matrix basis in which these are the components of the $\{e^a\}$. The reducible algebra $\mathbf{C}^{\mathbb{C}}_{3,0}$ is projected into simple components by the mutually commuting pair of central idempotents $P_{\pm} = \frac{1}{2}(1 \pm \mathrm{i}e^{123})$. Since $e^{123}P_{\pm} = \mp\mathrm{i}P_{\pm}$, giving $e^{12}P_{\pm} = \mp\mathrm{i}e^{3}P_{\pm}$, if we want $\boldsymbol{\sigma}^1\boldsymbol{\sigma}^2 = \mathrm{i}\boldsymbol{\sigma}^3$ then the $\{\boldsymbol{\sigma}^i\}$ must be the components of the $\{e^i\}$ in $\mathbf{C}^{\mathbb{C}}_{3,0}P_{-}$. To start the construction of the matrix basis we seek a pair of mutually orthogonal idempotents that are invariant under the involution ξ^*: these will form the diagonals of a basis in which ξ^* induces Hermitian conjugation. We choose

$$\begin{aligned} \mathbf{e}_{11} &= \tfrac{1}{2}(1 + e^3)P_{-} \\ \mathbf{e}_{22} &= \tfrac{1}{2}(1 - e^3)P_{-}. \end{aligned} \tag{2.8.3}$$

and since $\mathbf{e}_{22} = e^2\mathbf{e}_{11}e^2$ we may complete the basis with

$$\begin{aligned} \mathbf{e}_{12} &= \mathbf{e}_{11}e^2 = e^2\mathbf{e}_{22} \\ \mathbf{e}_{21} &= e^2\mathbf{e}_{11} = \mathbf{e}_{22}e^2 \end{aligned} \tag{2.8.4}$$

where

$$\mathbf{e}_{21}{}^{\xi^*} = \mathbf{e}_{12}.$$

If

$$e^iP_{-} = \sum_{\alpha,\beta=1}^{2} \sigma^i{}_{\alpha\beta}\boldsymbol{\varepsilon}_{\alpha\beta}$$

then

$$\sigma^i{}_{\alpha\beta} = \sum_{\lambda=1}^{2} \boldsymbol{\varepsilon}_{\lambda\alpha}e^i\boldsymbol{\varepsilon}_{\beta\lambda}.$$

For example,

$$\begin{aligned} \sigma^1{}_{12} &= \boldsymbol{\varepsilon}_{11}e^1\boldsymbol{\varepsilon}_{21} + \boldsymbol{\varepsilon}_{21}e^1\boldsymbol{\varepsilon}_{22} \\ &= -\mathrm{i}\boldsymbol{\varepsilon}_{11}e^{23}\boldsymbol{\varepsilon}_{21} - \mathrm{i}\boldsymbol{\varepsilon}_{21}e^{23}\boldsymbol{\varepsilon}_{22} \end{aligned}$$

since $e^{123}P_{-} = \mathrm{i}P_{-}$

$$= \mathrm{i}\boldsymbol{\varepsilon}_{11}e^2\boldsymbol{\varepsilon}_{21} + \mathrm{i}\boldsymbol{\varepsilon}_{21}e^2\boldsymbol{\varepsilon}_{22}$$

where the e^2 has been absorbed into $\boldsymbol{\varepsilon}_{21}$ and $\boldsymbol{\varepsilon}_{22}$. From the definition of $\boldsymbol{\varepsilon}_{21}$

$$\sigma^1{}_{12} = \mathrm{i}(\boldsymbol{\varepsilon}_{11} + \boldsymbol{\varepsilon}_{22}) = \mathrm{i}P_{-}$$

where, we recall, P_{-} is the identity in this simple algebra. In this way we construct the following:

$$\boldsymbol{\sigma}^1 = \begin{pmatrix} 0 & \mathrm{i} \\ -\mathrm{i} & 0 \end{pmatrix} \qquad \boldsymbol{\sigma}^2 = \begin{pmatrix} 0 & 1 \\ 1 & 0 \end{pmatrix} \qquad \boldsymbol{\sigma}^3 = \begin{pmatrix} 1 & 0 \\ 0 & -1 \end{pmatrix}. \tag{2.8.5}$$

We may use these matrices in (2.8.2) to obtain a standard representation of the γ-matrices. At this point we reverse the reasoning and construct the matrix basis corresponding to these components. From the diagonal γ^0 we construct a pair of (non-primitive) idempotents,

$$\tfrac{1}{2}(\mathbf{I} + \mathrm{i}\gamma^0) = \begin{pmatrix} 0 & & & \\ & 0 & & \\ & & 1 & \\ & & & 1 \end{pmatrix} \qquad \tfrac{1}{2}(\mathbf{I} - \mathrm{i}\gamma^0) = \begin{pmatrix} 1 & & & \\ & 1 & & \\ & & 0 & \\ & & & 0 \end{pmatrix}.$$

Putting (2.8.5) into (2.8.2) enables another pair of idempotent matrices to be constructed

$$\tfrac{1}{2}(\mathbf{I} + \mathrm{i}\gamma^1\gamma^2) = \begin{pmatrix} 0 & & & \\ & 1 & & \\ & & 0 & \\ & & & 1 \end{pmatrix} \qquad \tfrac{1}{2}(\mathbf{I} - \mathrm{i}\gamma^1\gamma^2) = \begin{pmatrix} 1 & & & \\ & 0 & & \\ & & 1 & \\ & & & 0 \end{pmatrix}.$$

Primitives are obtained from the four products of these two pairs of idempotents, for example

$$\tfrac{1}{2}(\mathbf{I} - \mathrm{i}\gamma^0)\tfrac{1}{2}(\mathbf{I} - \mathrm{i}\gamma^1\gamma^2) = \begin{pmatrix} 1 & & & \\ & 0 & & \\ & & 0 & \\ & & & 0 \end{pmatrix}.$$

If then

$$e^a = \sum_{i,\, j=1}^{4} \gamma^a{}_{ij}\mathbf{e}_{ij}$$

we have

$$\begin{aligned} \mathbf{e}_{11} &= \tfrac{1}{2}(1 - \mathrm{i}e^0)\tfrac{1}{2}(1 - \mathrm{i}e^{12}) \\ \mathbf{e}_{22} &= \tfrac{1}{2}(1 - \mathrm{i}e^0)\tfrac{1}{2}(1 + \mathrm{i}e^{12}) \\ \mathbf{e}_{33} &= \tfrac{1}{2}(1 + \mathrm{i}e^0)\tfrac{1}{2}(1 - \mathrm{i}e^{12}) \\ \mathbf{e}_{44} &= \tfrac{1}{2}(1 + \mathrm{i}e^0)\tfrac{1}{2}(1 + \mathrm{i}e^{12}). \end{aligned} \tag{2.8.6}$$

In exactly the same way we take products of the γ-matrices to produce a matrix of zeroes except for a 1 in the i, j entry, for all i and j. As may readily be checked this leads to the conclusion that the remainder of the matrix basis must be as shown in table 2.19.

In odd dimensions there are two inequivalent representations of the γ-matrices: these being the matrix components of the $\{e^a\}$ projected into either of the simple component algebras. We now show how a standard representation can be constructed for $\mathbf{C}^{\mathbb{C}}_{p,1}$, where now p is even. In this case

$$\mathbf{C}^{\mathbb{C}}_{p,1} = \mathcal{M}_{2^{p/2}}(\mathbb{C}) \oplus \mathcal{M}_{2^{p/2}}(\mathbb{C})$$

whereas $\mathbf{C}^{\mathbb{C}}_{p,0} = \mathcal{M}_{2^{p/2}}(\mathbb{C})$. The involution ξ^* is equivalent to Hermitian conjugation on $\mathbf{C}^{\mathbb{C}}_{p,0}$, whereas it swaps the components of $\mathbf{C}^{\mathbb{C}}_{p,1}$. We choose a matrix basis $\{\mathbf{e}_{ij}\}$ for $\mathbf{C}^{\mathbb{C}}_{p,0}$ in which ξ^* coincides with Hermitian conjugation. If $P_\pm$ are the central idempotents that project $\mathbf{C}^{\mathbb{C}}_{p,1}$ into simple components, and $\mathbf{e}_{ij}{}^{\pm} = \mathbf{e}_{ij}P_\pm$, then the $\mathbf{e}_{ij}{}^{\pm}$ form matrix bases for these component algebras. The involution $\dagger$ is defined on $\mathbf{C}^{\mathbb{C}}_{p,1}$ by the requirement that it conjugate the complex factor and satisfy the following properties on the generators of the real subalgebra: $e^{i\dagger} = e^i$, $i = 1, \ldots, p$, $e^{0\dagger} = -e^0$. Thus $\dagger$ coincides with ξ^* on $\mathbf{C}^{\mathbb{C}}_{p,0}$ and so is indeed Hermitian conjugation in the basis $\{\mathbf{e}_{ij}\}$. If $z = e^1 \ldots e^p e^0$ then for $p = 2 \mod 4$, $P_\pm = \frac{1}{2}(1 \pm \mathrm{i}z)$, whereas for $p = 0 \mod 4$, $P_\pm = \frac{1}{2}(1 \pm \mathrm{i}z)$. Now certainly $z^\dagger = -z^{\xi^*}$, and as we have remarked ξ^* swaps the simple components of $\mathbf{C}^{\mathbb{C}}_{p,1}$, that is $P_\pm{}^{\xi^*} = P_\mp$, thus $P_\pm{}^\dagger = P_\pm$. It follows that, as the notation suggests, $\dagger$ induces Hermitian conjugation in the simple component algebras in the bases that we have constructed, $\{\mathbf{e}_{ij}{}^{\pm}\}$. In such bases $e^iP_\pm$ are represented by Hermitian matrices, whereas $e^0P_\pm$ is represented by an antiHermitian matrix. In fact for $p = 0 \mod 4$ $e^0P_\pm = \mp\mathrm{i}e^1 \ldots e^pP_\pm$, whereas for $p = 2 \mod 4$, $e^0P_\pm = \mp e^1 \ldots e^pP_\pm$.

Table 2.19 A matrix basis for $\mathbf{C}^{\mathbb{C}}_{3,1}$.

$\mathbf{e}_{ij}\rightarrow$ $\downarrow$				
	$\mathbf{e}_{11}$	$-e^{23}\mathbf{e}_{22}$	$e^3\mathbf{e}_{33}$	$-\mathrm{i}e^1\mathbf{e}_{44}$
	$e^{23}\mathbf{e}_{11}$	$\mathbf{e}_{22}$	$\mathrm{i}e^1\mathbf{e}_{33}$	$-e^3\mathbf{e}_{44}$
	$e^3\mathbf{e}_{11}$	$-\mathrm{i}e^1\mathbf{e}_{22}$	$\mathbf{e}_{33}$	$-e^{23}\mathbf{e}_{44}$
	$\mathrm{i}e^1\mathbf{e}_{11}$	$-e^3\mathbf{e}_{22}$	$e^{23}\mathbf{e}_{23}$	$\mathbf{e}_{44}$

The representation-independent operator trace, Tr, projects a matrix algebra onto the subspace spanned by the identity. There is therefore a relation between the projection of the Clifford algebra onto the space of 0-forms, $\mathcal{S}_0$, and the trace of the γ-matrices. In even dimensions any element can be expanded in a matrix basis

$$a = \sum_{i,j}^{2^{n/2}} a_{ij}\mathbf{e}_{ij}.$$

Since the a_{ij} are (complex) 0-forms

$$\mathcal{S}_0(a) = \sum_{i,j}^{2^{n/2}} a_{ij}\mathcal{S}_0(\mathbf{e}_{ij}).$$

and since products can be reversed under S_0, (2.1.17),

$$\mathcal{S}_0(\mathbf{e}_{ij}) = \mathcal{S}_0(\mathbf{e}_{ii}\mathbf{e}_{ij}\mathbf{e}_{jj}) = \mathcal{S}_0(\mathbf{e}_{ij}\mathbf{e}_{jj}\mathbf{e}_{ii}) = \mathcal{S}_0(\mathbf{e}_{ii})\delta_{ij}.$$

The diagonals in the matrix basis are a set of pairwise orthogonal primitive idempotents, and these are all similar. So

$$\mathscr{S}_0(\mathbf{e}_{jj}) = \mathscr{S}_0(s\mathbf{e}_{ii}s^{-1})$$

for some s, thus

$$\mathscr{S}_0(\mathbf{e}_{jj}) = \mathscr{S}_0(\mathbf{e}_{ii}).$$

By writing the identity as a sum of primitives

$$1 = \mathbf{e}_{11} + \mathbf{e}_{22} + \ldots + \mathbf{e}_{rr} \qquad \text{with } r = 2^{n/2}$$

we have $S_0(\mathbf{e}_{ii}) = 1/(2^{n/2})$. Thus

$$\mathscr{S}_0(a) = \frac{1}{2^{n/2}} \sum_i a_{ii}$$

that is,

$$\mathscr{S}_0(a) = \frac{1}{2^{n/2}} \operatorname{Tr} a. \tag{2.8.7}$$

In odd dimensions we let $P_\pm$ denote the central idempotents. If, for example, $\{\mathbf{e}_{ij}{}^+\}$ is a matrix basis for the simple algebra whose identity is P_+ then

$$P_+ = \mathbf{e}_{11}{}^+ + \ldots + \mathbf{e}_{rr}{}^+ \qquad r = 2^{(n-1)/2}.$$

Since $\mathscr{S}_0(P_+) = \frac{1}{2}$ we have $\mathscr{S}_0(\mathbf{e}_{ii}{}^+) = 1/[2(2^{(n-1)/2})]$ giving $\mathscr{S}_0(aP_+) = \{1/[2(2^{(n-1)/2})]\} \operatorname{Tr}(aP_+)$. Thus

$$\mathscr{S}_0(a) = \frac{1}{2(2^{(n-1)/2})} [\operatorname{Tr}(aP_+) + \operatorname{Tr}(aP_-)]. \tag{2.8.8}$$

In calculating cross sections in quantum theory one uses various trace theorems for the γ-matrices. The following illustrative properties of S_0 are equivalent to some of the most important. If $\{a_1, \ldots, a_n\}$ is a set of 1-forms then $\mathscr{S}_0(a_1a_2 \ldots a_n) = 0$ for n odd, and $\mathscr{S}_0(a_1a_2 \ldots a_n) = \mathscr{S}_0(a_n \ldots a_2a_1)$. These follow from the more general relations $\eta\mathscr{S}_p = \mathscr{S}_p\eta$, $\xi\mathscr{S}_p = S_p\xi$. The 0-form component of a product of n 1-forms, with n even, can be related to that of products of $n-2$ terms, from (2.1.7)

$$\begin{aligned}
\mathscr{S}_0(a_1a_2 \ldots a_n) &= \mathscr{S}_0\{a_1 \wedge (a_2 \ldots a_n) + \mathrm{i}_{\tilde{a}_1}(a_2 \ldots a_n)\} \\
&= \mathscr{S}_0\{\mathrm{i}_{\tilde{a}_1}a_2a_3 \ldots a_n - a_2\mathrm{i}_{\tilde{a}_1}a_3a_4 \ldots a_n + \ldots + \\
&\qquad a_2 \ldots a_{n-1}\mathrm{i}_{\tilde{a}_1}a_n\} \\
&= g(a_1, a_2)\mathscr{S}_0(a_3a_4 \ldots a_n) - g(a_1, a_3)\mathscr{S}_0(a_2a_4 \ldots \\
&\qquad a_n) + \ldots + g(a_1, a_n)\mathscr{S}_0(a_2a_3 \ldots a_{n-1}).
\end{aligned}$$

In physics, elements of the vector space carrying an irreducible representation of the complexified Clifford algebra are termed Dirac spinors. Thus whilst in even dimensions this accords with what we have simply called a spinor of the complexified Clifford algebra, in odd dimensions a Dirac spinor is what we have called a semi-spinor. In n dimensions Dirac spinors are obviously elements of a $2^{[n/2]}$-dimensional complex vector space which we will identify with some minimal left ideal. If n is even the different minimal left ideals all carry equivalent representations, whilst for n odd the two inequivalent representations are carried by minimal left ideals lying in different simple component algebras. Any minimal left ideal can be taken as the first column in some matrix basis. If $\psi \in \mathbf{C}_n(\mathbb{C})P$ with P primitive then we may form a matrix basis, $\{\mathbf{e}_{ij}\}$, with $\mathbf{e}_{11} = P$, giving $\psi = \Sigma_i \psi_i \mathbf{e}_{i1}$. If $\psi' = \psi S^{-1}$ for some invertible S then ψ' lies in the first column of the matrix basis $\{\mathbf{e}'_{ij}\}$ where $\mathbf{e}'_{ij} = S\mathbf{e}_{ij}S^{-1}$. If we write $\psi' = \Sigma_i \psi'_i \mathbf{e}'_{ii}$ then if $S^{-1} = \Sigma_{p,q} S^{-1}_{pq} \mathbf{e}'_{pq}$ we have $\psi'_i = \Sigma_j S^{-1}_{ij} \psi_j$. Thus although a change of minimal left ideal is effected by Clifford multiplication from the right, the components in matrix bases for which the spinors form the first columns are related by matrix multiplication from the left.

The Dirac adjoint spinor, $\bar{\psi}$, is a 'row' spinor which enables spin-invariant products to be defined. Thus $\bar{\psi}$ is the adjoint of ψ with respect to some spin-invariant product, it being an element of the dual space carrying a contragradient representation. From table 2.18 we see that unless p is odd and q is even the involution $\xi\eta*$ is the adjoint involution of a pseudo-Hermitian product. When p is odd with q even then ξ^* is the adjoint involution of such a product. We consider the former case first. For some choice of matrix basis let $\dagger$ be the involution of Hermitian conjugation. (In odd dimensions $\dagger$ induces Hermitian conjugation in the simple component algebras.) Then $\dagger$ is related to $\xi\eta*$ as follows,

$$a^{\xi\eta^*} = Aa^\dagger A^{-1} \qquad \forall\, a \in \mathbf{C}^{\mathbb{C}}_{p,q} \tag{2.8.9}$$

with $A^{\xi\eta^*} = A$ (equivalently $A^\dagger = A$). If φ and ψ are Dirac spinors, lying in the first column in the matrix basis in which $\dagger$ is Hermitian conjugation, then we may define a spin-invariant product

$$(\varphi, \psi)_{\xi\eta^*} = A^{-1}\varphi^{\xi\eta^*}\psi. \tag{2.8.10}$$

This product, which having $\xi\eta^*$ as its adjoint involution is invariant under ${}^+\Gamma^\pm$, is a special case of (2.6.2). As such it takes values in the algebra of complex numbers whose identity is the primitive $\mathbf{e}_{11}$. We can trivially obtain a product with values in the underlying complex field. For if $(\varphi, \psi)_{\xi\eta^*} = \langle \varphi, \psi \rangle \mathbf{e}_{11}$ then

$$\langle \varphi, \psi \rangle = \mathrm{Tr}(\varphi, \psi)_{\xi\eta^*}. \tag{2.8.11}$$

The adjoint of ψ with respect to the product in (2.8.10) is the Dirac adjoint, that is

$$\bar{\psi} = A^{-1}\psi^{\xi\eta^*} = \psi^\dagger A^{-1}. \tag{2.8.12}$$

The defining relation for A, (2.8.9), involves Hermitian conjugation which is defined in some matrix basis, $\{\mathbf{e}_{ij}\}$. If $\{\mathbf{e}'_{ij}\}$ is another matrix basis with $\mathbf{e}'_{ij} = S\mathbf{e}_{ij}S^{-1}$ then $\mathbf{e}'^{\dagger}_{ij} = S^{-1\dagger}\mathbf{e}_{ji}S^\dagger$. Thus if $S^\dagger = S^{-1}$ then $\mathbf{e}'^{\dagger}_{ij} = \mathbf{e}_{ji}$ and the involution † also induces Hermitian conjugation in this basis. So in fact the involution †, and hence the relation (2.8.9), involves a class of bases the elements of which are related by unitary transformations. Suppose that we consider that class of matrix basis for $\mathbf{C}^{\mathbb{C}}_{p,1}$ in which $e^{0\dagger} = -e^0$, $e^{i\dagger} = e^i$, $i = 1, \ldots, p$. Equation (2.8.9) is equivalent to $e^{a\dagger} = -A^{-1}e^aA$, and so in such a basis we may choose $A^{-1} = \mathrm{i}e^0$. This gives the familiar relation

$$\bar{\psi} = \mathrm{i}\psi^\dagger e^0. \tag{2.8.13}$$

(The factor of i is absent in the case of $\mathbf{C}^{\mathbb{C}}_{1,q}$.) It is this relation (in component form) that is usually taken as the definition of the Dirac adjoint. It is the choice of $A^{-1} = \mathrm{i}e^0$ that arbitrarily restricts the representations of the γ-matrices to be related by unitary transformations. There is no need for this restriction. The notable exception to this restrictive definition of the Dirac adjoint is the book by Jauch and Rohrlich [4].

In the above we excluded the case in which p is odd and q is even. In this case it is $\xi*$, rather than $\xi\eta*$, that is the adjoint involution of a pseudo-Hermitian product. Unless p is even and q is odd in analogy with (2.8.9) we may define

$$a^{\xi^*} = Ba^\dagger B^{-1} \qquad \forall\, a \in \mathbf{C}^{\mathbb{C}}_{p,q} \tag{2.8.14}$$

with $B^{\xi^*} = B^\dagger = B$. Instead of (2.8.12) we define

$$\bar{\psi} = B^{-1}\,\psi^{\xi^*}. \tag{2.8.15}$$

Here the Dirac adjoint is defined with respect to a $_+\Gamma^\pm$-invariant product.

A Dirac spinor and its adjoint are used to form the so-called bilinear covariants. If φ and ψ are Dirac spinors then, as explained at the end of §2.5, the spinor representation gives rise to a representation τ. We define

$$\tau(s)(\varphi\bar{\psi}) = s\varphi(\overline{s\psi}).$$

When, for example, the Dirac adjoint is defined as in (2.8.12) then $\tau(s)(\varphi\bar{\psi}) = s(\varphi\bar{\psi})s^{\xi\eta^*}$. Thus for $s \in {}^+\Gamma^\pm$ the representation τ coincides with the vector representation, that is

$$\tau(s)(\varphi\bar{\psi}) = s(\varphi\bar{\psi})s^{-1}.$$

When q is odd the image of ${}^{+}\Gamma^{\pm}$ under the vector representation is the timelike-orientation-preserving subgroup of the orthogonal group; whilst for q even it is the spacelike-orientation-preserving subgroup. As pointed out in §2.4, the p-forms transform irreducibly under the vector representation of the Clifford group. Using (2.1.18) we expand $\varphi\bar{\psi}$ as a sum of p-forms,

$$\begin{aligned}\varphi\bar{\psi} &= \sum_A \mathcal{S}_0(\varphi\bar{\psi}e_A{}^{\xi})e^A \\ &= \sum_A \mathcal{S}_0(\bar{\psi}e_A{}^{\xi}\varphi)e^A \qquad \text{(by (2.1.17)).}\end{aligned}$$

This gives the p-form components in terms of the product (2.8.10), or (2.8.11),

$$\varphi\bar{\psi} = \sum_A \langle \psi, e_A{}^{\xi}\varphi\rangle \mathcal{S}_0(\mathbf{e}_{11})e^A. \tag{2.8.16}$$

For the particular case of $\mathbf{C}^{\mathbb{C}}_{3,1}$ we have

$$\begin{aligned}4\mathcal{S}_0(\psi\bar{\psi}) &= \mathrm{Tr}(\psi\bar{\psi}) \\ 4\mathcal{S}_1(\psi\bar{\psi}) &= \mathrm{Tr}(\bar{\psi}e^a\psi)e_a \\ 4\mathcal{S}_2(\psi\bar{\psi}) &= \tfrac{1}{2}\mathrm{Tr}(\bar{\psi}e^{ab}\psi)e_{ba} \\ 4\mathcal{S}_3(\psi\bar{\psi}) &= \mathrm{Tr}(\bar{\psi}e^a z\psi)e_a z \\ 4\mathcal{S}_4(\psi\bar{\psi}) &= -\mathrm{Tr}(\bar{\psi}z\psi)z.\end{aligned} \tag{2.8.17}$$

The components of these homogeneous forms are the familiar scalar, vector, tensor, pseudo-vector and pseudo-scalar. As was noted above, the spinor representation on ψ induces the representation τ on these bilinears. In particular, the spinor representation of ${}^{+}\Gamma^{\pm}$ induces the vector representation on the bilinears, the image of ${}^{+}\Gamma^{\pm}$ under the vector representation being the group of orthochronous orthogonal transformations. It is the behaviour under the parity transformation that, for example, distinguishes between the scalar and the pseudoscalar. The vector representation of the elements of the Clifford group which change time orientation cannot be induced on these bilinears from the spinor representation. The Wigner time-reversal operator on spinors is not a representation of the Clifford group, neither does it induce on these bilinears the transformations one would expect from the nomenclature of 'vector'. It is, however, a symmetry of the Maxwell–Dirac equations, as will be discussed in §10.3. In the physics literature the action of the spinor representation of that element of the Clifford group whose vector representation gives time reversal is called the Racah time reversal on spinors. It is not a symmetry of the Maxwell–Dirac equations, which accounts for its infrequent mention these days.

The Dirac adjoint is associated with the pseudo-Hermitian product for which $\xi\eta*$ or $\xi*$, is the adjoint involution. We also have the spin-invariant products for which the $\mathbb{C}$-linear involutions $\xi\eta$ and ξ are the adjoints. From table 2.19 we see that unless $n = 1 \bmod 8$ or $5 \bmod 8$ the involution $\xi\eta$ induces an involution on the simple components of the reducible Clifford algebras. If $\mathcal{T}$ denotes transposition in some matrix basis then, excepting the dimensions mentioned, we have

$$a^{\xi\eta} = \mathrm{C}a^{\mathcal{T}}\mathrm{C}^{-1} \qquad \forall\, a \in \mathbf{C}^{\mathbb{C}}_{p,q} \tag{2.8.18}$$

with $\mathrm{C}^{\xi\eta} = \mathrm{C}^{\mathcal{T}} = \pm\mathrm{C}$. The symmetry of C determines the symmetry of the complex bilinear product defined by

$$(\varphi, \psi)_{\xi\eta} = \mathrm{C}^{-1}\varphi^{\xi\eta}\psi. \tag{2.8.19}$$

Here φ and ψ are Dirac spinors lying in the first column of the matrix basis in which $\mathcal{T}$ is the transposition. The symmetry of this product, for which $\xi\eta$ is the adjoint involution, is given in table 2.1. The defining property of C, (2.8.18), is equivalent to

$$e^{a\,\mathcal{T}} = -\mathrm{C}^{-1}e^{a}\mathrm{C}$$

the matrix components of which are usually taken as the definition of the charge conjugation matrix. If $\widetilde{\psi}$ is the adjoint of ψ with respect to the product in (2.8.19) then

$$\widetilde{\psi} = \mathrm{C}^{-1}\psi^{\xi\eta} = \psi^{\mathcal{T}}\mathrm{C}^{-1}. \tag{2.8.20}$$

This adjoint spinor is often called the Majorana conjugate.

Except for $n = 3 \bmod 8$ or $7 \bmod 8$ the involution ξ induces an involution on the simple components of the reducible Clifford algebras. We may define

$$a^{\xi} = Da^{\mathcal{T}}D^{-1} \qquad \forall\, a \in \mathbf{C}^{\mathbb{C}}_{p,q} \tag{2.8.21}$$

with $D^{\xi} = D^{\mathcal{T}} = \pm D$. This gives

$$e^{a\,\mathcal{T}} = D^{-1}e^{a}D.$$

The symmetry of the product defined by

$$(\varphi, \psi)_{\xi} = D^{-1}\varphi^{\xi}\psi \tag{2.8.22}$$

is given in table 2.18. We shall also use $\widetilde{\psi}$ to denote the adjoint with respect to this product, specifying the relevant product whenever confusion is likely.

In §2.7 we were careful to distinguish the automorphism *, referred to as complex conjugation, from the automorphism $^\#$. Complex conjugation leaves invariant the real subalgebra generated by the real orthogonal space with signature p, q, whilst $^\#$ is defined to complex conjugate the matrix components in some matrix basis. Thus the definition of $^\#$

depends on the choice of some matrix basis. In (2.7.9) we showed that, excepting the case in which the real subalgebra is isomorphic to the algebra of complex matrices, these two automorphisms are related by $a^* = ma^{\#}m^{-1}\ \forall a$. When the real subalgebra is a real matrix algebra, or a sum of two such algebras, we may choose $mm^* = 1$. When the real subalgebra is the tensor product of a matrix algebra with the quaternions, or a sum of two such algebras, we may choose $mm^* = -1$. The real subalgebra is isomorphic to the algebra of complex matrices when $p - q = 3$ or $7 \bmod 8$. In this case complex conjugation of the complexified algebra swaps the simple components. Save for this exceptional case we use this relation between the two automorphisms to define the charge conjugate spinor ψ^{c}

$$\psi^{\mathrm{c}} = \psi^* m. \tag{2.8.23}$$

This can be rewritten in terms of the Dirac adjoint and the charge conjugation matrix. Unless $n = 1$ or $5 \bmod 8$, or p is odd with q even, we may use (2.8.9) and (2.8.18) to produce

$$(a^{\xi\eta^*})^{\xi\eta} = (A^{-1})^{\xi\eta}\mathrm{C}a^{\dagger\mathcal{T}}\mathrm{C}^{-1}A^{\xi\eta}.$$

Since complex conjugation commutes with the involution $\xi\eta$ and $\dagger\mathcal{T} = \mathcal{T}\dagger = \#$ we have

$$a^* = ma^{\#}m^{-1} \qquad \text{with } m = A^{-1*}\mathrm{C} \tag{2.8.24}$$

where we have used $A^{\xi\eta^*} = A$. We know that we can scale m such that $mm^* = \pm 1$, which can be accomplished by choosing C suitably. With m given by (2.8.24) equation (2.8.23) becomes

$$\psi^{\mathrm{c}} = \mathrm{C}\bar{\psi}^{\mathcal{T}}. \tag{2.8.25}$$

In exactly the same way, except for the case of $n = 3$ or $7 \bmod 8$ or p even with q odd, (2.8.23) can be written as

$$\psi^{\mathrm{c}} = D\bar{\psi}^{\mathcal{T}} \tag{2.8.26}$$

where now $\bar{\psi}$ is given by (2.8.15). The only cases in which we can use neither (2.8.25) nor (2.8.26) are for $p + q = 3$ or $7 \bmod 8$ with q even, or $p + q = 1$ or $5 \bmod 8$ with q odd. These cases can only occur for $p - q = 3$ or $7 \bmod 8$, which is the case we excluded from the definition of the charge conjugate spinor.

When the Dirac spinors carry a reducible representation of the real subalgebra, elements of the irreducible subspaces are called Majorana spinors. As was pointed out in §2.7 this ocurs when the real subalgebra is a real matrix algebra, or a sum of two such algebras, and this occurs when $p - q = 0,\ 1,\ 2 \bmod 8$, as is seen from table 2.8. In these dimensions reference to §2.7 shows how the space of Dirac spinors can be decomposed into eigenspaces of the charge conjugation operator.

Thus a Majorana spinor is an eigenspinor of the charge conjugation operation

$$\psi = \pm\psi^{c}. \tag{2.8.27}$$

This can be written in terms of the Dirac and Majorana conjugates by using (2.8.25) or (2.8.26).

In an even number of dimensions the irreducible representations of the complex Clifford algebra induce a reducible representation of the even subalgebra; the spinor representation splitting into two inequivalent semi-spinor representations of the even subalgebra. The central idempotents that project the even subalgebra into simple components are $P_{\pm} = \frac{1}{2}(1 \pm \check{z})$, where either $\check{z} = z$ or $\check{z} = \mathrm{i}z$ ensuring $\check{z}^2 = 1$, z denoting the volume n-form. If ψ is a Dirac spinor then it may be decomposed into subspaces that transform irreducibly under the even subalgebra,

$$\psi = \psi_+ + \psi_- \qquad \text{where } \psi_{\pm} = P_{\pm}\psi. \tag{2.8.28}$$

The semi-spinors $\psi_{\pm}$ are called Weyl spinors, or chiral spinors. The Weyl spinors can carry a reducible representation of the real even subalgebra. From table 2.10 this is seen to occur when $p - q = 0 \bmod 8$. In this case the real even subalgebra is the direct sum of two real matrix algebras, having the real central idempotents $P_{\pm} = \frac{1}{2}(1 \pm z)$. The 'Majorana condition' (2.8.27), can be consistently imposed together with the 'Weyl condition', (2.8.28), to decompose a Dirac spinor into subspaces transforming irreducibly under the real even subalgebra. The resulting spinors are called Majorana–Weyl spinors.

In an odd number of dimensions irreducible representation of the complexified Clifford algebra induce irreducible representations of the even subalgebra. These can induce a reducible representation of the real, even subalgebra. Obviously this is the case for $p - q = 1 \bmod 8$ where, as we have noted, Dirac spinors carry a reducible representation of the whole real subalgebra. From table 2.10 we see that for $p - q = 7 \bmod 8$ Dirac spinors carry irreducible representations of the real subalgebra and the even subalgebra. However they carry a reducible representation of the real even subalgebra. For $p - q = 7 \bmod 8$

$$\mathbf{C}_{p,q}(\mathbb{R}) \simeq \mathbb{C}\otimes\mathcal{M}_{2^{(n-1)/2}}(\mathbb{R})$$

and

$$\mathbf{C}^{+}_{p,q}(\mathbb{R}) \simeq \mathcal{M}_{2^{(n-1)/2}}(\mathbb{R})$$

where $p + q = n$. We may thus choose a matrix basis for the Clifford algebra in which the automorphism η simply complex conjugates the components. The complexified algebra $\mathbf{C}^{\mathbb{C}}_{p,q}$ is reducible, with η interchanging the simple components. Complex conjugation, *, also swaps the component algebras. The automorphism η^* will certainly preserve the simple components, and in a suitable basis we see from (2.6.17) that it coincides with #, the operation that complex conjugates the matrix components. A Dirac spinor ψ can be decomposed into spinors transforming irreducibly under the real even subalgebra

$$\psi = \psi_+ + \psi_- \qquad \text{with } \psi_\pm = \tfrac{1}{2}(\psi \pm \psi^{\eta^*}). \tag{2.8.29}$$

(Such spinors have attracted no special terminology in the physics literature.)

Of importance in many calculations, especially those involving supersymmetric theories, is the Fierz rearrangement formula. This allows products of bilinears to be rewritten in terms of different bilinears. Many similar results can be given, we illustrate the basic result below. Let α, β, ψ, φ be Dirac spinors lying in some minimal left ideal projected by the primitive P. If M and N are arbitrary elements of the Clifford algebra then

$$\begin{aligned} \bar{\alpha}M\beta\bar{\psi}N\varphi &= \bar{\alpha}S_0(M\beta\bar{\psi}Ne_A{}^\xi)e^A\varphi \qquad \text{(by 2.1.18)} \\ &= \bar{\alpha}e^A\varphi S_0(M\beta\bar{\psi}Ne_A{}^\xi). \end{aligned}$$

The terms in the brackets can be reordered using (2.1.13), and since $\varphi = \varphi P$

$$\bar{\alpha}M\beta\bar{\psi}N\varphi = \bar{\alpha}e^A\varphi S_0(\bar{\psi}Ne_A{}^\xi M\beta)P.$$

Now the term in brackets is in $P\mathbf{C}^{\mathbb{C}}_{p,q}P$, which is isomorphic to the algebra of complex numbers with P as identity. That is, $PXP = \lambda P$ for λ a complex 0-form, giving

$$S_0(PXP) = \lambda S_0(P)$$

and

$$S_0(PXP)P = S_0(P)PXP$$

so

$$\bar{\alpha}M\beta\bar{\psi}N\varphi = \bar{\alpha}e^A\varphi\bar{\psi}Ne_A{}^\xi M\beta S_0(P).$$

In terms of the product in (2.8.11) we have

$$\langle\alpha, M\beta\rangle\langle\psi, N\varphi\rangle P = \langle\alpha, e^A\varphi\rangle\langle\psi, Ne_A{}^\xi M\beta\rangle S_0(P)P$$

whose 0-form component is the basic Fierz formula

$$\langle\alpha, M\beta\rangle\langle\psi, N\varphi\rangle = \langle\alpha, e^A\varphi\rangle\langle\psi, Ne_A{}^\xi M\beta\rangle S_0(P). \tag{2.8.30}$$

(The factor of $S_0(P)$ arises from our normalisation of the e^A.)

The approach to spinors that we have pursued is essentially algebraic. From the Clifford algebra we can define the spin groups, and from the representations of the algebra we induce representations of these groups. One can, however, start from a knowledge of the covering group of the connected component of the orthogonal group and introduce its irreducible representations as spinors. Representations of the component of the orthogonal group connected to the identity can then be found from the tensor product of these spinor representations. For the case of four dimensions, and Lorentzian signature, such an approach has developed its own rather specialised notation and conventions. That is the Infeld–van der Waerden formalism, or 'two-component spinor formalism'. Given that the double covering of $SO^+(3, 1)$ is $SL(2, \mathbb{C})$ one introduces 'two-component spinors' as carrying irreducible representations of $SL(2, \mathbb{C})$. The complex conjugate representations of this group are inequivalent, and a special notation is used to distinguish them. If u is a vector carrying an $SL(2, \mathbb{C})$ representation such that the components of u transform with a matrix m then, say, the components of u are labelled by a Greek superscript. If the vector v transforms with the complex conjugate matrix then the components of v are labelled by a Greek superscript with a dot above it. (It is perhaps significant that such a notation was introduced before the advent of frequent photocopying!) The vector spaces carrying these representations both admit $SL(2, \mathbb{C})$-invariant symplectic products, and the adjoint of u, say, with respect to such a product has its components with respect to a dual basis written as subscripts. A similar situation holds for v. Thus indices are 'lowered' with the symplectic matrix, which can be taken to have plus one in the top right-hand entry. Because of the antisymmetry of this matrix a convention must be adopted as to which side the matrix is multiplied from to lower an index. The tensor product of these two representations, with themselves and each other, gives a representation of $SO^+(3, 1)$. Thus $SO^+(3, 1)$ irreducible representations are identified with certain expressions written with two Greek indices, either with or without dots, up and down, or a mixture. Of course, starting with $SL(2, \mathbb{C})$ irreducible representations only produces $SO^+(3, 1)$ representations, not $O(3, 1)$ representations. One can extend the representations of $SL(2, \mathbb{C})$ to include other transformations so that the tensor representation extends to a representation of $O(3, 1)$. However, such extensions are not unique and there is certainly no universal convention for complex phase factors. Without being exhaustive we shall show the relation between the 'two-component formalism' and the algebraic approach.

We shall consider the Clifford algebra associated with a four-dimensional Lorentzian space. Starting with the real even subalgebra we shall construct a basis for the complexified Clifford algebra. We saw in §2.3 that $\mathbf{C}^+_{3,1}(\mathbb{R}) \simeq \mathbb{C} \otimes \mathcal{M}_2$ and that if $\{\boldsymbol{\varepsilon}_{\alpha\beta}\}$ is a matrix basis there exists a 2-form c relating transposition, t, to the involution ξ

$$u^{\xi} = c a^{t} c^{-1}$$

and a 1-form x that squares to one and commutes with the $\boldsymbol{\varepsilon}_{\alpha\beta}$ and c. Thus c must have real components and, since it is certainly anti-symmetric, we can choose it such that its components form the standard symplectic matrix, that is

$$c = \sum_{\alpha,\beta} c_{\alpha\beta}\boldsymbol{\varepsilon}_{\alpha\beta} \tag{2.8.31}$$

where the matrix of components $c_{\alpha\beta}$ is

$$c_{\alpha\beta} = \begin{pmatrix} 0 & 1 \\ -1 & 0 \end{pmatrix}. \tag{2.8.32}$$

The even subalgebra of the complexified algebra is the direct sum of two algebras of complex order-two matrices.

$$\mathbf{C}_{3,1}^{\mathbb{C}+} \simeq \mathcal{M}_2(\mathbb{C}) \oplus \mathcal{M}_2(\mathbb{C}).$$

If $P_\pm = \frac{1}{2}(1 \pm iz)$, with z the volume 4-form, then $\{\boldsymbol{\varepsilon}_{\alpha\beta}P_+\}$ and $\{\boldsymbol{\varepsilon}_{\alpha\beta}P_-\}$ are bases for the simple component algebras. The complexified Clifford algebra is isomorphic to the algebra of order-four complex matrices, and so we can choose a matrix basis in which the even subalgebra is block diagonal. In such a basis any odd element must have off-diagonal components. If, as usual, we identify the space of Dirac spinors with the minimal left ideal formed by the first column then the upper two components and the lower two components will transform irreducibly under the even subalgebra. These are the even and odd parts of the spinor, forming the two inequivalent Weyl spinors. In this language one refers to a Dirac spinor as a bispinor, as it carrys a reducible representation of SL(2, $\mathbb{C}$). We can use the element x to form the off-diagonal elements in a matrix basis for $\mathbf{C}_{3,1}^{\mathbb{C}}$, $\{\mathbf{e}_{ij}\}$. We can schematically display the basis we have constructed as follows:

$$\mathbf{e}_{ij} : \begin{pmatrix} \boldsymbol{\varepsilon}_{\alpha\beta}P_+ & x\boldsymbol{\varepsilon}_{\alpha\beta}P_- \\ x\boldsymbol{\varepsilon}_{\alpha\beta}P_+ & \boldsymbol{\varepsilon}_{\alpha\beta}P_- \end{pmatrix}. \tag{2.8.33}$$

It can be checked that this is indeed an ordinary matrix basis. In this basis the diagonal blocks are related by complex conjugation, as are the off-diagonal blocks. If $\mathcal{T}$ denotes transposition in this basis then

$$(\boldsymbol{\varepsilon}_{\alpha\beta}P_\pm)^{\mathcal{T}} = \boldsymbol{\varepsilon}_{\beta\alpha}P_\pm.$$

But from the defining property of c, (2.3.2),

$$\begin{aligned} \boldsymbol{\varepsilon}_{\beta\alpha}P_\pm &= c^{-1}\boldsymbol{\varepsilon}_{\alpha\beta}{}^{\xi}cP_\pm \\ &= c^{-1}\boldsymbol{\varepsilon}_{\alpha\beta}{}^{\xi}P_\pm c \qquad \text{(since } c \text{ is even)} \\ &= c^{-1}(\boldsymbol{\varepsilon}_{\alpha\beta}P_\pm)^{\xi}c. \end{aligned}$$

Similarly $(x\boldsymbol{\varepsilon}_{\alpha\beta}P_\pm)^{\mathcal{T}} = x\boldsymbol{\varepsilon}_{\beta\alpha}P_\mp$

and

$$\begin{aligned} x\boldsymbol{\varepsilon}_{\beta\alpha}P_{\mp} &= xc^{-1}\boldsymbol{\varepsilon}_{\alpha\beta}{}^{\xi}cP_{\mp} \\ &= c^{-1}x\boldsymbol{\varepsilon}_{\alpha\beta}^{\xi}P_{\mp}c \qquad \text{(since } x \text{ commutes with } c\text{)} \\ &= c^{-1}P_{\pm}\boldsymbol{\varepsilon}_{\alpha\beta}{}^{\xi}xc \qquad \text{(since } xP_{\pm} = P_{\mp}x\text{)} \\ &= c^{-1}(x\boldsymbol{\varepsilon}_{\alpha\beta}P_{\pm})^{\xi}c \end{aligned}$$

and so

$$a^{\xi} = ca^{\mathcal{T}}c^{-1} \qquad \forall\, a \in \mathbf{C}_{3,1}^{\mathbb{C}}. \tag{2.8.34}$$

If ψ is a Dirac spinor we can write ψ in terms of its even and odd parts as $\psi = u + v$. If we introduce the notation

$$\begin{aligned} \boldsymbol{\varepsilon}_{\alpha 1}P_{+} &= \mathbf{b}_{\alpha} \\ x\boldsymbol{\varepsilon}_{\alpha 1}P_{+} &= \mathbf{b}_{\dot{\alpha}} \end{aligned} \tag{2.8.35}$$

then $u = u^{\alpha}\mathbf{b}_{\alpha}$, $v = v^{\dot{\alpha}}\mathbf{b}_{\dot{\alpha}}$, u^{α}, $v^{\dot{\alpha}} \in \mathbb{C}$. This accords with the conventional labelling since u and v carry complex conjugate representations of the real even subalgebra, and hence ${}_{+}\Gamma^{+}$ which is isomorphic to $\mathrm{SL}(2, \mathbb{C})$. The first row in the matrix basis is naturally identified with the dual space of the first column. If we define

$$\begin{aligned} \boldsymbol{\varepsilon}_{1\alpha}P_{+} &= \mathbf{B}^{\alpha} \\ x\boldsymbol{\varepsilon}_{1\alpha}P_{-} &= \mathbf{B}^{\dot{\alpha}} \end{aligned} \tag{2.8.36}$$

then $\mathbf{B}^{\alpha}\mathbf{b}_{\beta} = \delta^{\alpha}_{\beta}\boldsymbol{\varepsilon}_{11}P_{+}$, $\mathbf{B}^{\alpha}\mathbf{b}_{\dot{\beta}} = 0$, $\mathbf{B}^{\dot{\alpha}}\mathbf{b}_{\dot{\beta}} = \delta^{\dot{\alpha}}_{\dot{\beta}}\boldsymbol{\varepsilon}_{11}P_{+}$, $\mathbf{B}^{\dot{\alpha}}\mathbf{b}_{\beta} = 0$. We may define the Majorana conjugate of ψ as

$$\widetilde{\psi} = \psi^{\mathcal{T}}c^{-1} \tag{2.8.37}$$

(this is a special case of 8.21), then

$$\widetilde{\psi} = u^{\mathcal{T}}c^{-1} + v^{\mathcal{T}}c^{-1} = u^{\alpha}\mathbf{B}^{\alpha}c^{-1} + v^{\dot{\alpha}}\mathbf{B}^{\dot{\alpha}}c^{-1}.$$

We now introduce

$$\begin{aligned} \mathbf{B}_{\alpha} &= \mathbf{B}^{\alpha}c^{-1} \\ \mathbf{B}_{\dot{\alpha}} &= \mathbf{B}^{\dot{\alpha}}c^{-1} \end{aligned} \tag{2.8.38}$$

giving, by (2.8.31), $\mathbf{B}_{\alpha} = c^{-1}_{\alpha\nu}\mathbf{B}^{\nu}$ and $\mathbf{B}_{\dot{\alpha}} = c^{-1}_{\alpha\nu}\mathbf{B}^{\dot{\nu}}$. We can write the Majorana conjugate as

$$\widetilde{\psi} = u^{\alpha}\mathbf{B}_{\alpha} + v^{\dot{\alpha}}\mathbf{B}_{\dot{\alpha}} = u_{\alpha}\mathbf{B}^{\alpha} + v_{\dot{\alpha}}\mathbf{B}^{\dot{\alpha}}$$

where, for example, $u_{\alpha} = u^{\beta}c^{-1}_{\beta\alpha}$. That is, indices are lowered with the components of the symplectic matrix. If φ is another Dirac spinor written in even and odd parts as $\varphi = w + y$ then

$$\widetilde{\psi}\varphi = (u_{\alpha}w^{\alpha} + v_{\dot{\alpha}}y^{\dot{\alpha}})\boldsymbol{\varepsilon}_{11}P_{+}.$$

Thus this product on the Dirac spinors induces the SL(2, $\mathbb{C}$)-invariant symplectic products on the Weyl spinors. So far we have relabelled the first row and the first column of our matrix basis to facilitate a correspondence with the two-component formalism. Any element of the matrix basis can be written as a product of the first column by the first row, $\mathbf{e}_{ij} = \mathbf{e}_{i1}\mathbf{e}_{1j}$, and so we can apply this relabelling to the whole basis

$$\begin{aligned}\boldsymbol{\varepsilon}_{\alpha\beta}P_{+}c^{-1} &= \mathbf{b}_{\alpha}\mathbf{B}_{\beta}\\ \boldsymbol{\varepsilon}_{\alpha\beta}P_{-}c^{-1} &= \mathbf{b}_{\dot{\alpha}}\mathbf{B}_{\dot{\beta}}\\ x\boldsymbol{\varepsilon}_{\alpha\beta}P_{+}c^{-1} &= \mathbf{b}_{\dot{\alpha}}\mathbf{B}_{\beta}\\ x\boldsymbol{\varepsilon}_{\alpha\beta}P_{-}c^{-1} &= \mathbf{b}_{\alpha}\mathbf{B}_{\dot{\beta}}.\end{aligned} \tag{2.8.39}$$

The products on the right-hand side with no dots or two dots are even, whilst the terms with mixed indices are odd. Under complex conjugation a dotted index is replaced with an undotted one, and vice versa. These terms also have simple properties under the involution ξ, for example

$$\begin{aligned}(\mathbf{b}_{\alpha}\mathbf{B}_{\beta})^{\xi} &= -c^{-1}P_{+}\boldsymbol{\varepsilon}_{\alpha\beta}{}^{\xi}\\ &= -c^{-1}\boldsymbol{\varepsilon}_{\alpha\beta}{}^{\xi}cP_{+}c^{-1}\\ &= -\boldsymbol{\varepsilon}_{\beta\alpha}P_{+}c^{-1}\\ &= -\mathbf{b}_{\beta}\mathbf{B}_{\alpha}.\end{aligned}$$

Similarly we obtain for the full set

$$\begin{aligned}(\mathbf{b}_{\alpha}\mathbf{B}_{\beta})^{\xi} &= -\mathbf{b}_{\beta}\mathbf{B}_{\alpha}\\ (\mathbf{b}_{\dot{\alpha}}\mathbf{B}_{\dot{\beta}})^{\xi} &= -\mathbf{b}_{\dot{\beta}}\mathbf{B}_{\dot{\alpha}}.\\ (\mathbf{b}_{\dot{\alpha}}\mathbf{B}_{\beta})^{\xi} &= -\mathbf{b}_{\beta}\mathbf{B}_{\dot{\alpha}}\\ (\mathbf{b}_{\alpha}\mathbf{B}_{\dot{\beta}})^{\xi} &= -\mathbf{b}_{\dot{\beta}}\mathbf{B}_{\alpha}.\end{aligned} \tag{2.8.40}$$

If then n is any real odd form

$$\begin{aligned}n &= n^{\dot{\alpha}\beta}\mathbf{b}_{\dot{\alpha}}\mathbf{B}_{\beta} + n^{\dot{\alpha}\beta *}\mathbf{b}_{\alpha}\mathbf{B}_{\dot{\beta}}\\ n^{\xi} &= -n^{\dot{\alpha}\beta}\mathbf{b}_{\beta}\mathbf{B}_{\dot{\alpha}} - n^{\dot{\alpha}\beta *}\mathbf{b}_{\dot{\beta}}\mathbf{B}_{\alpha}.\end{aligned}$$

So if n is a 1-form, even under ξ, $n^{\dot{\alpha}\beta} = -n^{\dot{\beta}\alpha *}$. That is, the components can be arranged as an anti-Hermitian matrix (with different conventions the matrix of components is Hermitian). In particular the basis 1-forms, e^{a}, can be expanded in the matrix basis as

$$e^{a} = \sigma^{a\dot{\alpha}\beta}\mathbf{b}_{\dot{\alpha}}\mathbf{B}_{\beta} + \sigma^{a\dot{\alpha}\beta *}\mathbf{b}_{\alpha}\mathbf{B}_{\dot{\beta}}.$$

The anti-Hermitian matrices, $\sigma^{a\dot{\alpha}\beta}$, give the correspondence between a 'vector' and a 'rank-two spinor' (or a 'valence two spinor'). Similarly a real 3-form has components that form a Hermitian matrix. If m is real and even then

$$m = m^{\alpha\beta}\mathbf{b}_{\alpha}\mathbf{B}_{\beta} + m^{\alpha\beta *}\mathbf{b}_{\dot{\alpha}}\mathbf{B}_{\dot{\beta}}.$$

Requiring that m be odd under ξ is equivalent to it being a 2-form. From (2.8.40) it follows that this gives $m^{\alpha\beta} = m^{\beta\alpha}$. Obviously the components of a 0-form or 4-form must form an anti-symmetric matrix, but we still need to disentangle the two. We have

$$\begin{aligned}\mathcal{S}_0(\mathbf{b}_{\alpha}\mathbf{B}_{\beta}) &= \mathcal{S}_0(\mathbf{B}_{\beta}\mathbf{b}_{\alpha})\\ &= \mathcal{S}_0(\mathbf{B}^{\beta}c^{-1}\mathbf{b}_{\alpha})\\ &= \mathcal{S}_0(\boldsymbol{\varepsilon}_{1\beta}c^{-1}\boldsymbol{\varepsilon}_{\alpha 1}P_{+})\\ &= \mathcal{S}_0(c^{-1}_{\beta\alpha}\boldsymbol{\varepsilon}_{11}P_{+}).\end{aligned}$$

For the primitive $\boldsymbol{\varepsilon}_{11}P_{+}$ we have $\mathcal{S}_0(\boldsymbol{\varepsilon}_{11}P_{+}) = \frac{1}{4}$, giving

$$\mathcal{S}_0(\mathbf{b}_{\alpha}\mathbf{B}_{\beta}) = \tfrac{1}{4}c^{-1}_{\beta\alpha}$$

also

$$\mathcal{S}_4(\mathbf{b}_{\alpha}\mathbf{B}_{\beta}) = -\mathcal{S}_0(\boldsymbol{\varepsilon}_{1\beta}c^{-1}\boldsymbol{\varepsilon}_{\alpha 1}P_{+}z)z$$

and since $P_{+}z = -\mathrm{i}P_{+}$ it follows that $\mathcal{S}_4(\mathbf{b}_{\alpha}\mathbf{B}_{\beta}) = \frac{1}{4}\mathrm{i}c^{-1}_{\beta\alpha}z$. If the inverse matrix is introduced such that $c^{\beta\alpha}c^{-1}{}_{\alpha\lambda} = \delta^{\beta}{}_{\lambda}$ then, if $m^{\alpha\beta}$ is anti-symmetric, $m^{\alpha\beta} = \lambda c^{\alpha\beta}$ for some complex λ. It then immediately follows that $\mathcal{S}_0(m) = 2\,\mathrm{Re}\,\lambda$ whilst $\mathcal{S}_4(m) = -2\,\mathrm{Im}\,\lambda z$.

In this section we have established contact with the most usual notations and nomenclature used for spinors in physics. There is, however, yet one more impediment to multilingual fluency. For many applications in physics one works with 'anticommuting' spinors. That is, whenever the order of two spinor fields is reversed a minus sign is introduced. One rationale is that the components of the spinors take values in the odd part of some exterior algebra. Certain other fields are assigned values in the even part of this algebra; bilinears in the spinors being even, for example. In practice the rationale seems unimportant as the rules are easy to understand. The consequences are, for example, that certain expressions which are antisymmetric in 'commuting' spinors become symmetric in 'anticommuting' spinors. Thus although many of the results presented in this chapter are changed (for example the properties of the spin-invariant inner products) they are easily adapted to accommodate 'anticommuting' spinors.

Exercise 2.2 Abstract from tables 2.8, 2.15, 2.17 and 2.18 the information in the following tables. In part (*a*): C2 gives real dimension of irreducible algebra representation; C3 gives real-valued spin-invariant product associated with ξ; C4 gives real-valued spin-invariant product associated with $\xi\eta$. In part (*b*): C2 gives complex dimension of irreducible algebra representation; C3 gives complex-valued spin-invariant product associated with ξ; C4 gives complex-valued spin-invariant product associated with $\xi\eta$; C5 gives complex-valued spin-invariant product associated with ξ^*, C6 gives complex-valued spin-invariant product associated with $\xi\eta^*$.

(*a*)

Clifford algebra	C2	C3	C4
$\mathbf{C}_{4,4}(\mathbb{R}) \simeq \mathcal{M}_{16}(\mathbb{R})$	16 (Majorana spinors)	sym(index 8)	sym(index 8)
$\mathbf{C}^+_{4,4}(\mathbb{R}) \simeq \mathcal{M}_8(\mathbb{R}) \oplus \mathcal{M}_8(\mathbb{R})$	8 (Majorana–Weyl spinors)	sym(index 4)	
$\mathbf{C}_{8,0}(\mathbb{R}) \simeq \mathcal{M}_{16}(\mathbb{R})$	16 (Majorana spinors)	sym(index 0)	sym(index 8)
$\mathbf{C}^+_{8,0}(\mathbb{R}) \simeq \mathcal{M}_8(\mathbb{R}) \oplus \mathcal{M}_8(\mathbb{R})$	8 (Majorana–Weyl spinors)	sym(index 0)	
$\mathbf{C}_{0,8}(\mathbb{R}) \simeq \mathcal{M}_{16}(\mathbb{R})$	16 (Majorana spinors)	sym(index 8)	sym(index 0)
$\mathbf{C}^+_{0,8}(\mathbb{R}) \simeq \mathcal{M}_8(\mathbb{R}) \oplus \mathcal{M}_8(\mathbb{R})$	8 (Majorana–Weyl spinors)	sym(index 0)	
$\mathbf{C}_{9,1}(\mathbb{R}) \simeq \mathcal{M}_{32}(\mathbb{R})$	32 (Majorana spinors)	sym(index 16)	skew
$\mathbf{C}^+_{9,1}(\mathbb{R}) \simeq \mathcal{M}_{16}(\mathbb{R}) \oplus \mathcal{M}_{16}(\mathbb{R})$	16 (Majorana–Weyl spinors)	swap	
$\mathbf{C}_{1,1}(\mathbb{R}) \simeq \mathcal{M}_2(\mathbb{R})$	2 (Majorana spinors)	sym(index 1)	skew
$\mathbf{C}^+_{1,1} \simeq \mathbb{R} \oplus \mathbb{R}$	1 (Majorana–Weyl spinors)	swap	

(*b*)

Clifford algebra	C2	C3	C4	C5	C6
$\mathbf{C}^{\mathbb{C}}_{8,0}(\mathbb{R}) \simeq \mathcal{M}_{16}(\mathbb{C})$	16 (Dirac spinors)	sym	sym	Hermitian sym (index 0)	Hermitian sym (index 8)
$\mathbf{C}^{\mathbb{C}+}_{8,0}(\mathbb{R}) \simeq \mathcal{M}_8(\mathbb{C}) \oplus \mathcal{M}_8(\mathbb{C})$	8 (Weyl spinors)	sym		Hermitian sym (index 0)	
$\mathbf{C}^{\mathbb{C}}_{9,1}(\mathbb{R}) \simeq \mathcal{M}_{32}(\mathbb{C})$	32 (Dirac spinors)	sym	skew	Hermitian sym (index 16)	Hermitian sym (index 16)
$\mathbf{C}^{\mathbb{C}+}_{9,1}(\mathbb{R}) \simeq \mathcal{M}_{16}(\mathbb{C}) \oplus \mathcal{M}_{16}(\mathbb{C})$	16 (Weyl spinors)	swap		swap	

Bibliography

Chevalley C 1954 *The Algebraic Theory of Spinors* (New York: Columbia)
Coquereaux R 1982 *Phys. Lett.* **115B** 389
Crumeyrolle A 1969 *Ann. Inst. Henri Poincaré* A **11** 19
—— 1971 *Ann. Inst. Henri Poincaré* A **14** 309
—— 1972 *Ann. Inst. Henri Poincaré* A **16** 171
—— 1974 Algebres de Clifford et spineurs *Cours et Seminaires de l'Universite de Toulouse III*
Greub W 1978 *Multilinear Algebra* 2nd edn
Lounesto P 1980 *Ann. Inst. Henri Poincaré* **33** 53
Porteous I 1981 *Topological Geometry* 2nd edn (Cambridge: Cambridge University Press)

3

Pure Spinors and Triality

This chapter contains some further properties of Clifford algebras and spinors. They may be regarded as more advanced material and the presentation will be adapted accordingly. Some readers may prefer to defer a study of these topics until later: they are not essential prerequisites for understanding the bulk of the material that follows, although we shall briefly make reference to certain properties of pure spinors in the last chapter.

3.1 Pure Spinors

In certain cases spinors may have a rather direct geometrical interpretation. As was observed by Cartan [5] certain spinors of $\mathbf{C}(V, g)$ may be correlated with maximal totally isotropic subspaces of V: these spinors being called *pure*. (An isotropic subspace of V is one on which g induces the zero bilinear form.) The account of pure spinors that we shall give follows that given in Chevalley [6]. We shall only consider the case in which V is even-dimensional. It turns out that in four (and six) dimensions all complex Weyl spinors are pure. For the physically interesting Lorentzian case this gives a correlation between Weyl spinors (or Majorana spinors) and null planes. In the positive-definite case maximal isotropic subspaces, and hence pure spinors, can be put into correspondence with complex structures. In more than six dimensions not all spinors are pure. The possibility of constraining spinors to be pure in physical theories formulated in higher dimensions has been investigated ([7], [8]).

Let V be an F-linear space with $\dim_F V = 2r$, and g an F-valued F-bilinear form with maximal index. (Here F will be either $\mathbb{R}$ or $\mathbb{C}$.) We can express V in terms of maximal (r-dimensional) totally isotropic subspaces M and N as $V = M \oplus N$. A *Witt basis* for V is formed from the isotropic bases $\{x^i\}$ for M and $\{y^i\}$ for N such that

$$x^i y^j + y^j x^i = \delta^{ij}. \tag{3.1.1}$$

Since g is of maximal index the Clifford algebra is a total matrix algebra

$$\mathbf{C}(V, g) \simeq \mathcal{M}_{2^r}(F) \tag{3.1.2}$$

whilst the structure of the even subalgebra is given by

$$\mathbf{C}^+(V, g) \simeq \mathcal{M}_{2^{r-1}}(F) \oplus \mathcal{M}_{2^{r-1}}(F). \tag{3.1.3}$$

Let $\check{z}$ be the $2r$-form with $\check{z}^2 = 1$ so that the idempotents $P_\pm = \frac{1}{2}(1 \pm \check{z})$ reduce $\mathbf{C}^+(V, g)$ to simple ideals. In terms of a Witt basis for V we may choose

$$\check{z} = [x^1, y^1][x^2, y^2] \ldots [x^r, y^r] \tag{3.1.4}$$

the brackets denoting Clifford commutators.

Let z_M be the r-form product of some basis for M. Since M is totally isotropic its Clifford algebra is just its exterior algebra $\Lambda(M)$ and so the r-form product of a different basis will differ from z_M by the determinant of the general linear transformation relating the bases. Given the Witt decomposition $V = M \oplus N$ we can express any element of $\mathbf{C}(V, g)$ in terms of products of the x^i and the y^i. Using the relations (3.1.1) the elements of M can be positioned at the right-hand side of any terms so that we see that $\mathbf{C}(V, g)z_M = \mathbf{C}(N, g)z_M = \Lambda(N)z_M$. Thus the left ideal $\mathbf{C}(V, g)z_M$ has the dimension of the exterior algebra of N, 2^r, and is hence a minimal left ideal. We may take this minimal left ideal as the space of spinors. If $\psi \in \mathbf{C}(V, g)z_M$ then $\psi = Bz_M$ for $B \in \Lambda(N)$. Thus $\check{z}\psi = B^\eta \check{z} z_M$. We have

$$\begin{aligned} \check{z} z_M &= [x^1, y^1][x^2, y^2] \ldots [x^r, y^r]x^1 x^2 \ldots x^r \\ &= [x^1, y^1]x^1[x^2, y^2]x^2 \ldots [x^r, y^r]x^r \end{aligned}$$

and from (3.1.1)

$$[x^i, y^i]x^i = x^i y^i x^i = (1 - y^i x^i)x^i = x^i$$

so $\check{z} z_M = z_M$. Thus $\check{z}\psi = B^\eta z_M$ and the even and odd (under η) subspaces of $\mathbf{C}(V, g)z_M$ form the semi-spinor spaces of the even subalgebra. Just as a maximal totally isotropic subspace can be used to define a minimal left ideal it can also be used to define a minimal right ideal. Since the Clifford algebra is a total matrix algebra the intersection of a minimal left ideal with a minimal right ideal is a 1-dimensional F-linear space. (For if P and P' are primitive idempotents with $P' = SPS^{-1}$ then $P'\mathbf{C}(V, g)P = SP\mathbf{C}(V, g)P$ and $P\mathbf{C}(V, g)P = \lambda P$ for $\lambda \in F$.) So if we use a maximal totally isotropic subspace M to define our space of spinors any other maximal totally isotropic subspace T can be used to define a minimal right ideal and hence a one-dimensional subspace of the spinor space.

If M, T are maximal totally isotropic subspaces then any element of $z_T\mathbf{C}(V, g)z_M$ is a *representative spinor* for T (with respect to M). A spinor that represents some T is called *pure*. (3.1.5)

An immediate consequence of this definition is the following:

If $T = \chi(s).M$ for $s \in \Gamma$ then a representative for T is $u = sz_M$. (3.1.6)

The space of representative spinors for M is spanned by z_M, so if u is a representative for M then $xu = 0\ \forall x \in M$. If now u is any element of $\mathbf{C}(V, g)z_M$ then $u = Bz_M$ for some $B \in \Lambda(N)$. For any $y^i \in N$ we can write $B = y^iB_1 + B_2$ with B_1 and B_2 in the exterior algebra of the subspace of N spanned by the remaining y. So $x^iu = B_1u$ and $x^iu = 0$ only if B lies in the exterior algebra of the $(r - 1)$-dimensional subspace of N spanned by the remaining y. Thus u is a representative for M if and only if $xu = 0\ \forall x \in M$. Because of (3.1.6) this can be couched more generally.

A spinor u is a representative for T if and only if $xu = 0\ \forall x \in T$. (3.1.7)

Given the totally isotropic M there is no unique N such that $V = M \oplus N$. If T is a maximal totally isotropic subspace with $\dim(T \cap M) = h$ then we can always choose a Witt basis such that $\{x^i\}$ is a basis for M and $\{x^1, \ldots, x^h, y^{h+1}, \ldots, y^r\}$ is a basis for T. Starting with a basis $\{x^1, \ldots, x^h\}$ for $T \cap M$ the Witt basis can be completed by a Gram–Schmidt type of construction. If we adapt the Witt basis in this way to the isotropic subspaces M and T then a representative for T is $u = y^{h+1} \ldots y^rz_M$. It will often be useful to have this canonical form for a pure spinor.

In even dimensions all elements of the Clifford group are either even or odd. Thus, by (3.1.6), all pure spinors are either even or odd. This property of the representative spinors can be used to classify the maximal totally isotropic subspaces as either even or odd.

If T_1, T_2 are maximal totally isotropic subspaces then T_1 and T_2 are both even or odd if and only if $\dim(T_1 \cap T_2) = r \bmod 2$. (3.1.8)

There is some $s \in \Gamma$ such that $\chi(s).T_2 = M$. If u_1, u_2 are representatives for T_1 and T_2 then $z_M = su_2$ and if $u = su_1$ then u is a representative for $T = \chi(s).T_1$. Since s is either even or odd then u and z_M behave the same under η if and only if u_1 and u_2 do. Moreover, $T \cap M = \chi(s).(T_1 \cap T_2)$ so it is sufficient to prove that representatives for T and M are both even or odd if $\dim(T \cap M) = r \bmod 2$. If we adapt a Witt basis to T and M then a representative u for T has the canonical form $u = y^{h+1} \ldots y^r z_M$ where $\dim(T \cap M) = h$. So u and z_M are both even or odd if $r - h = 0 \bmod 2$, that is $h = r \bmod 2$.

In general not all spinors will be pure; whereas we can always choose a basis of pure spinors, linear combinations of pure spinors will not in general be pure. The following gives the conditions for the sum of two pure spinors to be pure.

If u_1, u_2 represent T_1 and T_2 then a necessary and sufficient condition for $u_1 + u_2$ to be pure is that $\dim(T_1 \cap T_2) = r$ or $r - 2$. If this is the case then non-trivial linear combinations of u_1 and u_2 represent all T such that $T \cap T_2 = T_1 \cap T_2$. (3.1.9)

As in the proof of (3.1.8) it is sufficient to consider representatives for T and M. We adapt a Witt basis to these subspaces. A non-trivial linear combination of representatives for these subspaces will be pure if u is, where

$$u = \lambda z_M + y^{h+1} \ldots y^r z_M \qquad \lambda \in F. \tag{3.1.10}$$

Now $x^i u = 0$ if and only if $i = 1, \ldots, h$ and $\Sigma_{j=h+1}^r \lambda_j x^j u = 0$ if and only if $\lambda_j = 0 \ \forall j = h+1, \ldots r$, so if u is pure, representing T' say, then $T' \cap M = T \cap M$. If this is the case then we can choose a Witt basis $\{x^i, y^{i\prime}\}$ $i = 1, \ldots, r$ with $\{x^1, \ldots, x^h, y^{h+1\prime}, \ldots, y^{r\prime}\}$ a basis for T'. In this basis representatives for T' will take the canonical form, so if u is pure

$$\lambda z_M + y^{h+1} \ldots y^r z_M = \mu y^{h+1\prime} \ldots y^{r\prime} z_M \tag{3.1.11}$$

for some $\mu \in F$. By repeatedly using (3.1.1) the Clifford products in $y^{h+1} \ldots y^r z_M$ can be written in terms of exterior products, $y^{h+1} \ldots y^r z_M$ having homogeneous $(h, h+2, \ldots, 2r-h)$-form components. Similarly $y^{h+1\prime} \ldots y^{r\prime} z_M$ will have homogeneous components of the same degrees. Equating h-form components in (3.1.11) gives

$$\mu = 1. \tag{3.1.12}$$

If $h + 2 \neq r$ then this can be used to equate $(h+2)$-forms in (3.1.11):

$$x^1 \ldots x^h (y^{h+1} \wedge x^{h+1} + \ldots + y^r \wedge x^r)$$
$$= x^1 \ldots x^h (y^{h+1\prime} \wedge x^{h+1} + \ldots + y^{r\prime} \wedge x^r). \tag{3.1.13}$$

The $\{y^{i\prime}\}$ can be written as linear combinations of the basis $\{x^i, y^i\}$. Since $\{x^i, y^{i\prime}\}$ is also a Witt basis we have

$$y^{i\prime} = y^i + \sum_{j=h+1}^{r} M^{ij} x^j + N^i \qquad i = h+1, \ldots, r \tag{3.1.14}$$

where $M^{ij} = -M^{ji}$ and N^i is a linear combination of $\{x^1, \ldots x^h\}$. Inserting (3.1.14) in (3.1.13) gives $M^{ij} = 0 \ \forall i, j = h+1, \ldots, r$ and hence $\{x^1, \ldots, x^h, y^{h+1'}, \ldots, y^{r'}\}$ is just a new basis for T; that is, $T' = T$. So the only non-trivial case is $h + 2 = r$. In this case (3.1.11) is seen to be satisfied by

$$y^{r-1'} = y^{r-1} + \lambda x^r + N^{r-1} \qquad y^{r'} = y^r - \lambda x^{r-1} + N^r.$$

Here λ is seen to parametrise all T' with $T' \cap M = T \cap M$. Although in general, as we have stated, not all spinors are pure, in sufficiently low dimensions the above result can be used to show that all semi-spinors are pure.

$$\text{If } r \leqslant 3 \text{ then all semi-spinors are pure.} \tag{3.1.15}$$

In general we can always choose a set of pure spinors as a basis for the spinor space. Any semi-spinor will be a linear combination of pure spinors that are all even or odd. From (3.1.8) we know that if u_1, u_2 are two such pure spinors representing T_1 and T_2 then $\dim(T_1 \cap T_2) = r \bmod 2$, whereas from (3.1.9) linear combinations of u_1 and u_2 will be pure if $\dim(T_1 \cap T_2) = r$ or $r - 2$. Thus if $r \leqslant 3$ linear combinations of any two even or odd pure spinors are pure and hence all semi-spinors are pure.

Through (3.1.7) a pure spinor is related to the maximal isotropic subspace that it represents. However, given a semi-spinor this does not give a very practical way of determining whether or not it is pure. Given a spin-invariant inner product then the tensor product of a spinor with its adjoint can be identified with an element of the Clifford algebra. Necessary and sufficient conditions for a spinor to be pure can be given in terms of these tensors on the space of spinors (or 'spinor bilinears'). These conditions give a practical way of determining whether any given spinor is pure or not, and, in the case in which it is, recovering the associated maximal totally isotropic subspace.

Let (,) be an F-valued, symmetric or skew, product on spinors with ξ as adjoint involution. Let $\tilde{u}$ be the spinor adjoint to u with respect to this product.

If u_1, u_2 represent T_1 and T_2 then $T_1 \cap T_2 \neq \varnothing$ if and only if $(u_1, u_2) = 0$. (3.1.16)

Suppose firstly that there is some x in $T_1 \cap T_2$. Then there is some y such that $xy + yx = 1$ and

$$(u_1, u_2) = (u_1, (xy + yx)u_2) = (u_1, xyu_2)$$

since $x \in T_2$. Since the spinor product has ξ as adjoint involution then $(u_1, xyu_2) = (xu_1, yu_2)$ and this is zero if $x \in T_1$. So $T_1 \cap T_2 \neq \varnothing$ implies that $(u_1, u_2) = 0$. To prove the converse we let M and N be any two maximal isotropic subspaces such that $V = M \oplus N$. Then a spinor basis, each element of which is pure, is given by $\{y^I z_M\}$ with I a multi-index. Now we have already shown that $(z_M, y^I z_M) = 0$ unless $y^I = z_N$, and since the spinor product is non-degenerate we must have $(z_M, z_N z_M) \neq 0$. But $z_N z_M$ is a spinor representing N which was any maximal totally isotropic subspace not intersecting with M. So if u_1 and u_2 represent T_1 and T_2 then (u_1, u_2) can only be zero if $T_1 \cap T_2 \neq \emptyset$.

If $\check{z}v = +(-)v$ then for u any spinor $\mathcal{S}_{2r-p}(u\tilde{v}) = +(-)(-1)^r \mathcal{S}_p(u\tilde{v})\,\check{z}$ for all p. (3.1.17)

Since $\check{z}^2 = 1$

$$\mathcal{S}_{2r-p}(u\tilde{v}) = \mathcal{S}_{2r-p}(u\tilde{v})\,\check{z}\,\check{z} = \mathcal{S}_p(u\tilde{v}\,\check{z})\,\check{z} = \mathcal{S}_p(u(\widetilde{\check{z}^{\xi}v}))\,\check{z}$$

as the adjoint spinor is defined with respect to a product with ξ as the adjoint involution. Since $\check{z}$ is a $2r$-form $\check{z}^{\xi} = (-1)^r\,\check{z}$ and $\mathcal{S}_{2r-p}(u\tilde{v}) = (-1)^r \mathcal{S}_p(u(\widetilde{\check{z}v}))\,\check{z}$ and the result follows. If $v = Bz_M$ for $B \in \Lambda(N)$ then $\check{z}v = B^{\eta} z_M$ and $(-1)^r\,\check{z}v = v^{\eta}$. So if $v^{\eta} = \pm v$ then $\mathcal{S}_{2r-p}(u\tilde{v}) = +(-)\mathcal{S}_p(u\tilde{v})\,\check{z}$.

If u_1, u_2 represent T_1 and T_2 with $\dim(T_1 \cap T_2) = h$ then $\mathcal{S}_p(u_2\tilde{u}_1) = 0$ if $p < h$ or $p > 2r - h$, whilst $\mathcal{S}_h(u_2\tilde{u}_1) = z_{T_1 \cap T_2}$. (3.1.18)

If $s \in \Gamma$ then $\mathcal{S}_p(su_2\tilde{s}\tilde{u}_1) = \lambda(s)s\mathcal{S}_p(u_2\tilde{u}_1)s^{-1}$ so without loss of generality we can assume that u_1 represents M with u_2 representing some T with $\dim(T \cap M) = h$. In an adapted Witt basis we need to consider $y^{h+1} \ldots y^r z_M \tilde{z}_M$. In the proof of (3.1.16) we showed that $(z_M, y^I z_M) = 0$ unless $y^I = z_N$. Now $z_M z_N z_M = \pm z_M$ so we can always normalise the spinor product such that $(z_M, z_N z_M) z_M = z_M z_N z_M$. The definition of $u_2\tilde{u}_1$ is that $u_2\tilde{u}_1 v = u_2(u_1, v)$, so $z_M \tilde{z}_M y^I z_M = (z_M, y^I z_M) z_M$. This is zero unless $y^I = z_N$ and for the normalisation just mentioned $(z_M, z_N z_M) z_M = z_M z_N z_M$. But $z_M y^I z_M = 0$ unless $y^I = z_N$ and so $(z_M \tilde{z}_M) y^I z_M = z_M y^I z_M$ for all multi-indices I and so $z_M \tilde{z}_M = z_M$. Hence $y^{h+1} \ldots y^r z_M \tilde{z}_M = y^{h+1} \ldots y^r z_M$. The form of lowest degree in $y^{h+1} \ldots y^r z_M$ is proportional to $x^1 \ldots x^h$, which is just the product of a basis for $T \cap M$. Since all pure spinors are semi-spinors it follows from (3.1.17) that there is no non-vanishing p-form for $p > 2r - h$.

A semi-spinor u is pure if and only if $\mathcal{S}_p(u\tilde{u}) = 0 \quad \forall p \neq r$. (3.1.19)

From (3.1.18) we see that if u is pure then certainly $\mathcal{S}_p(u\tilde{u}) = 0$ $\forall p \neq r$, so what we need to do is to show that this condition on a semi-spinor is sufficient for it to be pure. Any spinor u can be written as $u = Bz_M$ where $B \in \Lambda(N)$. There is some $s \in \Gamma$ such that $su = (1 + b)z_M$ where $b \in \Lambda(N)$ and $\mathcal{S}_0(b) = 0$. If u is a semi-spinor then so is su and hence b must be an even element of $\Lambda(N)$. Suppose that $\mathcal{S}_2(b) \neq 0$, then $\exp(-\mathcal{S}_2(b)) \in \Gamma \cap \Lambda(N)$. Now the Clifford algebra of N is just its exterior algebra and so

$$\begin{aligned}\mathcal{S}_2[\exp(-\mathcal{S}_2(b))b] &= \mathcal{S}_0[\exp(-\mathcal{S}_2(b))]\mathcal{S}_2(b) + \mathcal{S}_2[\exp(-\mathcal{S}_2(b))]\mathcal{S}_0(b) \\ &= \mathcal{S}_2(b)\end{aligned}$$

since $\mathcal{S}_0(b) = 0$. Thus $\exp(-\mathcal{S}_2(b))su = (1 + b')z_M$ where $b' \in \Lambda^+(N)$ with $\mathcal{S}_0(b') = \mathcal{S}_2(b') = 0$. Suppose that the non-vanishing homogeneous component of b' of lowest degree is an h-form. In an appropriate basis we assume that

$$\exp(-\mathcal{S}_2(b))su = (1 + \lambda y^1y^2 \ldots y^h + \ldots)z_M$$

where the extra terms are of degree h or higher. Multiplying by $y^ry^{r-1} \ldots y^{h+1}$ will annihilate these other terms so if

$$v \equiv x^{h+1} \ldots x^ry^r \ldots y^{h+1} \exp(-\mathcal{S}_2(b))su$$

then

$$v = (1 + \lambda y^1y^2 \ldots y^h)z_M. \tag{3.1.20}$$

Now we come to the point of this construction. If u is any spinor and $s \in \Gamma$ then $\mathcal{S}_p(su\widetilde{su}) = \lambda(s)s\mathcal{S}_p(u\tilde{u})s^{-1}$ and $\mathcal{S}_p(u\tilde{u}) = 0 \;\; \forall p \neq r \Leftrightarrow \mathcal{S}_p(su\widetilde{su}) = 0 \;\; \forall p \neq r$. If a is any element of V then $au\widetilde{au} = au\tilde{u}a$. By (2.1.7) and (2.1.8) $au\tilde{u}a = g(a, a)(u\tilde{u})^\eta - 2a \wedge \mathrm{i}_{\tilde{a}}(u\tilde{u})^\eta$ and $\mathcal{S}_p(au\widetilde{au}) = (-1)^pg(a, a)\mathcal{S}_p(u\tilde{u}) - 2(-1)^pa \wedge \mathrm{i}_{\tilde{a}}\mathcal{S}_p(u\tilde{u})$. So if $\mathcal{S}_p(u\tilde{u}) = 0 \;\forall p \neq r$ then $\mathcal{S}_p(au\widetilde{au}) = 0 \;\forall p \neq r \;\forall a \in V$. Thus if the semi-spinor u that we started with satisfies $\mathcal{S}_p(u\tilde{u}) = 0 \;\forall p \neq r$ then the v we have constructed in (3.1.20) also satisfies these conditions. We will now show that this can only hold if $\lambda = 0$; that is $\exp(-\mathcal{S}_2(b))su = z_M$ and hence u is pure. Now v is the sum of two pure spinors and, as we have already noted, a pure spinor will satisfy the conditions of the theorem. So if u satisfies these conditions then

$$\lambda\mathcal{S}_p\{z_M(\widetilde{y^1 \ldots y^hz_M}) + y^1 \ldots y^hz_M\tilde{z}_M\} = 0 \qquad \forall p \neq r.$$

As we noted in the proof of (3.1.18) $z_M\widetilde{z_M} = z_M$ and so

$$z_M(\widetilde{y^1 \ldots y^hz_M}) + y^1 \ldots y^hz_M\tilde{z}_M = z_My^h \ldots y^1 + y^1 \ldots y^hz_M.$$

We now rearrange these terms, remembering that h is even:

$$\begin{aligned}z_My^h \ldots y^1 + y^1 \ldots y^hz_M = \{&(x^1y^1) \ldots (x^hy^h) + (-1)^{h/2}(y^1x^1) \ldots \\ &(y^hx^h)\}x^{h+1} \ldots x^r.\end{aligned}$$

Now $x^iy^i = \frac{1}{2} + x^i \wedge y^i$ whereas $y^ix^i = \frac{1}{2} - x^i \wedge y^i$. So if $h/2$ is even there will be a non-vanishing 0-form in $\{\;\}$, whereas if $h/2$ is odd there will be a non-vanishing 2-form. Thus in the first case the total expression has a non-vanishing $(r - h)$-form, whilst in the second the

$(r - h + 2)$-form component is non-zero. Since $h > 2$ then in both cases there is a non-vanishing p-form with $p < r$, so

$$\mathcal{S}_p(u\tilde{u}) = 0 \ \forall p \neq r \qquad \Rightarrow \mathcal{S}_p(v\tilde{v}) = 0 \ \forall p \neq r \Rightarrow \lambda = 0.$$

As we have already noted this shows that u is pure.

Eight dimensions are interesting as the lowest number of dimensions in which not all semi-spinors are pure. If $\dim_F V = 8$ with $F = \mathbb{R}$ or $\mathbb{C}$ and g is of maximal index, then from tables 2.15 and 2.17 we see that $(\ ,\ )$ induces a symmetric product on the semi-spinors. Hence $(u, e_A u) = (e_A u, u) = (u, e_A{}^{\xi} u)$ and $\mathcal{S}_p(u\tilde{u}) = 0$ if $[\frac{1}{2}p]$ is odd. If u is a semi-spinor then $u\tilde{u} = \check{z}u(\widetilde{\check{z}u}) = \check{z}u\tilde{u}\,\check{z}^{\xi} = \check{z}u\tilde{u}\,\check{z}$ in eight dimensions, and so $u\tilde{u} = (u\tilde{u})^{\eta}$. So if u is any semi-spinor then $u\tilde{u} = \Sigma_{p=0,4,8}\mathcal{S}_p(u\tilde{u})$. The 0-forms and 8-forms are related by (3.1.17) so in eight dimensions a semi-spinor u is pure if and only if $\mathcal{S}_0(u\tilde{u}) = 0$, that is, $(u, u) = 0$.

In this section we have taken the space of spinors to be a particular minimal left ideal of the Clifford algebra. This is convenient, enabling a basis of spinors to be constructed so as to facilitate the various algebraic proofs. However, it is not essential. Indeed all we really need is that the spinor space carry an irreducible representation of the Clifford algebra. Then (3.1.7) can be taken as the definition of a pure spinor, the stated results for pure spinors then following from this. Of course in general it would make no sense to talk about the behaviour of a spinor under the involution η, but all references to 'even' and 'odd' spinors can be interpreted as referring to their behaviour under multiplication by $(-1)^r\check{z}$.

For a real (pseudo-) orthogonal space whose metric has maximal index the pure spinors of the real Clifford algebra have a direct geometrical interpretation. For the remaining real Clifford algebras we cannot apply the above theory of pure spinors directly. However, if V is any real even-dimensional orthogonal space we may correlate the pure spinors of $\mathbf{C}^{\mathbb{C}}(V, g)$ with maximal totally isotropic subspaces of $V^{\mathbb{C}}$. In certain cases these maximal totally isotropic subspaces of $V^{\mathbb{C}}$ can be interpreted in terms of structures on the real vector space V.

Of particular physical interest is the case in which V is a four-dimensional Lorentzian vector space (g has signature $(p, q)=(3, 1)$). Then if M is a maximal totally isotropic subspace of $V^{\mathbb{C}}$ we have $\dim_{\mathbb{C}} M = 2$. Suppose that u and v are respectively even and odd semi-spinors of $\mathbf{C}^{\mathbb{C}}(V, g)$ representing T_1 and T_2. Then because of (3.1.15) they are both pure. From (3.1.8) we see that $\dim_{\mathbb{C}}(T_1 \cap T_2)$ must be odd (for r is here even, namely two). Hence $\dim_{\mathbb{C}}(T_1 \cap T_2) = 1$. If z is the volume 4-form of V then $\check{z} = \mathrm{i}z$. So if superscript c denotes the conjugate-linear charge conjugation operation (here involutory) and u is even then u^{c} is odd. The intersection of the maximal totally isotropic subspaces of $V^{\mathbb{C}}$ represented by u and u^{c} is one dimensional, containing n say. Thus $nu = nu^{c} = 0$. But $nu = 0$ implies that $n^{*}u^{c} = 0$, and similarly $nu^{c} = 0$ implies that $n^{*}u = 0$. So n^{*} lies in the one-dimensional intersection of the subspaces represented by u and u^{c} and $n^{*} = \lambda n$ for some $\lambda \in \mathbb{C}$. Since complex conjugation is involutory, λ must satisfy $\lambda\lambda^{*} = 1$, that is $\lambda \in \mathrm{U}(1)$. There is some $\mu \in \mathrm{U}(1)$ such that $\lambda = \mu^{2}$, and if $x = \mu n$ it follows that $x^{*} = x$. Thus x is a real null vector such that

$$x(u + u^{c}) = 0. \tag{3.1.21}$$

This real null vector is determined up to multiplication by a real number.

Suppose that u represents T which has a basis $\{x, w\}$. Now $x(wu^{c}) = -wxu^{c} = 0$ since $xu^{c} = 0$: and certainly $w(wu^{c}) = 0$ since $w^{2} = 0$. Thus wu^{c} and u both represent T. Since the space of representative spinors for T is one dimensional there is some $\lambda \in \mathbb{C}$ such that $wu^{c} = \lambda u$. We cannot have $\lambda = 0$ since w does not lie in the subspace represented by u^{c}. So if $\omega \equiv \lambda^{-1}w$ then

$$\omega u^{c} = u. \tag{3.1.22}$$

The charge conjugate of this is $\omega^{*}u = u^{c}$. So $\omega\omega^{*}u = \omega u^{c} = u$, and since $\omega \in T$ we have $(\omega\omega^{*} + \omega^{*}\omega)u = u$, and thus

$$\omega\omega^{*} + \omega^{*}\omega = 1. \tag{3.1.23}$$

From the (complex) null 1-form ω we can construct a unit 1-form a:

$$a \equiv \omega + \omega^{*}. \tag{3.1.24}$$

We have because of (3.1.22)

$$a(u + u^{c}) = u + u^{c}. \tag{3.1.25}$$

The real unit 1-form a is determined up to the addition of an arbitrary multiple of the null 1-form x. So equivalently we have extracted from the complex semi-spinor u a real null 1-form x and a real decomposable 2-form F

$$F \equiv x \wedge a \tag{3.1.26}$$

both determined up to a real multiple. If $\psi \equiv u + u^{c}$ then ψ is a Majorana spinor and because of (3.1.21) and (3.1.25) we can equivalently think of the real forms x and F as being determined by ψ.

The theorems (3.1.18) and (3.1.19) enable the real forms x and F to be expressed in terms of u, u^{c} and their adjoint spinors. There is a freedom to scale the spinor product (,) whose adjoint is ξ by a complex number. In the Lorentzian case we can always choose a spinor basis such that charge conjugation simply conjugates the spinor components. Thus we can require that the spinor product satisfies

$$(u_1, u_2)^* = (u_1{}^{\mathrm{c}}, u_2{}^{\mathrm{c}}) \tag{3.1.27}$$

this leaving only a real scaling freedom. Taking a spinor product which satisfied (3.1.27) we turn to (3.1.18). The intersection of the subspaces represented by u and u^{c} is spanned by the real 1-form x. So (3.1.18) tells us that $\mathscr{S}_1(\mathrm{i}u\tilde{u}^{\mathrm{c}})$ is a complex multiple of x. The factor of i is inserted to ensure that this 1-form is in fact real. For

$$\mathscr{S}_1(\mathrm{i}u\tilde{u}^{\mathrm{c}}) = \mathscr{S}_0(\mathrm{i}u\tilde{u}^{\mathrm{c}}e_a)e^a = (u^{\mathrm{c}}, \mathrm{i}e_a u)e^a$$

and

$$\begin{aligned}\mathscr{S}_1(\mathrm{i}u\tilde{u}^{\mathrm{c}})^* &= -(u, \mathrm{i}e_a u^{\mathrm{c}})e^a && \text{(by (3.1.27))}\\ &= -(\mathrm{i}e_a u, u^{\mathrm{c}})e^a && \text{(since } \xi \text{ is the adjoint)}\\ &= (u^{\mathrm{c}}, \mathrm{i}e_a u)e^a && \text{(since the product is skew)}\\ &= \mathscr{S}_1(\mathrm{i}u\tilde{u}^{\mathrm{c}}).\end{aligned}$$

and thus

$$\mathscr{S}_1(\mathrm{i}u\tilde{u}^{\mathrm{c}}) = x \tag{3.1.28}$$

where x is, of course, only determined up to a real multiple. Let $\{x, \omega\}$ be a basis for T, represented by u, where ω is the complex 1-form satisfying (3.1.22). Then ω is determined up to the addition of a multiple of x. From (3.1.18) we know that $\mathrm{i}u\tilde{u}$ is a complex multiple of $x\omega$, say $\mathrm{i}u\tilde{u} = 2\exp(\mathrm{i}\theta)\,x\omega$ for an appropriately scaled x. So if $G = \frac{1}{2}(\mathrm{i}u\tilde{u} - \mathrm{i}u^{\mathrm{c}}\tilde{u}^{\mathrm{c}})$ then $G = \exp(\mathrm{i}\theta)x\omega + \exp(-\mathrm{i}\theta)x\omega^*$ and $G(\omega + \omega^*) = \cos\theta\, x + 2\mathrm{i}\sin\theta\, x \wedge \omega \wedge \omega^*$. This G will be nothing other than the F of (3.1.26) if in fact $\theta = 0$. We have

$$2G(\omega + \omega^*) = \mathrm{i}u[\widetilde{(\omega + \omega^*)u}] - \mathrm{i}u^{\mathrm{c}}[\widetilde{(\omega + \omega^*)u^{\mathrm{c}}}] = \mathrm{i}u\tilde{u}^{\mathrm{c}} - \mathrm{i}u^{\mathrm{c}}\tilde{u}$$

since $\omega u = 0$ and $\omega u^{\mathrm{c}} = u$. But if α, β, φ, ψ are any spinors then

$$\begin{aligned}(\alpha, (\varphi\tilde{\psi})^{\xi}\beta) &= ((\varphi\tilde{\psi})\alpha,\beta) = (\psi, \alpha)(\varphi, \beta) = -(\alpha, \psi)(\varphi, \beta)\\ &= -(\alpha, (\psi\tilde{\varphi})\beta).\end{aligned}$$

and so $(\varphi\tilde{\psi})^{\xi} = -\psi\tilde{\varphi}$. Thus $2G(\omega + \omega^*) = \mathrm{i}u\tilde{u}^{\mathrm{c}} + (\mathrm{i}u\tilde{u}^{\mathrm{c}})^{\xi}$ and since $\mathrm{i}zu = u$ then

$$(u\tilde{u}^{\mathrm{c}})^{\eta} = -zu\tilde{u}^{\mathrm{c}}z = -zu(\widetilde{zu^{\mathrm{c}}}) = -\mathrm{i}zu(\widetilde{\mathrm{i}zu})^{\mathrm{c}} = -u\tilde{u}^{\mathrm{c}}.$$

and so $u\tilde{u}^{\mathrm{c}} = \mathscr{S}_1(u\tilde{u}^{\mathrm{c}}) + \mathscr{S}_3(u\tilde{u}^{\mathrm{c}})$. Since 3-forms change sign under ξ we have $G(\omega + \omega^*) = \mathscr{S}_1(\mathrm{i}u\tilde{u}^{\mathrm{c}}) = x$ by (3.1.28). That is, the F of (3.1.26) can be written as

$$F = \mathrm{Re}(\mathrm{i}u\tilde{u}). \tag{3.1.29}$$

From (3.1.21) and (3.1.25) we see that x and F can equivalently be thought of as being associated with the Majorana spinor $\psi = u + u^{\mathrm{c}}$. We can also express x and F in terms of ψ and its adjoint. Since

$u = \mathrm{i}zu$ we have

$$\psi(\widetilde{z\psi}) = (-\mathrm{i}u\tilde{u} + \mathrm{i}u^{\mathrm{c}}\tilde{u}^{\mathrm{c}}) + \mathrm{i}(u\tilde{u}^{\mathrm{c}} + (u\tilde{u}^{\mathrm{c}})^{\xi})$$

where, since $(\varphi\tilde{\psi})^{\xi} = -\psi\tilde{\varphi}$, the first term is odd under ξ whilst the second is even. Comparison with (3.1.28) and (3.1.29) shows that

$$\tfrac{1}{2}\mathscr{G}_1(\psi(\widetilde{z\psi})) = x \tag{3.1.30}$$

$$-\tfrac{1}{2}\mathscr{G}_2(\psi(\widetilde{z\psi})) = F. \tag{3.1.31}$$

If u' is related to u by

$$u' = \exp(\mathrm{i}\theta)u \tag{3.1.32}$$

then, from (3.1.28), we see that u' determines the same null direction as u. If u' determines the 2-form F' then

$$F' = \mathrm{Re}(\cos 2\theta\, \mathrm{i}u\tilde{u} - \sin 2\theta\, u\tilde{u}).$$

and since $u = \mathrm{i}zu$

$$\begin{aligned} F' &= \mathrm{Re}(\cos 2\theta\, \mathrm{i}u\tilde{u} - \sin 2\theta\, z\, \mathrm{i}u\tilde{u}) \\ &= \cos 2\theta\, F - \sin 2\theta\, zF \\ &= \exp(-2\theta z)F. \end{aligned}$$

Since $zF = - * F$ we see that the 2-form determined by u' is related to that determined by u by a *duality rotation*.

We have established the relationship between a complex Lorentzian semi-spinor and the null direction x and 2-form F by using the previously established results on pure spinors. This correspondence between Weyl spinors and 'null flags' has been emphasised by Penrose and Rindler [9].

We now consider the case of V a real even dimensional orthogonal space with the metric g positive-definite. A *complex structure* on V is a 1-1 tensor (or linear transformation) J satisfying $J^2 = -1$. This complex structure is compatible with g if

$$g(a, b) = g(Ja, Jb) \qquad \forall\, a, b \in V \tag{3.1.33}$$

that is, J is an isometry of V. We will show that any such J is in one-to-one correspondence with a maximal totally isotropic subspace of $V^{\mathbb{C}}$. Hence the one-dimensional space of pure spinors of the complexified Clifford algebra is in one-to-one correspondence with a complex structure on V†.

Suppose firstly that we have such a J. Then by complex linearity J defines a tensor on $V^{\mathbb{C}}$. Define $M \subset V^{\mathbb{C}}$ by

†We thank G Segal for pointing this out to us.

$$x \in M \qquad \text{iff } Jx = \mathrm{i}x \tag{3.1.34}$$

and $y \in M^*$ iff $y^* \in M$. Then $V^{\mathbb{C}} = M \oplus M^*$. If J satisfies (3.1.33) then $g(x^1, x^2) = g(Jx^1, Jx^2)$, and for $x^1, x^2 \in M$ we have $g(x^1, x^2) = 0$. Hence M is a maximal totally isotropic subspace of $V^{\mathbb{C}}$. Conversely now suppose that we have a maximal totally isotropic subspace M. We can define J on elements of M by (3.1.34). Requiring $Jx^* = (Jx)^*$ defines J unambiguously on the whole of $V^{\mathbb{C}}$ and, by restriction, on V. Such a J certainly satisfies $J^2 = -1$. For any $a \in V^{\mathbb{C}}$ we can write $a = a^+ + a^-$ with $a^+ \in M$ and $a^- \in M^*$. Then $g(a, b) = g(a^+, b^-) + g(a^-, b^+)$ and it follows that if $Ja^+ = \mathrm{i}a^+$ and $Ja^- = -\mathrm{i}a^-$ then J satisfies (3.1.33).

This correspondence between pure spinors and complex structures will be used in Chapter 10.

3.2 Triality

Let V be an F-linear space with an F-bilinear symmetric metric g. If S is the space of spinors of $\mathbf{C}(V, g)$ then we may define a spin-invariant product on S. In certain cases (for $F = \mathbb{R}$ or $\mathbb{C}$) there is an F-bilinear symmetric product on S, h say. We can then ask 'when is $\mathbf{C}(V, g) \simeq \mathbf{C}(S, h)$?'. These algebras will be isomorphic when $\dim_F V = \dim_F S$ and the index of g is the same as that of h. If $S = S^+ \oplus S^-$, with S^+ and S^- semi-spinor spaces carrying inequivalent irreducible representations of $\mathbf{C}^+(V, g)$, with h inducing a product on the semi-spinor spaces, then we can also ask the question 'when is $\mathbf{C}(V, g) \simeq \mathbf{C}(S^+, h) \simeq \mathbf{C}(S^-, h)$?'. Again this will be when $\dim_F V = \dim_F S^\pm$ with the index of g the same as that of the metric induced by h on $S^\pm$. We now examine the possibility of this latter situation occurring. If $\dim_F V = n$ then n must be even if $\mathbf{C}^+(V, g)$ is to be reducible with S splitting into semi-spinor spaces. Then $\dim_F S = 2^{n/2}$ and for $\dim_F S^\pm$ to be equal to $\dim_F V$ we need $2^{n/2} = 2n$, which requires $n = 8$. If $F = \mathbb{C}$ then we see from table 2.17 that the situation we are looking for does occur in eight dimensions, with h being the spin-invariant spinor metric associated with the involution ξ. For $F = \mathbb{R}$ the situation depends on the signature of g. For given p and q the third entry in table 2.15 classifies the spin-invariant product associated with ξ, h say, on the irreducible representation spaces of the even subalgebra. If this entry is $1 \oplus 1$ then the even subalgebra has two semi-spinor representations, with an $\mathbb{R}$-bilinear symmetric product on each. Such entries occur for $(p, q) = (8, 0)$, $(0, 8)$ or $(4, 4)$. From (2.6.23) we see that in all these cases the index of h is the same as that of g. Actually a little care is needed in reaching this conclusion for the case of $C_{4,4}(\mathbb{R})$. We know that h on S has maximal index, but we could have h inducing a positive-definite product on S^+ and a negative-definite one on S^-. However, if $x \in V$ with $x^2 = 1$, then for $v \in S^-$ there is a $u \in S^+$ such that $v = xu$. Then $h(v, v) = h(xu, xu) = h(u, x^2u)$, since the adjoint involution of h is ξ, and the indices of the metrics induced by h on S^+ and S^- are the same. In the following V will either be a complex eight-dimensional vector space or a real eight-dimensional vector space with g having signature (8, 0), (0, 8) or (4, 4).

By taking the direct sum of the vector spaces V and S we form a 24-dimensional vector space E:

$$E = V \oplus S^+ \oplus S^-. \tag{3.2.1}$$

If elements Φ_i of E are decomposed into these subspaces as $\Phi_i = x_i + u_i + v_i$ then a bilinear form B is defined on E by

$$B(\Phi_1, \Phi_2) = g(x_1, x_2) + h(u_1, u_2) + h(v_1, v_2)\,. \tag{3.2.2}$$

(We shall frequently decompose an element Φ as above, the symbols x, u and v being reserved for the components of Φ in the subspaces V, S^+ and S^-.)

We can introduce a totally symmetric (3, 0) tensor T on E in terms of the inner product h. We define

$$\begin{aligned} T(\Phi_1, \Phi_2, \Phi_3) \equiv\ & h(u_1, x_2v_3) + h(u_1, x_3v_2) + h(u_2, x_1v_3) \\ & + h(u_2, x_3v_1) + h(u_3, x_1v_2) + h(u_3, x_2v_1). \end{aligned} \tag{3.2.3}$$

Each term on the right-hand side is linear in each Φ_i, thus T is indeed multilinear. By construction T is totally symmetric. We can use the bilinear B and trilinear T to define a bilinear map $\circ$:

$$\circ : E \times E \to E \qquad \text{such that } T(\Phi_1, \Phi_2, \Phi_3) = B(\Phi_1 \circ \Phi_2, \Phi_3)\,. \tag{3.2.4}$$

The non-degeneracy of B ensures that $\circ$ is indeed well defined. Its bilinearity follows from the trilinearity of T. Since T is totally symmetric $\Phi_1 \circ \Phi_2 = \Phi_2 \circ \Phi_1$. If Φ_1 and Φ_2 are both in the same subspace, either V, S^+ or S^-, then $T(\Phi_1, \Phi_2, \Phi_3) = 0$ from (3.2.3) and hence $\Phi_1 \circ \Phi_2 = 0$. For $x \in V$, $u \in S^+$ and $v \in S^-$ we have

$$B(x \circ u, v) = T(x, u, v) = h(u, xv) = h(xu, v) = B(xu, v)$$

and so

$$x \circ u = xu \tag{3.2.5}$$

similarly

$$x \circ v = xv. \tag{3.2.6}$$

If $\tilde{u}$ is the adjoint of u with respect to h then

$$\begin{aligned} B(u \circ v, x) &= T(u, v, x) = h(xu, v) = \widetilde{xu}v \\ &= \mathcal{S}_0(\tilde{u}xv) = \mathcal{S}_0(xv\tilde{u}) = \mathcal{S}_0(x\mathcal{S}_1(v\tilde{u})) = B(x, \mathcal{S}_1(v\tilde{u})) \end{aligned}$$

so

$$u \circ v = \mathscr{S}_1(v\tilde{u}). \tag{3.2.7}$$

The product $\circ$ is not associative, for example we have

$$x \circ (x \circ u) = x \circ xu = x^2u = g(x, x)u \tag{3.2.8}$$

whereas $x \circ x = 0$. The norm of the spinor $x \circ u$ is related to the norms of x and u by

$$h(x_1 \circ u, x_2 \circ u) = g(x_1, x_2)h(u, u). \tag{3.2.9}$$

This follows from (3.2.5), (3.2.6) and the fact that the adjoint of h is ξ. The 24-dimensional vector space E forms a non-associative algebra $\mathscr{A}$ under the $\circ$ product.

The spinor representation of the Clifford group, ρ on $S^{\pm}$, and the vector representation χ on V naturally induce a reducible representation Υ on E by

$$\Upsilon(s).(x + u + v) \equiv \chi(s).x + \rho(s).u + \rho(s).v. \tag{3.2.10}$$

Whereas g in invariant under $\chi(s)$ $\forall\, s \in \Gamma$, h is only invariant under $\rho(s)$ for $s \in {}_+\Gamma$ and so

$$B(\Phi_1, \Phi_2,) = B(\Upsilon(s).\Phi_1, \Upsilon(s).\Phi_2) \qquad \forall\, s \in {}_+\Gamma. \tag{3.2.11}$$

It readily follows that in addition

$$T(\Phi_1, \Phi_2, \Phi_3) = T(\Upsilon(s).\Phi_1, \Upsilon(s).\Phi_2, \Upsilon(s).\Phi_3) \qquad \Upsilon s \in {}_+\Gamma. \tag{3.2.12}$$

From these last two relations we can infer from (3.2.4) that

$$\Upsilon(s).(\Phi_1 \circ \Phi_2) = (\Upsilon(s).\Phi_1) \circ (\Upsilon(s).\Phi_2) \qquad \forall\, s \in {}_+\Gamma \tag{3.2.13}$$

that is, $\Upsilon(s)$ is in the automorphism group of the non-associative algebra $\mathscr{A}$. Conversely it follows that if σ is any automorphism of $\mathscr{A}$ that transforms V and S into themselves then $\sigma = \Upsilon(s)$ for some $s \in {}_+\Gamma$. (The starting point of the argument is that for $\psi \in S$ then $\sigma.\psi = s\psi$ for some regular element s of the Clifford algebra.)

The orthogonal space V under consideration has been carefully selected to ensure that V, S^+ and S^- are all isometric. The existence of an isometry that cyclicly permutes these three orthogonal spaces can be taken as being Cartan's 'principle of triality'. Such an isometric map will be constructed out of a mapping that interchanges two of these three spaces. Let $u_0 \in S^+$ be some unit-norm semi-spinor, $h(u_0, u_0) = 1$. Then a linear transformation $\tau(u_0)$ from V to S^- is defined by

$$\tau(u_0).x = x \circ u_0. \tag{3.2.14}$$

It immediately follows from (3.2.9) that $\tau(u_0)$ is in fact an orthogonal transformation from V to S^-. The linear transformation $\tau(u_0)$ is uniquely extended to an automorphism of period two on $V \oplus S^-$: that is, if $v \in S^-$ such that $v = \tau(u_0).x$ for some unique x then we define $\tau(u_0).v = x$. Finally we define $\tau(u_0)$ on S^+ by

$$\tau(u_0).u = 2h(u, u_0)u_0 - u. \qquad (3.2.15)$$

That is, $\tau(u_0)$ acts on S^+ by sending u to minus its reflection in the plane orthogonal to u_0. Thus $\tau(u_0)$ is an orthogonal transformation of S^+, and hence of E. In addition, $\tau(u_0)$ leaves T invariant. Note first that since the image under $\tau(u_0)$ of any of the three subspaces, V, S^+ or S^-, lies in only one subspace we need only consider T acting on elements lying in distinct subspaces. If $v = \tau(u_0).a$ for some a then

$$\begin{aligned}\tau(u_0).v\tau(u_0).x = axu_0 &= 2g(a, x)u_0 - xau_0 \\ &= 2h(\tau(u_0).a, \tau(u_0).x)u_0 - xv\end{aligned}$$

since $\tau(u_0)$ is an isometry from V to S^- and so

$$\tau(u_0).v\tau(u_0).x = 2h(v, xu_0)u_0 - xv = 2h(xv, u_0)u_0 - xv = (\tau(u_0).x)v.$$

Since

$$T(\tau(u_0).\Phi_1, \tau(u_0).\Phi_2, \tau(u_0).\Phi_3) = T(\tau(u_0).u_1, \tau(u_0).v_2, \tau(u_0).x_3) + \ldots$$

it follows from (3.2.3) that

$$T(\tau(u_0).\Phi_1, \tau(u_0).\Phi_2, \tau(u_0).\Phi_3) = T(\Phi_1, \Phi_2, \Phi_3). \qquad (3.2.16)$$

Whereas $\tau(u_0)$ is an orthogonal transformation of E that interchanges V and S^-, $\Upsilon(s)$ is an orthogonal transformation of E that interchanges S^+ and S^-. If $x_0 \in V$ is a unit vector, $g(x_0, x_0) = 1$, then $x_0 \in {}_+\Gamma$ and $\Upsilon(x_0)$ is of period two, $\Upsilon(x_0)^2 = 1$. Out of these two involutory transformations of E we construct an orthogonal transformation of period three. The *triality map* $\Xi(x_0, u_0)$ is defined by

$$\Xi(x_0, u_0) \equiv \Upsilon(x_0)\tau(u_0). \qquad (3.2.17)$$

To see that $\Xi(x_0, u_0)$ is of period three we want to show that

$$\tau(u_0)\Upsilon(x_0)\tau(u_0) = \Upsilon(x_0)\tau(u_0)\Upsilon(x_0). \qquad (3.2.18)$$

For example, if $x \in V$ then

$$\begin{aligned}\tau(u_0)\Upsilon(x_0)\tau(u_0).x &= \tau(u_0)\Upsilon(x_0).(xu_0) = \tau(u_0).(x_0xu_0) \\ &= 2h(x_0xu_0, u_0)u_0 - x_0xu_0 \\ &= 2h(x \circ u_0, x_0 \circ u_0)u_0 - x_0xu_0 \\ &= 2g(x, x_0)u_0 - x_0xu_0 \qquad \text{(by (3.2.9))} \\ &= xx_0u_0.\end{aligned}$$

On the other hand

$$\Upsilon(x_0)\tau(u_0)\Upsilon(x_0).x = \Upsilon(x_0)\tau(u_0).(x_0xx_0) = \Upsilon(x_0).(x_0xx_0u_0) = xx_0u_0.$$

The validity of (3.2.18) can be similarly demonstrated on elements from the other two subspaces. Given (3.2.18) we have

$$\Xi(x_0, u_0)^3 = (\Upsilon(x_0)\tau(u_0)\Upsilon(x_0))(\tau(u_0)\Upsilon(x_0)\tau(u_0)) = (\Upsilon(x_0)\tau(u_0)\Upsilon(x_0))^2$$

and since both $\Upsilon(x_0)$ and $\tau(u_0)$ are of period two

$$\Xi(x_0, u_0)^3 = 1. \tag{3.2.19}$$

Because $\Upsilon(x_0)$ and $\tau(u_0)$ both have these properties separately we have

$$B(\Phi_1, \Phi_2) = B(\Xi(x_0, u_0).\Phi_1, \Xi(x_0, u_0).\Phi_2) \tag{3.2.20}$$

and

$$T(\Phi_1, \Phi_2, \Phi_3) = T(\Xi(x_0, u_0).\Phi_1, \Xi(x_0, u_0).\Phi_2, \Xi(x_0, u_0).\Phi_3). \tag{3.2.21}$$

The three subspaces of E are permuted under $\Xi(x_0, u_0)$ as follows:

$$\Xi(x_0, u_0).V \subset S^+ \qquad \Xi(x_0, u_0).S^+ \subset S^- \qquad \Xi(x_0, u_0).S^- \subset V. \tag{3.2.22}$$

We have focused on a V such that $\mathbf{C}(V, g) \simeq \mathbf{C}(S^+, h) \simeq \mathbf{C}(S^-, h)$. The map $\Xi(x_0, u_0)$ isometrically permutes these three spaces. Any isometry between two orthogonal spaces uniquely extends to an isomorphism between their Clifford algebras. Let N be the isomorphism obtained from $\Xi(x_0, u_0)$:

$$N: \mathbf{C}(V, g) \longmapsto \mathbf{C}(S^+, h) \longmapsto \mathbf{C}(S^-, h) \longmapsto \mathbf{C}(V, g). \tag{3.2.23}$$

Because $S^+ \oplus S^-$ is the spinor space of $\mathbf{C}(V, g)$ the map N enables any two of the three spaces V, S^+ and S^- to be taken as the spinor space of the Clifford algebra of the third! For example, $S^- \oplus V$ can be taken as the spinor space of $\mathbf{C}(S^+, h)$. Let $\diamond$ denote the Clifford product of $\mathbf{C}(S^+, h)$. Then for $x \in V$ and $\psi \in S$

$$N(x\psi) = N(x) \diamond N(\psi).$$

That is, if $u \in S^+$ and $\psi' \in S' \equiv S^- \oplus V$ then

$$u \diamond \psi' = N((N^{-1}u)(N^{-1}\psi')). \tag{3.2.24}$$

Under this multiplication by u the spaces S^- and V are interchanged; these being the semi-spinor spaces of $\mathbf{C}^+(S^+, h)$.

Exercise 3.1

Show that if V is a complex vector space then $\mathbf{C}(V, g) \simeq \mathbf{C}(S, h)$ if $\dim_{\mathbf{C}} V = 2, 4$. In the real case what signatures can g have? (Remember that the spinor inner product could be associated with either ξ or $\xi\eta$.)

Bibliography

Chevalley C 1954 *The Algebraic Theory of Spinors* (New York: Columbia University Press)

4

Manifolds

Like many concepts in mathematics that of a manifold is based on intuitive ideas which require some sophistication to make precise. Perhaps the simplest example of a manifold is Euclidean three-space. Of necessity at this stage we must refrain from defining Euclidean space, but shall nevertheless assume that the reader has some intuitive ideas about this model description of our perceived three-dimensional world. (The term Euclidean space is not synonymous with Euclidean vector space. A Euclidean vector space is a real vector space with a positive-definite symmetric metric.) At an early age we all learnt how a Cartesian coordinate system can be introduced to put points in Euclidean space into correspondence with an ordered triple of real numbers, an element of $\mathbb{R}^3$. However, it is important that we distinguish Euclidean three space from $\mathbb{R}^3$. Euclidean space has no preferred coordinate system. Indeed we need not of course even be restricted to Cartesian coordinates. Despite our emphasis on the distinction between Euclidean three-space and $\mathbb{R}^3$ it is nonetheless in $\mathbb{R}^3$ that the familiar calculus of differentiation and integration is introduced. Through the introduction of a coordinate system one may then apply this calculus to Euclidean space. It is the correspondence of Euclidean space to $\mathbb{R}^n$, through the introduction of a coordinate system, that generalises to provide the definition of a manifold. This is defined, in a sense that will be made precise, to be locally like $\mathbb{R}^n$. Because we can define differentiation and integration on $\mathbb{R}^n$ we can extend these notions to a manifold.

Unlike Euclidean space, for an arbitrary manifold we cannot choose some origin to put all points on the manifold into a unique correspondence with points in $\mathbb{R}^n$. For example, we could take the two-dimensional outer surface of a hollow rubber ball. Whilst any cap of the ball could be put into one-to-one correspondence with points in a plane (by cutting the section out and flattening it), we cannot do this with the whole surface. (If we simply squashed the ball then two points on the surface would be mapped to the same point on the plane.) The fact that the surface is locally like $\mathbb{R}^2$ is sufficient to establish a differential calculus on the surface. This does not require a knowledge of embedding in three-space.

The intuitive examples of the Euclidean plane and the two-sphere convey ideas of more structure than that of an arbitrary manifold. Although locally any manifold resembles, in some sense, $\mathbb{R}^n$ this does not imply the existence of any metric or distance function on the manifold. Rather the resemblance relates to topology, this being an abstraction of the concept of 'nearness' from that given by distance.

We start by defining a topological space. By making precise the idea of being 'locally like $\mathbb{R}^n$' we arrive at the definition of a topological manifold. After reviewing differentiation on $\mathbb{R}^n$ we show how a system of coordinates on a topological manifold enables differentiation to be defined, giving a differentiable manifold. From its introduction in $\mathbb{R}^n$ the concept of a tangent vector will undergo a metamorphosis, the imago emerging in a form appropriate to the environment of an arbitrary differentiable manifold. This leads naturally to vector fields, and hence tensor fields. After introducing the computationally powerful exterior and Lie derivatives we define integration on manifolds. Similar to the case of differentiation, the definition reduces integration on manifolds to integration on $\mathbb{R}^n$. Only at the end of the chapter do we consider metric tensor fields. We are then equipped to apply our heavy artillery to the example of Euclidean three-space. This is done in Appendix B. Actually there is still an important facet of Euclidean space that will not be discussed until the following chapter, that of parallelism.

4.1 Topological Manifolds

The usual definition of continuity of a function $f: U \to W$ where U and W are subsets of $\mathbb{R}$ relies on the notion of 'nearness' of different elements of $\mathbb{R}$. Such 'nearness' is measured by a proximity function $\mathbf{d}: \mathbb{R} \times \mathbb{R} \to \mathbb{R}$ with the properties: $\mathbf{d}(x, y) = \mathbf{d}(y, x)$, $\mathbf{d}(x, y) = 0$ if and only if $x = y$, $\mathbf{d}(x, z) \leqslant \mathbf{d}(x, y) + \mathbf{d}(y, z)$. (Note $x, y \in \mathbb{R}$.) A natural proximity function for the real line that has these properties is the absolute value or modulus map, $(x, y) \to |x - y|$ and f is said to be continuous at $x \in \mathbb{R}$ if one can find a positive $\delta \in \mathbb{R}$ for any positive ε belonging to $\mathbb{R}$ such that if $\mathbf{d}(x, y) < \delta$ then $\mathbf{d}(f(x), f(y)) < \varepsilon$. Thus one probes the neighbourhood of the image of f induced by a neighbourhood about x in the domain of f. The first generalisation of this idea to arbitrary sets consists of defining a new set called the *neighbourhood*

$\text{nbh}(x, \delta) \subset S$ if $x \in S$. This is the set of elements $y \in S$ such that $\mathbf{d}(x, y) < \delta$, that is a set of all points that are within a 'distance' δ from x as measured by some proximity function $\mathbf{d}$. One often refers to $\mathbf{d}$ as a distance or metric function, although since we do not assume here that the set has any vector space structure it is logically distinct from the metric g defined earlier on vector spaces. Indeed what we have called a metric on a vector space would not in general define a distance function for a metric space. Here there is no requirement that $\mathbf{d}$ should be linear in either of its arguments. With this caveat in mind one refers to the pair $(S, \mathbf{d})$ as a metric space. The defining properties of the proximity function $\mathbf{d}$ of course remind one of the properties of distances between points in Euclidean space (for example, the triangle inequality) and indeed it is worth noting that if $\mathbb{R}^n$ is given a vector space structure one can choose $\mathbf{d}(x, y) = [g(x-y, x-y)]^{1/2}$ provided g is the positive-definite Euclidean metric. If one does use the Euclidean metric to define $\mathbf{d}$ then the set $\text{nbh}(x, \delta)$ in Euclidean $\mathbb{R}^n$ looks like an open ball (open because of the inequality $\mathbf{d}(x, y) < \delta$, $\forall y \in \text{nbh}(x, \delta)$. The triangle inequality property of $\mathbf{d}$ ensures that all points $y \in \text{nbh}(x, \delta)$ have some neighbourhoods that are contained in $\text{nbh}(x, \delta)$. In general the proximity function on $\mathbb{R}^n$ need not coincide with the metric on $\mathbb{R}^n$ regarded as a vector space.

A boundary element x of a set S' contained in the set S with distance function $\mathbf{d}$ is an element such that $\text{nbh}(x, \delta)$, for some positive $\delta \in \mathbb{R}$, contains both elements in S' and elements not in S'. The set of all boundary points of S' is called the *boundary* of S'. In particular if $S' = \text{nbh}(x, \delta) \subset S$ then S' does not contain its boundary and is called an *open set* in $(S, \mathbf{d})$. If any boundary points are not in the set then it is an open set. If all boundary points are in the set it is *closed*.

In general it is possible to find different distance functions that determine the same class of continuous functions. A valuable generalisation then is to concentrate on the open sets themselves as the primitive notions and reformulate 'nearness' directly in terms of them rather than in terms of any particular proximity function. The immediate usefulness of open sets is a reformulation of the definition of a continuous function $f: U \to W$. f is *continuous* at $p \in U$ if and only if, for any neighbourhood W' containing $f(p)$ there is a neighbourhood U' containing p whose image $f(U') \subseteq W'$. Such a notion of continuity relies on the open set structure of the spaces related by f and not on a particular choice of proximity function used in specifying these open sets. Consequently one attempts to bypass any mention of a proximity function and establish a more general definition of open sets on any space. The declaration of which subsets of a space are to be considered as open is called a definition of its *topology* provided such a family of subsets satisfy the following axioms.

(i) The whole space and the empty set belong to the family.

(ii) The intersection of any finite number from the family belong to the family.

(iii) The union of any number of sets from the family belong to the family.

With these definitions we now refer to any open set containing a point p in a topological space as a neighbourhood $\mathrm{Nbh}(p)$ and the definition of continuity of a function between topological spaces is now independent of any choice of proximity function; it has been replaced by the choice of open sets. The definition of boundary points of a set and the boundary generalises simply to arbitrary topologies by replacing $\mathrm{nbh}(p, \delta)$ by $\mathrm{Nbh}(p)$. A space with a topology defined on it is called a *topological space*. If a map between topological spaces is continuous with a continuous inverse then it is called a *homeomorphism*.

One further property defines the topology as being *Hausdorff*:

(iv) Disjoint neighbourhoods can be defined about distinct elements of the space.

That is, one may find open sets whose intersection is the empty set.

If a space has a proximity function $\mathbf{d}$ then we may if we wish define $\mathrm{Nbh}(p) = \mathrm{nbh}(p, \delta)$ and the space is said to have a *metric topology* (which is always Hausdorff). One of the commonest metric topologies is associated with $\mathbb{R}^n$ and $\mathbf{d}(x, y) = |x - y|$, $x, y \in \mathbb{R}^n$. With the above $\mathbf{d}(x, y)$ on $\mathbb{R}^n$ the open sets may be visualised as all possible open hypercubes in $\mathbb{R}^n$.

It is useful to have such examples of a natural metric topology in $\mathbb{R}^n$ since they can be used to induce topologies on subsets of $\mathbb{R}^n$. The induced topology on a subset $\tilde{S}$ of a topological space S is the collection of all sets formed by the intersection of $\tilde{S}$ with all open sets of S. These are then declared to be open in $\tilde{S}$ (they need not be open in S) and $\tilde{S}$ is called a *topological subspace* of S. Subsets of Euclidean $\mathbb{R}^3$ provide some of the simplest visualisable models of topological spaces. Thus the sphere S^2 is the subset of $\mathbb{R}^3$ defined by $|x| = 1$, $x \in \mathbb{R}^3$ with a topology induced from the metric topology of $\mathbb{R}^3$. It is topologically equivalent (homeomorphic) to the ellipsoid ($a^2x^2 + b^2y^2 + c^2z^2 = 1$, $a, b, c \in \mathbb{R}$) with the topology induced from that of $\mathbb{R}^3$; that is one can establish a homeomorphism between them. Neither is homeomorphic to the 2-torus, $S^1 \times S^1$. However all these examples (and indeed any two-surface) have points with neighbourhoods homeomorphic to the open disc $\{x \mid |x| < 1, x \in \mathbb{R}^2\}$. Such spaces are said to be locally homeomorphic. The fact that they need not be homeomorphic is sometimes phrased by saying that they have different global topologies.

If one exploits the vector space structure of $\mathbb{R}^3$ one can project any sufficiently small region of a two-surface onto a suitable two-plane in $\mathbb{R}^3$

to obtain a neighbourhood in $\mathbb{R}^2$ and a bijective map with a continuous inverse. This suggests the definition of an n-dimensional topological manifold. An n-dimensional *topological manifold* is a Hausdorf topological space, with a countable basis for its topology, that is locally homeomorphic to an open set of $\mathbb{R}^n$. A collection of open sets is a basis for a topology if every neighbourhood can be expressed as the union of members in the basis.

The elements of a topological manifold are often referred to as points. It is clear from the examples above that one cannot in general find a homeomorphism from the whole topological space to an open set of $\mathbb{R}^n$. The above definition of a topological manifold is sufficiently general that not all topological two-manifolds are subsets of $\mathbb{R}^3$.

4.2 Derivatives of Functions $\mathbb{R}^m \to \mathbb{R}^n$

Our discussion of continuity culminated in the definition of a topological manifold as being locally homeomorphic to $\mathbb{R}^n$. This local correspondence with $\mathbb{R}^n$ can be used to establish a criterion for differentiability of maps on manifolds. We first briefly review the differentiation of vector-valued functions on $\mathbb{R}^m$.

If f is a function from $\mathbb{R}^m$ to $\mathbb{R}^n$ then the derivative of f at $p \in \mathbb{R}^m$ in the direction of $V \in \mathbb{R}^m$ is given by

$$\mathrm{D}_V f(p) = \lim_{h \to 0} \left(\frac{f(p + hV) - f(p)}{h} \right) \tag{4.2.1}$$

where $h \in \mathbb{R}$. (Other commonly used notations for $\mathrm{D}_V f(p)$ are $\mathrm{d}f(p)V$, $\mathrm{d}f_p(V)$ and $f'_p V$.) Whereas the discussion of the continuity of f only involved the topology of $\mathbb{R}^m$ and $\mathbb{R}^n$, the right-hand side of this equation manifestly uses the vector space structure of these spaces. If all the directional derivatives of f exist at p then f is said to be differentiable at p. In this case $\mathrm{D}f(p)$ is a linear transformation from $\mathbb{R}^m$ to $\mathbb{R}^n$. $\mathrm{D}f(p) : V \mapsto \mathrm{D}_V f(p)$, determining the linear part of an approximation to f in the vicinity of p. The function f sends the point p to $f(p)$: if the point p starts to move in the direction of V then $f(p)$ will correspondingly start to move in the direction $\mathrm{D}_V f(p)$ (refer to figure 4.1).

Intuitively we think of the derivative of f as sending an 'arrow' in $\mathbb{R}^m$, with its tail at p and tip at $p + V$, to an 'arrow' in $\mathbb{R}^n$, with $f(p)$ as tail and $f(p) + \mathrm{D}_V f(p)$ as tip. We may formalise this by defining the *tangent space* to $\mathbb{R}^m$ at p, $T_p\mathbb{R}^m$, to be the set of pairs (p, V) for all $V \in \mathbb{R}^m$. These pairs (tangent vectors) form a vector space, isomorphic to $\mathbb{R}^m$, with the rule

$$\lambda(p, V) + \mu(p, U) = (p, \lambda V + \mu U) \qquad \lambda, \mu \in \mathbb{R}. \qquad (4.2.2)$$

We may now define the derivative of f at p, or tangent map, f_{*p}:

$$f_{*p} : T_p\mathbb{R}^m \longmapsto T_{f(p)}\mathbb{R}^n$$
$$(p, V) \longmapsto (f(p), \mathrm{D}_V f(p)). \qquad (4.2.3)$$

Since $\mathrm{D}f(p)$ is a linear transformation on $\mathbb{R}^m$ it follows that f_{*p} is a linear map on $T_p\mathbb{R}^m$. The tangent space of $\mathbb{R}^m$ at p is just a subspace of the direct sum of $\mathbb{R}^m$ with itself, and so there is a natural way of adding tangent vectors lying in different tangent spaces. This feature will not carry over to the following section where we generalise to the concept of a tangent space to a manifold. Since in general the manifold itself will have no vector space structure, there will be no natural way of adding vectors from tangent spaces associated with different points on the manifold.

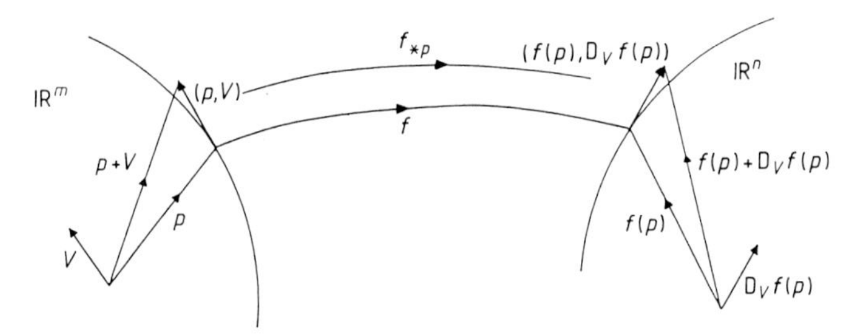

Figure 4.1 The tangent map of f:$\mathbb{R}^m \to \mathbb{R}^n$.

If $\{e_i\}$ and $\{e'_i\}$ are the natural bases for $\mathbb{R}^m$ and $\mathbb{R}^n$ then the component functions of f, $f^i : \mathbb{R}^m \mapsto \mathbb{R}$ $i = 1, \ldots, n$, are given by

$$f(p) = \sum_{i=1}^{n} f^i(p)e'_i. \qquad (4.2.4)$$

The directional derivatives of these component functions along the basis vectors for $\mathbb{R}^m$ are called the partial derivatives, and a special notation is customary:

$$\mathrm{D}_{e_j} f^i(p) \equiv (\partial f^i/\partial x^j)(p). \qquad (4.2.5)$$

For any $V \in \mathbb{R}^m$

$$\mathrm{D}_V f(p) = \sum_{i=1}^{n} \mathrm{D}_V f^i(p)e'_i = \sum_{j=1}^{m}\sum_{i=1}^{n} V^j \mathrm{D}_{e_j} f^i(p)e'_i$$

by the linearity of $\mathrm{D}f^i(p)$. Thus the matrix of the linear transformation $\mathrm{D}f(p)$ is formed by the partial derivatives. The $n \times m$ matrix $[(\partial f^i/\partial x^j)(p)]$, with i labelling the rows, is called the *Jacobian* and we have

$$\mathrm{D}_V f(p) = \sum_{j=1}^{m} \sum_{i=1}^{n} [(\partial f^i / \partial x^j)(p)] V^j e'_i. \tag{4.2.6}$$

The partial derivatives may be regarded as real functions of the point p and hence higher partial derivatives may be formed. A map between subsets of $\mathbb{R}^m$ and $\mathbb{R}^n$ for which all partial derivatives up to order k exist and are continuous is said to be a C^k map. A homeomorphism that is a C^k map with a C^k inverse is called a C^k *diffeomorphism*. We shall be primarily concerned with C^∞ maps, which will be called smooth.

Example 4.1
Let $f: \mathbb{R}^2 \mapsto \mathbb{R}^3, p \equiv (x^1, x^2) \mapsto f(p) = ((x^1)^2, x^1 x^2 + 1, x^2)$. Taking $V = (v^1, v^2)$ in (4.2.1) gives

$$\mathrm{D}_V f(p) = (2v^1 x^1, x^1 v^2 + x^2 v^1, v^2) = (v^1, v^2) \begin{pmatrix} 2x^1 & x^2 & 0 \\ 0 & x^1 & 1 \end{pmatrix}$$

where the entries in the matrix are recognised as the partial derivatives of the function f.

4.3 Differentiable Manifolds

With the notion of smooth maps between $\mathbb{R}^m$ and $\mathbb{R}^n$ established we proceed now to define a *differentiable manifold*. A topological manifold is locally homeomorphic to $\mathbb{R}^n$. By setting up a system of charts that map neighbourhoods of the manifold onto neighbourhoods of $\mathbb{R}^n$ we can use the differential structure on $\mathbb{R}^n$ to define the differential structure on topological manifolds.

In order to motivate the definition of a differentiable manifold let us first discuss the problem of coordinating a patch of a topological manifold by returning to the example of S^2 as a subset of $\mathbb{R}^3$. Suppose this subset is constructed from thin perspex and the boundary of a region is marked out by painting a closed curve on the perspex surface. Furthermore paint a fishnet of curves within and on this boundary so that distinct curves in the net intersect only once and each intersects the boundary image once also. Imagine a light is shone through this net of curves and examine the image shadow on any two-plane placed conveniently to collect the shadow. If each intersection in the net of painted curves casts a unique shadow on the two-plane then the neighbourhood chosen on the sphere yields a proper coordinate patch with respect to the projection scheme. Each intersection can be uniquely labelled by labelling all the curvilinear line shadows uniquely. If a lens of suitable material is placed between the image and perspex patch one can even arrange that the shadow lines appear orthogonal with respect to the induced Euclidean metric on the two-plane. Such a projection system establishes a homeomorphism from the open set U of S^2 containing the net onto the open set of $\mathbb{R}^2$ formed by the shadow. To each point $p \in U$ we assign two real coordinates $\varphi(p) = (\varphi^1(p), \varphi^2(p)) \in \mathbb{R}^2$. The set of images labelled $\varphi^j(p) = \text{constant}$ $(j = 1, 2)$ are sometimes called coordinate lines (or planes in general). There are many ways of establishing such an optical arrangement and equally many ways of painting lines on S^2 yielding alternative coordinate systems. Thus there is no unique way of assigning coordinate labels to points in U. We choose a projection system such that φ is a homeomorphism for then and only then will a sequence of points in the topological manifold with a limiting point (in the manifold topology) map into a sequence of coordinates with a corresponding limit.

To completely coordinate a topological manifold we shall in general need several overlapping patches, as the example of a sphere shows. We are then prompted to examine the relations between the different coordinates assigned to points in the region of overlap.

Returning to the general case of an n-dimensional topological manifold M we recall that by definition each point of M has a neighbourhood U_a homeomorphic to an open set of $\mathbb{R}^n$. If we label one such homeomorphism $\varphi_a : U_a \to \varphi_a(U_a)$ then the pair (U_a, φ_a) is called a *coordinate chart* for U_a (with the chart domain U_a). The image $\varphi_a(p)$ for $p \in U_a$ assigns to the point p the n real coordinates $(\varphi_a^1(p), \varphi_a^2(p), \ldots, \varphi_a^n(p))$. For each chart labelled by a the real-valued function $\varphi_a^j : U_a \to \mathbb{R}$, $(j = 1, \ldots, n)$ is called the jth coordinate function and is projected from φ_a by the j-projection map π^j

$$\pi^j : \mathbb{R}^n \longrightarrow \mathbb{R}, \ \varphi_a(p) \longmapsto \pi^j \circ \varphi_a(p) \equiv \varphi_a^j(p) \qquad (4.3.1)$$

for all $p \in U_a$. When we work in a prescribed chart we often drop the chart label 'a' on φ_a^j and a common notation for the set of n numbers $\{\varphi^j(p)\}$ is $\{x^j(p)\}$.

One of the most important hurdles to overcome when first working with general coordinates is to resist the instinct to infer any metric or distance properties of the manifold from the use of the symbol x^j. Whereas the coordinates $\{x^j(p)\}$ of p are elements of $\mathbb{R}^n$, regarded as a Euclidean vector space, the metric on $\mathbb{R}^n$ need not define any metric or distance function on the manifold. For example, x^1 and x^2 could be the 'usual' polar coordinates θ, φ for a neighbourhood of the two-sphere. Although the Euclidean metric is used on $(\theta(p), \varphi(p))$ to differentiate functions on the sphere this is not necessarily related to any metric on the sphere, certainly not to the standard metric.

A collection of charts (U_a, φ_a) $a = 1, 2, \ldots$ becomes an *atlas* for M

provided the union of all the U_a is M itself. Two charts (U_a, φ_a) and (U_b, φ_b) such that $U_a \cap U_b \neq \emptyset$ give rise to a homeomorphism between neighbourhoods of $\mathbb{R}^n$. If $U_a \cap U_b \equiv U_{ab}$ then we define (see figure 4.2)

$$h_{ab} \equiv \varphi_b \circ \varphi_a^{-1} : \ \varphi_a(U_{ab}) \longrightarrow \varphi_b(U_{ab}). \tag{4.3.2}$$

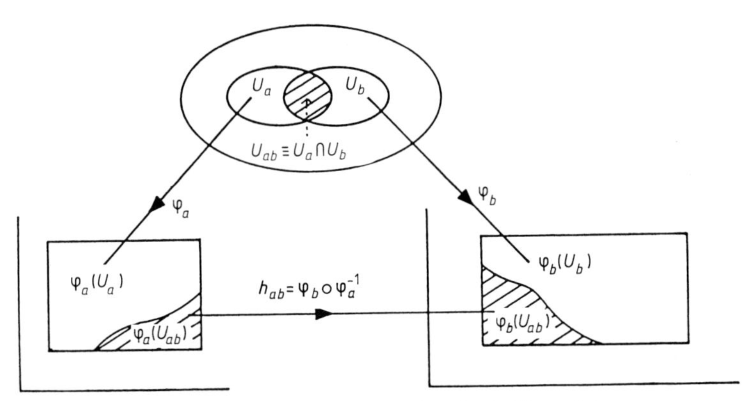

Figure 4.2 The chart maps for $U_a \cap U_b \subset M$.

Then $\varphi_b(p) = h_{ab} \circ \varphi_a(p)$ expresses the n coordinates $\varphi_b^i(p)$ of p in the 'b' chart in terms of n continuous functions h_{ab}^i of the coordinates $\varphi_a^j(p)$ of p in the 'a' chart, that is a coordinate transformation expresses the coordinates of p in one chart in terms of the coordinates of the same point in another overlapping chart. If as is often done we write $x^i \equiv \varphi_a^i(p)$ and $y^i \equiv \varphi_b^i(p)$ then $x^i = h_{ab}^i(y^1, y^2, \ldots, y^n)$ $i = 1, \ldots, n$. Similarly h_{ab}^{-1} is a homeomorphism from $\varphi_b(U_{ab})$ to $\varphi_a(U_{ab})$ and gives the inverse mapping between the coordinates. The maps $[h_{ab}]$ between all overlapping members of the atlas are called the *chart transformations*. If all these maps are differentiable the atlas is said to be differentiable. It is this new property that turns a topological manifold into a differentiable one. Since h_{aa} is the identity map and $h_{bc} \circ h_{ab} = h_{ac}$ then $h_{ab}^{-1} = h_{ba}$ and so the inverse chart transformations are differentiable; hence they are diffeomorphisms on $\mathbb{R}^n$. New charts (U, φ) can be added to the atlas $[(U_a, \varphi_a)]$ provided $\varphi \circ \varphi_a^{-1}$ and $\varphi_a \circ \varphi^{-1}$ are differentiable for all a, in which case (U, φ) is *compatible* with the atlas. If every member of one atlas is compatible with every member of another atlas then the two atlases are compatible. A *differentiable structure* on a topological manifold is specified by giving a differentiable atlas from the class of all compatible differentiable atlases for M. If a topological manifold can be provided with two differentiable atlases that are incompatible then the topological manifold is said to admit two different differentiable structures. An n-dimensional C^∞

manifold (or *smooth manifold*) is defined as an n-dimensional topological manifold together with a C^∞ differentiable structure.

As an example of how the topological space $\mathbb{R}$ (the real line) can be assigned different C^∞ structures consider the atlas with single chart $(\mathbb{R}, \varphi)$ with $\varphi:\mathbb{R} \mapsto \mathbb{R}$, $x \to x$. Consider another atlas for $\mathbb{R}$ with chart $(\mathbb{R}, \beta)$ where $\beta:\mathbb{R} \mapsto \mathbb{R}$, $x \to x^3$. Then $(\varphi \circ \beta^{-1})(x) = x^{1/3}$ which is not differentiable at $x = 0$. Hence $(\mathbb{R}, \varphi)$ and $(\mathbb{R}, \beta)$ are not compatible and each atlas defines a different C^∞ structure on the same underlying topological manifold. In what follows we shall always assume that our manifolds have been given a particular differentiable structure.

If the manifold admits a covering by charts such that each h_{ab} is orientation preserving (that is the determinant of the Jacobian of the map $(h_{ab})_*$ is everywhere of the same sign for all a, b) then the manifold is said to admit an *orientation*. Every *oriented differential manifold* admits two orientations corresponding to the two signs of the Jacobian determinant. The ribbon with one twist (Möbius band) is an example of a two-dimensional differential manifold that is non-orientable. If it is regarded as being a subset of Euclidean three-dimensional space one notices that it is not possible to assign unambiguously a smooth field of everywhere normal unit vectors to such a surface.

Having used the differentiability of functions on $\mathbb{R}^n$ to establish the notion of a smooth manifold we can now similarly define differentiable maps between smooth manifolds. A map f from a smooth manifold M_1 to a smooth manifold M_2 is said to be *differentiable* at $p \in M_1$ if, for some charts (U_1, φ_1) for M_1 and (U_2, φ_2) for M_2, the map $\varphi_2 \circ f \circ \varphi_1^{-1}$ is differentiable at $\varphi_1(p)$. Since a change of chart is a differentiable operation the differentiability of f does not depend on the chart used to represent it. A homeomorphism between smooth manifolds is a *diffeomorphism* if both it and its inverse are differentiable. A map f such that

$$p_2 = f(p_1) \qquad p_2 \in M_2,\ p_1 \in M_1$$

may be represented in local coordinates by writing

$$\varphi_2(p_2) = \varphi_2 \circ f(p_1) = \varphi_2 \circ f \circ \varphi_1^{-1} \circ \varphi_1(p_1) = f_{21} \circ \varphi_1(p_1)$$

where

$$f_{21} \equiv \varphi_2 \circ f \circ \varphi_1^{-1}.$$

If we write $x^i(p_2) \equiv \varphi_2^i(p_2) = \pi^i(\varphi_2(p_2))$ $i = 1, \ldots, n$, for the coordinates of p_2 in (U_2, φ_2) and $y^j(p_1) \equiv \varphi_1^j(p_1) = \pi^j(\varphi_1(p_1))$ $j = 1, \ldots, m$, for the coordinates of p_1 in (U_1, φ_1), then

$$x^i(p_2) = f_{21}^i(y^1(p_1), y^2(p_1), \ldots, y^m(p_1)). \qquad (4.3.3)$$

If we take M_2 to be $\mathbb{R}$ and write $M_1 = M$ then f is usually called simply a function on M. If f is defined on an open set W of M $f: W \to \mathbb{R}$, then in a local chart (U, φ) it defines a function

$$f_\varphi : \varphi(U \cap W) \to \mathbb{R} \tag{4.3.4}$$

by the rule $f_\varphi = f \circ \varphi^{-1}$, that is

$$\begin{aligned} f(p) &= (f_\varphi \circ \varphi)(p) \\ &= f_\varphi(\varphi^1(p), \varphi^2(p), \ldots, \varphi^n(p)) \qquad \forall p \in W. \end{aligned}$$

We define φ^* by the rule

$$(\varphi^* f_\varphi) = f_\varphi \circ \varphi. \tag{4.3.5}$$

Writing $f = f_\varphi \circ \varphi = \varphi^* f_\varphi$, the map f_φ is said to be *pulled back* from $\varphi(U \cap W)$ to $U \cap W$.

This notion generalises to any diffeomorphism ψ between the manifolds M and N. For $f: N \to \mathbb{R}$ we define

$$\psi^* f: M \to \mathbb{R} \qquad p \mapsto (\psi^* f)(p) = f(\psi(p)) \tag{4.3.6}$$

and say that the real-valued function f on N has been pulled back to the real-valued function $\psi^* f$ on M (see figure 4.3). It follows immediately that under a composition of diffeomorphisms:

$$(\varphi \circ \psi)^* = \psi^* \circ \varphi^*. \tag{4.3.7}$$

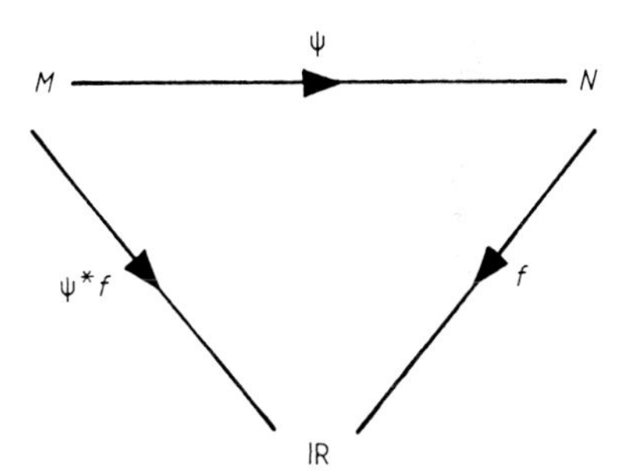

Figure 4.3 The pull-back map.

Suppose f is a smooth map from a manifold M to a manifold N. If $\dim(f_*(T_pM)) = r$ then f is said to have rank r at $p \in M$. The tangent map f_{*p} is said to be *injective* at p if $r = \dim M$ ($\dim M \leqslant \dim N$). If $r = \dim N$ then f_{*p} is said to be *surjective*. The mapping f for which f_{*p} is injective for all $p \in M$ is called an *immersion* and M is an *immersed submanifold* of N. When the immersion f is injective it is referred to as an *imbedding* and M is an (*imbedded*) *submanifold* of N. Unless specified otherwise by *submanifold* we shall mean an imbedded submanifold. In this case coordinate systems for N exist around $f(p)$ endowing

$f(M)$ with a smooth manifold structure.

As an example consider the map $f: S^1 \to \mathbb{R}^2$ where the image point traverses the figure 0 once without stopping. Then f is an injective immersion since both f and f_* are injective, and $f(S^1)$ is a one-dimensional imbedded submanifold of $\mathbb{R}^2$. If the map uniformly traverses the image set more than once it becomes an immersion, with f no longer injective. Similarly if the image $f(S^1)$ is the figure 8 traversed uniformly once the map is an immersion, since although again f_* is injective f is not. The map $f: [-1, 1] \to \mathbb{R}$, $x \mapsto x^3$, is neither an immersion nor an imbedding since although f is injective the map f_* fails to be injective at $x = 0$.

4.4 Parametrised Curves

Having defined real-valued functions on a manifold we now examine the generalisation of the directional derivative. We cannot simply apply the definition (4.2.1) since there is no vector space structure to enable points on a manifold to be added. By suitably defining curves on a manifold we can define differentiation of functions in the direction of a curve. Just as differentiation of maps between manifolds is defined by using the chart maps the derivative of a function along a curve will be defined by using a parametrisation of the curve; the derivative being defined for a real function of a real variable.

A *parametrised curve* C on a manifold M is a map from an open interval $I \subset \mathbb{R}$ to M. If p is any point on the image of C and (U, φ) is a chart for the neighbourhood of p then C may be specified in this neighbourhood by n real-valued functions

$$\pi^i \varphi[C(t)] \equiv \varphi^i \circ C(t) \qquad t \in I. \tag{4.4.1}$$

Thus denoting $\varphi^i \circ C$ by C^i we write in a local chart the representation of C

$$x^i(p) = C^i(t). \tag{4.4.2}$$

Different parametrised curves can have the same image on M. If h maps the open interval $J \subset \mathbb{R}$ into $I \subset \mathbb{R}$ then $C': J \mapsto M$ is said to be a reparametrisation of $C: I \to M$ if $C' = C \circ h$ (see figure 4.4). Whereas reparametrised curves have the same image, if we think of the parameter as a time, a change of parameter affects the rate at which that image evolves.

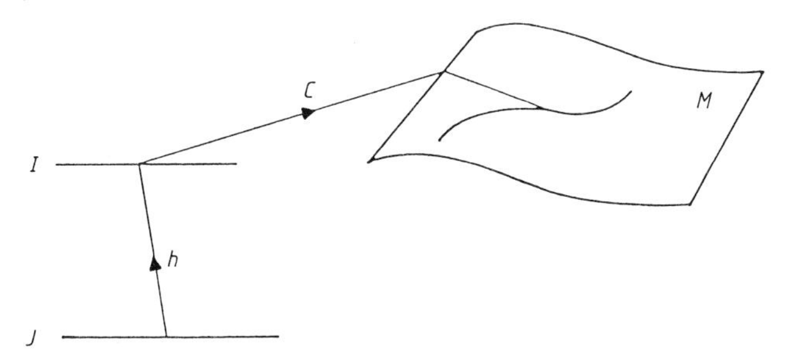

Figure 4.4 Different parametrised curves with the same image.

If f is a smooth function defined in the neighbourhood of $p_0 = C(t_0)$, with C smooth at t_0, then the derivative of f along C at p_0, $V^C_{p_0}(f)$ is defined to be

$$V^C_{p_0}(f) = \frac{\mathrm{d}}{\mathrm{d}t}(f \circ C)(t_0). \tag{4.4.3}$$

(The reason for adopting the notation $V^C_{p_0}(f)$ will be clear later.) Since $f \circ C$ is a map from I to $\mathbb{R}$, smooth at t_0, the derivative in (4.4.3) needs no further explanation. If (U, φ) is a chart for a neighbourhood of $p_0 = C(t_0)$ then the chart map φ can be used to express $V^C_{p_0}(f)$ in terms of the directional derivative of $f_\varphi = f \circ \varphi^{-1}$. We may write $f \circ C$ as the composition of maps from I to $\mathbb{R}^n$ and $\mathbb{R}^n$ to $\mathbb{R}$:

$$f \circ C = (f \circ \varphi^{-1}) \circ (\varphi \circ C).$$

The chain rule of differentiation then gives

$$\frac{\mathrm{d}}{\mathrm{d}t}(f \circ C)(t_0) = (\partial f_\varphi / \partial x^i)(\varphi(p_0)) \frac{\mathrm{d}C^i}{\mathrm{d}t}(t_0). \tag{4.4.4}$$

If $\mathscr{V}$ is the vector in $\mathbb{R}^n$ with components $\mathrm{d}C^i(t_0)/\mathrm{d}t$ then (4.4.4) expresses the derivative of f along C as the directional derivative of f_φ

$$V^C_{p_0}(f) = \mathrm{D}_{\mathscr{V}} f_\varphi(\varphi(p_0)). \tag{4.4.5}$$

Since this relation holds for all functions f we have a correspondence between the curve C, with image containing p_0, and the tangent vector to $\mathbb{R}^n$, $(\varphi(p_0), \mathscr{V})$. A curve C_1 with $C_1(\lambda_0) = p_0$ will be called equivalent to C at p_0 if $V^{C_1}_{p_0}(f) = V^C_{p_0}(f)$ for all functions f. Thus, for some choice of chart map, equivalent curves at p_0 correspond to the same tangent vector in $T_{\varphi(p_0)}\mathbb{R}^n$. By taking all curves passing through p_0 we obtain a one-to-one correspondence between equivalence classes of curves and vectors in $T_{\varphi(p_0)}\mathbb{R}^n$ (see figure 4.5).

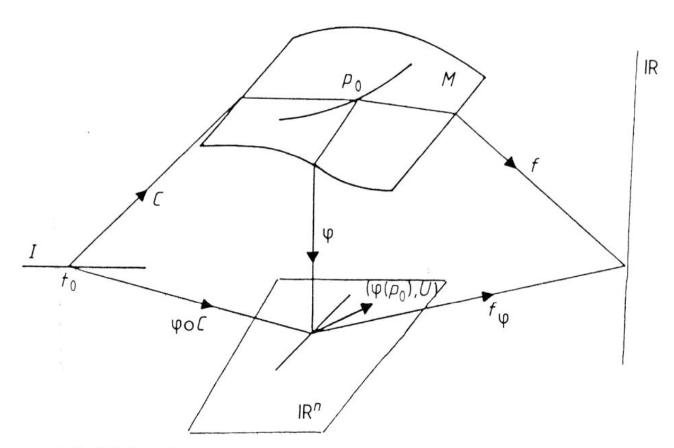

Figure 4.5 This diagram illustrates the relation between real functions on M and curves.

4.5 Tangent Vectors

In view of the previous section we could define a tangent vector to the manifold M at the point p_0 to be an equivalence class of curves passing through p_0. Such a class of curves defines a direction at the point p_0 and enables functions to be differentiated. Further, for any chart map we can put this class of curves into correspondence with a tangent vector in $\mathbb{R}^n$, this having been previously defined. It is most convenient (and usual) to adopt an equivalent definition of tangent vectors, modelled on the abstraction of differentiating along a curve. A tangent vector at p_0 will be defined to be a certain mapping from real-valued functions, defined in the neighbourhood of p_0. Such a mapping is given by any curve passing through p_0, namely the mapping to the derivative of the function along the curve. For this reason we used the notation $V_{p_0}^C(f)$ to denote the derivative of f along C at p_0: with the definition that we shall give $V_{p_0}^C$ will be identified with a tangent vector, the tangent to the curve C at p_0, whose action on f is given by (4.4.3). Similarly the definition of the tangent vector to $\mathbb{R}^n$, based on the intuitive idea of a directed line segment, is equivalent to the more abstract definition of being a derivation into $\mathbb{R}$ on functions. Given the tangent vector $(p, V) \in T_p\mathbb{R}^n$ we may take the directional derivative of the function f along V at p. In the following the reader should check that the properties we require of a tangent vector are satisfied by the derivative of a function along a curve. Later in this chapter we shall show, as is intuitively clear, that every tangent vector has a curve tangent to it.

The notion of a tangent vector at a point p on a manifold is a local one. Therefore it is convenient to classify together all maps in the neighbourhood of some point with similar properties. So we take the set of differentiable maps defined on some neighbourhood of $p \in M$ and say that two maps in this set are equivalent if their restrictions to a common neighbourhood agree. Maps satisfying this property belong to an equivalence class which is denoted $[f_{M_p}]$ and is called a (differentiable) *germ* of a map from M to N at p. The collection of all such equivalence classes is called the collection of germs of C^∞ maps at p. Clearly elements in $[f_{M_p}]$ yield the same image for p.

For example consider the germs of C^∞ maps $C : \mathbb{R} \to N$ at t. These 'path' germs yield the images of curves in N that all pass through $C(t)$ with the same velocity. Such curves were called equivalent in the previous section, and we expect the general notion of a tangent vector to be related to a germ $[C_{\mathbb{R}_t}]$ rather than to be related to a particular curve in this class.

If $f : M \to N$ and $g : N \to P$ are any representatives of the germs $[f_{M_p}]$ and $[g_{N_q}]$ then the composition $[g_{N_q}] \circ [f_{M_p}]$ is the germ obtained by composing representatives : namely $g \circ f$. Similarly we define the pull-back of germs in terms of any representitives

$$[f]^*[g] = [f^*g] = [g \circ f]. \tag{4.5.1}$$

It is convenient not to distinguish notationally between $[f]^*$ and f^* since no confusion need arise in practise.

We denote by $\mathscr{F}(M)$ the set of real-valued smooth functions on the manifold M. The elements of $\mathscr{F}(M)$ form a ring with $(f+g)(p) = f(p) + g(p)$ and $(fg)(p) = f(p)g(p)$. By identifying the constant functions with the real numbers the ring $\mathscr{F}(M)$ may be regarded as a real vector space, and hence an algebra. A *derivation* into $\mathbb{R}$ on $[\mathscr{F}(M)_p]$ is a linear map $X : [\mathscr{F}(M)_p] \to \mathbb{R}$ that obeys the Leibnitz rule

$$X(f_1 f_2) = X(f_1) f_2(p) + f_1(p) X(f_2). \tag{4.5.2}$$

Since linear combinations of derivations are derivations they form a vector space over $\mathbb{R}$ at p. If we set $f_1 = f_2 = 1$, the identity map, then (4.5.2) implies $X(1) = 0$ and hence, by linearity, X annihilates any element of $\mathbb{R}$. The vector space of derivations of the above germs at $p \in M$ is defined as the *tangent space* T_pM of the smooth manifold at p.

We introduced earlier the pull-back map f^* associated with the diffeomorphism $f : M \to N$, $p \mapsto q = f(p)$. The tangent map at p associated with f is denoted f_{*p} and is defined in terms of f^* by

$$f_{*p} : T_pM \longrightarrow T_qN \qquad X \longmapsto f_{*p}X = Xf^*. \tag{4.5.3}$$

Thus (see figure 4.6) $f_{*p}X$ is a derivation on elements $g \in [\mathscr{F}(N)_{f(p)}]$ obtained by pulling back g with f^* and then acting with X, that is

$$(f_{*p}X)(g) = X(f^*(g)) = X(g \circ f). \tag{4.5.4}$$

From this point on we shall also apply the definition of a tangent vector being a derivation into $\mathbb{R}$ on functions, to tangent vectors to $\mathbb{R}^m$. We must therefore show the equivalence with the previous definition of a tangent vector being an ordered pair of elements from $\mathbb{R}^m$. Let $X \equiv (p, V) \in T_p\mathbb{R}^m$. If h is a real-valued function on $\mathbb{R}^m$ then we define X to map h to $\mathbb{R}$ by taking the directional derivative, that is

$$X(h) = \mathrm{D}_V h(p).$$

With this rule the tangent vector X is a derivation on functions in the neighbourhood of p. It also ensures the consistency of the definition of the tangent map given in (4.5.3) with the earlier definition (4.2.3), as will be explicitly demonstrated in a moment.

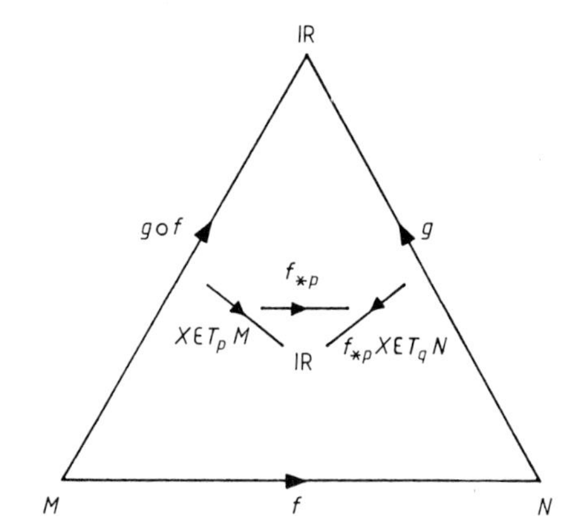

Figure 4.6 The tangent map $f_{*p} : T_pM \to T_qN$.

We now construct a local basis for T_pM in terms of a local chart germ at p, $[\varphi_p] : M \to \mathbb{R}^n$, that assigns the point $p \in M$ to the origin in $\mathbb{R}^n$. As usual let x^ν, $\nu = 1, \ldots, n$ denote the coordinate maps $\varphi^\nu : U_M \to \mathbb{R}$. Then φ^* is a map from function germs in $\mathbb{R}^n$ to function germs in M and φ_{*p} maps tangent vectors from T_pM to $T_0\mathbb{R}^n$. Of all the derivations on real-valued functions on $\mathbb{R}^n$ we denote by $X_\nu \in T_0\mathbb{R}^n$, the partial derivative:

$$X_\nu : [\mathscr{F}(\mathbb{R}^n)_0] \longrightarrow \mathbb{R} \qquad [f] \longmapsto [(\partial f/\partial x^\nu)(0)]. \tag{4.5.5}$$

Suppose $a^\nu X_\nu = 0$ for some n real numbers a^ν, then since $(X_\nu(x^\mu))(0) = \delta^\mu_\nu$, acting on x^μ gives $a^\mu = 0$. Thus the X_ν are linearly independent and the n tangent vectors $\{X_\nu\}$ form a local basis for the n-dimensional vector space $T_0\mathbb{R}^n$.

We may express any tangent vector $X \in T_pM$ in terms of

$\varphi_{*p}X \in T_0\mathbb{R}^n$. If $f \in \mathscr{F}(M)$ then by writing $f = \varphi^* f_\varphi$ we have

$$X(f) = X(\varphi^* f_\varphi) = (\varphi_{*p}X)f_\varphi. \tag{4.5.6}$$

In a natural basis associated with the chart (U_M, φ)

$$\varphi_{*p}X = \sum_{\nu=1}^{n} a^\nu(\partial/\partial x^\nu), \tag{4.5.7}$$

where it is to be understood that the derivative acts at $x^\nu = 0$, this gives

$$X(f) = \left[\left(\sum_{\nu=1}^{n} a^\nu(\partial/\partial x^\nu)\right)f_\varphi\right](x^i(p), \ldots, x^n(p)). \tag{4.5.8}$$

Often for computations, real-valued maps f on M are specified locally in terms of their local representatives $f_\varphi = f \circ \varphi^{-1}$ on $\mathbb{R}^n$ and the details of the chart φ are suppressed. However it may be important when dealing with global properties of manifolds to remember the distinction between f and f_φ since for a general manifold it is not possible to find an atlas consisting of a single chart. Just as the charts are often suppressed when discussing real-valued maps, in a similar way the representative $\psi \circ f \circ \varphi^{-1}$ of a map between manifolds is often written with the charts ψ and φ omitted. In the following we shall denote such a map by $\tilde{f}$. We may specify any $X \in T_pM$ by giving $\varphi_{*p}X$, as in (4.5.7). It is common not to distinguish $\varphi_{*p}X$ from X, identifying $(\partial/\partial x^\nu)$ with a tangent vector to M. Having pointed out the distinction we shall nevertheless employ this abuse of notation in the following sections.

Consider the expression for the tangent map f_{*p} where f is a representative of a germ at p from some n-dimensional manifold M to some m-dimensional manifold N. Suppose $(x^1, \ldots, x^n)$ are local chart functions that assign to $p \in M$ the origin of $\mathbb{R}^n$ and $(y^1, \ldots, y^m)$ are local chart functions that assign to $f(p) \in N$ the origin of $\mathbb{R}^m$. Thus f may be specified in terms of the m real-valued functions $(f^1, \ldots, f^m)$ and we represent it by the map

$$\tilde{f}: U(\mathbb{R}^n) \longrightarrow \mathbb{R}^m,$$

$$(x^1, \ldots, x^n) \longmapsto (y^1 = f^1(x^1, \ldots, x^n), \ldots, y^m = f^m(x^1, \ldots, x^n)).$$

We recall that $\{X_\nu\} = \{(\partial/\partial x^\nu)\}$ is a basis for $T_0\mathbb{R}^n$ in this chart. If g is any element of $[\mathscr{F}(\mathbb{R}^m)_0]$ then

$$\begin{aligned}(\tilde{f}_{*0}(\partial/\partial x^\nu))g &= (\partial/\partial x^\nu)(\tilde{f}^* g) = (\partial/\partial x^\nu)(g \circ \tilde{f}) \\ &= \sum_{\mu=1}^{m}(\partial g/\partial y^\mu)(0)(\partial f^\mu/\partial x^\nu)(0)\end{aligned}$$

or more simply

$$\tilde{f}_{*0}(\partial/\partial x^\nu) = \sum_{\mu=1}^{m}(\partial f^\mu(0)/\partial x^\nu)(\partial/\partial y^\mu). \tag{4.5.9}$$

The action of $\tilde{f}_{*0}$ on an arbitrary vector in $T_0\mathbb{R}^n$ now follows directly since $\tilde{f}_{*0}$ is linear:

$$\tilde{f}_{*0}(a^\nu(\partial/\partial x^\nu)) = a^\nu \tilde{f}_{*0}(\partial/\partial x^\nu) = a^\nu(\partial f^\mu/\partial x^\nu)(0)(\partial/\partial y^\mu). \quad (4.5.10)$$

The Jacobian matrix gives a representation of the linear map between T_pM and $T_{f(p)}N$.

Equation (4.5.10) expresses the chain rule of differentiation and establishes the equivalence of definitions (4.2.3) and (4.5.3) for the tangent map on $T_p\mathbb{R}^n$. If $A \in T_0\mathbb{R}^n$ is regarded as an ordered pair, $A = (0, a)$ with $a = \Sigma_{\nu=1}^n a^\nu e_\nu$ in the natural basis for $\mathbb{R}^n$ then A is equivalent to the derivation $a^\nu(\partial/\partial x^\nu)|_0$. The effect of $\tilde{f}_{*0}$ on this derivation is given in (4.5.10). The derivation on the right-hand side of (4.5.10) is equivalent to the ordered pair $(f(0), a^\nu(\partial f^\mu/\partial x^\nu)(0)e'_\mu)$ where $\{e'_\mu\}$ is the natural basis for $\mathbb{R}^m$. From (4.2.6) we recognise this as $(f(0), \mathrm{D}_a f(0))$, which is the form of $\tilde{f}_{*0}A$ given in (4.2.3).

Figure 4.7 summarises the relationship between φ and f and the maps that they induce. Let us next observe that if $\psi : U_1(\mathbb{R}^n) \to U_2(\mathbb{R}^n)$

$$x^\mu \longrightarrow x'^\mu = \psi^\mu(x^1, \ldots, x^n) \quad (4.5.11)$$

we may infer from the above that

$$\psi_{*0}(\partial/\partial x^\nu) = (\partial \psi^\mu/\partial x^\nu)(0)(\partial/\partial x'^\mu). \quad (4.5.12)$$

The tangent vector X at $p \in U_M$, that was represented in the chart (U_M, φ) by $\varphi_{*p}X = a^\nu(\partial/\partial x^\nu)$, will have a different representation in the chart $(U_M, \psi \circ \varphi)$, since

$$\begin{aligned}(\psi \circ \varphi)_{*p}X &= \psi_{*0} \circ \varphi_{*p}X = \psi_{*0}(a^\nu(\partial/\partial x^\nu)) \\ &= a^\nu(\partial \psi^\rho/\partial x^\nu)(0)(\partial/\partial x'^\rho) \qquad \text{(from (4.5.12))} \\ &\equiv a'^\rho(\partial/\partial x'^\rho)\end{aligned}$$

where $a'^\rho = (\partial \psi^\rho/\partial x^\nu)(0)a^\nu$.

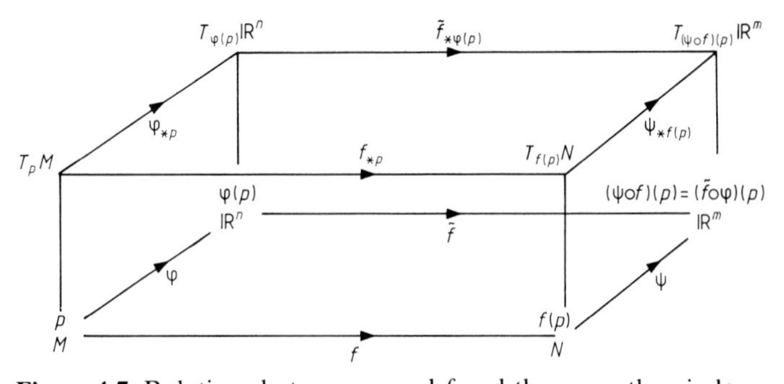

Figure 4.7 Relations between φ and f and the maps they induce.

This representation of the same tangent vector $X \in T_pM$ at p in a different chart should be distinguished from the tangent vector $f_{*p}X \in T_{f(p)}M$. The latter is induced from a differentiable germ $f: M \to M$ at p: the former from a change of coordinates in the neighbourhood of $p \in M$. The relation between the natural (or chart-induced) components $\{a'^{\rho}\}$ of X in the basis $\{(\partial/\partial x'^{\rho})\}$ at p to the natural components $\{a^{\nu}\}$ of X in the basis $\{(\partial/\partial x^{\nu})\}$ of a different chart about p, may be recognised as a $\mathrm{Gl}(n, \mathbb{R})$ basis-induced transformation. (Recall coordinate transformations are invertible.) Historically this was one of the characterisations of a 'contravariant' vector. It prescribed how the components of a vector were to be related to a change of coordinates.

In the previous section we motivated the definition of a tangent vector by considering differentiation along a curve. Having now defined tangent vectors we can return and define the tangent vector to a curve. If $C: I \to M$ is a smooth curve with $C(t_0) = p_0$ then the tangent vector to C at p_0 is

$$V^C_{p_0} \equiv C_{*t_0}(\partial/\partial t) \tag{4.5.13}$$

so for $f \in \mathcal{F}(M)$

$$V^C_{p_0}(f) = (C_{*t_0}(\partial/\partial t))(f) = (\partial/\partial t)(f \circ C)(t_0).$$

Thus the tangent vector to C at p_0 maps functions to their derivative along the curve at p_0, as was anticipated by the choice of notation in (4.4.3)

$$(C_*)_{t_0}(\partial/\partial t) = ((\partial C^i/\partial t))(t_0)(\partial/\partial x)^i \in T_{p_0}M. \tag{4.5.14}$$

As an illustration consider $C: (0,1) \to \mathbb{R}^2$ given by

$$C^1(t) = a \sin bt$$

$$C^2(t) = a \cos bt \qquad a, b \in \mathbb{R}.$$

If $\{(\partial/\partial x^1), (\partial/\partial x^2)\}$ is a natural basis for $T_{p_0}\mathbb{R}^2$ then

$$C_{*t_0}(\partial/\partial t) = (\partial C^i/\partial t)(t_0)(\partial/\partial x^i).$$

From the above we have

$$(\partial/\partial t)C^1(t_0) \equiv \dot{C}^1(t_0) = ab \cos bt_0 = bC^2(t_0)$$

$$(\partial C^2/\partial t)(t_0) \equiv \dot{C}^2(t_0) = -ab \sin bt_0 = -bC^1(t_0).$$

4.6 Vector Fields

So far tangent vectors have been associated with points on the manifold. By smoothly assigning a tangent vector to each point we define a vector field. Thus a vector field maps functions to functions. In fact this is a convenient starting point for the definition of a vector field, it being a consequence that a vector field assigns a tangent vector to each point.

A vector field X on a manifold M is a derivation on the algebra of smooth functions

$$X : \mathscr{F}(M) \longrightarrow \mathscr{F}(M)$$

$$X(\lambda f + \mu g) = \lambda X(f) + \mu X(g) \qquad \lambda, \mu \in \mathbb{R}; f, g \in \mathscr{F}(M)$$

$$X(fg) = X(f)g + f\,X(g). \tag{4.6.1}$$

(In the previous section we used capital letters to denote *tangent vectors*; in the following capital letters will be used for *vector fields*. Tangent vectors will henceforth be labelled by the point with which they are associated.) Whereas tangent vectors are derivations into $\mathbb{R}$, vector fields are derivations that map the algebra of smooth functions into itself. A vector field X is called smooth if, for every smooth $f \in \mathscr{F}(M)$, $X(f)$ is smooth. The set of smooth vector fields on M will be denoted $T^1(M)$. Given an $X \in T^1(M)$ we may define a vector $X_p \in T_pM$, for any $p \in M$, by

$$(Xf)(p) = X_p f. \tag{4.6.2}$$

It is clear from the derivation properties of X and X_p that this does indeed define a tangent vector. Since vector fields map functions to functions we may define a product in an obvious way. For X, $Y \in T^1(M)$

$$XY : \mathscr{F}(M) \longrightarrow \mathscr{F}(M)$$

$$f \longmapsto X(Y(f)). \tag{4.6.3}$$

This composed mapping will not, however, be a vector field. It will not satisfy the Leibnitz property (4.6.1) required of a derivation. In fact

$$(XY)(fg) = (XY)(f)g + f(XY)(g) + X(f)Y(g) + Y(f)X(g).$$

From this it is clear that we can obtain a new vector field from the commutator of two vector fields

$$[X, Y] = XY - YX. \tag{4.6.4}$$

Being the commutator of an associative product this bracket operation on vector fields is antisymmetric and satisfies the *Jacobi identity*

$$[[X, Y], Z] + [[Y, Z], X] + [[Z, X], Y] = 0. \tag{4.6.5}$$

Smooth vector fields form a module (see Appendix A) over $\mathscr{F}(M)$, and hence a vector space over $\mathbb{R}$ identified with the constant functions. The commutator then turns the vector fields into an (infinite-dimensional) Lie algebra. The commutator is also called the *Lie bracket*.

If $f: M \to N$ is a smooth map between manifolds then, for any $p \in M$, the tangent map f_{*p} sends T_pM to $T_{f(p)}N$. If X and Y are smooth vector fields on M and N respectively, with X_p and Y_p given by (4.6.2), such that

$$Y_{f(p)} = f_{*p}X_p \qquad \forall\, p \in M \tag{4.6.6}$$

then X and Y are said to be f-related. We will often simply write $Y = f_*X$. This notation does not imply that any smooth map $f: M \to N$ enables a smooth vector field on M to be mapped to one on N. If f is not one to one, with $f(p) = f(q)$ say, then for an arbitrary X, $f_{*p}X \neq f_{*q}X$. If f is not onto then smooth vector fields on N that are f-related to $X \in T^1(M)$ can differ outside the image of f. An important example is that of a smooth curve $C: I \to M$. Different smooth vector fields on M can be tangent to all the points on the image of C. For the special case in which f is a diffeomorphism for every $X \in T^1(M)$ there is a unique $Y \in T^1(N)$ such that

$$Y = f_*X.$$

As we noted in the previous section it is common not to distinguish $X_p \in T_pM$ from its coordinate representation $\varphi_{*p}X$. Thus if (U, φ) is a chart for the neighbourhood of p, with coordinate functions $\{x^i\}$, one identifies $\{(\partial/\partial x^i)|_p\}$ with a basis for T_pM. If $X \in T^1(M)$ then in the neighbourhood of p we can express X as $X = X^i(\partial/\partial x^i)$, where $X^i \in \mathscr{F}(M)$ are not distinguished from their representations in this chart. The elements $(\partial/\partial x^i)$ form a basis for $T^1(U)$, the $\mathscr{F}$-module of smooth vector fields on U. They form the *natural local basis* or *local coordinate basis*. Since the ring of smooth functions is not a division ring there is no reason why the $\mathscr{F}$-module $T^1(M)$ should have a basis, and in general it will not have. This is because for a general manifold there are no vector fields that do not vanish somewhere. (The two-sphere, for example, is such a manifold.)

4.7 The Tangent Bundle

One way of formalising the way a vector field on an n-dimensional manifold M assigns a tangent vector to each point is to construct a new $2n$-dimensional manifold TM by collecting together all the tangent spaces T_pM from all points of M:

$$TM = \bigcup_p T_pM. \tag{4.7.1}$$

An element of TM is a tangent vector X_p, labelled by the point p and its components in some basis for T_pM. Moreover the construction of TM must satisfy certain smoothness criteria with respect to these assignments. If a tangent vector $X_p \in T_pM$ is represented in a local chart (U_M, φ_M), with coordinate maps (x^i), by $\varphi_{*p}X_p = y^j \partial/\partial x^j$ then we define $(x^i(p), y^j(p)) \in \mathbb{R}^{2n}$ as the coordinates of a point in TM. That is, the chart (U_M, φ_M) for M induces a chart (U_{TM}, φ_{TM}) for TM by

$$(\varphi_{TM})(X_p) \equiv (x^i(p), y^i(p))$$

where $\varphi_M(p) = x^i(p)e_i$ and $(\varphi_M)_{*p}X_p = y^i(p)(\partial/\partial x^i)\,|_{(\varphi_M)(p)}$, $\{e_i\}$ being the natural basis for $\mathbb{R}^n$. As we have remarked earlier a tangent vector to $\mathbb{R}^n$ is equivalent to an element of $\mathbb{R}^{2n}$: the derivative in the direction V at p being equivalent to (p, V). Thus φ_{TM} assigns to X_p the element of $\mathbb{R}^{2n}$ equivalent to $(\varphi_M)_{*p}X_p \in T_{\varphi(p)}\mathbb{R}^n$.

Since M is a differentiable manifold it is possible to give a topology and differentiable manifold structure to TM. If (U_{TM}, φ_{TM}) is a local chart for TM, induced by (U_M, φ_M), then the map specifying a change of coordinates in TM: $(\varphi_{TM})_2 \circ (\varphi_{TM}^{-1})_1 : \mathbb{R}^{2n} \to \mathbb{R}^{2n}$, is given in terms of the map specifying a change of coordinates on M

$$(\varphi_M)_2 \circ (\varphi_M^{-1})_1 : \mathbb{R}^n \longrightarrow \mathbb{R}^n \qquad x^i(p) \longmapsto x'^i(p).$$

The tangent map is

$$((\varphi_M)_2 \circ (\varphi_M^{-1}))_{1*(\varphi_M)_1(p)} : T_{(\varphi_M)_1(p)}\mathbb{R}^n \longrightarrow T_{(\varphi_M)_2(p)}\mathbb{R}^n$$
$$y^k \partial/\partial x^k \longmapsto y^k((\partial x'^i/\partial x^k))\partial/\partial x'^i$$

(where we are using summation convention) so that

$$((\varphi_{TM})_2 \circ (\varphi_{TM}^{-1})_1)(x^i, y^i)(q) = (x'^i(p), y^k(p)(\partial x'^i/\partial x^k)(p)). \quad (4.7.2)$$

These maps define (see figure 4.8) a diffeomorphism $(\varphi_{TM})_{12}$ of

$$(\varphi_{TM})_1((U_{TM})_1 \cap (U_{TM})_2)$$

onto

$$(\varphi_{TM})_2((U_{TM})_1 \cap (U_{TM})_2).$$

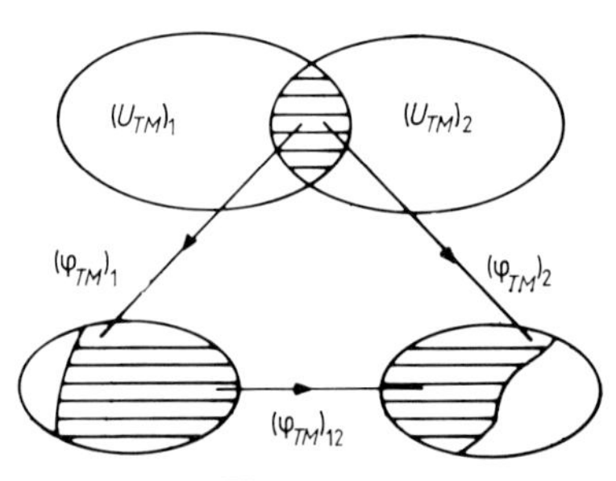

Figure 4.8

From its construction U_{TM} is diffeomorphic to $U_M \times \mathbb{R}^n$, but globally TM need not be a product manifold. A product manifold $M \times N$ is formed from ordered pairs of elements from the manifolds M and N. If $\{(U_a, \varphi_a)\}$ and $\{(V_b, \psi_b)\}$ are atlases for M and N respectively then an atlas for $M \times N$ is defined by the collection of charts

$$\varphi_a \times \psi_b : U_a \times V_b \longrightarrow \mathbb{R}^{\dim M + \dim N} \qquad (p, q) \longmapsto (\varphi_a(p), \psi_b(q)).$$

Such a collection of maps satisfies the criteria for being an atlas. The local product structure of TM allows the definition of a natural projection map

$$\Pi : TM \longrightarrow M, \; X_p \longmapsto p \tag{4.7.3}$$

which identifies the point on M to which the tangent vector in TM is attached. It is convenient to picture U_{TM}, with its local product structure exposed, as a space over U_M (see figure 4.9). All the tangent vectors at p are drawn as the space T_pM associated by the projection Π to a point p of M. The local coordinate representative of Π is usually given the same name, $\Pi : \mathbb{R}^{2n} \to \mathbb{R}^n$, $(x^i, y^i) \mapsto x^i$. (The inverse image set T_pM is sometimes denoted $\Pi^{-1}(p)$ and U_{TM} denoted $\Pi^{-1}(U_M)$ although this notation should not be confused with the notion of an inverse map!).

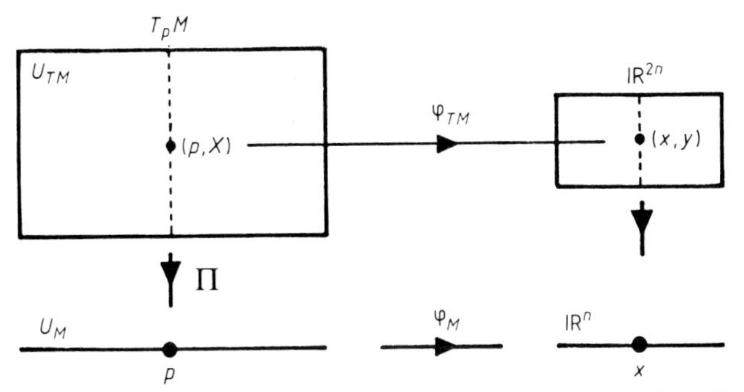

Figure 4.9 The local product structure of the tangent bundle.

The existence of a projection map makes TM into a fibred space, the elements related to p by Π being the *fibre* over p. The manifold TM together with Π is called the *tangent bundle* of M. We have here an example of a fibre bundle. Although in all fibre bundles the fibre spaces are fused together by giving the bundle the structure of a product manifold locally, bundles with different global topologies can be constructed by relating fibres in overlapping neighbourhoods $(U_{TM})_1 \cap (U_{TM})_2$ in different ways. This is like the difference between a cylindrical ribbon with a twist and one without a twist. In both cases the twist can be eliminated from any neighbourhood but is an essential characteristic distinguishing one ribbon from the other.

A smooth *section* of TM is a C^∞ map

$$\sigma : M \longrightarrow TM \tag{4.7.4}$$

such that $\Pi \circ \sigma = (id)_M$. Thus $\sigma(p) \in T_pM$ for all $p \in M$. It may be represented in local charts $(U_{TM}, \varphi_{TM}),(U_M, \varphi_M)$ by

$$\tilde{\sigma}(x) = (x^i, y^i \equiv \sigma^i(x)) \tag{4.7.5}$$

where the $\{\sigma^i\}$ are real functions on $U_{\mathbb{R}^n}$ (see figure 4.10). Thus σ smoothly assigns a tangent vector to each point $p \in M$. We may identify a smooth section σ with a smooth vector field X by

$$(Xf)(p) = \sigma(p)f \qquad \forall f \in \mathscr{F}(M). \tag{4.7.6}$$

In this way every smooth vector field on M is equivalent to a smooth section of TM. If ΓTM is the space of smooth sections of TM we will henceforth use the above to identify $T^1(M)$ with ΓTM.

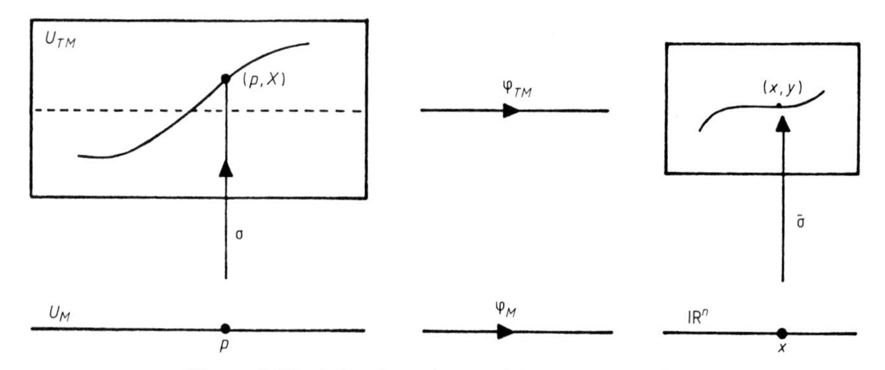

Figure 4.10 A local section and its representation.

4.8 Differential 1-Forms

The smooth vector fields on M form a module over the commutative ring of smooth functions, and hence inherit a vector space structure over $\mathbb{R}$ identified with the constant functions. We shall frequently need to distinguish maps that are linear with respect to the module structure from those that are only linear with respect to this vector space structure. Thus we refer to maps as being $\mathscr{F}(M)$-*linear* (or more simply $\mathscr{F}$-linear) or $\mathbb{R}$-linear. A 1-form field (or 1-form on M) is an element of the module dual to $T^1(M)$; that is, an $\mathscr{F}$-valued $\mathscr{F}$-linear map on vector fields. A 1-form is smooth if it maps smooth vectors to smooth functions. A smooth 1-form on M will also be called a differential 1-form. The space of smooth 1-forms on M is denoted $T_1(M)$. If $X \in T^1(M)$ assigns $X_p \in T_pM$ to the point p then for $\omega \in T_1(M)$ we define ω_p by

$$(\omega(X))(p) = \omega_p(X_p). \tag{4.8.1}$$

Clearly ω_p is a linear map from T_pM to $\mathbb{R}$, that is, an element of the dual space T^*_pM. Elements of T^*_pM are called co-vectors or 1-forms at p. Thus ω smoothly assigns an element of T^*_pM to every point p of M. In analogy to the construction of TM we may collect together all the cotangent spaces and form a new space

$$T^*M = \bigcup_p T^*_pM \tag{4.8.2}$$

Like TM the space T^*M inherits a manifold structure from that of M, with a natural projection from T^*M to M. With this structure T^*M becomes the *cotangent bundle*. We may identify a smooth 1-form on M with a smooth section of T^*M. So if ΓT^*M is the space of smooth sections we have a natural equivalence between elements of ΓT^*M and $T_1(M)$.

For every $f \in \mathcal{F}(M)$ we may associate an element $\mathrm{d}f \in T_1(M)$ by the rule

$$X(f) = (\mathrm{d}f)(X) \qquad \forall X \in T^1(M). \tag{4.8.3}$$

That is, $\mathrm{d}f \in \Gamma T^*M$ assigns $(\mathrm{d}f)_p \in T^*_pM$ to the point p with

$$X_p(f) = (\mathrm{d}f)_p(X_p). \tag{4.8.4}$$

The element $(\mathrm{d}f)_p$ which maps T_pM to $\mathbb{R}$ is related to f_{*p} which maps T_pM to $T_{f(p)}\mathbb{R}$: in fact they are naturally isomorphic. If g is a real-valued function on $\mathbb{R}$, $\lambda \mapsto g(\lambda)$, then from (4.5.4)

$$(f_{*p}X_p)(g) = X_p(g \circ f).$$

By the chain rule

$$X_p(g \circ f) = X_p(f)\,\frac{\mathrm{d}g}{\mathrm{d}\lambda}(f(p)).$$

Thus $f_{*p}X_p \in T_{f(p)}\mathbb{R}$ is equivalent to the ordered pair

$$(f(p), X_pf) = (f(p), (\mathrm{d}f)_p(X_p)).$$

The existence and linearity of f_{*p} ensures that (4.8.3) really does define a 1-form. Despite this natural isomorphism we shall distinguish the maps $(\mathrm{d}f)_p$ and f_{*p}.

If x^i is one of the coordinate functions and $(\partial/\partial x^j)$ is a vector from the natural local basis then (4.8.3) gives

$$\mathrm{d}x^i(\partial/\partial x^j) = (\partial x^i/\partial x^j) = \delta^i_j. \tag{4.8.5}$$

Thus $\{dx^i\}$ is a local basis for $T_1(M)$ naturally dual to the basis $\{(\partial/\partial x^j)\}$. In some coordinate neighbourhood, for any $f \in \mathscr{F}(M)$, df can be expanded in a local basis $df = df(\partial/\partial x^i)dx^i$, giving the classical expression

$$df = (\partial f/\partial x^i)dx^i \tag{4.8.6}$$

from (4.8.3). It is worth emphasising that in this expression the dx^i are not 'infinitesimal increments of the coordinates' but linear mappings on the tangent vectors. By evaluating this expression on a vector tangent to some curve we obtain the derivative of f along the curve: in this way df encodes the way in which the value of f changes as the point in M begins to move. The components of a 1-form with respect to the natural basis $\{dx^i\}$, associated with the chart (U_M, φ), are used to coordinate the bundle T^*M. If $\alpha \in T_1(M)$ with $\alpha = \alpha_\mu dx^\mu$, $\alpha_\mu \in \mathscr{F}(M)$, then α is associated with the smooth section ρ

$$\rho : M \longrightarrow T^*M$$

represented in a local chart by

$$x^\mu(p) \longmapsto (x^\mu(p), \alpha_\mu(p)).$$

If $f : M \to N$ is a smooth map then we have already defined the pull-back map f^* that takes a smooth function g on N to a smooth function f^*g on M, $f^*g = g \circ f$. Thus f^*g is evaluated at p by using f to send p from M to N where it is evaluated with g. In the same spirit we can define the pull-back of a 1-form ω on N to a 1-form $f^*\omega$ on M. If $X_p \in T_pM$ we define

$$(f^*\omega)_p X_p = \omega_{f(p)}(f_{*p}X_p). \tag{4.8.7}$$

We need to check that for a smooth assignment of X_p to T_pM and a smooth ω on N this rule assigns $(f^*\omega)_p$ smoothly to T^*_pM. This can be seen from the local coordinate expression for (4.8.7). Firstly we note that for $g \in \mathscr{F}(M)$

$$\begin{aligned}(f^*(g\omega))_p X_p &= (g\omega)_{f(p)}(f_{*p}X_p) \\ &= (g \circ f)(p)\omega_{f(p)}(f_{*p}X_p) = (f^*g)(p)(f^*\omega)_p X_p\end{aligned}$$

thus

$$f^*(g\omega) = (f^*g)(f^*\omega). \tag{4.8.8}$$

If $\{x^i\}$ $i = 1, \ldots, m$ and $\{y^j\}$ $j = 1, \ldots, n$ are local coordinates for M and N such that the coordinate representation of f is given by $y^j = f^j(x^i)$, then if $X = X^i(\partial/\partial x^i)$

$$f_{*p}X_p = X^i(p)(\partial f^j/\partial x^i)(p)(\partial/\partial y^j)|_{f(p)}$$

and for $\mathrm{d}x^j \in T^*_{f(p)}N$

$$\mathrm{d}x^j(f_{*p}X_p) = X^i(p)(\partial f^j/\partial x^i)(p) = (\partial f^j/\partial x^i)(p)\ \mathrm{d}x^i(X_p).$$

It follows from (4.8.8) that if $\omega = \omega_j\ \mathrm{d}x^j$

$$f^*\omega = (\omega_j \circ f)(\partial f^j/\partial x^i)\ \mathrm{d}x^i \tag{4.8.9}$$

and the smoothness of f and the component functions ω_j ensure that $f^*\omega$ is smooth.

If $\omega \in T^*_pM$ then we can use any chart for the neighbourhood of p to represent ω, using the natural local basis. Given two different charts we can compare the representations of ω by using the pull-back of the map that relates the charts. Suppose (U_M, φ) is a chart for the neighbourhood of p with $\varphi(U_M) = U_1$. Given a diffeomorphism $\psi : U_1(\mathbb{R}^n) \to U_2(\mathbb{R}^n)$ we have a new chart $(U_M, \psi \circ \varphi)$. If ψ is specified by

$$\begin{aligned} \psi : U_1(\mathbb{R}^n) &\longrightarrow U_2(\mathbb{R}^n) \\ x^\mu &\longmapsto x'^\mu = \psi(x^1, \ldots, x^n) \end{aligned}$$

then we have the inverse map

$$\begin{aligned} \psi^{-1} : U_2(\mathbb{R}^n) &\longrightarrow U_1(\mathbb{R}^n) \\ x'^\mu &\longmapsto x^\mu = \psi^{-1\mu}(x'^1, \ldots, x'^n). \end{aligned}$$

In §4.5 we showed that if $X \in T_pM$ is represented in the (U_M, φ) chart by

$$\varphi_{*p}X = X^\nu(\partial/\partial x^\nu)|_{\varphi(p)}$$

then the representative in $(U_M, \psi \circ \varphi)$ is

$$(\psi \circ \varphi)_{*p}X = X^\nu(\partial\psi^\mu/\partial x^\nu)(\varphi(p))(\partial/\partial x'^\mu)|_{(\psi\circ\varphi)(p)}.$$

When representing a 1-form we have to remember that the pull-back map acts in the opposite direction to the map itself. Since chart maps are invertible $\omega \in T^*_pM$ is represented in (U_M, φ) by $\varphi^{-1*}_{\varphi(p)}\omega$, with

$$\varphi^{-1*}_{\varphi(p)}\omega = \omega_\nu\ \mathrm{d}x^\nu|_{\varphi(p)}$$

say. In the chart $(U_M, \psi \circ \varphi)$ the representation of ω is $(\psi \circ \varphi)^{-1*}_{(\psi\circ\varphi)(p)}\ \omega$, where

$$\begin{aligned} (\psi \circ \varphi)^{-1*}_{(\psi\circ\varphi)(p)}\omega &= \psi^{-1*}_{(\psi\circ\varphi)(p)}\varphi^{-1*}_{\varphi(p)}\ \omega = \psi^{-1*}_{(\psi\circ\varphi)(p)}\omega_\nu \mathrm{d}x^\nu \\ &= \omega_\nu(\partial\psi^{-1\nu}/\partial x'^\mu)((\psi \circ \varphi)(p))\ \mathrm{d}x'^\mu|_{(\psi\circ\varphi)(p)} \\ &= \omega'_\mu\ \mathrm{d}x'^\mu|_{(\psi\circ\varphi)(p)} \end{aligned}$$

from (4.8.9). Thus whereas the components of the tangent vector X are transformed with the Jacobian matrix representing ψ, the components of the 1-form ω transform with the inverse matrix since

$$(\partial\psi^{\mu}/\partial x^{\alpha})(\partial\psi^{-1\alpha}/\partial x'^{\nu}) = \delta^{\mu}_{\nu}.$$

That is, the components of ω transform contragradiently to those of X. The behaviour of the change in the components of ω induced by changing the coordinate basis is the historical characterisation of a covariant vector (see figure 4.11).

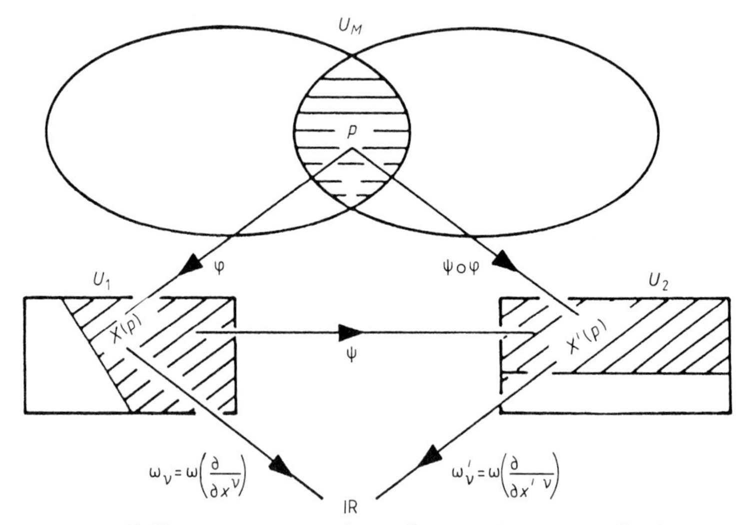

Figure 4.11 Different representations of a covariant vector field on U_M.

4.9 Tensor Fields

In Chapter 1 we introduced the tensor algebra associated with an arbitrary vector space. We may now apply this to the particular case when that vector space is the cotangent space at any point of a manifold. Thus elements of $T^s_r(T^*_pM)$ are called tensor fields at p of covariant degree r and contravariant degree s.

It is purely for convenience that we have selected the cotangent space rather than the tangent space, the notation of Chapter 1 having been chosen such that taking the arbitrary vector space V to be T^*_pM gives the conventional labelling for mixed tensors. It is for this reason that it was convenient in Chapter 1 to think of elements of V as acting on V rather than the other way around. Clearly we have* $T^s_r(T^*_pM) = T^r_s(T_pM)$.

Whereas the cotangent space at any point is a real vector space the set of 1-form fields forms an $\mathscr{F}$-module. In the same way as we constructed the tensor product of vector spaces we may construct the tensor product of the $\mathscr{F}$-module of 1-form fields with itself and the dual module of smooth vector fields. Elements of the tensor product module, $T^s_r(M)$, are called *tensor fields* of covariant degree r and contravariant degree s. We identify $T^0_0(M)$ with $\mathcal{F}(M)$. Thus an element of $T^s_r(M)$ smoothly assigns an element of $T^s_r(T^*_pM)$ to each point p in M. As for the case of vector and 1-form fields this way of regarding tensor fields is formalised in terms of a fibre bundle, the bundle of mixed tensors T^s_rM. Thus $T^s_rM = \cup_p T^s_r(T^*_pM)$, with a coordinate system induced from that of M. The natural projection of the bundle maps each tensor field to the point in M at which it is attached. Smooth sections can be defined in an obvious way, allowing the identification of the set of smooth tensor fields $T^s_r(M)$ with the space of smooth sections ΓT^s_rM.

If (U, φ) is a chart for some neighbourhood of M, with chart maps $\{x^i\}$, then $\{(\partial/\partial x^i)\}$ and $\{\mathrm{d}x^i\}$ are bases for $T^1(U)$ and $T_1(U)$ respectively. Thus locally any tensor field $T \in T^s_r(M)$ can be written as

$$T = T_{i_1 i_2 \ldots i_r}{}^{j_1 j_2 \ldots j_s} \mathrm{d}x^{i_1} \otimes \mathrm{d}x^{i_2} \otimes \ldots \otimes \mathrm{d}x^{i_r} \otimes (\partial/\partial x^{j_1}) \otimes (\partial/\partial x^{j_2}) \otimes \ldots \otimes (\partial/\partial x^{j_s}). \tag{4.9.1}$$

This is just a formula from §1.5 rewritten with $\mathrm{d}x^{i_1}$ replacing e^{i_1} and $(\partial/\partial x^{j_1})$ replacing X_{j_1}. The summation convention is employed. The indices are staggered in anticipation of the introduction of a metric tensor field when we shall use the raising and lowering conventions introduced in Chapter 1. Whereas one can always use a local coordinate basis in which to expand tensor fields such a basis is not always the most convenient. In particular, when we have a metric tensor it is often useful to employ a suitably adapted basis.

The submodule of $T^s_r(M)$ formed by all totally antisymmetric covariant tensor fields forms the exterior algebra of differential forms, $\Lambda(M)$, under the exterior product of (1.2.2). We shall identify $\Lambda_0(M)$ with $\mathcal{F}(M)$. Thus a smooth differential form is associated with a smooth section of the exterior bundle $\Lambda M = \cup_p \Lambda(T^*_pM)$. Whereas an element of the exterior algebra of an arbitrary vector space is called an exterior form, the term differential form is reserved for an element of the $\mathcal{F}$-module $\Lambda(M)$. If $\beta \in \Gamma\Lambda_rM$, section of the bundle of exterior r-forms we may use a local coordinate basis to write

$$\beta = \sum_{\mu_1 \leq \mu_2 \leq \ldots \mu_r} \beta_{\mu_1 \mu_2 \ldots \mu_r} (\mathrm{d}x^{\mu_1}) \wedge (\mathrm{d}x^{\mu_2}) \wedge \ldots \wedge (\mathrm{d}x^{\mu_r}) \tag{4.9.2}$$

equivalently

$$\beta = \frac{1}{n!} \beta_{\mu_1 \mu_2 \ldots \mu_r} (\mathrm{d}x^{\mu_1}) \wedge (\mathrm{d}x^{\mu_2}) \wedge \ldots \wedge (\mathrm{d}x^{\mu_r})$$

where the summation convention is used. These formulae are transcribed from §1.2 with the substitution of $\mathrm{d}x^{\mu_i}$ for e^{μ_i}.

Given a smooth map f between two manifolds the induced maps on the tangent and cotangent spaces can, to some extent, be extended to tensor fields. If $f : M \to N$ then we extend the map f_{*p} to an $\mathbb{R}$-linear map on contravariant tensors at p

$$f_{*p} : T^s(T^*_pM) \longrightarrow T^s(T^*_{f(p)}N)$$

$$X_1 \otimes X_2 \otimes \ldots \otimes X_s \longmapsto f_{*p}X_1 \otimes f_{*p}X_2 \otimes \ldots \otimes f_{*p}X_s \qquad X_i \in T_pM. \quad (4.9.3)$$

As for the case of vector fields we cannot in general use this map to obtain a smooth tensor field on N from one on M. We have, of course, the obvious generalisation to f-related contravariant tensor fields. The smooth map f does, however, give rise to a map f^* which enables smooth 1-form fields on N to be pulled back to smooth 1-forms on M. This pull-back map may be extended to an $\mathbb{R}$-linear map on smooth covariant tensors on N

$$f^* : T_r(N) \longrightarrow T_r(M)$$

$$\omega^1 \otimes \omega^2 \otimes \ldots \otimes \omega^r \longmapsto f^*\omega^1 \otimes f^*\omega^2 \otimes \ldots \otimes f^*\omega^r \qquad \omega^i \in T_1(N). \quad (4.9.4)$$

For such a definition to make sense it is important that we have (4.8.8), that is

$$f^*(g\omega) = (f^*g)(f^*\omega) \qquad g \in \mathscr{F}(N),\ \omega \in T_1(N).$$

For $\beta \in T_r(N)$ and $\{X_i\} \in T_pM$ $i = 1, \ldots, r$ we have

$$(f^*\beta)_p(X_1, X_2, \ldots, X_r) = \beta_{f(p)}(f_{*p}X_1, f_{*p}X_2, \ldots, f_{*p}X_r).$$

In general the smooth map $f : M \to N$ does not induce a map on smooth contravariant tensor fields on M; nor on mixed tensor fields, the maps f_{*p} and f^*_p acting in opposite directions. For the special case of a diffeomorphism, however, there is an induced map on smooth vector fields as was noted in §4.6, and the problem of the maps acting in different directions is readily overcome since diffeomorphisms are invertible. If $\varphi : M \to N$ is a diffeomorphism then we define $\hat{\varphi}$ by

$$\hat{\varphi} : T^s_r(M) \longrightarrow T^s_r(N)$$

$$\hat{\varphi}(\omega^1 \otimes \omega^2 \otimes \ldots \otimes \omega^r \otimes X_1 \otimes \ldots \otimes X_s)$$

$$= \varphi^{-1*}\omega^1 \otimes \varphi^{-1*}\omega^2 \otimes \ldots \otimes \varphi^{-1*}\omega^r \otimes \varphi_*X_1 \otimes \ldots \otimes \varphi_*X_s$$

$$\omega^i \in T_1(M),\ X_i \in T^1(M). \quad (4.9.5)$$

Again we require (4.8.8) for consistency. Equivalently

$$\hat{\varphi}(\omega^1 \otimes \omega^2 \otimes \ldots \otimes \omega^r \otimes X_1 \otimes \ldots \otimes X_s)(Y_1, Y_2, \ldots, Y_r, \alpha^1, \ldots, \alpha^s)$$
$$= \omega^1 \otimes \omega^2 \otimes \ldots \otimes \omega^r \otimes X_1 \otimes \ldots \otimes X_s(\varphi^{-1}{}_*Y_1, \varphi^{-1}{}_*Y_2, \ldots, \varphi^{-1}{}_*Y_r, \varphi^*\alpha^1, \ldots, \varphi^*\alpha^s). \tag{4.9.6}$$

Example 4.2
For the smooth map

$$\varphi : \mathbb{R}^2 \longrightarrow \mathbb{R}^2 \qquad p \longmapsto \varphi(p)$$
$$(a, b) \longmapsto (\varphi^1(p), \varphi^2(p)) = (a\cos t + b\sin t, b\cos t - a\sin t)$$

with t a constant, the inverse is given by

$$\varphi^{-1} : \mathbb{R}^2 \longrightarrow \mathbb{R}^2 \qquad p \longmapsto \varphi^{-1}(p)$$
$$(a, b) \longmapsto ((\varphi^{-1})^1(p), (\varphi^{-1})^2(p)) = (a\cos t - b\sin t, b\cos t + a\sin t).$$

For a vector field Y, $\hat{\varphi}Y = \varphi_* Y$. Taking $Y = x^2(\partial/\partial x) + xy(\partial/\partial y)$, with x and y the standard coordinates on $\mathbb{R}^2$, gives

$$\varphi_{*p}Y_p = x^2(p)\{(\partial\varphi^1/\partial x)(p)(\partial/\partial x)|_{\varphi(p)} + (\partial\varphi^2/\partial x)(p)(\partial/\partial y)|_{\varphi(p)}\}$$
$$+ x(p)y(p)\{(\partial\varphi^1/\partial y)(p)(\partial/\partial x)|_{\varphi(p)} + (\partial\varphi^2/\partial y)(p)(\partial/\partial y)|_{\varphi(p)}\}$$
$$(\varphi_*Y)_{\varphi(p)} = (x^2\cos t + xy\sin t)(p)(\partial/\partial x)|_{\varphi(p)} + (xy\cos t - x^2\sin t)(p)(\partial/\partial y)|_{\varphi(p)}$$

We may use φ^{-1} to express the coordinates of p in terms of those of $\varphi(p)$, giving

$$(\varphi_*Y)_{\varphi(p)} = (x^2\cos t - xy\sin t)(\varphi(p))(\partial/\partial x)|_{\varphi(p)} + (xy\cos t - y^2\sin t)(\varphi(p))(\partial/\partial y)|_{\varphi(p)}.$$

so

$$\varphi_*Y = (x^2\cos t - xy\sin t)(\partial/\partial x) + (xy\cos t - y^2\sin t)(\partial/\partial y).$$

We consider now a 1-form $\alpha = x^2\,\mathrm{d}x + xy\,\mathrm{d}y$

$$(\hat{\varphi}\alpha)_{\varphi(p)} = (\varphi^{-1*}\alpha)_{\varphi(p)} = \varphi^{-1*}\alpha_p$$
$$= x^2(p)\{(\partial(\varphi^{-1})^1/\partial x)(\varphi(p))\mathrm{d}x|_{\varphi(p)} + (\partial(\varphi^{-1})^1/\partial y)(\varphi(p))\mathrm{d}y|_{\varphi(p)}\}$$
$$+ (xy)(p)\{(\partial(\varphi^{-1})^2/\partial x)(\varphi(p))\mathrm{d}x|_{\varphi(p)} + (\partial(\varphi^{-1})^2/\partial y)(\varphi(p))\mathrm{d}y|_{\varphi(p)}\}$$
$$= \{x^2(p)\cos t + (xy)(p)\sin t\}\mathrm{d}x|_{\varphi(p)} + \{(xy)(p)\cos t - x^2(p)\sin t\}\mathrm{d}y|_{\varphi(p)}$$
$$= (x^2\cos t - xy\sin t)(\varphi(p))\mathrm{d}x|_{\varphi(p)} + (xy\cos t - y^2\sin t)(\varphi(p))\mathrm{d}y|_{\varphi(p)}$$

as in the previous example, so

$$\hat{\varphi}\alpha = (x^2 \cos t - xy \sin t)\mathrm{d}x + (xy \cos t - y^2 \sin t)\mathrm{d}y.$$

We have

$$\begin{aligned}
&(\hat{\varphi}\alpha)(\hat{\varphi}Y)(p)\\
&= (x^2 \cos t - xy \sin t)^2(p) + (xy \cos t - y^2 \sin t)^2(p)\\
&= \{(x \cos t - y \sin t)^4 + (x \cos t - y \sin t)^2(y \cos t + x \sin t)^2\}(p)\\
&= (x^4 + x^2y^2)(\varphi^{-1}(p)) = (\alpha(Y))(\varphi^{-1}(p))\\
&= (\alpha(Y) \circ \varphi^{-1})(p)
\end{aligned}$$

so

$$(\hat{\varphi}\alpha)(\hat{\varphi}Y) = \hat{\varphi}(\alpha(Y)).$$

4.10 Exterior Derivatives

In §4.8 we associated with every $f \in \mathscr{F}(M)$ an element $\mathrm{d}f \in \Gamma T^*M$. Thus we have an operator mapping functions to 1-forms. We may extend this operator to an $\mathbb{R}$-linear map on $\Gamma\Lambda M$:

$$\mathrm{d}: \Gamma\Lambda_p M \longrightarrow \Gamma\Lambda_{p+1} M \tag{4.10.1}$$

with the properties:

$$\mathrm{d}f(X) = Xf \qquad X \in \Gamma\Lambda M,\ f \in \mathscr{F}M \tag{4.10.2a}$$

$$\mathrm{d}(\alpha \wedge \beta) = \mathrm{d}\alpha \wedge \beta + (-1)^p \alpha \wedge \mathrm{d}\beta \qquad \alpha \in \Gamma\Lambda_p M,\ \beta \in \Gamma\Lambda M \tag{4.10.2b}$$

$$\mathrm{dd} \equiv \mathrm{d}^2 = 0. \tag{4.10.2c}$$

The operator d is called the exterior derivative. Its existence and uniqueness are most easily demonstrated using a local chart and the properties of the exterior algebra. In any coordinate neighbourhood of M an element of $\Gamma\Lambda M$ can be expressed in a local natural basis. Since d is $\mathbb{R}$-linear it is sufficient to consider its effect on an element of the form

$$\omega = g\ \mathrm{d}x^{i_1} \wedge \ldots \wedge \mathrm{d}x^{i_k} \qquad g \in \mathscr{F}(M).$$

From properties (4.10.2b) and (4.10.2c)

$$\mathrm{d}\omega = \mathrm{d}g \wedge \mathrm{d}x^{i_1} \wedge \ldots \wedge \mathrm{d}x^{i_k}$$

with dg given by property (4.10.2a). So for the assumed form of ω we have the unique form for dω. The defining properties of d enable dω to be evaluated on a set of vector fields, for any $\omega \in \Gamma\Lambda M$. We consider first a 1-form, it being sufficient to assume

$$\omega = g\,\mathrm{d}x \qquad x, g \in \mathscr{F}(M)$$

thus

$$\begin{aligned}\mathrm{d}\omega(X_1, X_2) &= (\mathrm{d}g \wedge \mathrm{d}x)(X_1, X_2)\\ &= \tfrac{1}{2}\{\mathrm{d}g(X_1)\mathrm{d}x(X_2) - \mathrm{d}g(X_2)\mathrm{d}x(X_1)\}\end{aligned}$$

from the definition of the exterior product. From property (4.10.2a)

$$\begin{aligned}2\,\mathrm{d}\omega(X_1, X_2) &= X_1(g)\,\mathrm{d}x(X_2) - X_2(g)\,\mathrm{d}x(X_1)\\ &= X_1(g\,\mathrm{d}x(X_2)) - gX_1(\mathrm{d}x(X_2)) - X_2(g\,\mathrm{d}x(X_1)) + gX_2(\mathrm{d}x(X_1))\\ &= X_1(\omega(X_2)) - X_2(\omega(X_1)) + g[X_2, X_1](x).\end{aligned}$$

Using this property once again in the last term gives

$$2\,\mathrm{d}\omega(X_1, X_2) = X_1(\omega(X_2)) - X_2(\omega(X_1)) - \omega([X_1, X_2]).$$

It follows that for any $\alpha \in \Gamma\Lambda_1 M$

$$(\mathrm{d}\alpha)(X, Y) = (1/2)\{X(\alpha(Y)) - Y(\alpha(X)) - \alpha([X, Y])\}. \qquad (4.10.3)$$

Similarly if $\alpha \in \Gamma\Lambda_2 M$

$$\begin{aligned}(\mathrm{d}\alpha)(X, Y, Z) = (1/3)\{&X(\alpha(Y, Z)) + Y(\alpha(Z, X)) + Z(\alpha(X, Y))\\ &- \alpha([X, Y], Z) - \alpha([Y, Z], X) - \alpha([Z, X], Y)\}\\ &\forall X, Y, Z \in \Gamma TM. \qquad (4.10.4)\end{aligned}$$

For the general case of $\alpha \in \Gamma\Lambda_r M$

$$\begin{aligned}(\mathrm{d}\alpha)(X_0, X_1, \ldots, X_r) &= \frac{1}{r+1}\sum_{j=0}^{r}(-1)^j X_j(\alpha(X_0, \ldots, \hat{X}_j, \ldots, X_r))\\ &+ \frac{1}{r+1}\sum_{0\leq j\leq k\leq r}(-1)^{j+k}\alpha([X_j, X_k], X_0, \ldots, \hat{X}_j, \ldots, \hat{X}_k, \ldots, X_r)\\ &\forall X_0, X_1, \ldots, X_r \in \Gamma TM \qquad (4.10.5)\end{aligned}$$

where $\hat{X}_j$ means omit this term from the argument list.

An important property of d is that it commutes with the pull-back map $f^* : \Gamma\Lambda N \to \Gamma\Lambda M$ induced from a diffeomorphism $f : M \to N$. First observe that if $g \in \mathscr{F}(M)$, $X \in \Gamma TM$, then

$$\begin{aligned}(f^*\mathrm{d}g)(X) &= \mathrm{d}g(f_*X) = (f_*X)(g) &&\text{(by 4.10.2}a\text{)}\\ &= X(f^*g)\\ &= \mathrm{d}(f^*g)(X) &&\text{(using property (4.10.2}a\text{) again)}\end{aligned}$$

giving

$$f^*\mathrm{d}g = \mathrm{d}(f^*g) \qquad (4.10.6)$$

Now consider

$$\begin{aligned}
&\mathrm{d}\{f^*(g\mathrm{d}x^{i_1} \wedge \mathrm{d}x^{i_2} \wedge \ldots \wedge \mathrm{d}x^{i_k})\} \\
&= \mathrm{d}\{f^*(g\mathrm{d}x^{i_1}) \wedge f^*(\mathrm{d}x^{i_2}) \wedge \ldots \wedge f^*(\mathrm{d}x^{i_k})\} \\
&= \mathrm{d}\{f^*(g\mathrm{d}x^{i_1}) \wedge \mathrm{d}(f^*x^{i_2}) \wedge \ldots \wedge \mathrm{d}(f^*x^{i_k})\} \qquad \text{from above} \\
&= \mathrm{d}(f^*(g\mathrm{d}x^{i_1})) \wedge \mathrm{d}(f^*x^{i_2}) \wedge \ldots \wedge \mathrm{d}(f^*x^{i_k}) \qquad \text{as } \mathrm{d}^2 = 0 \\
&= \mathrm{d}((f^*g)(f^*\mathrm{d}x^{i_1})) \wedge f^*(\mathrm{d}x^{i_2} \wedge \ldots \wedge \mathrm{d}x^{i_k}) \\
&= \mathrm{d}(f^*g) \wedge f^*\mathrm{d}x^{i_1} \wedge f^*(\mathrm{d}x^{i_2} \wedge \ldots \wedge \mathrm{d}x^{i_k}) \qquad \text{as } \mathrm{d}(f^*\mathrm{d}x^i) = \mathrm{dd}(f^*x^i) = 0 \\
&= f^*\mathrm{d}g \wedge f^*\mathrm{d}x^{i_1} \wedge f^*(\mathrm{d}x^{i_2} \wedge \ldots \wedge \mathrm{d}x^{i_k}) \\
&= f^*\mathrm{d}(g\mathrm{d}x^{i_1} \wedge \mathrm{d}x^{i_2} \wedge \ldots \wedge \mathrm{d}x^{i_k}).
\end{aligned}$$

It follows since d and f^* are $\mathbb{R}$-linear maps that

$$f^*\mathrm{d} = \mathrm{d}f^* \tag{4.10.7}$$

on arbitrary elements of $\Gamma\Lambda M$.

4.11 One-Parameter Diffeomorphisms and Integral Curves

In many situations in theoretical physics one is concerned with situations that can be described in terms of 'flows on a manifold'. This technical term is borrowed from what is perhaps the simplest case to visualise, the laminar flow of a fluid around a smooth surface. The motion of a fluid around a vortex is another familiar example of a flow. If each element of the medium experiencing such a flow is followed in time it traces out the image of a curve. Hence for a smooth flow one can establish a correspondence between local fluid flow and a local vector field. The notion of a flow in time is naturally associated with a bijective mapping, the flow taking a neighbourhood $U(p)$ of a point p on a manifold M to a neighbourhood $U(p')$ in some fixed interval of time. For some fixed interval t we describe such an evolution by

$$\varphi_t : U(p) \longrightarrow U(p'). \tag{4.11.1}$$

For a chosen $U(p)$ we have a diffeomorphism for $t \in I_p$, where $I_p \subset \mathbb{R}$ is an open interval about 0. To describe what happens in an arbitrary time interval we define φ in terms of φ_t by

$$\varphi : W \subset (I \times M) \longrightarrow M, \ (t, p) \longmapsto \varphi(t, p) = \varphi_t(p) \tag{4.11.2}$$

where, for each $t \in I$, φ_t is a local diffeomorphism from some $U(p) \subset M$ to $U(p') \subset M$. Conversely, for every $U(p) \subset M$ there is an

$I_p \subset I$ such that φ_t is a diffeomorphism from $U(p)$ to $U(p')$ for $t \in I_p$. Motivated by the example of fluid flows, in which the configuration of fluid elements at any time can be obtained from the successive composition of evolution maps, we demand that

$$\varphi_{t_2} \circ \varphi_{t_1} = \varphi_{t_1+t_2} \qquad \forall t_1, t_2 \in I \text{ such that } t_1 + t_2 \in I$$

and that

$$\varphi_0(p) = p \qquad \forall p \in M. \tag{4.11.3}$$

In particular, $\varphi_t^{-1} = \varphi_{-t}$. Families of diffeomorphisms of this type are called *local one-parameter diffeomorphisms* on M. A one-parameter family of local diffeomorphisms gives rise to a vector field on M. For every $p \in M$ the map φ defines a curve $\varphi(p)$ starting at p

$$\begin{aligned} \varphi(p) : I_p &\longrightarrow M \\ t &\longmapsto \varphi_t(p). \end{aligned} \tag{4.11.4}$$

To say that the curve starts at p means that $\varphi_0(p) = p$. Using the definition (4.5.13) we have a tangent vector defined at every point of the image of the curve. By taking the set of all such curves we define a tangent vector at each point of M. Since different curves have image points in common it is necessary to check that this rule gives an unambiguous assignment of tangent vectors. Suppose that $\varphi_{t_0}(p) = \varphi_{t_0'}(p')$ for some (t_0, p) and (t_0', p'), then

$$\begin{aligned} \varphi_t(p') &= \varphi_{(t-t_0')+t_0'}(p') = (\varphi_{t-t_0'} \circ \varphi_{t_0'})(p') \qquad \text{by (4.11.3)} \\ &= \varphi_{t-t_0'}(\varphi_{t_0'}(p')) = \varphi_{t-t_0'}(\varphi_{t_0}(p)) \\ &= (\varphi_{t-t_0'} \circ \varphi_{t_0})(p) = \varphi_{t-t_0'+t_0}(p) \end{aligned}$$

by (4.11.3) again. Thus if the curves $\varphi(p')$ and $\varphi(p)$ have image points in common then $\varphi(p')$ is a reparametrisation of $\varphi(p)$. The parametrisations merely differ by the addition of a constant and so $\varphi(p')_{*t} = \varphi(p)_{*(t-t_0'+t_0)}$, and the tangent vectors agree where the image points coincide. Hence the one-parameter family of local diffeomorphisms defines a tangent vector at each point of M; the smoothness of φ_t ensures that the assignment of tangent vectors is smooth and we have a smooth vector field. For the example of a fluid flow this vector field is everywhere tangential to the flow lines.

In the above we showed how a one-parameter family of local diffeomorphisms defined a set of curves, enabling a vector field to be introduced that was everywhere tangential to these curves. We now show how the argument can be reversed. If X is a vector field on M then a curve $C : I \to M$, $t \mapsto p(t)$, is called an integral curve of X if X is C-related to $(\partial/\partial t)$. That is, if C is specified by $C : t \mapsto x^\mu = \lambda^\mu(t)$, giving $C_*(\partial/\partial t) = \dot{\lambda}^\mu(t)(\partial/\partial x^\mu)|_{\lambda(t)}$, and in local coordinates $X = f^\mu(\partial/\partial x^\mu)$, then C is an integral curve of X if

$$\dot{\lambda}^\mu(t) = f^\mu(\lambda^1(t), \ldots, \lambda^n(t)) \tag{4.11.5}$$

for $\mu = 1, \ldots, n$. It follows from the theory of ordinary differential equations that solutions to (4.11.5) always exist, being uniquely determined by the initial conditions $x^\mu(p) = \lambda^\mu(0)$. The smoothness of the f^μ ensures that such solutions are not only smooth functions of t, for t in some interval $I \subset \mathbb{R}$, but are also smooth functions of the initial point $x^\mu(p)$, for p in some neighbourhood $U \subset M$. Thus if $C: I \to M$ and $C': I' \to M$ are integral curves of X starting at p we must have $I' \subset I$ say, with C equal to C' on the restriction to I'. By taking the largest such interval we have a uniquely determined *maximal integral curve* of X starting at p.

Example 4.3
Suppose $X = x(\partial/\partial y) - y(\partial/\partial x) \in \Gamma T\mathbb{R}^2$. Let $C: I \to \mathbb{R}^2$, $t \mapsto (\lambda^1(t), \lambda^2(t))$ be an integral curve of X that starts at the point $(a, b) \in \mathbb{R}^2$. Solving $\dot{\lambda}^1 = -\lambda^2$, $\dot{\lambda}^2 = \lambda^1$ subject to this condition gives: $\lambda^1(t) = a\cos t - b\sin t$, $\lambda^2(t) = b\cos t + a\sin t$. Here we may take $I = \mathbb{R}$, the maximal integral curve mapping the whole real line into the circle, the curve being periodic with period 2π.

A vector field whose maximal integral curves starting at p are defined on all of $\mathbb{R}$, for every $p \in M$, is called *complete*. In general this will not be the case, the domain of the maximal integral curves depending on which point they start at. Introducing a suggestive notation we denote by $\varphi(p)$ the maximal integral curve of $X \in \Gamma TM$ starting at p

$$\varphi(p): I_p \longrightarrow M \qquad t \longmapsto \varphi_t(p).$$

If $t_0 \in I_p$ with $\varphi_{t_0}(p) = q$ then setting

$$h: J_q \longrightarrow I_p \qquad t \longmapsto t + t_0$$

gives a curve $\psi(q) = \varphi(p) \circ h$. The images of $\psi(q)$ and $\varphi(p)$ coincide, as do their tangent vectors since the reparametrisation merely involves the addition of a constant. Thus $\psi(q)$ is certainly an integral curve of X, starting at q. If $I_p = (a, b)$ then $J_q = (a - t_0, b - t_0)$ and since $a < 0 < b$ we have $-t_0 \in J_q$, giving $\psi_{-t_0}(q) = p$. If $\psi(q)$ were not maximal, with $J_q \subset I_q$, then reversing the argument would contradict I_p being the maximal domain of integral curves starting at p. So maximal integral curves with image points in common are all related by reparametrisations that translate the domain of definition along the real line. It then follows that if φ_t is defined by

$$\varphi_t: p \longmapsto \varphi_t(p) \qquad \forall p \text{ with } t \in I_p$$

then φ_t is an invertible map satisfying (4.11.3). Above each point $p \in M$ we erect the fibre I_p and denote the space formed by all the fibres by W. If $I_p \subseteq I \; \forall p$ then $W \subset (I \times M)$, and φ is defined by

$$\varphi: W \longrightarrow M, \; (t, p) \longmapsto \varphi_t(p).$$

Each φ_t is an invertible map on some domain contained in M. Furthermore these maps are smooth, the solutions to the differential equations for an integral curve being smooth functions of the starting point. It follows that every smooth vector field X on M generates a one-parameter family of local diffeomorphisms: each point of M being mapped along an integral curve of X. In local coordinates the transformation $p \longmapsto \varphi_t(p)$ is represented by

$$x^\mu(p) \longmapsto \varphi^\mu(t, x^1(p), \ldots, x^n(p)) \qquad \mu = 1, \ldots, n$$

where

$$\varphi^\mu(0, x^1(p), \ldots, x^n(p)) = x^\mu(p)$$

and

$$\begin{aligned}\varphi^\mu(t_1 + t_2, x^1(p), \ldots, x^n(p)) = \varphi^\mu\{t_2, \; & \varphi^1(t_1, x^1(p), \ldots, x^n(p)), \\ & \varphi^2(t_1, x^1(p), \ldots, x^n(p)), \ldots, \\ & \varphi^n(t_1, x^1(p), \ldots, x^n(p))\}. \quad (4.11.6)\end{aligned}$$

We may use the smoothness of the functions φ^μ in the variable t to obtain a linear approximation of φ^μ for small t

$$\begin{aligned}\varphi^\mu(t, x^1(p), \ldots, x^n(p)) = \; & \varphi^\mu(0, x^1(p), \ldots, x^n(p)) \\ & + t\dot{\varphi}^\mu(0, x^1(p), \ldots, x^n(p)) + \ldots \quad (4.11.7)\end{aligned}$$

where $\dot{\varphi}^\mu$ denotes the derivative with respect to t. Since $\varphi(p)$ is an integral curve of X, starting at p, if in local coordinates $X = f^\mu(\partial/\partial x^\mu)$ we have

$$\varphi^\mu(0, x^1(p), \ldots, x^n(p)) = x^\mu(p)$$

and

$$\dot{\varphi}^\mu(0, x^1(p), \ldots, x^n(p)) = f^\mu(p). \qquad (4.11.8)$$

Thus for t sufficiently small (4.11.6) may be approximated by

$$x^\mu(p) \longmapsto x^\mu(p) + t\, f^\mu(p) + \ldots. \qquad (4.11.9)$$

Example 4.4
If x coordinates $\mathbb{R}$ then a smooth vector field on $\mathbb{R}$ is $X = x^2(\partial/\partial x)$. If $\varphi(t, p)$ is the maximal integral curve starting at p we require

$$\dot{\varphi}(t, p) = \varphi(t, p)^2$$
$$\varphi(0, p) = p.$$

The solution is $\varphi(t, p) = p/(1 - tp)$. If $p > 0$ we must have $t \in (-\infty, p^{-1})$, if $p = 0$, $t \in (-\infty, \infty)$ whilst for $p < 0$, $t \in (p^{-1}, \infty)$. The domain $W = \cup_p I_p$ is the region of $\mathbb{R}^2$ bounded by hyperbolae in the bottom-left and upper-right quadrants. This is shown in figure 4.12. We can verify that indeed $\varphi_{t_2} \circ \varphi_{t_1} = \varphi_{t_2+t_1}$.

$$\varphi_{t_2}(\varphi_{t_1}p) = \frac{p/(1 - t_1 p)}{1 - t_2 p/(1 - t_1 p)} = \frac{p}{1 - (t_1 + t_2)p} = \varphi_{t_1+t_2}(p).$$

We have shown in figure 4.12 the effect of one of the local diffeomorphisms φ_t.

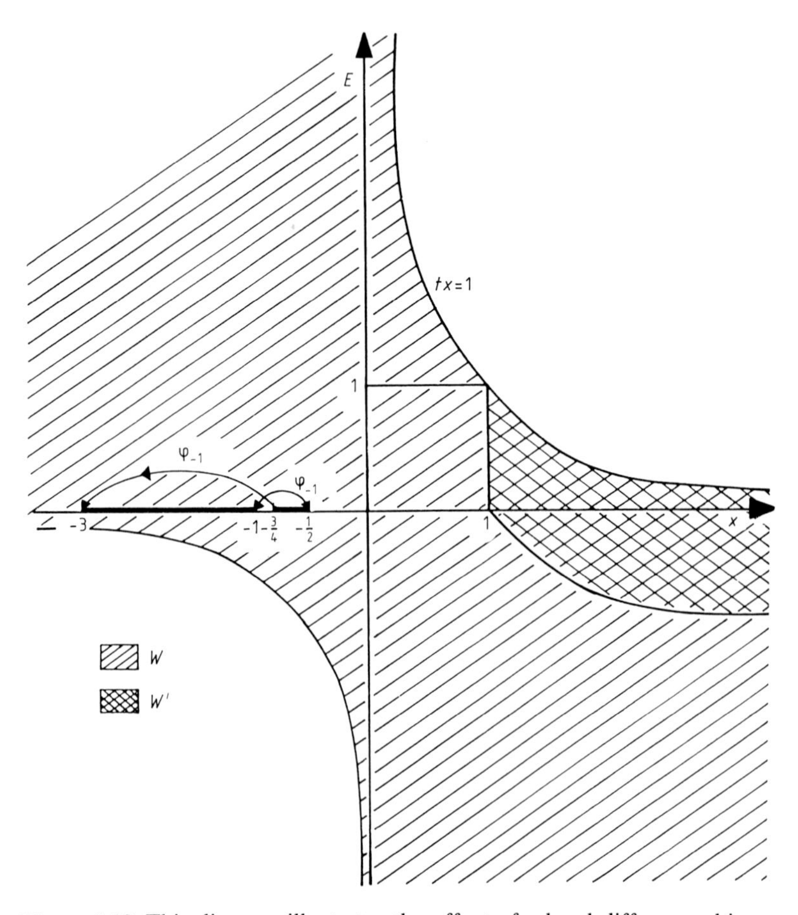

Figure 4.12 This diagram illustrates the effect of a local diffeomorphism φ_t.

If we modify the above example by restricting X to the manifold M consisting of the open interval $(1, \infty)$, then the maximal integral curve starting at p has domain $I'_p = (p^{-1} - 1, p^{-1})$. So in this case φ_t is defined somewhere if $t \in (-1, 1)$. The domain $W' = \cup_p I'_p \ \forall p \in M$ is shown in figure 4.12.

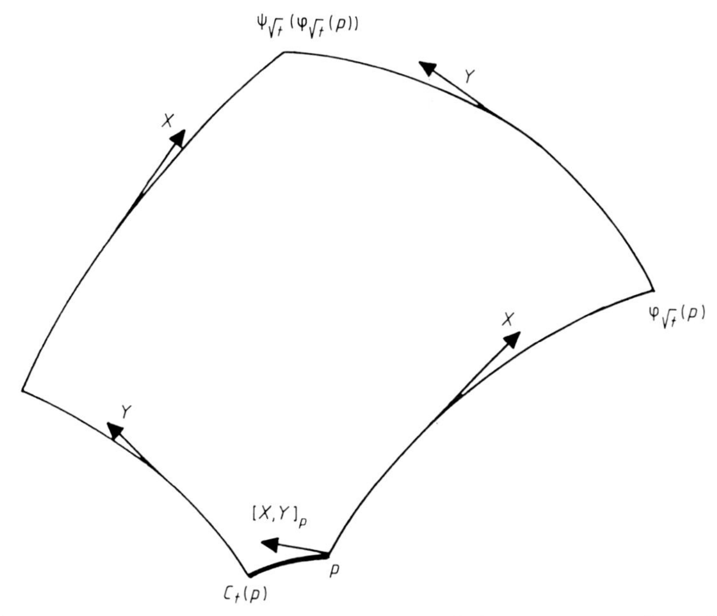

Figure 4.13 The geometrical interpretation of the commutator $[X, Y]$ of the vector fields X and Y.

Exercise 4.1
Let X and Y be vector fields with $\varphi(p)$ and $\psi(p)$ the respective integral curves starting at p, and φ_t and ψ_t the associated local diffeomorphisms (see figure 4.13). For t sufficiently small and positive a one-parameter family of local diffeomorphisms is given by $C_t = \psi_{-\sqrt{t}} \circ \varphi_{-\sqrt{t}} \circ \psi_{\sqrt{t}} \circ \varphi_{\sqrt{t}}$, with $C(p) : t \mapsto C_t(p)$ a smooth curve starting at p. If $\dot{C}_0(p)$ is the tangent vector to $C(p)$ at the point p show that $\dot{C}_0(p) = [X, Y]_p$. Hint: For $f \in \mathscr{F}(M) f \circ \varphi_t = f + tXf + t^2/2X^2f + O(t^3)$, where $X^2f = X(Xf)$.

4.12 Lie Derivatives

In §4.4 we motivated the concept of a tangent vector by introducing differentiation of functions along a curve. Having arrived at the definition by which a vector field is a derivation on the algebra of smooth functions we showed in that section how the action of any vector field on a function is the derivative of that function along a curve; namely the integral curve of the vector field. That is, if φ is an integral curve of X, starting at p

$$\begin{aligned}(Xf)(p) &= \frac{\mathrm{d}}{\mathrm{d}t}(f \circ \varphi(p))(0) \qquad \forall f \in \mathscr{F}(M) \\ &= \lim_{t\to 0} t^{-1}\{f(\varphi_t(p)) - f(p)\}. \end{aligned} \tag{4.12.1}$$

Since X is a smooth vector field then associated with the curve $\varphi(p)$ starting at p, is the local diffeomorphism φ_t on the neighbourhood of p. It is instructive to rewrite the above in terms of the pull-back φ_t^*

$$(Xf)(p) = \lim_{t\to 0} t^{-1}\{(\varphi_t^* f)(p) - f(p)\}. \tag{4.12.2}$$

This form of the derivative, X on f, suggest a generalisation to a derivative on an arbitrary tensor field $T \in \Gamma T^r_s M$. We may use the map $\hat{\varphi}_{-t}$, associated with the vector field X, to map $T_{\varphi_t(p)}$ back to the point p where it can be compared with T_p. If the limit as t tends to zero of the difference between the two tensors divided by the parameter t exists, it is called the *Lie derivative* at p of T with respect to X, denoted by $(\mathscr{L}_X T)(p)$ (see figure 4.14)

$$(\mathscr{L}_X T)(p) = \lim_{t\to 0} t^{-1}\{(\hat{\varphi}_{-t}T)_p - T_p\} \qquad \forall p \in M. \tag{4.12.3}$$

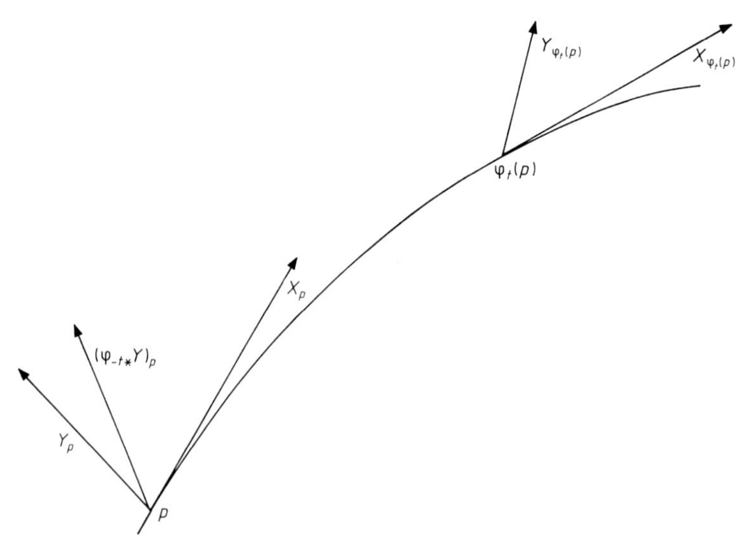

Figure 4.14 This diagram illustrates the vectors used in the definition of the Lie derivative of a vector field Y with respect to the vector field X (tangent to some integral curve).

The definition of $\hat{\varphi}_{-t}$ is given in (4.9.5). In the particular case of $f \in \mathscr{F}(M)$

$$\hat{\varphi}_{-t}f = \varphi_{-t}^{-1*}f = \varphi_t^* f$$

and we have

$$\mathscr{L}_X f = Xf \qquad \forall f \in \mathscr{F}(M). \tag{4.12.4}$$

It follows from (4.9.5) that $\mathscr{L}_X$ is a derivation on the algebra of tensor fields

$$\mathscr{L}_X(S \otimes T) = \mathscr{L}_X S \otimes T + S \otimes \mathscr{L}_X T \tag{4.12.5}$$

in particular if $f \in \mathscr{F}(M)$

$$\mathscr{L}_X(fT) = (Xf)T + f\mathscr{L}_X T. \tag{4.12.6}$$

From (4.9.6) we may deduce that $\mathscr{L}_X$ commutes with contractions:

$$\begin{aligned}\mathscr{L}_X(T(\alpha_1, \ldots, \alpha_r, X_1, \ldots, X_s)) &= (\mathscr{L}_X T)(\alpha_1, \ldots, \alpha_r, X_1, \ldots, X_s) \\ &+ \sum_{k=1}^{r} T(\alpha_1, \ldots, \mathscr{L}_X \alpha_k, \ldots, \alpha_r, X_1, \ldots, X_s) \\ &+ \sum_{k=1}^{s} T(\alpha_1, \ldots, \alpha_r, X_1, \ldots, \mathscr{L}_X X_k, \ldots, X_s).\end{aligned} \tag{4.12.7}$$

Applying the general definition of a Lie derivative to a vector field Y gives

$$(\mathscr{L}_X Y)(p) = \lim_{t \to 0} \frac{1}{t}\{(\varphi_{-t*}Y)_p - Y_p\}. \tag{4.12.8}$$

For any $f \in \mathscr{F}(M)$

$$\begin{aligned}(\mathscr{L}_X Y)_p f &= \lim_{t \to 0} t^{-1}\{[(\varphi_{-t})_*(Y_{\varphi_t(p)})]f - Y_p f\} \\ &= \lim_{t \to 0} t^{-1}\{Y_{\varphi_t(p)}(\varphi^*_{-t}f) - Y_p f\}.\end{aligned}$$

Now from (4.12.2)

$$\varphi_t^* f = f + tXf + \mathrm{O}(t^2)$$

hence

$$\begin{aligned}(\mathscr{L}_X Y)_p f &= \lim_{t \to 0} t^{-1}\{Y_{\varphi_t(p)}[f - tXf + \mathrm{O}(t^2)] - Y_p f\} \\ &= -Y_p(Xf) + \lim_{t \to 0} t^{-1}\{Y_{\varphi_t(p)}f - Y_p f\} \\ &= -Y_p(Xf) + \lim_{t \to 0} t^{-1}\{(Yf)(\varphi_t(p)) - (Yf)(p)\}.\end{aligned}$$

Since $Yf \in \mathscr{F}(M)$ we see from (4.12.1) that the last term is $X_p(Yf)$, therefore

$$(\mathcal{L}_X Y)_p f = X_p(Yf) - Y_p(Xf)$$
$$((\mathcal{L}_X Y)f)(p) = (X(Yf) - Y(Xf))(p) \quad \forall p$$

or

$$\mathcal{L}_X Y = [X, Y] \tag{4.12.9}$$

where $[X, Y] \equiv XY - YZ$ is the commutator of the two vector fields. From this example we note that $\mathcal{L}_{fX} \neq f\mathcal{L}_X$ where $f \in \mathcal{F}(M)$. We say that $\mathcal{L}_X$ is not $\mathcal{F}$-linear in X. This reflects the fact that the Lie derivative along a curve depends on the parametrisation of the curve and not just on its image. When using a coordinate basis to evaluate $\mathcal{L}_X$ on tensor fields with contravariant components it is useful to note from (4.12.9) that

$$\mathcal{L}_{(\partial/\partial x^i)}(\partial/\partial x^j) = [(\partial/\partial x^i), (\partial/\partial x^j)] = 0.$$

The properties of $\mathcal{L}_X$ that we have established are sufficient to determine it completely; it being the unique type-preserving derivation on tensor fields that satisfies (4.12.5), (4.12.6), (4.12.7) and (4.12.9). A consequence of these uniqueness properties is

$$[\mathcal{L}_X, \mathcal{L}_Y] = \mathcal{L}_{[X,Y]}. \tag{4.12.10}$$

The commutator of two derviations that commute with contractions is a derivation that commutes with contractions. Certainly both sides of (4.12.10) agree when evaluated on a function, so we need only confirm that they agree when evaluated on an arbitrary vector field. This follows from the Jacobi identity (4.6.6). Whereas the established properties of the Lie derivative completely specify it, these being used in any practical calculation, the definition (4.12.3) conveys the geometrical significance: $\mathcal{L}_X T = 0$ if and only if the tensor field T is invariant under the local diffeomorphisms generated by X.

Example 4.5

We shall evaluate $\mathcal{L}_X Y$ for $X, Y \in \Gamma T\mathbb{R}^2$ given by $X = x(\partial/\partial y) - y(\partial/\partial x)$, $Y = x^2(\partial/\partial x) + xy(\partial/\partial y)$. First we shall apply the definition (4.12.8) directly. In the first example of §4.11 we found the integral curves of X starting at $p = (a, b)$. This gives the diffeomorphism φ_t

$$(a, b) \longmapsto \varphi_t(a, b) = (a\cos t - b\sin t, b\cos t + a\sin t).$$

We have already computed $\varphi_{-t*}Y$, in example 4.2 (the map φ there being called φ_{-t} here). So (4.12.8) becomes

$$\mathscr{L}_X Y = \lim_{t\to 0} t^{-1}\,[(x^2\cos t - xy\sin t - x^2)(\partial/\partial x)$$

$$+ (xy\cos t - y^2\sin t - xy)(\partial/\partial y)]$$

$$= \lim_{t\to 0} t^{-1}(\cos t - 1)[x^2(\partial/\partial x) + xy(\partial/\partial y)]$$

$$- \lim_{t\to 0} t^{-1}\sin t[xy(\partial/\partial x) + y^2(\partial/\partial x)]$$

$$= -xy(\partial/\partial x) - y^2(\partial/\partial y).$$

We now evaluate $\mathscr{L}_X Y$ more practically, using the derived properties of $\mathscr{L}_X$

$$\begin{aligned}\mathscr{L}_X Y &= X(x^2)(\partial/\partial x) + x^2\mathscr{L}_X(\partial/\partial x) + X(xy)(\partial/\partial y) + xy\mathscr{L}_X(\partial/\partial y)\\ &= X(x^2)(\partial/\partial x) - x^2\mathscr{L}_{(\partial/\partial x)}X + X(xy)(\partial/\partial y) - xy\mathscr{L}_{(\partial/\partial y)}X\\ &= X(x^2)(\partial/\partial x) - x^2(\partial/\partial y) + X(xy)(\partial/\partial y) + xy(\partial/\partial x)\end{aligned}$$

since $\mathscr{L}_{(\partial/\partial x)}(\partial/\partial y) = 0$

$$= -xy(\partial/\partial x) - y^2(\partial/\partial y).$$

Since the Lie derivative is a derivation on the tensor algebra it is also a derivation on the exterior algebra of differential forms. There are a number of useful properties of the Lie derivative acting on differential forms. First, since the exterior derivative on forms commutes with φ^* for any smooth map φ, it follows that

$$\mathscr{L}_X \mathrm{d} = \mathrm{d}\mathscr{L}_X. \tag{4.12.11}$$

This is a very useful property for calculations involving Lie derivations of covariant tensor fields expressed in a natural coordinate basis. In Chapter 1 we gave the definition of the interior operator on exterior forms with respect to a vector from the dual space. The interior operator i_X on a differential form ω, with respect to a vector field X, is naturally defined to satisfy

$$(\mathrm{i}_X\omega)_p = \mathrm{i}_{X_p}\omega_p.$$

Thus the graded derivation i_X is $\mathscr{F}$-linear in X. Since $\mathscr{L}_X$ commutes with contractions it follows that

$$[\mathscr{L}_X, \mathrm{i}_Y] = \mathrm{i}_{[X,Y]}. \tag{4.12.12}$$

When acting on differential forms the Lie derivative can be expressed in terms of the exterior and interior derivatives

$$\mathscr{L}_X = \mathrm{d}\,\mathrm{i}_X + \mathrm{i}_X\,\mathrm{d} \qquad \forall X \in \Gamma TM. \tag{4.12.13}$$

The equality of these expressions is most readily seen by noting that both are derivations on the exterior algebra, commuting with d and agreeing on functions. For if $f \in \mathscr{F}(M)$

$$(\mathrm{d}\,\mathrm{i}_X + \mathrm{i}_X\,\mathrm{d})f = \mathrm{i}_X\,\mathrm{d}f = \mathrm{d}f(X) = Xf = \mathscr{L}_X f.$$

If $\alpha \in \Gamma\Lambda_p M$ and $\beta \in \Gamma\Lambda M$ then

$$\begin{aligned}(\mathrm{d}\,\mathrm{i}_X + \mathrm{i}_X\,\mathrm{d})(\alpha \wedge \beta) &= \mathrm{d}[\mathrm{i}_X\alpha \wedge \beta + (-1)^p \alpha \wedge \mathrm{i}_X\,\beta] \\ &\quad + \mathrm{i}_X[\mathrm{d}\alpha \wedge \beta + (-1)^p \alpha \wedge \mathrm{d}\beta] \\ &= \mathrm{d}\mathrm{i}_X\alpha \wedge \beta + (-1)^{p-1}\mathrm{i}_X\alpha \wedge \mathrm{d}\beta + (-1)^p\,\mathrm{d}\alpha \wedge \mathrm{i}_X\beta + \alpha \wedge \mathrm{d}\mathrm{i}_X\beta \\ &\quad + \mathrm{i}_X\mathrm{d}\alpha \wedge \beta + (-1)^{p+1}\mathrm{d}\alpha \wedge \mathrm{i}_X\beta + (-1)^p\mathrm{i}_X\alpha \wedge \mathrm{d}\beta \\ &\quad + \alpha \wedge \mathrm{i}_X\mathrm{d}\beta \\ &= (\mathrm{d}\,\mathrm{i}_X + \mathrm{i}_X\,\mathrm{d})\alpha \wedge \beta + \alpha \wedge (\mathrm{d}\,\mathrm{i}_X + \mathrm{i}_X\,\mathrm{d})\beta.\end{aligned}$$

It is straightforward to see that $(\mathrm{d}\,\mathrm{i}_X + \mathrm{i}_X\,\mathrm{d})$ commutes with d since $\mathrm{d}^2 = 0$. The existence of a local coordinate basis for M ensures that the above properties are sufficient to establish (4.12.13).

Example 4.6

For $X \in \Gamma T\mathbb{R}^2$ and $\alpha \in \Gamma T^*\mathbb{R}^2$ we shall evaluate $\mathscr{L}_X\alpha$, first from the definition then, as will always be done in practice, from the established properties of $\mathscr{L}_X$. We take $X = x(\partial/\partial y) - y(\partial/\partial x)$, $\alpha = x^2\,\mathrm{d}x + xy\,\mathrm{d}y$. As was noted in the previous example we may use earlier examples to proceed from the definition

$$\begin{aligned}&\mathscr{L}_X\alpha \\ &\quad = \lim_{t\to 0} t^{-1}\,[(x^2\cos t - xy\sin t - x^2)\mathrm{d}x + (xy\cos t - y^2\sin t - xy)\mathrm{d}y] \\ &\quad = -xy\,\mathrm{d}x - y^2\,\mathrm{d}y.\end{aligned}$$

alternatively,

$$\begin{aligned}\mathscr{L}_X\alpha &= X(x^2)\mathrm{d}x + x^2\mathscr{L}_X\mathrm{d}x + X(xy)\,\mathrm{d}y + xy\mathscr{L}_X\mathrm{d}y \\ &= X(x^2)\mathrm{d}x + x^2\,\mathrm{d}(Xx) + X(xy)\mathrm{d}y + xy\mathrm{d}(Xy) \\ &= -2xy\,\mathrm{d}x + x^2\,\mathrm{d}(-y) + (x^2 - y^2)\,\mathrm{d}y + xy\,\mathrm{d}(x) \\ &= -xy\,\mathrm{d}x - y^2\,\mathrm{d}y.\end{aligned}$$

For the vector field Y of the previous example

$$(\mathscr{L}_X\alpha)(Y) = -x^3y - xy^3$$

$$\alpha(\mathscr{L}_X Y) = -x^3y - xy^3$$

whereas

$$\alpha(Y) = x^4 + x^2y^2.$$

These are indeed related by

$$\mathscr{L}_X(\alpha(Y)) = (\mathscr{L}_X\alpha)(Y) + \alpha(\mathscr{L}_XY).$$

4.13 Integration On Manifolds

The differential forms derive a certain prominence amongst the tensor fields on a manifold from the fact that they give rise to a theory of integration, generalising the Riemann integral in $\mathbb{R}^r$. We recall that such integrals may be defined as the limit attained by a Riemann sum of terms, each consisting of a measure associated with some (usually cubical) subdivision of a domain multiplied by the value taken by the function to be integrated at some point within each cell of the subdivision. We shall assume that the reader is familiar with the methods of evaluating multiple integrals in $\mathbb{R}^r$ by means of iterated integrals. The classical notation for a Riemann integral suggests a natural definition for the integral of an r-form on $\mathbb{R}^r$ over an oriented domain. With such a definition a mapping from $\mathbb{R}^r$ to an n-dimensional manifold M enables an r-form on M to be integrated: we use the map to pull it back to $\mathbb{R}^r$ where the integration is defined. Properly formulated the above idea gives the theory of integration of differential forms over oriented chains.

Let $[0, 1]^r$ be the set of points $p \in \mathbb{R}^r$ that satisfy $0 \leqslant \sigma^k(p) \leqslant 1$, $k = 1, \ldots, r$ in any natural chart $\{\sigma^k\}$ for $\mathbb{R}^r$. Thus $[0, 1]^r$ is the *unit cube* in $\mathbb{R}^r$. Introduce Ω^r for the natural 'volume' r-form $d\sigma^1 \wedge d\sigma^2 \wedge \ldots \wedge d\sigma^r$ which serves to orient $[0, 1]^r$. An *oriented r-cube* on an n-dimensional manifold M is the pair (C^r, Ω^r) where C^r is a C^∞ map C^r: $[0, 1]^r \to M$. (To say that C^r is C^∞ on the closed set means that there is a C^∞ map $\mathscr{C}^r$ between open sets containing the domain and image of C^r such that C^r is obtained from $\mathscr{C}^r$ by restriction.) In a local chart (U, x) we may represent the map $C^r : p \in [0, 1]^r \mapsto q \in M$ by its components, $(\lambda^1, \ldots, \lambda^r)$

$$x^i(q) = \lambda^i(\sigma^1(p), \ldots, \sigma^r(p)) \qquad i = 1, \ldots, n. \qquad (4.13.1)$$

Every oriented r-cube gives rise to $2r$ oriented $(r - 1)$-cubes called its oriented $(r - 1)$-*faces*. Each face is defined by restricting the map C^r to points p for which $\sigma^j(p) = \varepsilon$, where $\varepsilon = 0, 1$. Denoting the $(r - 1)$-faces by $C^{r-1}_{(j,\varepsilon)} : [0, 1]^{r-1} \to M$ we have then

$$\begin{aligned} C^{r-1}_{(j,\varepsilon)}(\sigma^1(p), \ldots, \sigma^{j-1}(p), \sigma^{j+1}(p), \ldots, \sigma^r(p)) \\ = C^r(\sigma^1(p), \ldots, \sigma^{j-1}(p), \varepsilon, \sigma^{j+1}(p), \ldots, \sigma^r(p)) \\ j = 1, \ldots, r; \varepsilon = 0, 1 \qquad (4.13.2) \end{aligned}$$

Each $(r-1)$-face may be given a unique orientation $\Omega^{(r-1)}_{(j,\varepsilon)}$, induced from the orientation of C^r:

$$\Omega^{r-1}_{(j,\varepsilon)} = (-1)^{\varepsilon+1}\, \mathrm{i}_{(\partial/\partial\sigma^j)}\Omega^r \qquad j = 1, \ldots, r;\ \varepsilon = 0, 1 \quad (4.13.3)$$

from which it follows that the faces labelled by $\varepsilon = 0, 1$ have opposite induced orientations. An oriented 1-cube has two oppositely oriented 0-faces (its end points or vertices) each of which is assigned an orientation $+$ or $-$. We may recursively define k-faces of C^r, for $k = r-2, r-3, \ldots, 0$; these being the k-cubes obtained by similarly restricting the $(k+1)$-cubes. The 2^r 0-faces (or vertices) of C^r are the 0-cubes obtained by restricting the map C^r with all $\sigma^i(p)$ equal to zero or one. For $b_j \in \mathbb{R}$ the finite sum $\Sigma_j b_j C^r_j$, that maps some set $\{C^r_j, \Omega^r_j\}$ of oriented r-cubes into M, is called an *oriented r-chain* (with real coefficients).

The oriented r-cube (C^r, Ω^r) has a *boundary* $(r-1)$-chain denoted by $\partial(C^r, \Omega^r)$ which is defined as

$$\partial(C^r, \Omega^r) = \sum_{i=1}^{r} \sum_{\varepsilon=0,1} (C^{r-1}_{(i,\varepsilon)}, \Omega^{r-1}_{(i,\varepsilon)}). \qquad (4.13.4)$$

The boundary operator ∂ extends naturally to all r-chains:

$$\partial\Big(\sum_j b_j(C^r_j, \Omega^r_j)\Big) = \sum_j b_j \partial(C^r_j, \Omega^r_j).$$

It follows directly from the definition of C^{r-2} that $\partial\partial = 0$ since the $(r-2)$-faces cancel pairwise.

From an r-form α, defined on the image of C^r, we can use the map C^r to 'pull back' α to $[0, 1]^r$. The r-form $(C^r)^*\alpha$ has the representation $h\mathrm{d}\sigma^{i_1} \wedge \mathrm{d}\sigma^{i_2} \wedge \ldots \wedge \mathrm{d}\sigma^{i_r}$, $h \in \mathscr{F}(\mathbb{R}^r)$. The orientation Ω^r of C^r is now used to define $\varepsilon_r = \pm 1$ by

$$\Omega^r = \varepsilon_r\, \mathrm{d}\sigma^{i_1} \wedge \mathrm{d}\sigma^{i_2} \wedge \ldots \wedge \mathrm{d}\sigma^{i_r}.$$

We define the integral of $C^{r*}\alpha$ over $[0, 1]^r$ in terms of the Riemann integral of h

$$\int_{[0,1]^r} C^{r*}\alpha = \varepsilon_r \int_{[0,1]^r} h\, \mathrm{d}\sigma^{i_1} \ldots \mathrm{d}\sigma^{i_r}.$$

This may be evaluated as the iterated integral

$$\varepsilon_r \int_0^1 \ldots \left[\int_0^1 \left(\int_0^1 h(\sigma^1, \sigma^2, \ldots, \sigma^r)\, \mathrm{d}\sigma^1\right) \mathrm{d}\sigma^2\right] \ldots \mathrm{d}\sigma^r.$$

We may now define the integral of an r-form on M over an oriented r-cube

$$\int_{C^r} \alpha = \int_{[0,1]^r} (C^r)^*\alpha. \qquad (4.13.5)$$

This definition is extended to include 0-forms by defining the integral of a 0-form over a 0-cube to be the difference between the values of the 0-form taken at the two end points. If $\varphi : [0, 1]^r \mapsto [0, 1]^r$ is a smooth reparametrisation that preserves orientations and $C'^r = C^r \circ \varphi$ then

$$\begin{aligned}\int_{C'^r} \alpha &= \int_{(C^r \circ \varphi)} \alpha \\ &= \int_{[0,1]^r} (C^r \circ \varphi)^* \alpha = \int_{[0,1]^r} \varphi^*(C^{r*} \alpha) \\ &= \int_{\varphi[0,1]^r} C^{r*} \alpha = \int_{[0,1]^r} C^{r*} \alpha\end{aligned}$$

since the last equality follows from a change of variable $\sigma \mapsto \sigma' = \varphi(\sigma)$ in the iterated integral. Hence

$$\int_{C'^r} \alpha = \int_{C^r} \alpha$$

and we say that the oriented r-cubes C^r and C'^r are *equivalent*. The integral of α over the r-chain $C = \Sigma_j b_j C_j^r$ is defined to be

$$\int_{C^r} \alpha = \sum_j b_j \int_{C_j^r} \alpha.$$

The culmination of this treatment of r-form integration over oriented r-chains is the elegant generalisation of Stokes's theorem afforded by this formalism. For any smooth $r - 1$ form β defined in the range of the r-chain $C (r \geqslant 1)$ we have

$$\int_C \mathrm{d}\beta = \int_{\partial C} \beta. \tag{4.13.6}$$

The definitions are such that this follows immediately from the result in $\mathbb{R}^r$. First we observe that (4.13.6) will hold for an arbitrary chain if it is true for any r-cube; then we use definition (4.13.5) to relate the integrals to Riemann integrals. Since $C^*\mathrm{d} = \mathrm{d}C^*$ the proof of (4.13.6) reduces to that of Stokes's theorem in $\mathbb{R}^r$. Since the Riemann integral can be written as a repeated integral the proof finally rests on the fundamental theorem of calculus; the integral of a real function is the anti-derivative.

An immediate consequence of Stokes's theorem is the generalisation of the rule for 'integration by parts' to exterior products of forms on a manifold. If $\alpha \in \Gamma\Lambda_r M$, $\beta \in \Gamma\Lambda_q M$ then

$$\mathrm{d}(\alpha \wedge \beta) = \mathrm{d}\alpha \wedge \beta + (-1)^r \alpha \wedge \mathrm{d}\beta.$$

Consequently for some $(r + q + 1)$-chain C

$$\int_C \mathrm{d}(\alpha \wedge \beta) = \int_C \mathrm{d}\alpha \wedge \beta + (-1)^r \int_C \alpha \wedge \mathrm{d}\beta = \int_{\partial C} \alpha \wedge \beta \tag{4.13.7}$$

by Stokes's theorem. If $\partial C = 0$ or $\alpha \wedge \beta = 0$ on ∂C we have the simple result

$$\int_C \mathrm{d}\alpha \wedge \beta = (-1)^{r+1} \int_C \alpha \wedge \mathrm{d}\beta. \tag{4.13.8}$$

Example 4.7
We consider the chain C:

$$C: [0, 1]^2 \longrightarrow \mathbb{R}^3$$
$$(\tau, \sigma) \longmapsto (\sin \pi\tau \cos 2\pi\sigma, \sin \pi\tau \sin 2\pi\sigma, \cos \pi\tau).$$

If (r, θ, φ) are the standard polar coordinates for $\mathbb{R}^3$ then this map sends (τ,σ) to the point on the unit sphere with polar coordinates $(1, \pi\tau, 2\pi\sigma)$. The spherical polar coordinates (θ, φ) do not cover the sphere, there are coordinate singularities at $\theta = 0, \pi$ and $\varphi = 0, 2\pi$ (see figure 4.15). Thus the C^∞ chain C is a diffeomorphism from the interior of its domain onto its image, whilst the boundary of the cube is mapped onto the points at which the coordinates are singular. We will integrate the 2-form $\omega = r^3 \sin\theta \, \mathrm{d}\theta \wedge \mathrm{d}\varphi$ over C. Note first that ω is smooth on the whole of $\mathbb{R}^3$. This can be seen by changing to Cartesian coordinates that cover all of $\mathbb{R}^3$, giving $\omega = x \, \mathrm{d}y \wedge \mathrm{d}z + y \, \mathrm{d}z \wedge \mathrm{d}x + z \, \mathrm{d}x \wedge \mathrm{d}y$. We have $C^*\mathrm{d}\theta = \pi \mathrm{d}\tau$, $C^*\mathrm{d}\varphi = 2\pi \mathrm{d}\sigma$ giving $C^*\omega = 2\pi^2 \sin(\pi\tau)\mathrm{d}\tau \wedge \mathrm{d}\sigma$ and

$$\int_{[0,1]^2} C^*\omega = 2\pi^2 \int_0^1 \left(\int_0^1 \sin(\pi\tau)\mathrm{d}\tau \right) \mathrm{d}\sigma = 4\pi.$$

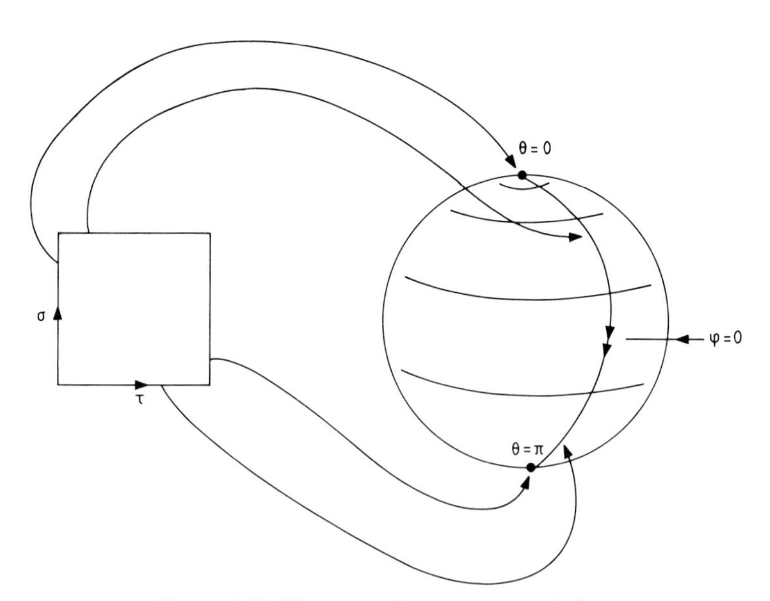

Figure 4.15 The two-sphere as a two-chain.

In the above example it is tempting to say that we have integrated 'over the surface of the unit sphere', although we can so far attach no meaning to this statement, our integrals of forms being over chains. However, a class of chains (a member of which was considered in the example above) can be put into correspondence with subsets of an oriented manifold N, such that we can unambiguously refer to integration over the subset. An oriented r-cube C^r is said to *parametrise* a region S of an oriented r-dimensional manifold N if $C^r([0,1]^r) = S$, C^r is a diffeomorphism on the interior of its domain and the orientation of the cube is compatible with that of the image. That is, if $\{(\partial/\partial\sigma^a)\}$ is an oriented basis for the cube then $\{C_{*p}(\partial/\partial\sigma^a)\}$ is positively oriented with respect to the orientation of N for all points p for which C_{*p} is a non-singular linear transformation. (These conditions are met in the above example with N the 2-sphere with orienting 2-form ω.) We can certainly parametrise a region S with more than one r-cube, the crucial result being that if ω is an r-form on N which is parametrised by both C^r and C'^r then $\int_{C^r}\omega = \int_{C'^r}\omega$. It is therefore meaningful to define

$$\int_S \omega = \int_{C^r} \omega$$

where C^r parametrises S. Although we shall not prove the above we observe that it is certainly reasonable. On the interior of their domains C^r and C'^r are invertible, and hence $(C^r)^{-1} \circ C'^r$ is an orientation-preserving diffeomorphism between the interiors of the domains. We have already shown that integrals are invariant under changes of chain that are related by orientation-preserving diffeomorphisms, and so to prove the above result it is necessary to show (as one would expect) that the boundary does not contribute to the integral. (Such an argument shows that parametrising cubes can be a little more general than defined here.)

An r-chain $C = \Sigma_s C^r{}_s$ parametrises a region S if the image of C is S, each $C^r{}_s$ parametrises its image and the images of the interiors of the cubes are non-intersecting. Again one can show that the integrals of any smooth r-form over any two parametrising chains are equal. The proof that one can parametrise certain regions (for example, compact manifolds and compact manifolds with boundary) is not simple and we refer the interested reader to the literature.

4.14 Metric Tensor Fields

A metric tensor field g on manifold M is a section of a second-rank tensor bundle over M. Restricted to a point $p \in M$ it provides a metric tensor on the space T_pM. If g is a symmetric positive-definite non-degenerate metric tensor field the manifold is said to be a *Riemannian* manifold. If g is a symmetric but indefinite non-degenerate metric tensor field the manifold is said to be a *pseudo-Riemannian* or (*semi-Riemannian*) one. For the special case of signature $(p, 1)$ a pseudo-Riemannian manifold is called *Lorentzian*.

Let us develop the description of a (pseudo-) Riemannian metric in a local chart (U_M, φ_M). If $\{dx^\mu\}$ is a local basis for 1 forms for T^*_pM we may write the tensor field g as

$$g = g_{\mu\nu} dx^\mu \otimes dx^\nu \qquad (4.14.1)$$

where the $n(n+1)/2$ real-valued functions $g_{\mu\nu} = g(\partial/\partial x^\mu, \partial/\partial x^\nu)$ satisfy $g_{\mu\nu} = g_{\nu\mu}$ $(\mu, \nu = 1, \ldots, n)$. A g-orthonormal basis $\{X_a\}$ of T_pM is one that satisfies

$$g(X_a, X_b) = \eta_{ab} = \pm 1 \qquad a, b = 1, \ldots, n. \qquad (4.14.2)$$

An ordered basis of local vector fields defines a local frame on M and an ordered basis of 1-forms a local co-frame. The components η_{ab} of g in a g-orthonormal co-frame are real constants and we may write

$$g = \eta_{ab} e^a \otimes e^b$$

where $\{e^a\} \in \Gamma T^*M$ is a g-orthonormal co-frame satisfying

$$e^a(X_b) = \delta^a_b \qquad \forall a, b = 1, \ldots, n. \qquad (4.14.3)$$

Fields of frames are sometimes called moving frames. As described in Appendix A the metric tensor enables T_pM and T^*_pM to be related. If $\alpha \in \Gamma T^*M$ then $\tilde{\alpha} \in \Gamma TM$ is defined by

$$g(\tilde{\alpha}, X) = \alpha(X) \qquad \forall X \in \Gamma TM. \qquad (4.14.4)$$

The contravariant (pseudo-Riemannian) metric g^* is a tensor field on M that when restricted to a point $p \in M$ provides a metric on the vector space T^*_pM, defined by

$$g^*(\alpha, \beta) = g(\tilde{\alpha}, \tilde{\beta}) \qquad \forall \alpha, \beta \in \Gamma T^*M. \qquad (4.14.5)$$

In a local chart we may write

$$g^* = g^{\mu\nu} \partial/\partial x^\mu \otimes \partial/\partial x^\nu = \eta^{ab} X_a \otimes X_b$$

where $g^{\nu\mu} = g^{\mu\nu} \in \mathcal{F}(M)$ and

$$g^{\mu\nu} g_{\nu\rho} = \delta^\mu_\rho$$

$$\eta^{ab} \eta_{bc} = \delta^a_c.$$

The $\mathrm{Gl}(n, \mathbb{R})$ elements e^a_μ relating natural and g-orthonormal co-frame fields,

$$e^a = e^a_\mu \mathrm{d}x^\mu \tag{4.14.6}$$

are now functions on M. Some authors refer to the co-frame $\{e^a\}$ as an n-bein, others reserve the term n-beins for the n^2 functions $e^a_\mu \in \mathscr{F}(M)$. It should be noticed that, unlike the natural co-basis, in general $\mathrm{d}e^a \neq 0$, $a = 1, \ldots, n$.

If the 1-form ω is written locally as $\omega = \omega_\mu \mathrm{d}x^\mu = \omega_a e^a$ then the metric dual is $\widetilde{\omega} = \omega^\mu \partial/\partial x^\mu = \omega^a X_a$, where $\omega^\mu = g^{\mu\nu}\omega_\nu$ and $\omega^a = \eta^{ab}\omega_b$. Similarly, if locally $X = \xi^\mu \partial/\partial x^\mu = \xi^a X_a$, then $\widetilde{X} = \xi_\mu \mathrm{d}x^\mu = \xi_a e^a$ where $\xi_\mu = g_{\mu\nu}\xi^\nu$ and $\xi_a = \eta_{ab}\xi^b$ (see Appendix A). The index notation is doing double duty here, the Greek and Roman alphabets indicating that the components are with respect to a natural and orthonormal basis respectively. The symbols $\omega^\mu = \omega(\mathrm{d}x^\mu)$ and $\omega^a = \omega(e^a)$ obviously represent different functions on M. Thus it is potentially hazardous when working with components to give μ and a a numerical value. Clearly a safer (but rarely used) procedure would be to write unambiguously

$$\omega = \omega(\partial/\partial x^\mu)\mathrm{d}x^\mu = \omega(X_a)e^a$$

$$X = \mathrm{d}x^\mu(X)\partial/\partial x^\mu = e^a(X)X_a.$$

We discussed in Chapter 1 how to use a metric on co-vectors to construct a metric on p-forms. That procedure can now be generalised to construct a metric on differential forms. If M is an n-dimensional orientable manifold with a fixed atlas, specifying a positive orientation say, then one may smoothly assign an orientation to T_pM for all $p \in M$. Equivalently, if (U_a, φ_a) and (U_b, φ_b) are any overlapping charts in this atlas, with coordinate functions $\{x^\mu\}$ are $\{y^\mu\}$ respectively, then the real-valued function f on $U_a \cap U_b$, defined by $\mathrm{d}x^1 \wedge \mathrm{d}x^2 \wedge \ldots \mathrm{d}x^n = f\mathrm{d}y^1 \wedge \mathrm{d}y^2 \wedge \ldots \wedge \mathrm{d}y^n$, is everywhere positive since f is just the Jacobian of the transition map between charts. Thus we are assured of a non-vanishing n-form on any orientable differential manifold. If such a manifold admits a (pseudo-)Riemannian metric tensor field then a canonical choice of orienting n-form is $z = e^1 \wedge e^2 \wedge \ldots \wedge e^n$ where $\{e^a\}$ is a g-orthonormal moving co-frame. We may now extend the construction of the Hodge map given earlier to M with $*1 = z$. This enables the domain of the Hodge map to be generalised to sections of ΛM.

If $\varphi : M \mapsto N$ is a smooth diffeomorphism between (pseudo)-Riemannian manifolds M and N such that the metric tensor fields g_M on M and g_N on N are related by

$$g_M = \varphi^* g_N$$

then φ is said to be a smooth *isometry*. As a special case if $M = N$ then φ is a smooth isometry of M. If $\{\varphi_i\}$ is a set of such maps on M

then they form the isometry group of M under composition. The set of vector fields $\{K_i\}$ that generate these isometries are known as *Killing vectors*. Because the commutator of Lie derivatives is the Lie derivative with respect to a commutator of vector fields, in the neighbourhood of any point in M the Killing vector fields form a Lie algebra under the commutator; $[K_i, K_j] = c_{ij}{}^k K_k$ where $\{c_{ij}{}^k\}$ are the *structure constants* in this basis. The isometry group defines a *Killing symmetry* of the (pseudo)-Riemannian structure on M; the metric tensor field satisfying

$$\mathcal{L}_K g = 0$$

for any vector field K in the algebra of Killing vectors. In general a (pseudo)-Riemannian manifold will admit no isometries, and hence possess no Killing vectors. Furthermore, there is a maximum number, $\frac{1}{2}n(n+1)$, of Killing fields that can exist for any metric on M.

Example 4.8: Euclidean Manifolds

The topological space whose points consist of the n-tuples in $\mathbb{R}^n$ may be given a manifold structure by adopting an atlas consisting of the identity chart that assigns a unique element of $\mathbb{R}^n$ to each point. On any open sets U, V on this manifold one may adopt 'local curvilinear coordinates', $\varphi_U : U \mapsto \mathbb{R}^n$, $\varphi_V : V \mapsto \mathbb{R}^n$ provided $\varphi_U \circ \varphi_V^{-1}$ is smooth and 1 : 1 with a non-zero Jacobian on $U \cap V$. This manifold has a natural Riemannian structure. In a global chart $\{x^i, \mathbb{R}^n\}$ the metric tensor field takes the form $g = \Sigma_{i=1,\ldots n} \mathrm{d}x^i \otimes \mathrm{d}x^i$. The manifold $\mathbb{R}^n$ with this Riemannian structure is a model for an n-dimensional Euclidean manifold. Any n-dimensional Riemannian manifold isometric to this one under a (smooth) diffeomorphism provides a model for the space of Euclid. Such manifolds admit $\frac{1}{2}n(n-1)$ *rotational* isometries (the integral curves of the Killing vectors lying on an $(n-1)$-sphere) together with n *translational* isometries (with the Killing vectors having open integral curves). The group of these isometries is known as the *Poincaré* group of n-dimensional Euclidean space.

Some of the ideas in this chapter are illustrated in Appendix B where the familiar vector calculus of three-dimensional Euclidean space is reformulated.

Bibliography

Abramhams R, Marsden J and Ratiu T 1983 *Tensor Analysis and Applications* (New York: Addison-Wesley)

Bishop R L and Goldberg S I 1980 *Tensor Analysis on Manifolds* (New York: Pitman)

Clarke C 1979 *Elementary General Relativity* (London: Edward Arnold)

Dodson C T J and Poston T 1977 *Tensor Geometry* (London: Pitman)
Hawking S and Ellis G 1973 *The Large Scale Structure of Space–Time* (Cambridge: Cambridge Unversity Press)
Kobayashi S and Nomizu K 1963 *Principles of Differential Geometry* (New York: Interscience)
Poor W A 1981 *Differential Geometric Structures* (New York: McGraw-Hill)
Thirring W E 1978 *A Course in Mathematical Physics: 2. Classical Field Theory* (Heidelberg: Springer)

5

Applications in Physics

5.1 Galilean Spacetimes

Since the time of Aristotle the evolution of the language for physics has to a large extent been governed by the choice of an appropriate event space. One may formulate the Galilean relativistic description of physics in terms of a four-dimensional fibre bundle in which each fibre is a Euclidean three-space and the projection is onto a one-dimensional oriented Euclidean time manifold. Events in this Galilean bundle are assigned a standard time point by this projection and the one-dimensional Euclidean metric on the base may be used to measure time differences between such events. Such elapsed times are unambiguous up to an arbitrary scaling corresponding to a choice of time units. If the time difference is zero the events are considered to be simultaneous and it is then possible to use the standard Euclidean metric on the corresponding fibre to define their spatial separation.

A family of curves, members of which intersect each fibre only once such that each point of every fibre lies on one and only one curve *foliates* the bundle.

Any two non-simultaneous events that lie on the same curve can be regarded as having the same spatial position with respect to this family. Each such family defines a coordinate system. The Galilean bundle is provided with a preferred class of families of curves; the trajectories of freely falling particles moving with uniform Newtonian velocities. They define the class of *inertial reference systems*. This dynamical structure endows the bundle with a preferred *parallelism*. We shall return to its mathematical formulation when we encounter the Newtonian connection. (The bundle may be given alternative parallelisms, for example, one might single out those reference frames in which particles have a uniform velocity when *falling freely* in some Newtonian gravitational field.)

In addition to the maximal set of six Euclidean Killing vectors on each fibre and the time translation symmetry, the existence of the preferred class of inertial frames endows the Galilean bundle with another three-parameter symmetry group corresponding to the transformation between inertial frames that differ by a uniform Newtonian three-velocity. The complete 10-parameter Galilean group is the relativistic group for Galilean physics (see figure 5.1).

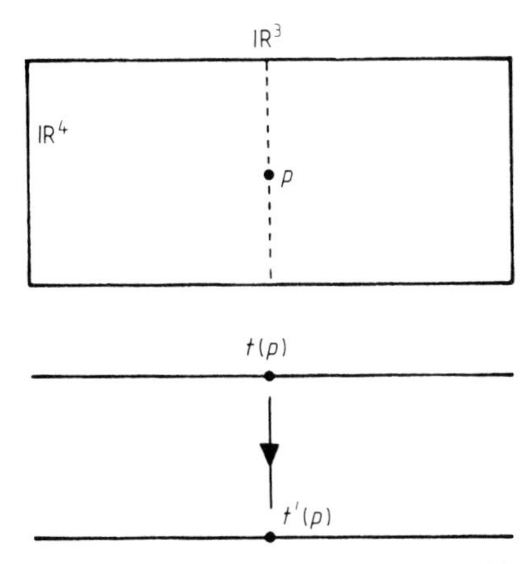

Figure 5.1 The Galilean bundle with a Euclidean three-space assigned an arbitrary time coordinate by projection.

The existence of the above structure for *Galilean relativistic* spacetime is a basic tenet of Newtonian dynamics. Physical descriptions prior to the introduction of a 'Lorentzian relativistic' structure for spacetime implicitily assume such a time-preferred fibre pattern for the spacetime manifold.

Two clocks at rest in a Galilean inertial system may assign different time parameters and even run at different rates relative to each other. However, it is a fundamental postulate of Galilean relativistic physics that the behaviour of all good clocks is independent of their relative state of motion. (By a good clock one means a clock that is robust and whose behaviour in external fields of force can in principle be compensated for.) It is further assumed that all good clocks may in principle be synchronised in an inertial system and used to calibrate the evolution rates of all physical processes. In Newtonian physics observers may also be equipped with measuring rods as well as clocks synchronisable with a hypothetical universal time. Rigid rods are used to construct rigid pieces of apparatus such as standard metres, telescopes, oscilloscopes etc and the Newtonian description of phenomena relies fundamentally on such a framework.

However if, as Einstein did, one builds a world picture based on a spacetime geometry with a Lorentzian-signatured metric structure such 'commonsense' operations as length and time measurement cannot be taken as primitive concepts. Thus a more appropriate notion of a clock is required and one must relinquish measuring processes based on extended rigid structures since they are strictly undefined as primitive operations. With any new set of measurement definitions associated with classical observers in a refined spacetime picture we must expect to be able to recover in some approximation the valuable global Newtonian spacetime notions. Einsteinian relativity has sharpened the notion of a good clock and made redundant the concept of a preferred time projection. Physical clocks that approximate the ideal clocks of a non-Galilean description measure the elapsed time between events in $\mathbb{R}^4$ as a function of their relative motions, and it is only for clocks moving with uniform relative Newtonian velocities, small compared with the Newtonian velocity of light, that the notion of elapsed time between events can be divorced from the relative state of motion of the measuring clocks. Such a reformulation is often referred to as a relativistic description. In the following we are motivated towards one particular relativistic formulation: that inherent in a reformulation of Maxwell's equations on a four-dimensional manifold possessing a Lorentzian metric structure and a *Poincaré* isometry group.

We shall follow the historical path that led Einstein to this elegant (and physically more accurate) world structure by examining one of the most successful of all physical theories: classical electrodynamics.

5.2. Maxwell's Equations and Minkowski Spacetime

Physical theories are usually formulated in terms of quantities with *physical dimensions*. The assignment of a physical dimension to a quantity often follows from its operational definition in terms of some measuring process, a coherent choice of units often facilitating the expression of a physical law. Our mathematical introduction of tensor fields is based upon an underlying manifold where chart coordinates and components of all tensors may be regarded as physically dimensionless numbers. However, in order to compare such a tensor field description with a physical theory written in terms of dimensioned quantities one must effect a transformation. If a physical theory is formulated in terms of tensors over the real field one may restore all physical dimensions appropriately as follows. The dimensionless tensor field equations describing the theory are initially expressed in a local chart with dimensionless spacetime event coordinate maps, say (t, x^1, x^2, x^3). Chart transformations are then performed to some standard coordinates with assigned physical dimensions. If necessary, new tensors with physical dimensions can be defined by scaling dimensionless ones by some constant parameter with appropriate dimensions. The numerical values chosen for such dimensioned parameters establish the choice of units for the system. If one wants to work with coordinates having the standard dimensions of time and length, say $(\underline{t}, \underline{x}^1, \underline{x}^2, \underline{x}^3)$, one may introduce three standard dimensioned units such as c, a standard speed, $\hbar$ a standard unit of action and a reference mass m_0. The restoration of physical units follows from the simple chart transformations

$$t = (m_0 c^2/\hbar)\underline{t}$$
$$x^k = (m_0 c/\hbar)\underline{x}^k \qquad k = 1, 2, 3. \tag{5.2.1}$$

A dimensionless tensor field will have components with dimensions when referred to a basis induced from a local chart with dimensioned coordinates.

It is a fundamental property of matter that it can exert a long-range influence on other matter by both the effect of its mass (the gravitational interaction) and its electrical charge (the electromagnetic interaction). The latter is a property that comes in two opposite varieties or *polarities* that are responsible for the 'attractive' and 'repulsive' forces of electrostatic interaction. (No analogous 'repulsive' long-range Newtonian gravitational interaction between matter has been observed.) After the pioneering efforts of Faraday and Maxwell the electromagnetic interaction between matter is described in terms of an intermediary physical field. This field was originally conceived to consist of a pair of vector fields $(\boldsymbol{E}, \boldsymbol{B})$ on Euclidean $\mathbb{R}^3$ parametrised by a universal time t. If we denote by the (time-dependent) function $\rho: \mathbb{R}^3 \to \mathbb{R}$ the electrical charge density in $\mathrm{C\,m^{-3}}$ and by $\boldsymbol{j}$ the (time-dependent) vector field on $\mathbb{R}^3$ describing the charge crossing normally a unit area (the current density in $\mathrm{A\,m^{-2}}$) then, in MKS dimensioned units, (mass in kilogrammes (kg), time in seconds, length in metres (m)) the electric $\boldsymbol{E}$ and magnetic $\boldsymbol{B}$ vector fields satisfy Maxwell's equations:

$$\operatorname{div}\boldsymbol{E} = \rho/\varepsilon_0 \qquad \operatorname{curl}\boldsymbol{E} = -\partial\boldsymbol{B}/\partial\underline{t}$$
$$\operatorname{div}\boldsymbol{B} = 0 \qquad \operatorname{curl}\boldsymbol{B} = \mu_0\boldsymbol{j} + \frac{1}{c^2}\frac{\partial\boldsymbol{E}}{\partial\underline{t}}. \tag{5.2.2}$$

We are assuming that the sources $(\rho, \boldsymbol{j})$ exist in a free space or 'vacuum' environment. If $\boldsymbol{E} = \underline{E}_i(\partial/\partial\underline{x}^i) \in \Gamma T\mathbb{R}^3$ then by $(\partial\boldsymbol{E}/\partial\underline{t})$ one means $(\partial\underline{E}_i/\partial\underline{t})(\partial/\partial\underline{x}^i)$ where, in the chart $(\underline{x}^1, \underline{x}^2, \underline{x}^3)$ for $\mathbb{R}^3$, the Euclidean metric tensor field has the representation $\underset{\sim}{\hat{g}} = \Sigma_{i=1}^{3} \mathrm{d}\underline{x}^i \otimes \mathrm{d}\underline{x}^i$. The constants ε_0, μ_0 and $c \equiv (\varepsilon_0 \mu_0)^{-1/2}$ ensure that the equations are dimensionally coherent. They are assigned dimensions as follows

$$[\mu_0] = \left[\frac{\mathrm{ML}}{\mathrm{Q}^2}\right] \qquad [\varepsilon_0] = \left[\frac{\mathrm{T}^2\mathrm{Q}^2}{\mathrm{ML}^3}\right].$$

The functions $(\underline{E}_i, \underline{B}_i) : \mathbb{R}^3 \to \mathbb{R}$, each depending on the time parameter t, will be called the MKS Cartesian components of the electric and magnetic field respectively. The Cartesian components of the electric field have dimensions $[\mathrm{ML/T^2Q}]$, with MKS units of $\mathrm{N\,C^{-1}}$, whilst those of the magnetic field have dimensions $[\mathrm{M/TQ}]$ with MKS units of Teslas (or $\mathrm{Wb\,m^{-2}}$).

The structure of this system of coupled partial differential equations permits one to construct a remarkable synthesis between the fields $(\boldsymbol{E}, \boldsymbol{B})$. This may be achieved by reformulating the system in terms of a pair of tensor equations on the event manifold $\mathbb{R}^4$ endowed with a particular metric structure. Instead of associating the Cartesian components of $\boldsymbol{E}, \boldsymbol{B}$ with vector fields on $\mathbb{R}^3$, they are used to construct a 2-form $\underline{F}$ on $\mathbb{R}^4$. Using a local chart $(\underline{t}, \underline{x}^1, \underline{x}^2, \underline{x}^3)$ we define

$$\underline{F} = \underline{B} + \mathrm{d}\underline{t} \wedge \underline{E} \tag{5.2.3}$$

where

$$\underline{B} = \underline{B}_1 \mathrm{d}\underline{x}^2 \wedge \mathrm{d}\underline{x}^3 + \underline{B}_2 \mathrm{d}\underline{x}^3 \wedge \mathrm{d}\underline{x}^1 + \underline{B}_3 \mathrm{d}\underline{x}^1 \wedge \mathrm{d}\underline{x}^2$$

$$\underline{E} = \underline{E}_1 \mathrm{d}\underline{x}^1 + \underline{E}_2 \mathrm{d}\underline{x}^2 + \underline{E}_3 \mathrm{d}\underline{x}^3.$$

In a similar way we unify the components of the current and charge density to construct the 3-form $\underline{\mathcal{J}}$:

$$\underline{\mathcal{J}} = c\mu_0 \underline{J} \wedge \mathrm{d}\underline{t} + (\rho / c\varepsilon_0) \mathrm{d}\underline{x}^1 \wedge \mathrm{d}\underline{x}^2 \wedge \mathrm{d}\underline{x}^3 \tag{5.2.4}$$

where

$$\underline{J} = \underline{j}_1 \mathrm{d}\underline{x}^2 \wedge \mathrm{d}\underline{x}^3 + \underline{j}_2 \mathrm{d}\underline{x}^3 \wedge \mathrm{d}\underline{x}^1 + \underline{j}_3 \mathrm{d}\underline{x}^1 \wedge \mathrm{d}\underline{x}^2.$$

The $\{\underline{j}_i\}$ are the components of the vector current $\boldsymbol{j}$. The choice of dimensioned coefficients ensures that $\underline{\mathcal{J}}$ and $\underline{F}$ have the same dimensions, namely $[\hbar/\mathrm{Q}]$. Note that, for any form f, $\mathrm{d}f$ and f have the same physical dimensions: the exterior derivative does not change the physical dimensions of the form on which it acts.

The metric tensor field adopted on $\mathbb{R}^4$ is given in this chart by

$$\underset{\sim}{g} = -c^2 \mathrm{d}\underline{t} \otimes \mathrm{d}\underline{t} + \underset{\sim}{\hat{g}}. \tag{5.2.5}$$

Hence $(c\mathrm{d}\underline{t}, \mathrm{d}\underline{x}^k)$ is an orthonormal co-frame with respect to this $\underset{\sim}{g}$. In terms of the Hodge map $\underline{*}$ associated with this Lorentzian-signatured metric Maxwell's equations may be expressed elegantly as the exterior equations

$$\mathrm{d}\underset{\sim}{*}\, \underset{\sim}{F} = \underset{\sim}{\mathcal{J}} \tag{5.2.6}$$

$$\mathrm{d}\underset{\sim}{F} = 0. \tag{5.2.7}$$

One further and desirable simplification can be made: the set can be written entirely in terms of dimensionless tensors. First it is trivial to define dimensionless forms F and j by scaling each with any convenient parameters having the dimensions $[\hbar/\mathrm{Q}]$. We choose to write

$$F = (e_0/\hbar)\underset{\sim}{F}$$

$$j = (e_0/\hbar)\underset{\sim}{\mathcal{J}}$$

where e_0 is the elementary charge on the electron. In general, equations involving the Hodge map make reference to a specific metric. The equations (5.2.6) and (5.2.7), however, remain unchanged if we replace $\underset{\sim}{g}$ by $2\lambda\underset{\sim}{g}$ where λ is any positive-definite real-valued function on $\mathbb{R}^4$. This follows since $\underset{\sim}{F}$ is a 2-form in four dimensions. It is convenient for us to exploit this freedom here to rescale $\underset{\sim}{g}$ by any constant with the dimensions of $[\mathrm{L}]^2$ and use a dimensionless metric tensor field $g = L^{-2}\underset{\sim}{g}$. We shall denote the Hodge map associated with g as simply $*$ and rewrite the Maxwell equations:

$$\mathrm{d}*F = j \tag{5.2.8}$$

$$\mathrm{d}F = 0. \tag{5.2.9}$$

One is of course free to use either dimensioned or dimensionless coordinates in extracting component equations from this set. We have spelt out in detail the straightforward manner in which one can make contact with the conventional MKS dimensioned field and source components. Henceforth we shall work with dimensionless coordinates and tensors. It is worth stressing that although we have built up these equations from the traditional Cartesian-oriented approach the equations are now fully tensorial on the four-dimensional manifold with metric tensor g. We have extricated ourselves from a particular chart including a particular time map. This is a major achievement and may be regarded as the cornerstone development in Einstein's 'relativistic' world view.

A metric such as g that has a signature with one minus sign is called *Lorentzian*. A four-dimensional manifold with Lorentzian metric will be called a *spacetime*. Tangent vectors in a Lorentzian spacetime may be classified into spacelike (positive-norm), timelike (negative-norm) or null (zero-norm) vectors. The tangent space is said to possess a *light cone* structure conferred on it by such a metric. Furthermore, timelike tangent vectors may be classified into future-pointing and past-pointing. If X_p is assigned a future-pointing role then $-X_p$ is defined to be past pointing at p. If this assignment can be made unambiguously over the whole manifold then the spacetime is said to be time orientable. It would be rather difficult to interpret physical phenomena on a manifold that was not time orientable.

The spacetime modelled on $\mathbb{R}^4$ with metric as in (5.2.5) is called *Minkowski spacetime*. Thus Minkowski spacetime admits a chart with coordinates (t, x^1, x^2, x^3) in which the metric tensor field is given by

$$g = -\mathrm{d}t \otimes \mathrm{d}t + \sum_{i=1}^{3} \mathrm{d}x^i \otimes \mathrm{d}x^i. \tag{5.2.10}$$

We observe that the vector field $(\partial/\partial t)$ has a negative norm whilst $(\partial/\partial x^i)$ has a positive norm for $i = 1, 2, 3$

$$\begin{aligned} g((\partial/\partial t), (\partial/\partial t)) &= -1 \\ g(\partial/\partial x^i, \partial/\partial x^i) &= 1 \qquad \text{(no sum)}. \end{aligned} \tag{5.2.11}$$

Minkowski space M possesses a 10-parameter group of isometries. In a chart in which the metric is given by (5.2.10) these isometries are generated by the following Killing vector fields

$$\begin{aligned} T_0 &= (\partial/\partial t),\ T_k = (\partial/\partial x^k) \qquad k = 1, 2, 3 \\ K_3 &= x^1(\partial/\partial x^2) - x^2(\partial/\partial x^1) \\ K_2 &= x^3(\partial/\partial x^1) - x^1(\partial/\partial x^3) \\ K_1 &= x^2(\partial/\partial x^3) - x^3(\partial/\partial x^2) \\ B_k &= t(\partial/\partial x^k) + x^k(\partial/\partial t) \qquad k = 1, 2, 3. \end{aligned} \tag{5.2.12}$$

The isometry group of Minkowski space is called the *Poincaré* group. The vectors T_μ, $\mu = 0, 1, 2, 3$, generate *translations*; the integral curves being open lines. The K_i, $i = 1, 2, 3$ generate *rotations*; the integral curves lying on the surface of a sphere. The B_k, $k = 1, 2, 3$, generate *boosts*, the integral curves being open, forming hyperbolae.

Exercise 5.1

Verify that if X is any of the vector fields in (5.2.12) then

$$\mathcal{L}_X g = 0.$$

The structure of Maxwell's equations motivated the introduction of Minkowski space. In fact the form of Maxwell's equations arrived at, (5.2.8) and (5.2.9), is immediately valid in any Lorentzian spacetime (one not necessarily having the large number of isometries present for Minkowski space). Such a generalisation is the essence of Einstein's incorporation of arbitrary gravitational interactions into the underlying geometry of spacetime.

5.3 Observer Curves

The classical physical interpretation of the components of a tensor field on spacetime is associated with the notion of an observer curve. To introduce the notion of local observer time into the spacetime manifold M we exploit the lightcone structure of the Lorentzian metric. A curve C whose image passes through $p \in \mathrm{M}$ is said to be timelike at p if its tangent vector is timelike there. Next consider the physical interpretation of the parametrisation of C: $[0, 1] \rightarrow \mathrm{M}$. If (t, x^k) are local chart maps for M we represent C parametrically by the equations $t(p) = C^0(\tau)$, $x^k(p) = C^k(\tau)$ and we restrict ourselves to monotonic functions of τ that make C a future timelike curve:

$$g(C_*\partial_\tau, C_*\partial_\tau) < 0. \tag{5.3.1}$$

The *length* of C is defined to be the real number

$$s = \int_0^1 |g(C_*\partial_\tau, C_*\partial_\tau)|^{1/2} \mathrm{d}\tau. \tag{5.3.2}$$

Under a change of parametrisation $\tau \mapsto \tau'(\tau)$ mapping $[0, 1] \rightarrow [0, 1]$ with $(\partial\tau'/\partial\tau) > 0 \; \forall \, \tau$ then $C_*\partial_\tau \mapsto (\partial_\tau \tau')(C'_*\partial_{\tau'})$ and $\mathrm{d}\tau \mapsto (\partial\tau/\partial\tau')\mathrm{d}\tau'$, so we see that the integral is invariant under such a reparametrisation. A parameter τ is said to provide a *proper-time parametrisation* for C if

$$g(C_*\partial_\tau, C_*\partial_\tau) = -1. \tag{5.3.3}$$

An *ideal observer* is defined to be a proper-time parametrised future-pointing timelike curve on spacetime. The observer image is represented as a history or *world line* on the manifold. Elapsed time between events on the world line, as measured by such an observer curve, is determined by the difference between the affine parameter assigned to each event. It is a fundamental assumption that there exist standard clocks that operationally determine such an affine parametrisation along their histories. For such curves (5.3.2) implies that the time between events linked by an observer curve is equal to the length of world line linking them; it is measured by a standard clock accompanying the ideal observer. This time measure is often called the *proper time* measured by C. It does appear that many natural processes (for example, decaying particles) can be used as standard clocks registering *proper time*. Once one is convinced of the existence of microscopic natural clocks for *proper time*, macroscopic clocks (assemblies of microscopic clocks) can then be synchronised using light signals, or any other physical mechanism that supports a formulation in terms of a locally Lorentzian geometry. Once this definition of a good clock is adopted it becomes evident that there is no unique *proper time* interval between two events that can be joined by a family of timelike observer curves.

Each curve will in general measure a different time interval since each curve has a different arc length.

A timelike vector field V is called a world velocity (or four-velocity) vector field if $g(V, V) = -1$. As an example consider the vector field

$$V = k\left(\partial_t + \sum_{j=1}^{3} v^j \partial_{x^j}\right) \tag{5.3.4}$$

in a local chart (t, x^j) in which the Minkowski metric g takes the form (5.2.10). The field is labelled by real constants k, v^1, v^2, v^3. V is a velocity vector if

$$k = [1 - (v^1)^2 - (v^2)^2 - (v^3)^2]^{-1/2}. \tag{5.3.5}$$

What observer curve C has a tangent vector that coincides with V at each p on its image? For this we require

$$C_*\partial_\tau|_p = (\partial t/\partial\tau)\partial_t|_p + \sum_{j=1}^{3}(\partial x^j/\partial\tau)\partial_{x^j}|_p = V|_p$$

that is $((\partial t/\partial\tau), (\partial x^j/\partial\tau)) = k(1, v^j)$. These equations fix the parametrisation of C up to an additive constant for τ. For C labelled by the triplet $\boldsymbol{v} = (v^1, v^2, v^3) \in \mathbb{R}^3$ a family of observer curves through the origin of the (t, x^k) chart has the representation $t(p) = k\tau$, $x^j(p) = kv^j\tau$ or $x^j(p) = v^j t$, $j = 1, 2, 3$. For arbitrary constant $\boldsymbol{v}$ the vector field V is a Killing field. We define a *stationary observer* to be a proper-time parametrised integral curve of a timelike Killing vector. Thus the vector field V, with arbitrary constant $\boldsymbol{v}$, yields a three-parameter family of stationary observers in Minkowski space.

Any global Minkowski space chart in which the metric takes the form (5.2.10) is often referred to as an inertial chart. The chart maps define a global co-frame of exact 1-forms. The vector field V defines a congruence of ideal observers, each ideal observer being an integral curve of V. One often sees the phrase 'an inertial frame' or 'an inertial system' in this context. Care will be exercised in not adopting this phrase too readily: we have not assigned a frame of vectors along any observer curve so cannot at this stage, strictly speaking, make reference to an observer's inertial frame. However, the frame $\{\partial_t, \partial_{x^j}\}$ associated with the inertial chart is an example of an inertial frame along the integral curves of ∂_t. We shall return to the general definition of observer frames after we have introduced the concept of vector transport.

The equation of the world line of a *stationary* observer in an inertial chart suggests that the triplet $\boldsymbol{v}$ be identified with the components of a Newtonian velocity three-vector. However, we would prefer to identify such a notion in the context of a general observer, not necessarily a *stationary* one. Since we now contemplate arbitrary observers we concentrate on T_pM rather than the whole history of the arbitrary observer world line. A point $p \in \mathrm{M}$ together with a future-pointing timelike vector with norm -1 will be called an *instantaneous* observer at p.

Let $\dot{C}$ be such an *instantaneous* observer associated with the general observer C and let λ be any timelike future-pointing 1-chain (not necessarily another observer) with tangent vector $\dot{\lambda}$ at p. We wish to define the Newtonian velocity of λ observed by C at p. Since $\dot{\lambda}, \dot{C} \in T_p\mathrm{M}$ we have a unique orthogonal decomposition

$$\dot{\lambda} = \boldsymbol{P} + \mathscr{E}\dot{C} \tag{5.3.6}$$

where $\mathscr{E} \in \mathbb{R}$ and $g(\boldsymbol{P}, \dot{C}) = 0$. This latter condition implies $\mathscr{E} = -g(\dot{\lambda}, \dot{C})$ since $g(\dot{C}, \dot{C}) = -1$, hence

$$\dot{\lambda} = \boldsymbol{P} - g(\dot{\lambda}, \dot{C})\dot{C}. \tag{5.3.7}$$

The *Newtonian velocity* of λ observed by C at p is now defined with respect to this orthogonal decomposition as $\boldsymbol{v} = \boldsymbol{P}/\mathscr{E}$, or

$$\boldsymbol{v} = -\boldsymbol{P}/g(\dot{\lambda}, \dot{C}) \tag{5.3.8}$$

showing that $\boldsymbol{v}$ depends on both $\dot{\lambda}$ and $\dot{C}$. The vector $\boldsymbol{P}$ is spacelike and is said to lie in an *instantaneous three-space* of $\dot{C}$ at p. This is defined as the orthogonal complement of $\dot{C}$ in $T_p\mathrm{M}$.

We next consider the case of a null 1-chain Γ observed by C. The condition $g(\dot{\Gamma}, \dot{\Gamma}) = 0$ inserted into $\dot{\Gamma} = \boldsymbol{P} - g(\dot{\Gamma}, \dot{C})\dot{C}$ gives, with the aid of $g(\boldsymbol{P}, \boldsymbol{P}) = g(\dot{\Gamma}, \boldsymbol{P})$,

$$\dot{\Gamma} = \mathscr{E}(\dot{C} - \boldsymbol{N}) \tag{5.3.9}$$

where $\mathscr{E} \equiv -g(\dot{\Gamma}, \dot{C})$ and $\boldsymbol{N} \equiv (g(\dot{\Gamma}, \dot{C})/g(\dot{\Gamma}, \boldsymbol{P}))\boldsymbol{P}$. $\mathscr{E}$ is called the *energy* that C observes for Γ at p whilst $\boldsymbol{N}$ is the *spatial direction* observed for Γ. Note that $\boldsymbol{N}$ is spacelike with $g(\boldsymbol{N}, \boldsymbol{N}) = 1$. It is a fundamental result that there exist propagating solutions to Maxwell's equations corresponding to the phenomenon of electromagnetic waves. Such waves propagate *in vacuo* without dispersion and have null vector fields associated with them. Thus null curves may model the flow of electromagnetic radiation, or photons.

The images of timelike future-pointing curves are models for either massive point particles or the streamlines of mass–energy flows. A point particle of mass m is modelled by a future-pointing curve ρ with $g(\dot{\rho}, \dot{\rho}) = -m^2$. Then $\dot{\rho} = \boldsymbol{P} + \mathscr{E}\dot{C}$ implies $g(\boldsymbol{P}, \boldsymbol{P}) + m^2 = \mathscr{E}^2$; $\boldsymbol{P} = \mathscr{E}\wedge$ therefore implies

$$g(\boldsymbol{P}, \boldsymbol{P}) = \mathscr{E}^2 g(\boldsymbol{v}, \boldsymbol{v}) = \mathscr{E}^2 - m^2. \tag{5.3.10}$$

Hence $\mathscr{E}$ and $\boldsymbol{P}$ may be expressed in terms of $\boldsymbol{v}$ as

$$\mathscr{E} = \frac{m}{[1 - g(\boldsymbol{v}, \boldsymbol{v})]^{1/2}} \tag{5.3.11}$$

$$\boldsymbol{P} = \frac{m\boldsymbol{v}}{[1 - g(\boldsymbol{v}, \boldsymbol{v})]^{1/2}}. \tag{5.3.12}$$

Clearly if $m \neq 0$, $g(\boldsymbol{v}, \boldsymbol{v}) = 1 - (m/\mathscr{E})^2 < 1$: that is, massive particles are observed to have bounded Newtonian velocities.

If, for example, $\dot{C} = \partial_t|_p$ and $\dot{\lambda} = \dot{x}^j(\tau)\partial_{x^j}|_p + \dot{t}(\tau)\partial_t|_p$ in an inertial Minkowski chart then $g(\dot{C}, \dot{\lambda}) = -\dot{t}$ and $\dot{\lambda} = \boldsymbol{P} + \dot{t}(\tau)\partial_t|_p$ gives $\boldsymbol{P} = \dot{x}^j(\tau)\partial_{x^j}$. Hence in this chart $\boldsymbol{v} = (\dot{x}^j(\tau)/\dot{t}(\tau))\partial_{x^j}|_p$ is the Newtonian velocity of $\dot{\lambda}$ observed by C at p.

If the projection onto the instantaneous three-space orthogonal to $\dot{C}$ at p is effected by the projection operator $\Pi_p : T_p\mathrm{M} \to (\dot{C})_p^{\perp}$, $\Pi_p = (1 - \{\widetilde{C}(\dot{C})\}^{-1}\dot{C} \otimes \widetilde{C})_p$, then the *Newtonian length* of any spacelike vector $V \in T_p\mathrm{M}$ observed by C is defined as $(g(\Pi_p V, \Pi_p V))^{1/2}$. If W is a second spacelike vector in $T_p\mathrm{M}$ then the *Newtonian angle* between V and W observed by C is given by

$$\cos\theta = \frac{g(\Pi_p V, \Pi_p W)}{[g(\Pi_p V, \Pi_p V) g(\Pi_p W, \Pi_p W)]^{1/2}}. \tag{5.3.13}$$

The presence of the projector Π_p in these formulae, defined by the observer curve, means that the Newtonian length and angles specified in this way depend on the observer as well as on the vectors being observed. For the general future-pointing vector $\dot{\lambda} = \boldsymbol{P} + \mathscr{E}\dot{C}$ we see that $\boldsymbol{V}$ has *Newtonian length* $(g(\boldsymbol{v}, \boldsymbol{v}))^{1/2} = [g(\boldsymbol{P}, \boldsymbol{P})]^{1/2}/\mathscr{E}$. If $\dot{\lambda}$ is null, $g(\boldsymbol{P}, \boldsymbol{P}) = \mathscr{E}^2$ and hence all null vectors are always observed to have unit length Newtonian velocities. We have already noted that $g(\boldsymbol{v}, \boldsymbol{v}) < 1$ if $\boldsymbol{v}$ is the Newtonian velocity of a particle with $m \neq 0$. If $g(\boldsymbol{v}, \boldsymbol{v}) \ll 1$ we may expand (4.3.11), (4.3.12) using the binomial expansion

$$\mathscr{E} = m + \tfrac{1}{2}mg(\boldsymbol{v}, \boldsymbol{v}) + \ldots \tag{5.3.14}$$

$$\boldsymbol{P} = m\boldsymbol{v} + \ldots. \tag{5.3.15}$$

These formulae reinforce our identification of the *instantaneous* energy and three-momentum for a point particle. We see that the Newtonian kinetic energy of such a particle differs from the relativistic energy $\mathscr{E}$ by the constant m. This difference between Newtonian and Einsteinian relativistic kinematics has had a profound effect in the subsequent development of relativistic physics.

The images of different observer curves may be related by a diffeomorphism of spacetime: in particular a diffeomorphism from the isometry group. We here consider a 'boost' diffeomorphism from the Poincaré group. We first compute part of an integral curve of the 'boost' vector field

$$X = x^1\partial_t + t\partial_{x^1} \tag{5.3.16}$$

passing through a point p_0 with coordinates

$$(t(p_0), x^1(p_0), x^2(p_0), x^3(p_0))$$

in an inertial chart. We shall take p_0 to lie outside the 'light cone of (0, 0, 0, 0)', defined as the set L of points p satisfying

$$\sum_{j=1}^{3} (x^j(p))^2 - (t(p))^2 = 0.$$

This ensures that at p_0 X is timelike. For definiteness we shall assume $t(p_0) > 0$, $x^j(p_0) > 0$ $j = 1, 2, 3$. The integral curve is given parametrically as $t(p) = \lambda^0(\tau)$, $x^j(p) = \lambda^j(\tau)$ $j = 1, 2, 3$, where the functions $\lambda^\mu:[0, \infty) \to \mathbb{R}$, $\mu = 0, 1, 2, 3$ satisfy

$$d\lambda^1/d\tau = \lambda^0, \ d\lambda^0/d\tau = \lambda^1$$

$$d\lambda^2/d\tau = 0, \ d\lambda^3/d\tau = 0.$$

Thus the curve is given by the solution

$$\begin{aligned} \lambda^1(\tau) &= \lambda^1(0)\cosh\tau + \lambda^0(0)\sinh\tau \\ \lambda^0(\tau) &= \lambda^0(0)\cosh\tau + \lambda^1(0)\sinh\tau \\ \lambda^2(\tau) &= \lambda^2(0) \\ \lambda^3(\tau) &= \lambda^3(0). \end{aligned} \tag{5.3.17}$$

Eliminating τ between $\lambda^1(\tau)$ and $\lambda^0(\tau)$ gives part of a hyperbola through p_0 and p

$$(\lambda^1(\tau))^2 - (\lambda^0(\tau))^2 = (\lambda^1(0))^2 - (\lambda^0(0))^2. \tag{5.3.18}$$

If we relabel the functions λ^μ with coordinate names (with p_0 specified by $\tau = 0$), equations (5.3.17) may be rewritten as

$$x^1(p) = \frac{x^1(p_0) + vt(p_0)}{(1 - v^2)^{1/2}} \tag{5.3.19}$$

$$t(p) = \frac{t(p_0) + vx^1(p_0)}{(1 - v^2)^{1/2}} \tag{5.3.20}$$

where $\cosh\tau = 1/(1 - v^2)^{1/2}$ and $\sinh\tau = v/(1 - v^2)^{1/2} > 0$. These familiar equations relate the point p_0 to the point p labelled by the parameter $v = \tanh\tau$ along the *boost orbit* (5.3.18).

For a fixed v we have a diffeomorphism, generated by X, that may be used to relate two observer fields. Define the map

$$\varphi_v : M \to M, \ p \to p'$$

$$t(p') = (t(p) + vx^1(p))/(1 - v^2)^{1/2}$$

$$x^1(p') = (x^1(p) + vt(p))/(1 - v^2)^{1/2}$$

$$x^k(p') = x^k(p) \qquad k = 2, 3$$

then

$$(\varphi_v)_* : \partial_t|_p \longmapsto (\partial_t + v\partial_x)|_{p'}/(1 - v^2)^{1/2}.$$

Thus the fixed parameter v can be identified as the Newtonian velocity of $(\varphi_v)_*\partial_t|_{p'}$ as measured by $\partial_t|_{p'}$ for all p'. Note that for all τ, $v = \tanh \tau < 1$. It is of interest to note that since two successive diffeomorphisms of the above type parametrised by τ_1 and τ_2 respectively produce a diffeomorphism parametrised by $\tau_1 + \tau_2$:

$$\varphi_{\tau_1} \circ \varphi_{\tau_2} = \varphi_{\tau_1 + \tau_2}$$

we obtain as Newtonian velocity parameter v_{12} corresponding to $\tau_1 + \tau_2$

$$\begin{aligned} v_{12} &= \tanh(\tau_1 + \tau_2) \\ &= \frac{\tanh \tau_1 + \tanh \tau_2}{1 + \tanh \tau_1 \tanh \tau_2} = \frac{v_1 + v_2}{1 + v_1 v_2}. \end{aligned}$$

For all $v_1, v_2 < 1$, $v_{12} = v_{21} < 1$, that is successive 'boost' transformations applied to observer curves can never give rise to observer curves with a Newtonian velocity in excess of 1 relative to all observers.

5.4 Electromagnetism

In §5.2 we used the structure of Maxwell's equations to motivate the introduction of a four-dimensional Lorentzian spacetime. We here examine some further properties of these equations.

If α is a p-form on U, $U \subset \mathrm{M}$, satisfying the equation $\mathrm{d}\alpha = 0$ it is said to be *closed* on U. Then there exist some region $W \subset U$ for which $\alpha = \mathrm{d}\beta$, for β a $(p-1)$-form on W. The p-form α is then said to be *exact* on W. It is an important result that the global topology of U determines whether or not all closed forms are exact on U. For our local discussion, however, we can assert that the Maxwell equation $\mathrm{d}F = 0$ implies that in some neighbourhood of every point on M there exists a 1-form A such that $F = \mathrm{d}A$. Clearly given such an A there exists an equivalence class satisfying the same condition. Two members of this class differ by an exact 1-form $\mathrm{d}\lambda$ where $\lambda \in \mathcal{F}(U)$. The freedom to choose a 1-form *potential* from such a class is known as local electromagnetic gauge invariance. Two potentials in this class are said to be *co-homologous*. In a local Minkowski chart (t, x^k) we may write

$$A = \sum_{k=1}^{3} A_k \mathrm{d}x^k + \varphi \mathrm{d}t$$

and hence relate the real-valued functions A_k, φ to some electrodynamic 'vector' and 'scalar' potentials. Introducing a local potential A means that (5.2.9) is satisfied identically and the other equation (5.2.8) becomes

$$\mathrm{d}{*}\mathrm{d}A = j. \tag{5.4.1}$$

The above equation may be written in terms of the Laplace–Beltrami operator. To define this we need to introduce the co-derivative. On a general n-dimensional (pseudo-) Riemannian manifold we define the *co-derivative*

$$\delta : \Gamma\Lambda_p \mathrm{M} \to \Gamma\Lambda_{p-1}\mathrm{M}$$

by

$$\delta = *^{-1}\mathrm{d}{*}\eta. \tag{5.4.2}$$

(Recall from (1.1.2) that if φ is a p-form $\eta\varphi \equiv \varphi^{\eta} = (-1)^p\varphi$.) Since on p-forms

$$\begin{aligned} ** &= (-1)^{p(n-p)}\frac{\det g}{|\det g|} \\ &\equiv \eta^{n-1}\frac{\det g}{|\det g|} \end{aligned} \tag{5.4.3}$$

it follows immediately that δ has the property $\delta\delta = 0$, in common with d. The signs in the definition of the co-derivative are chosen to ensure that it is the adjoint operator to the exterior derivative, with respect to a certain inner product on differential forms on a compact Riemannian manifold. If M is a compact Riemannian manifold ($\partial\mathrm{M} = 0$) then a symmetric product on p-forms is defined by

$$(\alpha, \beta) \equiv \int_{\mathrm{M}} \alpha \wedge *\beta \qquad \alpha, \beta \in \Gamma\Lambda_p\mathrm{M}. \tag{5.4.4}$$

An 'integration by parts' gives, with Stokes's theorem and the compactness of M,

$$(\varphi, \mathrm{d}\psi) = (\delta\varphi, \psi) \qquad \varphi \in \Gamma\Lambda_p\mathrm{M},\ \psi \in \Gamma\Lambda_{p-1}\mathrm{M}.$$

That is, δ is the adjoint of d with respect to this product. The Laplace–Beltrami operator Δ is defined by

$$\Delta \equiv -(\mathrm{d}\delta + \delta\mathrm{d}). \tag{5.4.5}$$

Note that since d(δ) increases (decreases) the degree of a form by one the Laplace–Beltrami operator preserves the degree of a form. With our conventions the Laplace–Beltrami operator has negative eigenvalues on a compact Riemannian manifold. In terms of the product of (5.4.4):

$$\begin{aligned} (\varphi, \Delta\varphi) &= -(\varphi, \mathrm{d}\delta\varphi) - (\varphi, \delta\mathrm{d}\varphi) \\ &= -(\delta\varphi, \delta\varphi) - (\mathrm{d}\varphi, \mathrm{d}\varphi). \end{aligned}$$

The positivity of the Riemannian metric ensures that the right-hand side is negative-definite, thus so are any eigenvalues.

The equation (5.4.1) can be written in terms of Δ as

$$(\Delta + \mathrm{d}\delta)A = -*j. \qquad (5.4.6)$$

It is possible to select a representative potential from the class of co-homologous 1-forms such that $\delta A = 0$. Such a choice is called selecting a Lorentz gauge. In this gauge the potential satisfies a Helmholtz wave equation: $\Delta A = -*j$. (Note that the potential is not uniquely fixed by the Lorentz gauge condition. If A is changed to $A' = A + \mathrm{d}\lambda$, $\lambda \in \mathcal{F}(\mathrm{M})$, then $\delta A' = \delta\mathrm{d}\lambda = 0$ also if λ is chosen to be *harmonic*, that is satisfy $\Delta\lambda = 0$.)

Let us examine some solutions to Maxwell's equations in a region of Minkowski spacetime free of sources. Suppose we seek a solution to (5.4.1) of the form $A = f\,\mathrm{d}t$, $f \in \mathcal{F}(\mathrm{M})$ using a polar chart (t, r, θ, φ) in which

$$g = -\mathrm{d}t\otimes\mathrm{d}t + \mathrm{d}r\otimes\mathrm{d}r + r^2\mathrm{d}\theta\otimes\mathrm{d}\theta + r^2\sin^2\theta\mathrm{d}\varphi\otimes\mathrm{d}\varphi.$$

We shall look for a static 'spherically symmetric' solution satisfying the symmetry condition $\mathcal{L}_{K_i}F = 0$ where the timelike Killing vector is

$$K_0 = (\partial/\partial t)$$

and the rotational Killing vectors take the form

$$\begin{aligned} K_1 &= \sin\varphi\,\partial_\theta + \cot\theta\cos\varphi\,\partial_\varphi \\ K_2 &= -\cos\varphi\,\partial_\theta + \cot\theta\sin\varphi\,\partial_\varphi \qquad (5.4.7) \\ K_3 &= \partial_\varphi. \end{aligned}$$

This can be achieved if the function f involves only the coordinate map r. A convenient orthonormal co-frame is $\{\mathrm{d}t, \mathrm{d}r, r\mathrm{d}\theta, r\sin\theta\,\mathrm{d}\varphi\}$. Then $\mathrm{d}A = \partial_r f\,\mathrm{d}r \wedge \mathrm{d}t = \partial_r f e^1 \wedge e^0$, so if $*1 = e^1 \wedge e^2 \wedge e^3 \wedge e^0$ then $*\mathrm{d}A = (\partial_r)fe^3 \wedge e^2 = (\partial_r)fr^2\sin\theta\,\mathrm{d}\varphi \wedge \mathrm{d}\theta$. Thus $\mathrm{d}*\mathrm{d}A = \partial_r(\partial_r fr^2)\mathrm{d}r \wedge \sin\theta\,\mathrm{d}\varphi \wedge \mathrm{d}\theta$. This is zero if $f = k/r$ for some constant k. The solution $A = k\mathrm{d}t/r$ yields the electric 2-form $F = \mathrm{d}A = -(k/r^2)\mathrm{d}r \wedge \mathrm{d}t$. This is the Coulomb solution. The frame-dependent electric field 1-form $E \equiv \mathrm{i}_{\partial_t}F = (k/r^2)\mathrm{d}r$ gives the electric field vector $\widetilde{E} = (k/r^2)\partial_r$, the integral curves of which give the familiar radial Coulomb pattern associated with a stationary charge in this frame.

For a general F we define $\int_C *F$ as the electric charge Q contained in the interior of the sphere which is the image of C. (If the charge is non-zero then this S^2 cannot be the boundary of a source-free region!) (Restoring dimensioned variables,

$$\int_{S^2} \underline{*\ F} = (\varepsilon_0/\mu_0)^{1/2}\underline{Q}$$

determines a charge Q in Coulombs.) A class of 2-chains will determine the same electric charge. We define an equivalence relation on 2-chains as follows:

$$C_1 \simeq C_2 \qquad \text{iff } C_1 = C_2 + \partial\Sigma, \text{ where } \Sigma \text{ is any source-free region.}$$

Equivalent chains are said to be *homologous*. Since in source-free regions $*F$ is closed, Stokes's theorem ensures that the charge Q only depends on the class of chain chosen. As an example, we take C to be the 2-chain in Minkowski spacetime whose image is the sphere t = constant, r = constant. Then for the Coulomb solution

$$\begin{aligned}\int_C *F &= k\int_C \sin\theta \, \mathrm{d}\theta \wedge \mathrm{d}\varphi \\ &= 2k\int_0^{\pi/2} \sin\theta \, \mathrm{d}\theta \int_0^{2\pi} \mathrm{d}\varphi = 4\pi k.\end{aligned}$$

Since a Lie derivative with respect to a Killing vector K commutes with the Hodge map, $\mathscr{L}_K * = *\mathscr{L}_K$, and all Lie derivatives commute with d, we may deduce that if F satisfies the Maxwell equations with source 3-form j then $\mathscr{L}_K F$ satisfies them with the source $\mathscr{L}_K j$. The existence of an underlying isometry group of spacetime is often used implicitly in constructing new solutions of Maxwell's equations from simpler ones. If we recall the definition of the Lie derivative, and compare it with the elementary textbook calculation used to construct the electric dipole solution as a limit of two equal and opposite Coulomb solutions, we indeed expect the following potential to provide a source-free solution

$$A = \left(\frac{\mu}{k}\right)\mathscr{L}_{(\partial/\partial x^1)}\frac{k}{r}\mathrm{d}t = -\frac{\mu}{r^2}(\mathscr{L}_{(\partial/\partial x^1)}r)\mathrm{d}t = -\frac{\mu x^1}{r^3}\mathrm{d}t.$$

The vector field $(\partial/\partial x^1)$ represents a Minkowski space Killing vector in an inertial chart. Since the Lie derivative commutes with d,

$$F = \frac{\mu}{k}\mathscr{L}_{(\partial/\partial x^1)}\left(\frac{-k}{r^2}\mathrm{d}r \wedge \mathrm{d}t\right)$$

is the field of a static electric dipole with moment μ. In general for positive integers p, q, r a 'p, q, r'-type electric multipole solution follows from Poincaré covariance as

$$F = \{\mathscr{L}_{(\partial/\partial x^i)}\}^p\{\mathscr{L}_{(\partial/\partial x^j)}\}^q\{\mathscr{L}_{(\partial/\partial x^k)}\}^r\left\{\frac{q}{r^2}\mathrm{d}r \wedge \mathrm{d}t\right\}$$

$i, j, k = 1, 2, 3$.

There is one further symmetry of Maxwell's equations that deserves mentioning. A spacetime is said to admit local conformal isometries, generated by a vector field C, if the metric g is such that

$$\mathscr{L}_C g = \lambda_C g \tag{5.4.8}$$

for some scale function λ_C. For any n-dimensional space (n even) it then follows that if $F \in \Gamma\Lambda_{n/2}M$ then

$$\mathcal{L}_C(*F) = *(\mathcal{L}_C F). \tag{5.4.9}$$

Hence in a spacetime ($n = 4$) with a metric g, if such a C exists and the Maxwell 2-form F solves Maxwell's equations with source j then $\mathcal{L}_C F$ will be a solution, in the same metric, with source $\mathcal{L}_C j$. In particular if $j = 0$ the source-free Maxwell equations exhibit a local conformal covariance in spaces admitting conformal isometries. Clearly, as a special case, all Killing vectors generate such symmetries, corresponding to the zero scale function. It turns out that in Minkowski space there are five further vector fields which are given in an inertial chart, with their scale functions, below

$$\begin{gathered} D = x^\mu(\partial/\partial x^\mu) \qquad \lambda_D = 2 \\ K_\mu = g(D, D)(\partial/\partial x^\mu) - 2x_\mu D \qquad \lambda_{K_\mu} = -4x_\mu \\ \mu = 1, 2, 3, 0. \end{gathered} \tag{5.4.10}$$

These vector fields along with the 10 Killing vectors generating the Poincaré group, generate the 15-parameter local conformal group of Minkowski space. The source-free Maxwell equations are said to be conformally covariant in Minkowski space. Such a symmetry will generalise to any space with a metric admitting local conformal isometries and the vector C in (5.4.8) is referred to as a conformal Killing vector of the metric g. The local conformal symmetry may generalise to a global symmetry if the topology of the spacetime manifold can accommodate a complete conformal Killing vector field.

In §5.2 our introduction to Minkowski spacetime was motivated by the elegant reformulation of Maxwell's equations into a four-dimensional form. We now reverse the argument and show how these four-dimensional electromagnetic fields can be broken down into electric and magnetic fields in the instantaneous three-space of an arbitrary observer. Given any velocity vector field V, whose integral curves coincide with a set of observer curves, we use the Minkowski metric to define the associated dual 1-form $\widetilde{V}$ and write any F uniquely as

$$F = \widetilde{\boldsymbol{E}} \wedge \widetilde{V} + B \tag{5.4.11}$$

where B is a 2-form satisfying $i_V B = 0$ and $\widetilde{\boldsymbol{E}}$ a 1-form satisfying $i_V \widetilde{\boldsymbol{E}} = 0$. (Note: $\widetilde{\partial/\partial t} = -dt$.) One refers to $B \in \Gamma\Lambda_2 M$ as the magnetic 2-form associated with $\widetilde{V}$ and F, and $\widetilde{\boldsymbol{E}} \in \Gamma\Lambda_1 M$ as the associated electric 1-form. The electric field observed by this class of observers is

$$\boldsymbol{E} = \widetilde{i_V F}. \tag{5.4.12}$$

The magnetic vector field observed by this class can be related to F as follows. We use the velocity vector to define a metric $\hat{g}$ on the instantaneous three spaces

$$g = -\ \widetilde{V} \otimes \widetilde{V} + \hat{g}. \tag{5.4.13}$$

We may factor the volume four-form as

$$*1 = \widetilde{V} \wedge \hat{*}1. \tag{5.4.14}$$

Any p-form ω can be '3 + 1 decomposed' with respect to the velocity vector V:

$$\omega = \alpha + \widetilde{V} \wedge \beta \tag{5.4.15}$$

with $\mathrm{i}_V \alpha = \mathrm{i}_V \beta = 0$. If $\hat{*}$ is the Hodge map associated with $\hat{g}$ then

$$*\omega = -(\hat{*}\alpha) \wedge \widetilde{V} - \hat{*}\beta. \tag{5.4.16}$$

Applying this result to (5.4.11) gives

$$*F = -(\hat{*}B) \wedge \widetilde{V} + \hat{*}\widetilde{\boldsymbol{E}}. \tag{5.4.17}$$

But $\mathrm{i}_V(\hat{*}\widetilde{\boldsymbol{E}}) = 0$ so $\mathrm{i}_V * F = -\hat{*}B$. We define the vector field $\boldsymbol{B} = \widetilde{\hat{*}B}$ as the magnetic field associated with V; hence in terms of F

$$\boldsymbol{B} = -\widetilde{\mathrm{i}_V * F}. \tag{5.4.18}$$

If $\{Y_a\}$ is a frame on the instantaneous three-space, orthonormal with respect to $\hat{g}$, then the electric and magnetic field components in such a basis are given in terms of F as

$$\widetilde{\boldsymbol{E}}(Y_a) = (\mathrm{i}_V F)(Y_a) = 2F(V, Y_a)$$

$$(\hat{*}B)(Y_a) = -(\mathrm{i}_V * F)(Y_a) = -2*F(V, Y_a).$$

As an example consider the Coulomb solution:

$$F = \frac{q}{r^2}\mathrm{d}r \wedge \mathrm{d}t$$

$$r^2 = x^2 + y^2 + z^2$$

with observer curves tangent to $V = (\partial/\partial t)$ and $W = \gamma((\partial/\partial t) + v(\partial/\partial x^1))$, $\gamma = (1 - v^2)^{-1/2}$. With respect to V:

$$\boldsymbol{E} = \frac{-q}{r^2}(\partial/\partial r) \qquad \boldsymbol{B} = 0.$$

On the other hand, since $r\mathrm{d}r = x^1\mathrm{d}x^1 + x^2\mathrm{d}x^2 + x^3\mathrm{d}x^3$, W observes

$$\boldsymbol{E}' = \frac{-q\gamma}{r^2}\left((\partial/\partial r) + \frac{vx^1}{r}(\partial/\partial t)\right)$$

$$\boldsymbol{B}' = \frac{-q\gamma v}{r^3}\left(x^2(\partial/\partial x^3) - x^3(\partial/\partial x^2)\right)$$

instead of $\boldsymbol{E}$ and $\boldsymbol{B}$ at p.

It is worth stressing that although observers in Minkowski space experiencing arbitrary motion do not have world lines that can be naturally associated with the Poincaré group (their world lines are not integral curves of Killing vectors) the local definition of electric and magnetic fields for such observers follows as before since only a local frame and its dual are of relevance.

During the historical development of classical electromagnetism it became apparent that a number of related properties could be assimilated into a single idea once the spacetime description of Maxwell's theory was recognised. These properties became particularly succinct in terms of a second-rank tensor known as the Maxwell stress tensor. Historically the components of this tensor, with respect to a basis with physical dimensions, were associated with the properties of mechanical systems. This was a consequence of the role played by such components in equations which coupled together the behaviour of fields and matter. We shall discuss such equations later. At this point we shall be content with introducing this tensor in the guise of a 3-form associated with every Maxwell field and arbitrary vector field, and proving that such a 3-form associated with a conformal Killing vector is closed in source-free regions.

Define for any vector field V and Maxwell solution F the 3-form

$$\tau_V = \tfrac{1}{2}\{\mathrm{i}_V F \wedge {*F} - \mathrm{i}_V {*F} \wedge F\}. \tag{5.4.19}$$

Applying the exterior derivative and using Maxwell's equations for F produces

$$\mathrm{d}\tau_V = \tfrac{1}{2}\{\mathrm{d}\mathrm{i}_V F \wedge {*F} - \mathrm{i}_V F \wedge j - \mathrm{d}\mathrm{i}_V {*F} \wedge F\}. \tag{5.4.20}$$

Recall the identity $\mathscr{L}_X = \mathrm{d}\mathrm{i}_X + \mathrm{i}_X \mathrm{d} \quad \forall X$: hence

$$\mathrm{d}\mathrm{i}_V F = \mathscr{L}_V F \tag{5.4.21}$$

as $\mathrm{d}F = 0$. Similarly $\mathrm{d}\mathrm{i}_V {*F} = \mathscr{L}_V {*F} - \mathrm{i}_V j$. Inserting this in (5.4.20) gives

$$\mathrm{d}\tau_V = \tfrac{1}{2}\{\mathscr{L}_V F \wedge {*F} - \mathscr{L}_V {*F} \wedge F - \mathrm{i}_V F \wedge j + \mathrm{i}_V j \wedge F\}. \tag{5.4.22}$$

If C is a conformal Killing vector then $\mathscr{L}_C {*F} \wedge F = {*\mathscr{L}_C F} \wedge F = F \wedge {*\mathscr{L}_C F} = \mathscr{L}_C F \wedge {*F}$. Hence specialising to the case of a conformal Killing vector

$$\mathrm{d}\tau_C = -\tfrac{1}{2}\mathrm{i}_C F \wedge j + \tfrac{1}{2}\mathrm{i}_C j \wedge F.$$

Since $\mathrm{i}_C(j \wedge F) = \mathrm{i}_C j \wedge F - j \wedge \mathrm{i}_C F$ and, being a 5-form, $j \wedge F$ is zero we have

$$\mathrm{d}\tau_C = -\mathrm{i}_C F \wedge j. \tag{5.4.23}$$

For each conformal Killing vector these equations describe a 'local conservation equation' in a source free region ($j = 0$). The identification of a closed 3-form $\mathcal{J}$ with a local conservation law is appropriate in an arbitrary spacetime. For consider a region described by some 4-chain U whose boundary may be written

$$\partial U = \Sigma_1 + \Sigma_2 + \Pi \tag{5.4.24}$$

with the image of each Σ_j a spacelike hypersurface (each tangent vector to Σ_j being spacelike). For $\mathcal{J}$ closed

$$\int_{\partial U} \mathcal{J} = \int_U \mathrm{d}\mathcal{J} = 0 \tag{5.4.25}$$

by Stokes's theorem, thus

$$\int_{\Sigma_1} \mathcal{J} = \int_{-\Sigma_2} \mathcal{J} - \int_{\Pi} \mathcal{J}. \tag{5.4.26}$$

In cases where U may be chosen so that $\int_\Pi \mathcal{J} = 0$ one recognises that the flux of $\mathcal{J}$ through Σ_1 equals the flux of $\mathcal{J}$ through Σ_2 (see figure 5.2).

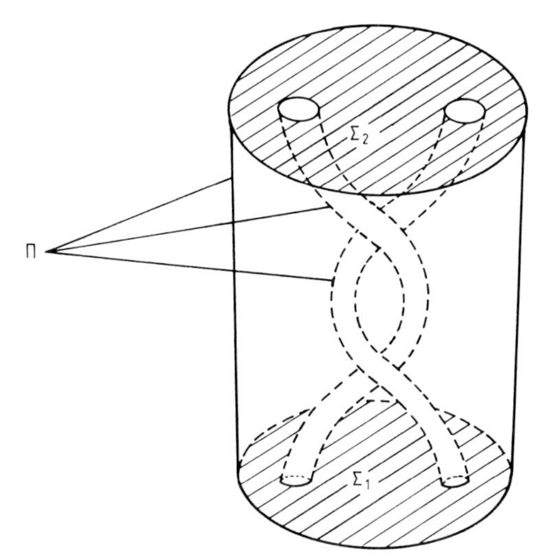

Figure 5.2 This diagram illustrates the equation $\partial U = \Sigma_1 + \Sigma_2 + \Pi$.

Suppose that we have a field system describing a simply connected source-free region U of Minkowski space. If τ is the proper time of some inertial observer passing through this region then in an adapted chart $\{\tau, \rho^1, \rho^2, \rho^3\}$ we take Σ_j to lie in the hypersurface $\tau(p) = c_j$, for some constant c_j. If the electromagnetic field vanishes at large spatial distances from the observer then we may take Π to complete the boundary of U such that the electromagnetic field vanishes on Π. Thus in this case the flux of $\mathcal{J}$ through the instantaneous three-space is time independent. If we write $\mathcal{J}$ in terms of a 2-form current $\hat{\mathcal{J}}$ and an associated 3-form density $\hat{\rho}$, $\mathcal{J} = \hat{\mathcal{J}} \wedge \mathrm{d}\tau + \hat{\rho}$ with $\mathrm{i}_{(\partial/\partial\tau)}\hat{\mathcal{J}} = 0$ and $\mathrm{i}_{(\partial/\partial\tau)}\hat{\rho} = 0$, then clearly $\Sigma^*\mathcal{J} = \hat{\rho}$ and

$$\int_\Sigma \mathcal{J} = \int_\Sigma \hat{\rho}. \tag{5.4.27}$$

It is tempting to reinterpret the conservation of $\mathcal{J}$-flux associated with U in terms of a local flow of current $\hat{\mathcal{J}}$ and an associated variation of density $\hat{\rho}$. Certainly the 3-form equation $\mathrm{d}\mathcal{J} = 0$ implies a local continuity equation. In the above chart we may write d when acting on $\mathcal{J}$ as $\mathrm{d} = \underline{\mathrm{d}} + \mathrm{d}\tau \wedge \mathcal{L}_{(\partial/\partial\tau)}$ where $\underline{\mathrm{d}}$ is the exterior derivative associated with the instantaneous three-space. Hence (as $\underline{\mathrm{d}}\,\hat{\rho} = 0$)

$$\underline{\mathrm{d}}\hat{\mathcal{J}} - \mathcal{L}_{(\partial/\partial\tau)}\hat{\rho} = 0. \tag{5.4.28}$$

If we express $\hat{\mathcal{J}}$ and $\hat{\rho}$ in a basis adapted to Σ:

$$\hat{\mathcal{J}} = \hat{\mathcal{J}}_1 \mathrm{d}\rho^2 \wedge \mathrm{d}\rho^3 + \hat{\mathcal{J}}_2 \mathrm{d}\rho^3 \wedge \mathrm{d}\rho^1 + \hat{\mathcal{J}}_3 \mathrm{d}\rho^1 \wedge \mathrm{d}\rho^2$$
$$\hat{\rho} = \rho\,\mathrm{d}\rho^1 \wedge \mathrm{d}\rho^2 \wedge \mathrm{d}\rho^3$$

(5.4.28) is equivalent to

$$\sum_{j=1}^{3} (\partial\hat{\mathcal{J}}_j/\partial\rho^j) - (\partial\rho/\partial\tau) = 0. \tag{5.4.29}$$

The interpretation of this local continuity equation must, however, be treated with caution. If $\mathcal{J}$ is a closed 3-form on U then so is $\mathcal{J}' = \mathcal{J} + \mathrm{d}\mathcal{K}$ where $\mathcal{K}$ is any smooth 2-form. If $\mathcal{K}$ is chosen such that $\int_{\partial\Sigma}\mathcal{K} = 0$ then $\mathcal{J}$ and $\mathcal{J}'$ both have the same flux through Σ, although $\mathcal{J}'$ will redistribute the local density.

Returning to (5.4.23) we see that there are 15 closed 3-forms, one for each of the 15 conformal generators of the Minkowski space conformal group. It is instructive to examine the currents associated with some of these Killing vectors. If V is a timelike Killing vector field generating time translations along its open integral curve then, using (5.4.12) and (5.4.18) to define $\boldsymbol{E}$ and $\boldsymbol{B}$ with respect to such a field, we easily find:

$$\tau_V = -\widetilde{\boldsymbol{E}} \wedge \widetilde{\boldsymbol{B}} \wedge \widetilde{V} + \tfrac{1}{2}(\widetilde{\boldsymbol{E}} \wedge \hat{*}\widetilde{\boldsymbol{E}} + \widetilde{\boldsymbol{B}} \wedge \hat{*}\widetilde{\boldsymbol{B}}). \tag{5.4.30}$$

The physically dimensioned components of the vector obtained by taking the metric dual of the 2-form $\widetilde{\boldsymbol{E}} \wedge \widetilde{\boldsymbol{B}}$ with respect to $\hat{g}$ was identified by Poynting as the local field energy transmitted 'normally' across unit area per second (that is the local field energy current). Similarly the $\hat{g}$-dual of the 3-form $\frac{1}{2}(\widetilde{\boldsymbol{E}} \wedge \hat{*}\widetilde{\boldsymbol{E}} + \widetilde{\boldsymbol{B}} \wedge \hat{*}\widetilde{\boldsymbol{B}})$ may, after restoring physical dimensions, be interpreted as a local field energy density. Since, for example $\widetilde{\boldsymbol{E}} \wedge \hat{*}\widetilde{\boldsymbol{E}} = \hat{g}(\boldsymbol{E}, \boldsymbol{E})\hat{*}1$, the signature of $\hat{g}$ ensures that this density is positive-definite. This interpretation has persisted although with the caveats above we would prefer to identify the oriented integral $\int_\Sigma \tau_V$, in a source-free region of spacetime, as the field energy associated with the spacelike 3-chain Σ and $\int_{S^2} \mathrm{i}_V \mathrm{d}\tau_V$ as a power flux across an oriented spacelike 2-chain S^2.

Suppose we consider a spacelike Killing vector field X generating spacelike translations along *open* integral curves and decompose τ_X according to

$$\tau_X = \mu_X \wedge \widetilde{V} + \mathcal{G}_X \tag{5.4.31}$$

with $\mathrm{i}_V \mu_X = \mathrm{i}_V \mathcal{G}_X = 0$. The Maxwell stress 2-form μ_X may be used to identify mechanical Newtonian forces produced by a 'flow' of a Newtonian field momentum density 3-form $\mathcal{G}_X$. In an analogous manner one may construct torque forms (angular momentum currents) using a Killing vector field that generates rotations along *closed* integral curves.

As promised we now relate the stress 3-forms to an associated second-rank tensor. Given any local frame $\{X_a\}$ $a = 0, 1, 2, 3$ in spacetime, with natural dual co-frame $\{e^b\}$, we may obtain 16 real functions T_{ab} defined by $*\tau_{X_b} = T_{bc}e^c$ or $T_{ab} = (*\tau_{X_a})(X_b)$. These may be used to define a second-rank tensor

$$T = T_{ab} e^a \otimes e^b \tag{5.4.32}$$

which is referred to as the *stress tensor*.

Exercise 5.2
Show that if $\tau_{X_a} \wedge e_b = \tau_{X_b} \wedge e_a$ then $T_{ab} = T_{ba}$: the stress tensor is *symmetric*. Show that if $\tau_{X_a} \wedge e^a = 0$ then $T_b{}^b = 0$: the stress tensor is *traceless*.

These properties are satisfied for the Maxwell stress tensor as follows directly from the definition. We shall meet these properties again at a later stage in the context of a Clifford representation for this tensor.

Exercise 5.3
If $F = \frac{1}{2} F_{ab} e^a \wedge e^b$ show that

$$T_{ab} = -\tfrac{1}{4} g_{ab} F^{cd} F_{cd} - F_{ac} F^c{}_b.$$

Exercise 5.4
Use the three angular momentum 3-forms τ_{K_i} to evaluate the torque on an electric dipole in a uniform static electric field. (Hint: calculate the total electromagnetic 2-form and use this in (5.4.19) where the Killing currents are computed with the aid of the rotational Killing vectors.)

Bibliography

Misner C, Thorne K and Wheeler A 1973 *Gravitation* (San Francisco: W H Freeman)

Sachs R K and Wu H 1977 *General Relativity for Mathematicians* (New York: Springer)
Schutz B F 1985 *A First Course in General Relativity* (Cambridge: Cambridge University Press)

6

Connections

The differentiable structure on a manifold enabled us to define two important differential operators; the exterior and Lie derivatives. Whereas the former acted only on antisymmetric tensor fields (differential forms) the latter acted on any tensor field. However, whilst reducing to the directional derivative on functions the Lie derivative is not a suitable generalisation to a 'directional derivative on tensors'. This is because the Lie derivative of a tensor at p, along a curve C, does not just depend on the tangent to the curve at p but on the behaviour of tangent vectors in the vicinity of p. This feature of the Lie derivative is reflected in the fact that $\mathscr{L}_X T$ is not $\mathscr{F}$-linear in the vector field X.

Another differential operator, a tensor *covariant derivative*, will now be introduced. The introduction of this new structure is equivalent to choosing a *parallelism* for the manifold. The general notion of parallelism is easy to grasp. It is only necessary to recognise that in general there is no preordained way to map a vector at one point on a manifold to a new vector at another point. Defining a parallelism on a manifold requires specifying a rule that will provide a means of comparing vectors at different points by transporting one to the other along some prescribed path connecting the points. Whereas the parallel transport map will depend on the path chosen to connect the points we do not want it to depend on how the path is traversed. (Parallel transport depends on the route taken but not on how bumpy the ride!) Although this feature of path dependence of parallel transport does not accord with the intuitive Euclidean concept it is an essential feature, characterising the *curvature* of the manifold. Given a parallelism we can define a covariant derivative by comparing a vector with its parallel translate and taking a suitable limit. Conversely, by introducing a new rule for differentiating vectors, and establishing a *linear connection*, we can define a vector field to be parallel along a curve if its derivative with respect to the tangent vector is zero.

6.1 Linear Connections

A *linear connection* on a manifold M is a map $\nabla : \Gamma TM \times \Gamma TM \to \Gamma TM$ that satisfies the following, $\forall f, g \in \mathcal{F}(M)$, $\forall X, Y, Z \in \Gamma TM$:

$$\nabla_{fX + gY} Z = f\nabla_X Z + g\nabla_Y Z \tag{6.1.1}$$

$$\nabla_X(fY + gZ) = X(f)Y + f\nabla_X Y + X(g)Z + g\nabla_X Z. \tag{6.1.2}$$

Thus ∇_X is a linear mapping on vector fields which is also $\mathcal{F}$-linear in X: it is called *covariant differentiation* with respect to X.

From these properties it follows that we can specify ∇ by giving the components of the vector $\nabla_{X_a} X_b$ in any convenient basis $\{X_a\}$:

$$\nabla_{X_a} X_b = \Gamma_{ab}{}^c X_c. \tag{6.1.3}$$

The n^3 functions $\Gamma_{ab}{}^c$, where $n = \dim M$, are known as the *connection components*, or *connection coefficients* in this basis. These coefficients can be used to define a set of 1-forms, the *connection 1-forms*,

$$\omega^a{}_b = \Gamma_{cb}{}^a e^c \tag{6.1.4}$$

where $\{e^a\}$ is the co-frame dual to $\{X_a\}$. Thus we can write (6.1.3) equivalently as

$$\nabla_{X_a} X_b = \omega^c{}_b(X_a) X_c. \tag{6.1.5}$$

If $\{Y_a\}$ is a new basis, related to $\{X_b\}$ by a general linear transformation $Y_a = A_a{}^b X_b$, then

$$\begin{aligned}\nabla_{Y_a} Y_b &= A_a{}^p \nabla_{X_p}(A_b{}^c X_c)\\ &= A_a{}^p A_b{}^c \Gamma_{pc}{}^q X_q + A_a{}^p X_p(A_b{}^c) X_c.\end{aligned}$$

If the inverse transformation is given by $A^{-1}{}_p{}^a A_a{}^b = \delta^b_p$, then the connection coefficients $\Gamma'{}_{ab}{}^r$ in the basis $\{Y_a\}$ are given by

$$\Gamma'_{ab}{}^r = A_a{}^p A_b{}^c \Gamma_{pc}{}^q A^{-1}{}_q{}^r + A_a{}^p X_p(A_b{}^c) A^{-1}{}_c{}^r. \tag{6.1.6}$$

Equivalently the connection 1-forms in this basis are given by

$$\omega'{}^a{}_b = A_b{}^q \omega^r{}_q A^{-1}{}_r{}^a + A^{-1}{}_q{}^a dA_b{}^q. \tag{6.1.7}$$

The 'inhomogeneous' term in this transformation represents a departure from the transformation of the components of a tensor, reflecting the fact that the map $X, Y \to \nabla_X Y$ is not $\mathcal{F}$-linear in Y.

As anticipated the covariant derivative of a vector field with respect to X, evaluated at the point p, depends only on the value of X at p. For if V is any vector field and $\{X_a\}$ is a basis in the neighbourhood of p then $(\nabla_V Z)|_p = V^a(p)(\nabla_{X_a} Z)|_p$. So if V vanishes at p then

$(\nabla_V Z)|_p = 0\ \forall Z$. Thus if X and Y are vector fields such that $X|_p = Y|_p$ then $(\nabla_X Z)|_p = (\nabla_Y Z)|_p\ \forall Z$. Hence for any $X_p \in T_pM$ we have a covariant derivative in the direction of X_p, $\nabla_{X_p} : \Gamma TM \to T_pM$.

Let C be a curve with tangent vector $\dot{C}$. Then if Y is a vector field we may covariantly differentiate Y in the direction of the tangent vector at any point $C(t)$ on the curve. An assignment of a vector $Y_{C(t)}$ to every $T_{C(t)}M$ is a (smooth) vector field along C if the map $t \mapsto Y_{C(t)}f$ is a smooth function of $t\ \forall f \in \mathscr{F}(M)$. Thus if C is a smooth curve $\nabla_{\dot{C}}Y$ is a smooth vector field along C. As will be seen below $\nabla_{\dot{C}}Y$ only depends on the value of Y along C, and so in fact any vector field along C can be covariantly differentiated with respect to the tangent vector to produce another vector field along C. (Some authors denote $\nabla_{\dot{C}}Y$ by $\mathrm{D}Y/\mathrm{d}t$ where t parametrises C.)

A vector field Y along a curve C is said to be *parallel along* C if it satisfies the equations

$$\nabla_{\dot{C}}Y = 0. \tag{6.1.8}$$

If we expand $Y = Y^j\partial_j$ in a local coordinate chart in which $x^j(p) = C^j(t)$ represents C, $j = 1, \ldots, n$ then $\dot{C} = C_*(\partial/\partial t) = \dot{C}^k(t)(\partial/\partial x^k)$. Hence $\nabla_{\dot{C}}(Y^j(\partial/\partial x^j)) = (\dot{C}Y^j)(\partial/\partial x^j) + Y^j\nabla_{\dot{C}}(\partial/\partial x^j)$. But $\dot{C}Y^j = [C_*(\partial/\partial t)]Y^j = \dot{C}^k(t)(\partial Y^j/\partial x^k) = \mathrm{d}(Y^j \circ C)/\mathrm{d}t$ and $\nabla_{\dot{C}}(\partial/\partial x^j) = \dot{C}^k(t)\nabla_{(\partial/\partial x^k)}(\partial/\partial x^j) = \dot{C}^k(t)\Gamma_{kj}{}^m(\partial/\partial x^m)$. Thus (6.1.8) gives the following differential equations for the components $Y^j \circ C$ of Y on C:

$$\frac{\mathrm{d}}{\mathrm{d}t}(Y^m \circ C) + (Y^j \circ C)\dot{C}^k(t)(\Gamma_{kj}{}^m \circ C) = 0. \tag{6.1.9}$$

For given functions $\dot{C}^k(t)$ and connection components $\Gamma_{kj}{}^m(C(t))$ these equations are known to have a unique solution $Y^m(C(t))$ specified by the choice of initial components $Y^m(C(0))$. (It is because these equations only depend on the components of Y along C that a vector field along C can be differentiated.) Because of the above uniqueness result a parallelism is established by the linear connection ∇. If $Y_{C(0)}$ is any vector in $T_{C(0)}M$ and Y is the unique vector field along C such that $\nabla_{\dot{C}}Y = 0$ then $Y_{C(t)}$ is called the parallel translate of $Y_{C(0)}$ along C.

Let Y be any smooth vector field along C with $Y_{C(0)} \neq 0$, and f the smooth function such that $(f \circ C)(t) = t$. For t sufficiently small Z is a smooth vector field on C

$$Z \equiv Y + \sum_{n=1}^{\infty} \frac{(-f)^n(\nabla_{\dot{C}})^n Y}{n!} \tag{6.1.10}$$

where $(\nabla_{\dot{C}})^2 Y = \nabla_{\dot{C}}(\nabla_{\dot{C}}Y)$ etc. We have

$$\nabla_{\dot{C}}Z = \nabla_{\dot{C}}Y - \sum_{n=1}^{\infty}\frac{n(-f)^{n-1}(\nabla_{\dot{C}})^n Y}{n!} + \sum_{n=1}^{\infty}\frac{(-f)^n(\nabla_{\dot{C}})^{n+1}Y}{n!} \text{(since } \dot{C}f = 1)$$

$$= -\sum_{n=2}^{\infty}\frac{n(-f)^{n-1}(\nabla_{\dot{C}})^{n}Y}{n!} + \sum_{n=1}^{\infty}\frac{(-f)^{n}(\nabla_{\dot{C}})^{n+1}Y}{n!}$$

$$= -\sum_{m=1}^{\infty}\frac{(m+1)(-f)^{m}(\nabla_{\dot{C}})^{m+1}Y}{(m+1)m!} + \sum_{n=1}^{\infty}\frac{(-f)^{n}(\nabla_{\dot{C}})^{n+1}Y}{n!} = 0.$$

So Z is parallel along C with $Z_{C(0)} = Y_{C(0)}$, thus $Z_{C(t)}$ must be the parallel translate of $Y_{C(0)}$ to $C(t)$. Note that *any* vector field Y satisfying $Y_{C(0)} = A$ can be taken in (6.1.10) to evaluate the parallel transport of $A \in T_{C(0)}M$ along C (see figure 6.1).

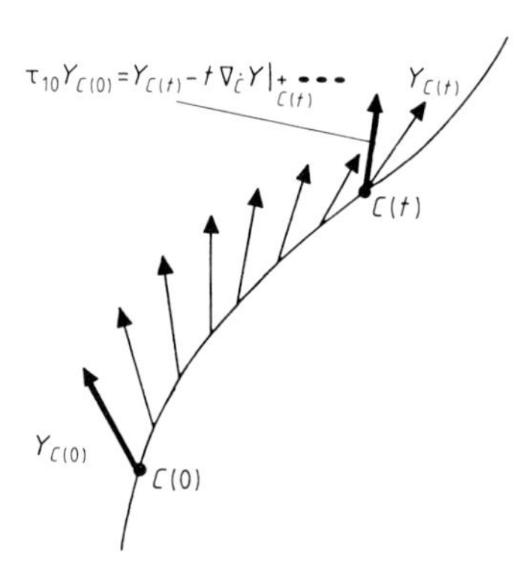

Figure 6.1 The parallel translation of Y along the curve C.

A vector field Y is said to be *parallel*, or covariantly constant, (with respect to ∇) if it satisfies the equation $\nabla_X Y = 0\ \forall X$. This implies that Y is parallel along all curves and thus the parallel transport map is independent of the path along which such a Y is transported.

Exercise 6.1

A connection on a two-dimensional manifold is specified in a local chart with coordinate maps (x^1, x^2) by $\Gamma_{11}{}^{1} = -x^1$ and $\Gamma_{22}{}^{2} = -x^2$ with all other connection components zero in this chart. Prove that for $a, b \in \mathbb{R}$ the vector field

$$Y = a\exp[(x^1)^2/2](\partial/\partial x^1) + b\exp[-(x^1)^2/2](\partial/\partial x^2)$$

is parallel along the curve

$$C:[0, 1] \longmapsto (x^1(p) = \sin t,\ x^2(p) = \cos t).$$

A linear connection enables us to define a 'straight line', generalising one of the intuitive properties of straight lines in Euclidean space. A curve C is an *autoparallel* (of ∇) if its tangent vector field is parallel along C. Such curves are given as solutions to the equation

$$\nabla_{\dot{C}}\dot{C} = 0. \tag{6.1.11}$$

(Autoparallels are more frequently called *geodesics* although we prefer to reserve this terminology for the autoparallels of a pseudo-Riemannian connection which will be discussed later. Students everywhere will be relieved to know that if by 'straight line' we mean autoparallel, then at least for the Riemannian connection 'straight lines' are (in a certain sense) the shortest curves connecting two points!) If an autoparallel C is given parametrically in a local chart by $x^j(p) = C^j(t)$ then the C^j must satisfy the system of differential equations

$$\frac{\mathrm{d}}{\mathrm{d}t}\dot{C}^m + (\Gamma_{kj}{}^m \circ C)\dot{C}^k(t)\dot{C}^j = 0$$

or

$$\ddot{C}^m + (\Gamma_{kj}{}^m \circ C)\dot{C}^k\dot{C}^j = 0. \tag{6.1.12}$$

It is important to note that the solution of (6.1.12) is a parametrised curve. Although a general reparametrisation of the solution will not change the image set on M of the reparametrised C, the corresponding map will not in general satisfy (6.1.12) and will not therefore be an autoparallel. If C is an autoparallel, with parameter t, then the reparametrised curve $C \circ h$ is also an autoparallel if and only if $h = at + b$ for $a, b \in \mathbb{R}$. For an arbitrary curve C we define the *acceleration* to be the vector field $\nabla_{\dot{C}}\dot{C}$ on C. (Thus the acceleration

Each autoparallel is fixed uniquely by specifying $(C^j, \dot{C}^j)$ for some initial t. That is, for every $X_p \in T_pM$ there is a unique maximal autoparallel starting at p in the direction of X_p. Let γ_{X_p} be this autoparallel. The *exponential mapping* at p, Exp_p, maps a subset of T_pM into M:$\mathrm{Exp}_pX_p = \gamma_{X_p}(1)$. Clearly Exp_p is defined on those X_p for which γ_{X_p} is defined on $[0, 1]$. Since for $\lambda \in \mathbb{R}$ $\gamma_{\lambda X_p}(t) = \gamma_{X_p}(\lambda t)$, if Exp_p is defined on X_p then it is also defined on λX_p for $\lambda \in [0, 1]$. It in fact follows from the nature of the differential equations (6.1.12) that for every $p \in M$ there is a neighbourhood of the origin in T_pM, $\mathcal{N}_0$, such that the exponential mapping is a diffeomorphism onto a neighbourhood of p, N_p. If such an $\mathcal{N}_0$ is star shaped then it is called a normal neighbourhood. (To say that $\mathcal{N}_0$ is star shaped means that if $v \in \mathcal{N}_0$ then $\lambda v \in \mathcal{N}_0$ $\forall \lambda \in [0, 1]$.) A *normal neighbourhood* of p is the image of a normal neighbourhood in T_pM under the exponential mapping. For every q in a normal neighbourhood of p, N_p, there is one and only one $Q \in T_pM$ such that $q = \mathrm{Exp}_pQ$. Thus if $\{X_i\}$ is any basis for T_pM with $Q = \Sigma_{i=1}^n Q^iX_i$ the mapping $q \mapsto \{Q^i\}$ provides a coordinate system for N_p (see figure 6.2). Such coordinates are called *normal coordinates* at p.

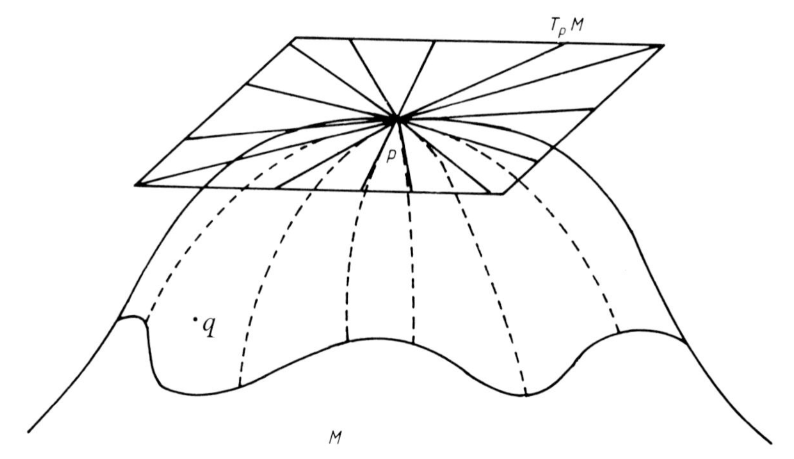

Figure 6.2 This diagram illustrates the exponential map and normal coordinates.

Exercise 6.2
If $\Gamma_{ij}{}^k$ are the connection coefficients with respect to a normal coordinate basis at p show that

$$\Gamma_{ij}{}^k(p) + \Gamma_{ji}{}^k(p) = 0.$$

Hint: Show that $\dot{\gamma}_v^i(t) = v^i$ where $v = v^i X_i$.

6.2 Examples and Newtonian Force

To gain some insight into covariant derivatives we turn to $\mathbb{R}^n$. This manifold has an absolute parallelism: the parallel-transport map is path independent. If $\{x^i\}$ are standard coordinates and $X_p = \Sigma_i c^i \partial_i|_p$ then the parallel translate of X_p at q is $X_q = \Sigma_i c^i \partial_i|_q$. Thus in such a standard chart the connection is defined by $\nabla_{\partial_i}\partial_j = 0$. Such a connection is referred to as the standard connection on $\mathbb{R}^n$.

It is of interest to compute the standard connection for $\mathbb{R}^2$ in a polar chart (r, θ) related to the standard one by

$$\begin{aligned} x^1 &= r\cos\theta \qquad & 0 < r < \infty \\ x^2 &= r\sin\theta \qquad & 0 < \theta \leqslant 2\pi. \end{aligned} \tag{6.2.1}$$

This induces a coordinate frame transformation:

$$\partial_r = (x^1/r)\partial_1 + (x^2/r)\partial_2 \tag{6.2.2}$$

and

$$\partial_\theta = -x^2\partial_1 + x^1\partial_2. \qquad (6.2.3)$$

Since ∂_1 and ∂_2 are parallel we have

$$\begin{aligned}
\nabla_{\partial_r}(\partial_r) &= [\partial_r(x^1/r)]\partial_1 + [\partial_r(x^2/r)]\partial_2 = 0 \\
\nabla_{\partial_r}(\partial_\theta) &= -[\partial_r(x^2)]\partial_1 + [\partial_r(x^1)]\partial_2 = (1/r)\partial_\theta \\
\nabla_{\partial_\theta}(\partial_\theta) &= -r\partial_r \\
\nabla_{\partial_\theta}(\partial_r) &= (1/r)\partial_\theta.
\end{aligned} \qquad (6.2.4)$$

Writing $\nabla_{\partial_r}(\partial_\theta)$ as $\Gamma_{r\theta}{}^r\partial_r + \Gamma_{r\theta}{}^\theta\partial_\theta$, etc we may read off the components of the connection in the polar chart; $\Gamma_{\theta r}{}^\theta = \Gamma_{r\theta}{}^\theta = (1/r)$, $\Gamma_{\theta\theta}{}^r = -r$ with all others zero.

If a curve C is given in the natural chart as $x^j(p) = C^j(t)$ then $\nabla_{\dot{C}}\dot{C} = \ddot{x}^j(t)\partial_j$. Let us evaluate the acceleration of the curve $C : [0, 1] \rightarrow \mathbb{R}^2$ given in the above polar chart by

$$(r \circ C)(t) = \rho(t),\ (\theta \circ C)(t) = \Theta(t)$$

for smooth real functions ρ and Θ of t. The tangent vector to C may be written

$$\dot{C} = C_*\partial_t = \dot{\rho}\partial_r + \dot{\Theta}\partial_\theta$$

hence

$$\nabla_{\dot{C}}\dot{C} = \ddot{\rho}\partial_r + \ddot{\Theta}\partial_\theta + \dot{\rho}\nabla_{\dot{C}}\partial_r + \dot{\Theta}\nabla_{\dot{C}}\partial_\theta.$$

But

$$\nabla_{\dot{C}}\partial_r = \dot{\rho}\nabla_{\partial_r}\partial_r + \dot{\Theta}\nabla_{\partial_\theta}\partial_r = (\dot{\Theta}/\rho)\partial_\theta$$

and

$$\begin{aligned}
\nabla_{\dot{C}}\partial_\theta &= \dot{\rho}\nabla_{\partial_r}\partial_\theta + \dot{\Theta}\nabla_{\partial_\theta}\partial_\theta \\
&= (\dot{\rho}/\rho)\partial_\theta - \rho\dot{\Theta}\partial_r.
\end{aligned}$$

Hence the natural $\mathbb{R}^2$ acceleration of C is

$$\nabla_{\dot{C}}\dot{C} = (\ddot{\rho} - \rho\dot{\Theta}^2)\partial_r + (\rho\ddot{\Theta} + 2\dot{\rho}\dot{\Theta})(1/\rho)\partial_\theta. \qquad (6.2.5)$$

With respect to the standard Euclidean metric on $\mathbb{R}^2$

$$g = \partial_1\otimes\partial_2 + \partial_2\otimes\partial_2 = \partial_r\otimes\partial_r + (1/r^2)\partial_\theta\otimes\partial_\theta \qquad (6.2.6)$$

and identifying the parameter t with Newtonian time, we recognise the orthonormal components of this acceleration in the polar frame as the radial and transverse components of Newtonian acceleration of a particle moving in two dimensions under the influence of some Newtonian force.

Exercise 6.3
Use the standard connection in $\mathbb{R}^3$ to compute the orthonormal components of the Newtonian acceleration of the curve $C : [0, 1] \to \mathbb{R}^3$ given by $(r \circ C)(t) = R(t)$, $(\theta \circ C)(t) = \Theta(t)$, $(\varphi \circ C)(t) = \Phi(t)$ where the maps (r, θ, φ) are standard polar coordinates in $\mathbb{R}^3$.

The above examples in $\mathbb{R}^2$ and $\mathbb{R}^3$ suggest that the Newtonian postulates describing the motion of a single point particle in space be rephrased in terms of the 'natural' connection as follows.

(1) A free particle is one that moves along the trajectory described by an autoparallel of the natural connection in Euclidean space, parametrised by universal time.

(2) A point particle of inertial mass m moving in a non-autoparallel curve C, parametrised by Newtonian time, experiences a force $\bar{\mathscr{F}}$ on C given by

$$\bar{\mathscr{F}} = \nabla_{\dot{C}}(m\dot{C}). \tag{6.2.7}$$

In many problems in physics $\bar{\mathscr{F}}$ arises as a restriction to C of a vector field on $\mathbb{R}^3$ determined from some field theory. If $\bar{\mathscr{F}}$ is prescribed, (6.2.7) may be used to determine a Newtonian trajectory. As an example, for motion of a particle under the gravitational force produced by a static spherically symmetric distribution of matter (with total *gravitational mass* M), we may use the Newtonian potential $\Phi = GM/r$ in a polar chart, where G is the Newtonian gravitational constant, to obtain

$$\bar{\mathscr{F}} = -m\widetilde{\mathrm{d}\Phi} = (GMm/r^2)\partial_r. \tag{6.2.8}$$

For a particle with electric charge q the Newtonian Lorentz force is $\bar{\mathscr{F}} = q\{\widetilde{\boldsymbol{E}} + \widetilde{\mathrm{i}_{\dot{C}}B}\}$. In standard coordinates $\{x^i\}$, $\boldsymbol{E} = E_i \mathrm{d}x^i$ and $B = B_1 \mathrm{d}x^2 \wedge \mathrm{d}x^3 + B_2 \mathrm{d}x^3 \wedge \mathrm{d}x^1 + B_3 \mathrm{d}x^1 \wedge \mathrm{d}x^2$ are 1-and 2-forms respectively on $\mathbb{R}^3$, parametrised by Newtonian universal time. The metric duals are taken with respect to the Euclidean metric. Solutions of (6.2.7) for particle trajectories subject to these force laws give an excellent description of the behaviour of matter in gravitational and electromagnetic fields provided the motion never approaches Newtonian speeds comparable with $10^8\ \mathrm{m\,s^{-1}}$.

6.3 Covariant Differentiation of Tensors

We have introduced the covariant derivative ∇_X as a map on vector fields. To extend the definition to its action on smooth 1-forms

$\beta \in \Gamma\Lambda_1 M$ we define $\nabla_X \beta$ by

$$(\nabla_X \beta)(Y) = -\beta(\nabla_X Y) + X(\beta(Y)) \qquad X, Y \in \Gamma TM. \quad (6.3.1)$$

If $f \in \mathscr{F}(M)$ it follows from this that

$$\nabla_X (f\beta) = f\nabla_X \beta + (Xf)\beta. \quad (6.3.2)$$

If $\{X_a\}$, $\{e^b\}$ are dual bases it follows from $e^b(X_a) = \delta^b_a$ that if $\omega^b{}_c$ are defined by (6.1.5) then

$$\nabla_{X_a} e^c = -\omega^c{}_b(X_a)e^b. \quad (6.3.3)$$

If for $f \in \mathscr{F}(M)$

$$\nabla_X f \equiv X(f) \quad (6.3.4)$$

we note that (6.3.1) is equivalent to adopting the rule

$$\nabla_X(\beta(Y)) = (\nabla_X \beta)(Y) + \beta(\nabla_X Y). \quad (6.3.5)$$

The covariant derivative is said to commute with contractions. Having defined the covariant derivative of 1-forms and vector fields we can extend the definition to arbitrary tensors by adopting this property of commuting with contractions

$$\nabla_X : \Gamma T^s_r M \longrightarrow \Gamma T^s_r M$$

$$\begin{aligned} \nabla_X T(X_1, \ldots, X_r, e^1, \ldots, e^s) &= -T(\nabla_X X_1, \ldots, X_r, e^1, \ldots, e^s) - \ldots \\ &\quad - T(X_1, \ldots, X_r, e^1, \ldots, \nabla_X e^s) \\ &\quad + \nabla_X(T(X_1, \ldots, X_r, e^1, \ldots, e^s)). \end{aligned} \quad (6.3.6)$$

Such a covariant derivative satisfies the Leibnitz property

$$\nabla_X(T \otimes W) = \nabla_X T \otimes W + T \otimes \nabla_X W \quad (6.3.7)$$

for all tensor fields T and W. That is, ∇_X becomes a type-preserving derivation on the algebra of tensor fields. If a mixed tensor has components $T^{a_1, \ldots, a_r}{}_{b_1, \ldots, b_s}$ in any basis it is conventional to denote the components of $\nabla_{X_k} T$ *in the same basis* by $T^{a_1, \ldots, a_r}{}_{b_1, \ldots, b_s; k}$. For any $T \in \Gamma T^s_r M$ the tensor field $\nabla T \in \Gamma T^s_{r+1} M$ defined by

$$(\nabla T)(X, X_1, \ldots, X_r, e^1, \ldots, e^s) = (\nabla_X T)(X_1, \ldots, X_r, e^1, \ldots, e^s)$$
$$\forall X, X_j \in \Gamma TM, \; e^a \in \Gamma T^* M \quad (6.3.8)$$

is called the *covariant differential* of T. Thus starting with a rule that defines a transport of vector fields along curves we have extended the covariant derivative to an operator on general tensor fields.

6.4 Curvature and Torsion Tensors of ∇

Whereas the lack of $\mathscr{F}$-linearity in the map $X, Y \to \nabla_X Y$ prevents ∇ itself from being identified with a tensor it may be used to construct two important tensors. First observe that for any function $f \in \mathscr{F}(M)$:

$$\nabla_X(fY) = (Xf)Y + f\nabla_X Y$$

and

$$\mathscr{L}_X(fY) \equiv [X, fY] = (Xf)Y + f[X, Y]$$
$$\forall X, Y \in \Gamma TM.$$

It follows that if we define

$$\boldsymbol{T}(X, Y) = \nabla_X Y - \nabla_Y X - [X, Y] \tag{6.4.1}$$

then $\boldsymbol{T}(X, fY) = f\boldsymbol{T}(X, Y)$. Since $\boldsymbol{T}(X, Y) = -\boldsymbol{T}(Y, X)$ by construction then $\boldsymbol{T}(X, Y)$ is $\mathscr{F}$-linear in both arguments. Consequently associated with $\boldsymbol{T}$ is a type (2, 1) tensor field T known as the *torsion tensor* of ∇:

$$T(X, Y, \beta) = \beta(\boldsymbol{T}(X, Y)). \tag{6.4.2}$$

Associated with any local basis is a set of *torsion 2-forms* T^a

$$T^a(X, Y) = \tfrac{1}{2}e^a(\boldsymbol{T}(X, Y)). \tag{6.4.3}$$

The torsion tensor can be written in terms of these 2-forms as

$$T = 2T^a \otimes X_a. \tag{6.4.4}$$

If $\{e^a\}$ is any co-frame, in which the connection 1-forms are $\{\omega^a{}_b\}$, then the torsion 2-forms are given by

$$T^a = \mathrm{d}e^a + \omega^a{}_b \wedge e^b. \tag{6.4.5}$$

This is called the *first structure equation*. It may be proved by contracting on a pair of arbitrary vectors. Using (4.10.3) have

$$\begin{aligned}
&2(\mathrm{d}e^a + \omega^a{}_b \wedge e^b)(X, Y)\\
&\quad = X(e^a(Y)) - Y(e^a(X)) - e^a([X, Y]) + \omega^a{}_b(X)e^b(Y) - \omega^a{}_b(Y)e^b(X)\\
&\quad = X(e^a(Y)) - \nabla_X e^a(Y) - Y(e^a(X)) + \nabla_Y e^a(X) - e^a([X, Y])
\end{aligned}$$

by (6.3.3). The right-hand side may be simplified by using (6.3.1), producing

$$(\mathrm{d}e^a + \omega^a{}_b \wedge e^b)(X, Y) = \tfrac{1}{2}e^a(\boldsymbol{T}(X, Y))$$

when (6.4.1) is used. Thus (6.4.5) follows from the definition (6.4.3).

The second important tensor constructed from ∇ involves two covariant differentiations. Again we note from the fundamental properties of

∇ that for any tensor field U

$$\nabla_X \nabla_{fY} U = f\nabla_X \nabla_Y U + (Xf)\nabla_Y U$$
$$\nabla_{fY} \nabla_X U = f\nabla_Y \nabla_X U \qquad \forall X, Y \in \Gamma TM.$$

If we define

$$\boldsymbol{R}(X, Y)U = \nabla_X \nabla_Y U - \nabla_Y \nabla_X U - \nabla_{[X, Y]} U \tag{6.4.6}$$

$\forall U, X, Y$, then again we have $\mathscr{F}$-linearity and antisymmetry in X, Y. Furthermore, for any smooth function f on M

$$\boldsymbol{R}(X, Y)(fU) = f\boldsymbol{R}(X, Y)U \tag{6.4.7}$$

and

$$\boldsymbol{R}(X, Y)f = 0. \tag{6.4.8}$$

Since ∇_X is a tensor derivation $\boldsymbol{R}(X, Y) \equiv [\nabla_X, \nabla_Y] - \nabla_{[X, Y]}$ is a type-preserving derivation on the algebra of tensor fields

$$\boldsymbol{R}(X, Y)(U \otimes W) = \boldsymbol{R}(X, Y)U \otimes W + U \otimes \boldsymbol{R}(X, Y)W \tag{6.4.9}$$

for all X Y, U and W. This derivation is called the *curvature operator* of ∇. The curvature operator may be used to define the (3, 1) *curvature tensor* R of ∇:

$$R(X, Y, Z, \beta) = \beta(\boldsymbol{R}(X, Y)Z). \tag{6.4.10}$$

Since $\boldsymbol{R}(X, Y) = -\boldsymbol{R}(Y, X)$ we may introduce a set of *curvature 2-forms* $R^d{}_c$ by

$$R = 2R^d{}_c \otimes e^c \otimes X_d. \tag{6.4.11}$$

In terms of the connection forms $\omega^a{}_b$ with respect to any co-frame $\{e^a\}$:

$$R^a{}_b = \mathrm{d}\omega^a{}_b + \omega^a{}_c \wedge \omega^c{}_b. \tag{6.4.12}$$

This is the *second structure equation*. For verification we contract on an arbitrary pair of vectors:

$$\begin{aligned}
&2(\mathrm{d}\omega^a{}_b + \omega^a{}_c \wedge \omega^c{}_b)(X, Y) \\
&\quad = X(\omega^a{}_b(Y)) - Y(\omega^a{}_b(X)) - \omega^a{}_b([X, Y]) + \omega^a{}_c(X)\omega^c{}_b(Y) \\
&\qquad - \omega^a{}_c(Y)\omega^c{}_b(X) \\
&\quad = X(e^a(\nabla_Y X_b)) - Y(e^a(\nabla_X X_b)) - e^a(\nabla_{[X, Y]} X_b) - \nabla_X e^a(X_c) e^c(\nabla_Y X_b) \\
&\qquad + \nabla_Y e^a(X_c) e^c(\nabla_X X_b) \\
&\quad = X(e^a(\nabla_Y X_b)) - Y(e^a(\nabla_X X_b)) - e^a(\nabla_{[X, Y]} X_b) - \nabla_X e^a(\nabla_Y X_b) \\
&\qquad + \nabla_Y e^a(\nabla_X X_b) \\
&\quad = e^a(\boldsymbol{R}(X, Y)X_b)
\end{aligned}$$

$$= R(X, Y, X_b, e^a)$$

$$= 2R^a{}_b(X, Y).$$

By contracting the (3, 1) curvature tensor we obtain a (2, 0) tensor: the *Ricci tensor*. That is,

$$\mathbf{Ric}(X, Y) = R(X_a, X, Y, e^a) \tag{6.4.13}$$

where the arbitrary bases $\{X_a\}$ and $\{e^a\}$ are dual. For a general connection '**Ric**' has no particular symmetry properties.

It is sometimes more convenient to work with the set of *Ricci 1-forms* $\{P_a\}$, elements of which are defined by

$$P_b = \mathrm{i}_{X_a} R^a{}_b \tag{6.4.14}$$

hence

$$P_a = \mathbf{Ric}(X_b, X_a)e^b. \tag{6.4.15}$$

Because of their $\mathscr{F}$-linearity the torsion and curvature operators can be evaluated on tangent vectors: they do not require vector fields. By suitably extending a pair of tangent vectors to vector fields we can construct figures out of segments of integral curves, giving a characterisation of the torsion and curvature operators.

Let N_p be a normal neighbourhood of p with X_p, $Y_p \in T_pM$. Each $q \in N_p$ lies on one and only one (up to a linear reparametrisation) geodesic radiating from p. We define $X_q \in T_qM$ by $X_q = \tau_{qp}X_p$ where τ_{qp} is the parallel translation map along the autoparallel. This assignment of a tangent vector to every $q \in N_p$ is smooth: we denote the resulting vector field by X. We similarly extend Y_p to a vector field Y. We have constructed X such that $\nabla_{Z_p}X = 0 \quad \forall Z_p \in T_pM$, thus $\boldsymbol{T}(Y, X)|_p = [X, Y]_p$. From exercise 4.1 at the end of §4.11 we see that $\boldsymbol{T}(Y_p, X_p)$ is the tangent at p to the curve formed from the integral curves of X and Y (see figure 6.3).

In considering the curvature we extend X_p and Y_p differently: this time to commuting vector fields X and Y. We could, for example, choose normal coordinates $\{x^i\}$ with

$$\left.\frac{\partial}{\partial x^1}\right|_p = X_p \qquad \text{and} \qquad \left.\frac{\partial}{\partial x^2}\right|_p = Y_p$$

with

$$X = \frac{\partial}{\partial x^1} \qquad \text{and} \qquad Y = \frac{\partial}{\partial x^2}.$$

If $\varphi(p)$ and $\psi(p)$ are the integral curves of X and Y respectively, starting at p, then we form the quadrilateral shown in figure 6.4.

We denote the parallel translation map from T_pM to T_qM, along the

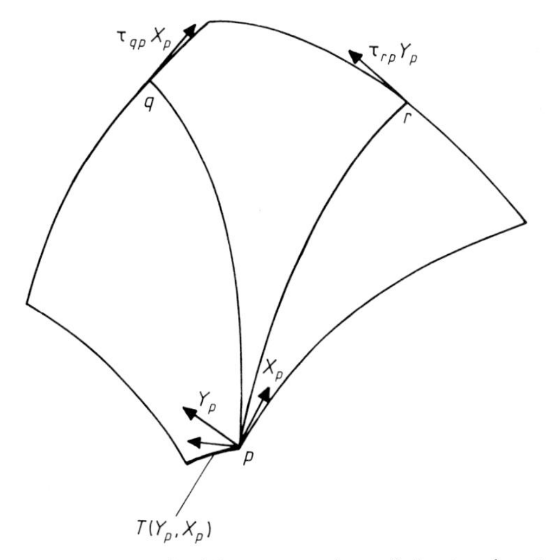

Figure 6.3 Geometrical interpretation of the torsion tensor.

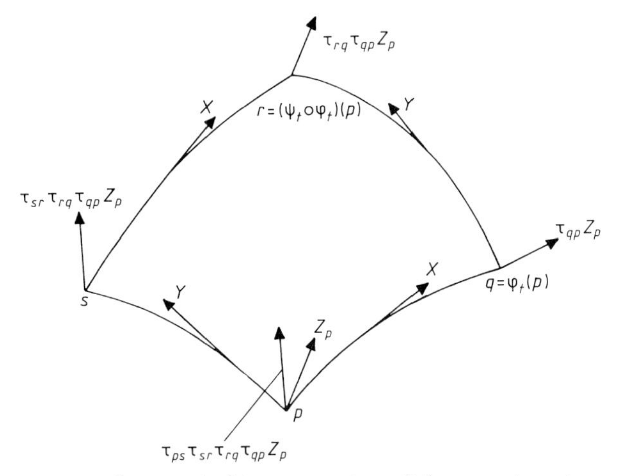

Figure 6.4 Geometrical interpretation of the curvature tensor.

curve shown, by τ_{qp}. If Z_p is any vector in T_pM then we calculate the parallel translate around the figure by using (6.1.10), dropping terms of order greater than t^2:

$$\tau_{qp}Z_p = \{Z - t\nabla_X Z + t^2/2\nabla_X{}^2 Z\}_q + O(t^3)$$

$$\tau_{rq}\tau_{qp}Z_p = \{Z - t(\nabla_X Z + \nabla_Y Z) + t^2/2(\nabla_X{}^2 Z + \nabla_Y{}^2 Z + 2\nabla_Y\nabla_X Z)\}_r + O(t^3).$$

Proceeding around the loop we compare $\tau_{ps}\tau_{sr}\tau_{rq}\tau_{qp}Z_p$ with Z_p:

$$\lim_{t\to 0}\frac{\tau_{ps}\tau_{sr}\tau_{rq}\tau_{qp}Z_p - Z_p}{t^2} = ([\nabla_Y, \nabla_X]Z)_p = \boldsymbol{R}(Y_p, X_p)Z_p.$$

since $[X, Y] = 0$. This expression shows that the curvature measures the path dependence of parallel translation.

6.5 Bianchi Identities

Because of the way in which the torsion and curvature tensors are constructed out of ∇ certain combinations of their covariant derivatives can be written back in terms of these two tensors. The resulting identities are called Bianchi identities.

The (1, 1) tensor field $(\nabla_X R)(Y, Z)$ is defined by $(\nabla_X R)(Y, Z)(W, \beta) = (\nabla_X R)(Y, Z, W, \beta)$. For any $X, Y, Z \in \Gamma TM$ consider the vector

$$\begin{aligned}\mathcal{Y} &= \{(\nabla_X R)(Y, Z) + (\nabla_Y R)(Z, X) + (\nabla_Z R)(X, Y)\}(V)\\ &\equiv \mathop{\mathcal{S}}_{X,Y,Z}\{(\nabla_X R)(Y, Z)\}(V).\end{aligned}$$

Here $\mathcal{S}_{X,\ Y,\ Z}$ denotes the cyclic sum of X, Y, Z. Now $(\nabla_X R)(Y, Z) = \nabla_X(R(Y, Z)) - R(\nabla_X Y, Z) - R(Y, \nabla_X Z)$, so we may write

$$\mathcal{Y} = \mathop{\mathcal{S}}_{X,Y,Z}\{A_{XYZ} - B_{XYZ}\}(V)$$

where

$$\begin{aligned}A_{XYZ}(V) &= \nabla_X(R(Y, Z))(V) = \nabla_X(R(Y, Z)(V)) - R(Y, Z)(\nabla_X V)\\ B_{XYZ}(V) &= (R(\nabla_X Y, Z) + R(Y, \nabla_X Z))(V).\end{aligned}$$

We may express A_{XYZ} in terms of the curvature operator

$$\begin{aligned}A_{XYZ}(V) &= \nabla_X(\boldsymbol{R}(Y, Z)(V)) - \boldsymbol{R}(Y, Z)(\nabla_X V)\\ &= [\nabla_X, [\nabla_Y, \nabla_Z] - \nabla_{[Y,\ Z]}]V.\end{aligned}$$

For any operators P, Q, R we have the (Jacobi) identity

$$\mathop{\mathcal{S}}_{P,Q,R}[P,[Q, R]] = 0$$

hence

$$\mathop{\mathcal{S}}_{X,Y,Z} A_{XYZ}(V) = -\mathop{\mathcal{S}}_{X,Y,Z}[\nabla_X, \nabla_{[Y,\ Z]}]V.$$

Writing out B_{XYZ} in terms of ∇

$$B_{XYZ}(V)$$

$$= (\nabla_{\nabla_X Y}\nabla_Z - \nabla_Z\nabla_{\nabla_X Y} - \nabla_{[\nabla_X Y,\, Z]} + \nabla_Y\nabla_{\nabla_X Z} - \nabla_{\nabla_X Z}\nabla_Y - \nabla_{[Y,\, \nabla_X Z]})V,$$

$$\underset{X,Y,Z}{\mathcal{S}} B_{XYZ}(V)$$

$$= \underset{X,Y,Z}{\mathcal{S}}(\nabla_{\nabla_Y Z}\nabla_X - \nabla_X\nabla_{\nabla_Y Z} - \nabla_{[\nabla_Y Z,\, X]} + \nabla_X\nabla_{\nabla_Z Y} - \nabla_{\nabla_Z Y}\nabla_X + \nabla_{[\nabla_Z Y,\, X]})V$$

$$= \underset{X,Y,Z}{\mathcal{S}}([\nabla_{[Y,\, Z]}, \nabla_X] - \nabla_{[[Y,\, Z],\, X]} + [\nabla_{T(Y,\, Z)}, \nabla_X] - \nabla_{[T(Y,\, Z),\, X]})V.$$

Using the Jacobi identity again gives

$$\underset{X,Y,Z}{\mathcal{S}} B_{XYZ}(V) = -\underset{X,Y,Z}{\mathcal{S}}\{[\nabla_X, \nabla_{[Y,\, Z]}] - \boldsymbol{R}(T(X,\, Y),\, Z)\}V.$$

Thus

$$\mathcal{Y} = -\underset{X,Y,Z}{\mathcal{S}}\{\boldsymbol{R}(T(X,\, Y),\, Z)\}V$$

and since this is valid for arbitrary V:

$$\underset{X,Y,Z}{\mathcal{S}}\{(\nabla_X R)(Y,\, Z) + R(T(X,\, Y),\, Z)\} = 0 \qquad (6.5.1)$$

$\forall X,\, Y,\, Z \in \Gamma TM$. This is known as *Bianchi's second identity*.

In a similar way we obtain an identity by covariantly differentiating the defining relation for the torsion tensor. We leave it as an exercise to prove *Bianchi's first identity*:

$$\underset{X,Y,Z}{\mathcal{S}}\{R(X,\, Y)(Z) - T(T(X,\, Y),\, Z) - (\nabla_X T)(Y,\, Z)\} = 0. \quad (6.5.2)$$

Because of the inherent antisymmetry of the exterior product these identities assume an elegant expression in terms of the torsion and curvature 2-forms. If we exteriorly differentiate the second structure equation and replace $\mathrm{d}\omega^a{}_c$ by $R^a{}_c - \omega^a{}_k \wedge \omega^k{}_c$ then the second Bianchi identity is expressed as

$$\mathrm{d}R^a{}_b + \omega^a{}_c \wedge R^c{}_b - R^a{}_c \wedge \omega^c{}_b = 0. \qquad (6.5.3)$$

Similarly by applying d to the first structure equation and expressing $\mathrm{d}\omega^a{}_c$ back in terms of $R^a{}_c$ and $\mathrm{d}e^b$ back in terms of T^b gives the first Bianchi identity as

$$\mathrm{d}T^a + \omega^a{}_b \wedge T^b = R^a{}_b \wedge e^b. \qquad (6.5.4)$$

6.6 Metric-Compatible Connections

The introduction of a connection on a manifold does not require any metric properties, and so far we have assumed none. However, when introducing a connection on a pseudo-Riemannian manifold we can impose relations between the connection and the pseudo-Riemannian structure. Parallel translation gives a map between the tangent spaces of any two points connected by some curve. On a pseudo-Riemannian manifold it is natural to require that this parallel-translation map be an isometry between the two tangent spaces. That is, parallel translation preserves the lengths of all vectors. A connection such that parallel translation has this property is called *metric compatible*.

Suppose that Y is a vector field parallel along the curve C. If ∇ is metric compatible then the length of Y will be constant along C, that is $\dot{C}(g(Y, Y)) = 0$. Since for $f \in \mathcal{F}(M)$ $\dot{C}(f) = \nabla_{\dot{C}} f$, and ∇ commutes with contractions

$$\begin{aligned}\dot{C}(g(Y, Y)) &= \nabla_{\dot{C}}(g(Y, Y)) \\ &= \nabla_{\dot{C}} g(Y, Y) + 2g(\nabla_{\dot{C}} Y, Y).\end{aligned}$$

If Y is parallel along C then the second term is zero. Requiring that the length of all parallel vectors along C be constant gives $\nabla_{\dot{C}} g = 0$. For a metric-compatible connection this holds for all C, so ∇ is metric compatible if and only if

$$\nabla g = 0. \tag{6.6.1}$$

If $\{X_i\}$ is any local basis then covariantly differentiating the functions $g_{ij} = g(X_i, X_j)$ gives

$$\begin{aligned}X(g_{ij}) &= \nabla_X g(X_i, X_j) + g(\omega^k{}_i(X)X_k, X_j) + g(X_i, \omega^k{}_j(X)X_k) \\ &= \nabla_X g(X_i, X_j) + \omega^k{}_i(X)g_{kj} + \omega^k{}_j(X)g_{ik}.\end{aligned}$$

If $\{X_i\}$ is orthonormal then the functions g_{ij} are constant. So in an orthonormal frame the connection forms of a metric-compatible connection satisfy the antisymmetry condition

$$\omega_{ji} + \omega_{ij} = 0 \tag{6.6.2}$$

where $\omega_{ji} \equiv g_{jk}\omega^k{}_i$.

Since the Hodge dual is defined by the metric it follows that covariant differentiation with respect to a metric-compatible connection commutes with this operation. First observe that the volume n-form is parallel

$$\nabla_X {*1} = 0 \qquad \forall X. \tag{6.6.3}$$

If $\{e^a\}$ is an orthonormal co-frame such that $*1 = e^1 \wedge e^2 \wedge \ldots \wedge e^n$

then $\nabla_X {*1} = -\omega^1{}_a(X)e^a \wedge e^2 \wedge \ldots \wedge e^n + \ldots + (-1)^n e^1 \wedge \ldots \wedge \omega^n{}_a(X)e^a$. Now $\omega^1{}_a(X)e^a \wedge e^2 \wedge \ldots \wedge e^n = \omega^1{}_1(X)e^1 \wedge e^2 \wedge \ldots \wedge e^n$ and so (6.6.3) follows from (6.6.2). It can now be seen from the definition (1.4.5) that if ∇ is metric compatible

$$\nabla_X * = *\nabla_X \qquad \forall X. \tag{6.6.4}$$

A metric-compatible connection is completely characterised by its torsion tensor. That is, the connection coefficients can be determined in terms of the metric and torsion tensors. For a metric-compatible ∇ we have $\nabla_U(g(V, W)) = g(\nabla_U V, W) + g(V, \nabla_U W)$ for any vector fields U, V and W. By cyclically permuting U, V and W we obtain three such expressions. Adding the first two and subtracting the third gives

$$\begin{aligned} U(g(V, W)) + V(g(W, U)) - W(g(U, V)) \\ = g(\nabla_U V, W) + g(V, \nabla_U W) + g(\nabla_V W, U) + g(W, \nabla_V U) \\ - g(\nabla_W U, V) - g(U, \nabla_W V). \end{aligned}$$

The definition of the torsion operator enables this to be rewritten as

$$\begin{aligned} 2g(\nabla_U V, W) = {} & U(g(V, W)) + V(g(W, U)) - W(g(U, V)) \\ & -g(U, [V, W]) + g(V, [W, U]) + g(W, [U, V]) \\ & -g(U, \boldsymbol{T}(V, W)) + g(V, \boldsymbol{T}(W, U)) + g(W, \boldsymbol{T}(U, V)). \end{aligned} \tag{6.6.5}$$

If $\{X_a\}$ is an arbitrary basis then the *structure functions* $C_{ab}{}^c$ of the basis are given by

$$[X_a, X_b] = C_{ab}{}^c X_c. \tag{6.6.6}$$

If three different basis vectors are inserted in (6.6.5) then we can solve for the connection coefficients:

$$\begin{aligned} \Gamma_{ab}{}^p = {} & \tfrac{1}{2} g^{cp}\{X_a(g_{bc}) + X_b(g_{ca}) - X_c(g_{ab}) - C_{bc}{}^d g_{ad} \\ & + C_{ca}{}^d g_{bd} + C_{ab}{}^d g_{cd} - T(X_b, X_c, \widetilde{X}_a) + T(X_c, X_a, \widetilde{X}_b) \\ & + T(X_a, X_b, \widetilde{X}_c)\}. \end{aligned} \tag{6.6.7}$$

Here g^{cp} is the inverse matrix to g_{ab}, $g_{pc}g^{cq} = \delta^q_p$. If $\{e^a\}$ is the dual basis then $\widetilde{X}_a = e_a = g_{ab}e^b$. There are two classes of bases in which this expression for the connection coefficients simplifies: in a coordinate basis the structure functions are zero, whilst in an orthonormal basis the metric components are constant. For the case of an orthonormal basis the above expression for the connection coefficients enables the connection 1-forms to be given as

$$2\omega_{ab} = e^d \mathrm{i}_{X_a}\mathrm{i}_{X_b}(de_d - T_d) + \mathrm{i}_{X_b}(de_a - T_a) - \mathrm{i}_{X_a}(de_b - T_b). \tag{6.6.8}$$

This formula is of great computational utility.

From now on we will only consider metric-compatible connections.

6.7 The Covariant Exterior Derivative

It is often convenient to work with sets of differential forms indexed with respect to some basis. The torsion and curvature forms provide an example. The Bianchi identities for these forms, (6.5.3) and (6.5.4), involve an exterior derivative plus 'correction terms' involving the connection 1-forms. Such combinations of terms can be efficiently encoded into a 'covariant exterior derivative'.

Given a mixed tensor that is totally antisymmetric in some subset of r vectors we can associate a set of r-forms with any basis $\{X_j\}$ with dual $\{e^j\}$. Suppose that S is such a tensor of type $(r + q, p)$. We define a set of r-forms $S^{i_1 \ldots i_p}{}_{j_1 \ldots j_q}$ by

$$S^{i_1 \ldots i_p}{}_{j_1 \ldots j_q}(X_1, \ldots, X_r) = S(X_1, \ldots, X_r, X_{j_1}, \ldots, X_{j_q}, e^{i_1}, \ldots, e^{i_p}). \tag{6.7.1}$$

We define the *covariant exterior derivative* D of the $S^{i_1 \ldots i_p}{}_{j_1 \ldots j_q}$ in terms of a connection ∇ by

$$\begin{aligned}(r + 1)\mathrm{D}S^{i_1 \ldots i_p}{}_{j_1 \ldots j_q}&(X_0, \ldots, X_r)\\ = \sum_{j=0}^{r}&(-1)^j \nabla_{X_j} S(X_0, \ldots, \hat{X}_j, \ldots, X_r, X_{j_1}, \ldots, X_{j_q}, e^{i_1}, \ldots, e^{i_p})\\ &- \sum_{0 \leqslant j < k \leqslant r} (-1)^{j+k} S(\boldsymbol{T}(X_j, X_k), X_0, \ldots, \hat{X}_j, \ldots, \hat{X}_k, \ldots,\\ &X_r, X_{j_1}, \ldots, X_{j_q}, e^{i_1}, \ldots, e^{i_p}).\end{aligned} \tag{6.7.2}$$

The 'hat' above a symbol indicates that that term is omitted from the sequence. $\boldsymbol{T}$ is the torsion operator of ∇. It follows from the above rather cumbersome expression that

$$\begin{aligned}\mathrm{D}S&^{i_1 \ldots i_p}{}_{j_1 \ldots j_q}\\ &= \mathrm{d}S^{i_1 \ldots i_p}{}_{j_1 \ldots j_q} + \omega^{i_1}{}_{i_s} \wedge S^{i_s \ldots i_p}{}_{j_1 \ldots j_q} + \ldots + \omega^{i_p}{}_{i_s} \wedge S^{i_1 \ldots i_s}{}_{j_1 \ldots j_q}\\ &\quad - \omega^{j_s}{}_{j_1} \wedge S^{i_1 \ldots i_p}{}_{j_s \ldots j_q} - \ldots - \omega^{j_s}{}_{j_q} \wedge S^{i_1 \ldots i_p}{}_{j_1 \ldots j_s}.\end{aligned} \tag{6.7.3}$$

This can be verified by using (4.10.5). For the special case in which $p = q = 0$ the covariant exterior derivative reduces to the ordinary exterior derivative. We can then infer from (6.7.2) that

$$e^a \wedge \nabla_{X_a} = \mathrm{d} - T^a \wedge \mathrm{i}_{X_a}. \tag{6.7.4}$$

(Alternatively this important relation can be verified on 0-and 1-forms; its general validity then following from the fact that both expressions are graded derivations.)

Repeated application of (6.7.3) gives the following Bianchi identity for D,

$$\begin{aligned} D^2 S^{i_1 \ldots i_p}{}_{j_1 \ldots j_q} = {} & R^{i_1}{}_{i_s} \wedge S^{i_s \ldots i_p}{}_{j_1 \ldots j_q} + \ldots + R^{i_p}{}_{i_s} \wedge S^{i_1 \ldots i_s}{}_{j_1 \ldots j_q} \\ & - R^{j_s}{}_{j_1} \wedge S^{i_1 \ldots i_p}{}_{j_s \ldots j_l} - \ldots \\ & - R^{j_s}{}_{j_q} \wedge S^{i_1 \ldots i_p}{}_{j_1 \ldots j_s}. \end{aligned} \tag{6.7.5}$$

It follows from (6.7.1) that under a change of basis the set of forms $S^{i_1 \ldots i_p}{}_{j_1 \ldots j_q}$ transform according to the classical tensor transformation rules. The $\mathscr{F}$-linearity in the arguments of the right-hand side of (6.7.2) ensures that the $DS^{i_1 \ldots i_p}{}_{j_1 \ldots j_q}$ transform like the $S^{i_1 \ldots i_p}{}_{j_1 \ldots j_q}$ under a change of basis. If S^I and T^J are sets of r-forms and s-forms respectively, labelled by the multi-indices I and J then, as may be seen from (6.7.3),

$$D(S^I \wedge T^J) = DS^I \wedge T^J + (-1)^r S^I \wedge DT^J. \tag{6.7.6}$$

The interior derivative with respect to a set of basis vectors maps a set of p-forms indexed with q indices into a set of $(p-1)$-forms indexed with $(q+1)$ indices. The anticommutator of this operator with D gives a useful relation. If

$$L_{X_a} \equiv D i_{X_a} + i_{X_a} D \tag{6.7.7}$$

then L_{X_a} maps a set of p-forms into a set of p-forms indexed by the extra index a. It acts as a derivation on exterior products

$$L_{X_a}(S^I \wedge T^J) = L_{X_a} S^I \wedge T^J + S^I \wedge L_{X_a} T^J. \tag{6.7.8}$$

First we consider a set of 1-forms, $A^{i_1 \ldots i_p}{}_{j_1 \ldots j_q}$. For A any 1-form (6.7.4) gives

$$i_{X_a} dA = \nabla_{X_a} A - e^b i_{X_a} \nabla_{X_b} A + i_{X_a} T^b i_{X_b} A.$$

Using this in (6.7.3) gives

$$\begin{aligned} & i_{X_a} D A^{i_1 \ldots i_p}{}_{j_1 \ldots j_q} \\ & = \nabla_{X_a} A^{i_1 \ldots i_p}{}_{j_1 \ldots j_q} - e^b i_{X_a} \nabla_{X_b} A^{i_1 \ldots i_p}{}_{j_1 \ldots j_q} + i_{X_a} T^b i_{X_b} A^{i_1 \ldots i_p}{}_{j_1 \ldots j_q} \\ & \quad \ldots + i_{X_a} \omega^{i_s}{}_{i_r} A^{i_1 \ldots \hat{i}_s i_r \ldots i_p}{}_{j_1 \ldots j_q} - \omega^{i_s}{}_{i_r} i_{X_a} A^{i_1 \ldots \hat{i}_s i_r \ldots i_p}{}_{j_1 \ldots j_q} \\ & \quad \ldots - i_{X_a} \omega^{j_r}{}_{j_s} A^{i_1 \ldots i_p}{}_{j_1 \ldots \hat{j}_s j_r \ldots j_q} + \omega^{j_r}{}_{j_s} i_{X_a} A^{i_1 \ldots i_p}{}_{j_1 \ldots \hat{j}_s j_r \ldots j_q}. \end{aligned}$$

Now if A is any 1-form

$$\nabla_{X_b} A = \nabla_{X_b} i_{X_c} A e^c - i_{X_c} A \omega^c{}_p(X_b) e^p$$

so

$$\mathrm{i}_{X_a}\nabla_{X_b}A = \nabla_{X_b}\mathrm{i}_{X_a}A - \omega^c{}_a(X_b)\mathrm{i}_{X_c}A$$

and

$$e^b\mathrm{i}_{X_a}\nabla_{X_b}A = \mathrm{di}_{X_a}A - \omega^c{}_a\mathrm{i}_{X_c}A \qquad \text{(by (6.7.4) again)}$$

so

$$\begin{aligned}\mathrm{i}_{X_a}\mathrm{D}A^{i_1\ldots i_p}{}_{j_1\ldots j_q}& \\ &= \nabla_{X_a}A^{i_1\ldots i_p}{}_{j_1\ldots j_q} - \mathrm{di}_{X_a}A^{i_1\ldots i_p}{}_{j_1\ldots j_q} + \omega^c{}_a\mathrm{i}_{X_c}A^{i_1\ldots i_p}{}_{j_1\ldots j_q}\\ &\quad + \mathrm{i}_{X_a}T^b\mathrm{i}_{X_b}A^{i_1\ldots i_p}{}_{j_1\ldots j_q} + \mathrm{i}_{X_a}\omega^{i_s}{}_{i_r}A^{i_1\ldots \hat{\imath}_s i_r\ldots i_p}{}_{j_1\ldots j_q}\\ &\quad - \omega^{i_s}{}_{i_r}\mathrm{i}_{X_a}A^{i_1\ldots \hat{\imath}_s i_r\ldots i_p}{}_{j_1\ldots j_q}\ \ldots\\ &\quad - \mathrm{i}_{X_a}\omega^{j_r}{}_{j_s}A^{i_1\ldots i_p}{}_{j_1\ldots \hat{\jmath}_s j_r\ldots j_q} + \omega^{j_r}{}_{j_s}\mathrm{i}_{X_a}A^{i_1\ldots i_p}{}_{j_1\ldots \hat{\jmath}_s j_r\ldots j_q}.\end{aligned}$$

Recognising the right-hand side as containing $\mathrm{Di}_{X_a}A^{i_1\ldots i_p}{}_{j_1\ldots j_q}$ enables this to be written as

$$\begin{aligned}\mathrm{L}_{X_a}A^{i_1\ldots i_p}{}_{j_1\ldots j_q}& \\ &= (\nabla_{X_a} + \mathrm{i}_{X_a}T^b\wedge\mathrm{i}_{X_b})A^{i_1\ldots i_p}{}_{j_1\ldots j_q} + \omega^{i_s}{}_{i_r}(X_a)A^{i_1\ldots \hat{\imath}_s i_r\ldots i_p}{}_{j_1\ldots j_q}\\ &\quad \ldots - \omega^{j_r}{}_{j_s}(X_a)A^{i_1\ldots i_p}{}_{j_1\ldots \hat{\jmath}_s j_r\ldots j_q}.\end{aligned}$$

We have obtained this expression for the $A^{i_1\ldots i_p}{}_{j_1\ldots j_q}$ 1-forms; but L_{X_a}, ∇_{X_a} and $\mathrm{i}_{X_a}T^b\wedge\mathrm{i}_{X_b}$ are all derivations on exterior products of multi-indexed p-forms so it is consequently valid on arbitrary p-forms:

$$\begin{aligned}\mathrm{L}_{X_a}S^{i_1\ldots i_p}{}_{j_1\ldots j_q}& \\ &= (\nabla_{X_a} + \mathrm{i}_{X_a}T^b\wedge\mathrm{i}_{X_b})S^{i_1\ldots i_p}{}_{j_1\ldots j_q} + \omega^{i_s}{}_{i_r}(X_a)S^{i_1\ldots \hat{\imath}_s i_r\ldots i_p}{}_{j_1\ldots j_q}\\ &\quad \ldots - \omega^{j_r}{}_{j_s}(X_a)S^{i_1\ldots i_p}{}_{j_1\ldots \hat{\jmath}_s j_r\ldots j_q}.\end{aligned} \tag{6.7.9}$$

If $S^{i_1\ldots i_p}{}_{j_1\ldots j_q} = S(e^{i_1}, \ldots, e^{i_p}, X_{j_1}, \ldots, X_{j_q})$ then this can be written as

$$\begin{aligned}&\mathrm{L}_{X_a}S^{i_1\ldots i_p}{}_{j_1\ldots j_q}\\ &= \nabla_{X_a}S(e^{i_1}, \ldots, e^{i_p}, X_{j_1}, \ldots, X_{j_q}) + \mathrm{i}_{X_a}T^b\wedge\mathrm{i}_{X_b}S^{i_1\ldots i_p}{}_{j_1\ldots j_q}.\end{aligned} \tag{6.7.10}$$

For the special case of φ, any $\mathscr{F}$-valued p-form, this reduces to

$$\mathrm{i}_{X_a}\mathrm{d}\varphi + \mathrm{Di}_{X_a}\varphi = \nabla_{X_a}\varphi + \mathrm{i}_{X_a}T^b\wedge\mathrm{i}_{X_b}\varphi. \tag{6.7.11}$$

The definition (6.7.2) can be applied to 0-forms where there is the simplification that the torsion terms do not enter. Since $g_{ab} = g(X_a, X_b)$ we have $\mathrm{D}g_{ab}(X) = \nabla_X g(X_a, X_b)$. Thus for a metric-compatible connection

$$\mathrm{D}g_{ab} = 0. \tag{6.7.12}$$

It follows that if indices labelling a set of forms are raised or lowered with the components of the metric then this operation commutes with the covariant exterior derivative. If the volume n-form is expanded as $*1 = (n!)^{-1}\varepsilon_{i_1 \dots i_n} e^{i_1} \wedge \dots \wedge e^{i_n}$ then $\varepsilon_{i_1 \dots i_n} = n! *1(X_{i_1}, \dots, X_{i_n})$. So for a metric-compatible connection

$$\mathrm{D}\varepsilon_{i_1 \dots i_n} = 0. \tag{6.7.13}$$

As anticipated the Bianchi identities (6.5.3) and (6.5.4) can now be written as

$$\mathrm{D}R^a{}_b = 0 \tag{6.7.14}$$

$$\mathrm{D}T^a = R^a{}_b \wedge e^b. \tag{6.7.15}$$

In an orthonormal basis the connection 1-forms of a metric-compatible connection are antisymmetric: they satisfy (6.6.2). It follows that the curvature 2-forms satisfy an analogous relation. Moreover, because of the tensorial nature of the transformation of the curvature 2-forms under a change of basis this antisymmetry is maintained in an arbitrary basis. Using this antisymmetry the second Bianchi identity (6.7.15) can be contracted to obtain various other identities. We leave it as an exercise to prove the following *contracted Bianchi identities*:

$$\underset{p,q,r,a}{\mathcal{S}} \mathrm{i}_{X_p}\mathrm{i}_{X_q}\mathrm{i}_{X_r}\mathrm{D}T_a = 2(\mathrm{i}_{X_a}\mathrm{i}_{X_q}R_{pr} - \mathrm{i}_{X_p}\mathrm{i}_{X_r}R_{aq}) \tag{6.7.16}$$

$$\mathrm{i}_{X_p}\mathrm{i}_{X_q}\mathrm{i}_{X_a}\mathrm{D}T^a = \mathrm{i}_{X_q}P_p - \mathrm{i}_{X_p}P_q \tag{6.7.17}$$

$$\mathrm{i}_{X_p}\mathrm{i}_{X_q}\mathrm{i}_{X_r}\mathrm{D}T_a e^a + \underset{p,q,r}{\mathcal{S}} \mathrm{i}_{X_p}\mathrm{i}_{X_q}\mathrm{D}T_r = 2\underset{p,q,r}{\mathcal{S}} \mathrm{i}_{X_q}R_{pr} \tag{6.7.18}$$

$$\mathrm{i}_{X_a}\mathrm{D}T^a = P_b \wedge e^b. \tag{6.7.19}$$

6.8 The Curvature Scalar and Einstein Tensor

The existence of a metric tensor enables 'type-changing' of the (3, 1) curvature tensor to various other fourth-rank tensors. We will normally denote all such tensors by the *same* symbol, making it clear in the *context* in which it appears exactly which tensor is meant. Similarly the Ricci tensor can be related to a (1, 1) tensor which can then be contracted to a scalar. That is, the *curvature scalar* $\mathcal{R}$ is given by

$$\mathcal{R} = \mathbf{Ric}(X_a, X^a) \tag{6.8.1}$$

where as usual $X^a = g^{ab}X_b$. In terms of the Ricci 1-forms P_a,

$$\mathcal{R} = \mathrm{i}_{X^a}P_a. \tag{6.8.2}$$

In n-dimensions the *Einstein $(n-1)$-forms* G_c are defined by

$$G_c = R_{ab} \wedge \mathrm{i}_{X_c} {*e^{ab}}. \tag{6.8.3}$$

These may be related to the Ricci forms; we have

$$\begin{aligned} G_c &= R_{ab} \wedge \mathrm{i}_{X_c}\mathrm{i}_{X^b} {*e^a} = R_{ba} \wedge \mathrm{i}_{X^b}\mathrm{i}_{X_c} {*e^a} \\ &= \mathrm{i}_{X^b}(R_{ba} \wedge \mathrm{i}_{X_c} {*e^a}) - P_a \wedge \mathrm{i}_{X_c} {*e^a} \\ &= \mathrm{i}_{X^b}(R_{ab} \wedge \mathrm{i}_{X^a}\mathrm{i}_{X_c} {*1}) - P_a \wedge \mathrm{i}_{X_c} {*e^a} \\ &= \mathrm{i}_{X^b}\{\mathrm{i}_{X^a}(R_{ab} \wedge \mathrm{i}_{X_c} {*1}) - P_b \wedge \mathrm{i}_{X_c} {*1}\} - P_a \wedge \mathrm{i}_{X_c} {*e^a}. \end{aligned}$$

Now $R_{ab} \wedge \mathrm{i}_{X_c} {*1} = 0$, since it is an $(n+1)$-form, so

$$G_c = -\mathcal{R}{*e_c} + P_b \wedge \mathrm{i}_{X^b}\mathrm{i}_{X_c} {*1} - P_a \wedge \mathrm{i}_{X_c} {*e^a} = -\mathcal{R}{*e_c} - 2P^a \wedge {*e_{ac}}$$

and

$$\begin{aligned} P^a \wedge {*e_{ac}} &= \mathrm{i}_{X^b} P^a e_b \wedge {*e_{ac}} \\ &= \mathrm{i}_{X^b} P^a \{-\mathrm{i}_{X_c}(e_b \wedge {*e_a}) + g_{bc} {*e_a}\} \\ &= \mathrm{i}_{X^b} P^a \{-g_{ba} {*e_c} + g_{bc} {*e_a}\} \\ &= -\mathrm{i}_{X_a} P^a {*e_c} + \mathrm{i}_{X_c} P^a {*e_a}. \end{aligned}$$

The contracted Bianchi identity (6.7.17) gives the antisymmetric part of the Ricci tensor in terms of the torsion, so

$$\begin{aligned} P^a \wedge {*e_{ac}} &= -\mathcal{R}{*e_c} + \mathrm{i}_{X^a} P_c {*e_a} + \mathrm{i}_{X^a}\mathrm{i}_{X_c}\mathrm{i}_{X_b} \mathrm{D}T^b {*e_a} \\ &= -\mathcal{R}{*e_c} + {*P_c} + {*\mathrm{i}_{X_c}\mathrm{i}_{X_b}\mathrm{D}T^b} \end{aligned}$$

thus

$$G_c = \mathcal{R}{*e_c} - 2{*P_c} - 2{*\mathrm{i}_{X_c}\mathrm{i}_{X_b}\mathrm{D}T^b}$$

or

$$*^{-1}G_c = \mathcal{R}e_c - 2P_c - 2\mathrm{i}_{X_c}\mathrm{i}_{X_b}\mathrm{D}T^b. \tag{6.8.4}$$

The set of Einstein forms are equivalent to a (2, 0) tensor. The *Einstein tensor* G is defined by

$$G = *^{-1}G_c \otimes e^c. \tag{6.8.5}$$

The antisymmetric part of the Einstein tensor is determined by the torsion. Using (6.7.17) once again gives

$$\mathrm{i}_{X_b} *^{-1}G_c - \mathrm{i}_{X_c} *^{-1}G_b = -2\mathrm{i}_{X_b}\mathrm{i}_{X_c}\mathrm{i}_{X_a}\mathrm{D}T^a. \tag{6.8.6}$$

The covariant exterior derivative of the Einstein forms can also be related to the torsion. Writing $G^c = R_{ab} \wedge {*e^{abc}}$ we have $\mathrm{D}G^c = \mathrm{D}R_{ab} \wedge {*e^{abc}} + R_{ab} \wedge \mathrm{D}{*e^{abc}}$. The first term is zero by the first

Bianchi identity. In n-dimensions we can expand the Hodge dual as

$$*e^{i_1i_2i_3} = \frac{1}{(n-3)!} e^{i_4} \wedge e^{i_5} \wedge \ldots \wedge e^{i_n} \mathrm{i}_{X_{i_n}} \ldots \mathrm{i}_{X_{i_5}} \mathrm{i}_{X_{i_4}} *e^{i_1i_2i_3}.$$

Now $\mathrm{i}_{X_{i_n}} \ldots \mathrm{i}_{X_{i_5}} \mathrm{i}_{X_{i_4}} *e^{i_1i_2i_3}$ is proportional to $*1$ contracted on n vectors, thus its covariant exterior derivative is zero, so

$$\begin{aligned} \mathrm{D}*e^{i_1i_2i_3} &= \frac{1}{(n-3)!}(T^{i_4} \wedge e^{i_5} \wedge \ldots \wedge e^{i_n} - e^{i_4} \wedge T^{i_5} \wedge \ldots \wedge e^{i_n} \\ &\quad + \ldots + (-1)^{n-4} e^{i_4} \wedge \ldots \wedge T^{i_n}) \mathrm{i}_{X_{i_n}} \ldots \mathrm{i}_{X_{i_5}} \mathrm{i}_{X_{i_4}} *e^{i_1i_2i_3} \\ &= \frac{1}{(n-4)!} T^{i_4} \wedge e^{i_5} \wedge \ldots \wedge e^{i_n} \mathrm{i}_{X_{i_n}} \ldots \mathrm{i}_{X_{i_5}} *e^{i_1i_2i_3}{}_{i_4} \\ &= T^{i_4} \wedge *e^{i_1i_2i_3}{}_{i_4} \end{aligned}$$

thus

$$\mathrm{D}G^c = R_{ab} \wedge T^p \wedge *e^{abc}{}_p. \tag{6.8.7}$$

Equivalently this relation can be written in terms of the (2, 0) Einstein tensor. The divergence of G, $\nabla.G$, is a 1-form defined by

$$(\nabla.G)(Y) = \nabla_{X_a} G(X^a, Y) \tag{6.8.8}$$

thus

$$\begin{aligned} \nabla.G &= (\mathrm{i}_{X^a} *^{-1} \nabla_{X_a} G_p - \omega^c{}_p(X_a) \mathrm{i}_{X^a} *^{-1} G_c) e^p \\ &= *^{-1}(e^a \wedge \nabla_{X_a} G_p - \omega^c{}_p \wedge G_c) e^p. \end{aligned}$$

We may now use (6.7.4) to give

$$\nabla.G = *^{-1}(\mathrm{D}G_p - T^a \wedge \mathrm{i}_{X_a} G_p) e^p$$

and (6.8.7) then gives

$$\nabla.G = -*^{-1}(T^q \wedge \mathrm{i}_{X_q} R_{ab} \wedge *e^{ab}{}_p) e^p. \tag{6.8.9}$$

6.9 The Pseudo-Riemannian Connection

Since a metric-compatible connection is completely characterised by its torsion tensor it follows that there is a unique torsion-free metric-compatible connection for any pseudo-Riemannian structure. This connection is called the *pseudo-Riemannian connection*. It is also sometimes associated with the names of Levi–Civita and Christoffel. From (6.6.7) we see that in a **coordinate basis** the condition of zero torsion is expressed as a symmetry of the connection coefficients, $\Gamma_{ab}{}^p = \Gamma_{ba}{}^p$. For this reason a torsion-free connection is often called 'symmetric'. The connection coefficients of the pseudo-Riemannian connection expressed in a coordinate basis are often called the *Christoffel symbols*. For actual computations it is often most efficient to use an orthonormal basis. In such a basis there are, by (6.6.2), $\frac{1}{2}n(n-1)$ independent 1-forms or $\frac{1}{2}n^2(n-1)$ independent connection coefficients. For a coordinate basis the zero-torsion condition cuts down the number of connection coefficients to $\frac{1}{2}n^2(n+1)$. Thus in an orthonormal basis there are n^2 fewer connection coefficients.

Because of the Bianchi identities the curvature tensor of a torsion-free connection has extra symmetries. Equation (6.7.16) reduces to an expression of the 'pairwise interchange' symmetry of the Riemann tensor. Equation (6.7.17) shows that for zero torsion the Ricci tensor is symmetric. For zero torsion the Einstein tensor is symmetric, by (6.8.6), and divergenceless by (6.8.9).

We can use (6.7.4) to write the exterior derivative in terms of any torsion-free connection. Since any metric-compatible connection satisfies (6.6.4) we obtain a useful relation between the pseudo-Riemannian connection and the co-derivative δ which was introduced in (5.4.2). If φ is a differential p-form then $\mathrm{i}_{X^a}\nabla_{X_a}\varphi$ is certainly a $(p-1)$-form. Introducing the Hodge map and its inverse:

$$\begin{aligned} \mathrm{i}_{X^a}\nabla_{X_a}\varphi &= \mathrm{i}_{X^a}**^{-1}\nabla_{X_a}\varphi = \mathrm{i}_{X^a}*\nabla_{X_a}*^{-1}\varphi \qquad \text{by (6.6.4)} \\ &= *(\nabla_{X_a}*^{-1}\varphi \wedge e^a) \qquad \text{(by (1.4.7))} \\ &= *(e^a \wedge \nabla_{X_a}\eta*^{-1}\varphi) \end{aligned}$$

where η is defined in (1.1.2). We now use (6.7.4):

$$\mathrm{i}_{X^a}\nabla_{X_a}\varphi = *\mathrm{d}\eta*^{-1}\varphi.$$

The inverse of the Hodge map is given in (5.4.3). By considering the cases of even and odd dimensions separately it can be seen that this can be rewritten as $\mathrm{i}_{X^a}\nabla_{X_a}\varphi = -*^{-1}\mathrm{d}*\eta\varphi$, that is

$$\mathrm{i}_{X^a}\nabla_{X_a}\varphi = -\delta\varphi. \tag{6.9.1}$$

From now on, unless we specify to the contrary, we shall restrict ourselves to the pseudo-Riemannian connection. For most of what follows it will be essential that the connection is metric compatible, whereas in most places torsion merely contributes extra terms.

Exercise 6.4

An *Einstein space* is one for which $\mathrm{Ric} = cg$ for some constant c. Show that if, in three or more dimensions, $\mathrm{Ric} = fg$ for $f \in \mathcal{F}(M)$ then:

(i) $f = \mathcal{R}/n$

(ii) $G_c = \dfrac{(n-2)}{n}\mathcal{R}*e_c$

(iii) $\mathrm{d}\mathcal{R} = 0$.

Example 6.1
Let g be the metric tensor of a four-dimensional spacetime: $g = -e^0 \otimes e^0 + \Sigma_{k=1}^3 e^k \otimes e^k$. In a local chart with coordinates $(t(p), r(p), \theta(p), \varphi(p))$ a class of spherically symmetric metrics may be parametrised by functions H_0, H_1, H_2 of $r(p)$ and a function λ of $t(p)$, by choosing a local orthonormal co-frame as

$$e^0 = H_0 \mathrm{d}t$$

$$e^1 = e^{\lambda} H_1 \mathrm{d}r$$

$$e^2 = e^{\lambda} H_2 \mathrm{d}\theta$$

$$e^3 = e^{\lambda} H_2 \sin\theta \mathrm{d}\varphi.$$

As an example of using (6.6.8) verify that the connection 1-forms ω_{ab} of the pseudo-Riemannian connection are given in this basis by table 6.1. Hence construct table 6.2 for $\nabla_{X_a} e^b$ where X_a is a dual orthonormal frame: $e^b(X_a) = \delta^b_a$.

6.10 Sectional Curvature

A two-dimensional subspace S of T_pM will be called a *tangent plane* to M at p. If $\{X, Y\}$ is any basis for S and

$$Q(X, Y) = g(X, X)g(Y, Y) - (g(X, Y))^2 \tag{6.10.1}$$

then $Q(X, Y) = 0$ if and only if g induces a degenerate metric on S. Such a tangent plane is called degenerate. If S is any non-degenerate tangent plane at p then the *sectional curvature* of M at p, along the plane section S, is $K(S)$:

$$K(S) = -\frac{g(\boldsymbol{R}(X, Y)X, Y)}{Q(X, Y)}. \tag{6.10.2}$$

Thus the sectional curvature at p is a real function of the tangent planes at p.

Exercise 6.5
Verify that the definition of $K(S)$ is independent of the basis chosen.

For the case in which M is Riemannian the sectional curvature generalises the intuitive notions of curvature of two-dimensional surfaces. If $\mathcal{N}_0$ is a normal neighbourhood of the origin in T_pM then $\mathrm{Exp}_p(\mathcal{N}_0 \cap S)$ is a two-dimensional Riemannian submanifold of M. Let $\mathcal{B}(r)$ be an open ball of radius r centred about the origin in $\mathcal{N}_0 \cap S$, with r sufficiently small that Exp_p is a diffeomorphism onto $B(r)$, an open ball centred about p. Let $\mathcal{A}(r)$ be the area of $\mathcal{B}(r)$ and $A(r)$ be

Table 6.1 The torsion-free orthonormal connection forms $\omega_{ab} = -\omega_{ba}$ for the metric of example 6.1.

a \ b	0	1	2	3
0	0	$-(H_0'/H_0H_1)e^{-\lambda}e^0 - (\dot{\lambda}/H_0)e^1$	$-(\dot{\lambda}/H_0)e^2$	$-(\dot{\lambda}/H_0)e^3$
1		0	$-(H_2'/H_1H_2)e^{-\lambda}e^2$	$-(H_2'/H_1H_2)e^{-\lambda}e^3$
2			0	$-(\cot\theta/H_2)e^{-\lambda}e^3$
3				0

$\dot{\lambda} \equiv d\lambda/dt$, $H_a' \equiv dH_a/dr$.

Table 6.2 Associated table of Levi–Cevita connection coefficients specified by $\nabla_{X_a}e^b$ in the dual bases satisfying $e^b(X_a) = \delta^b{}_a$.

	e^0	e^1	e^2	e^3
∇_{X_0}	$-(H_0'/H_0H_1)e^{-\lambda}e^1$	$-(H_0'/H_0H_1)e^{-\lambda}e^0$	0	0
∇_{X_1}	$-(\dot{\lambda}/H_0)e^1$	$-(\dot{\lambda}/H_0)e^0$	0	0
∇_{X_2}	$-(\dot{\lambda}/H_0)e^2$	$(H_2'/H_1H_2)e^{-\lambda}e^2$	$-(\dot{\lambda}/H_0)e^0$ $-(H_2'/H_1H_2)e^{-\lambda}e^1$	0
∇_{X_3}	$-(\dot{\lambda}/H_0)e^3$	$(H_2'/H_1H_2)e^{-\lambda}e^3$	$(\cot\theta/H_2)e^{-\lambda}e^3$	$-(\dot{\lambda}/H_0)e^0 - (H_2'/H_1H_2)e^{-\lambda}e^1$ $-(\cot\theta/H_2)e^{-\lambda}e^2$

$\dot{\lambda} \equiv d\lambda/dt$, $H_a' \equiv dH_a/dr$

the area of $B(r)$. Thus $\mathscr{A}(r)$ is determined by the Euclidean geometry of T_pM whilst $A(r)$ is determined by the Riemannian geometry of $\mathrm{Exp}_p(\mathcal{N}_0 \cap S)$. The sectional curvature is determined by a comparison of these two areas:

$$K(S) = \lim_{r \to 0} 12 \frac{\mathscr{A}(r) - A(r)}{r^2 \mathscr{A}(r)}. \tag{6.10.3}$$

The proof of these assertions can be found in, for example, Helgason (1978).

Exercise 6.6
Take M to be the two-sphere with the standard metric induced from $\mathbb{R}^3$ (see figure 6.5). Calculate the sectional curvature using (6.10.2). Verify that (6.10.3) gives the same result. (Note that $B(r)$ is a spherical cap with geodesic radius r (figure 6.5).)

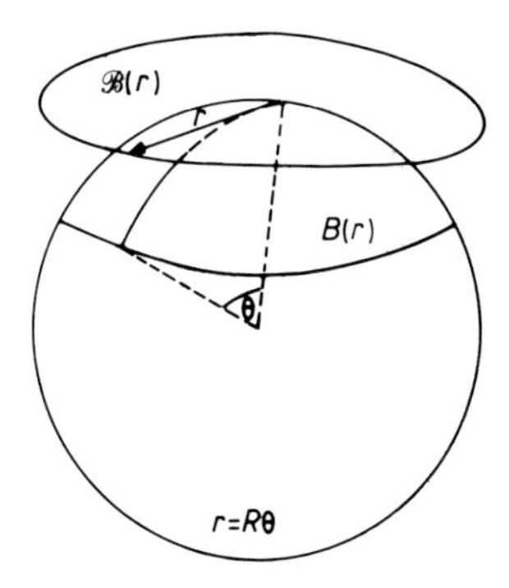

Figure 6.5

A manifold is said to have *constant curvature* if its sectional curvature is constant.

Exercise 6.7
Show that M has constant curvature c if and only if

$$R^{ab} = ce^{ab}. \tag{6.10.4}$$

6.11 The Conformal Tensor

Two metric tensor fields g and $\hat{g}$ such that $\hat{g} = \exp(2\lambda)g$ for some function λ are said to be *conformally related*. Whereas a conformal rescaling of the metric will change the curvature it is possible to construct a tensor out of the Riemann tensor that is invariant under such scalings. Let $\{e^a\}$ be a g-orthonormal co-frame, with dual $\{X_a\}$, and $\{\widehat{e^a}\}$ a $\hat{g}$-orthonormal co-frame, with dual $\{\widehat{X_a}\}$, where

$$\widehat{e^a} = \exp(\lambda)e^a \qquad \widehat{X_a} = \exp(-\lambda)X_a. \tag{6.11.1}$$

If $\hat{\nabla}$ is the pseudo-Riemannian connection of $\hat{g}$ with connection forms $\widehat{\omega_{ab}}$ with respect to $\{\widehat{e^a}\}$ then from (6.6.8)

$$(\widehat{\omega_{ab}}) = \omega_{ab} + X_b(\lambda)e^a - X_a(\lambda)e_b. \tag{6.11.2}$$

Similarly the curvature forms $\widehat{R_{ab}}$ of $\hat{\nabla}$ in the $\{\widehat{e^a}\}$ basis are

$$\begin{aligned}(\widehat{R_{ab}}) = R_{ab} &+ \nabla_{X_b}\mathrm{d}\lambda \wedge e_a - \nabla_{X_a}\mathrm{d}\lambda \wedge e_b + X_b(\lambda)e_a \wedge \mathrm{d}\lambda \\ &- X_a(\lambda)e_b \wedge \mathrm{d}\lambda - X_c(\lambda)X^c(\lambda)e_{ab}.\end{aligned} \tag{6.11.3}$$

We have used $\mathrm{D}X_a(\lambda) = \nabla_{X_a}\mathrm{d}\lambda$, which follows from (6.7.11). Contracting with $\widehat{X^a}$ gives the Ricci forms and curvature scalar of $\hat{\nabla}$:

$$\begin{aligned}\exp(\lambda)\widehat{P_b} = P_b &+ (2-n)\nabla_{X_b}\mathrm{d}\lambda + (n-2)X_b(\lambda)\mathrm{d}\lambda \\ &+ (2-n)X_c(\lambda)X^c(\lambda)e_b - \mathrm{i}_{X_a}\nabla_{X^a}\mathrm{d}\lambda e_b\end{aligned} \tag{6.11.4}$$

$$\exp(2\lambda)\hat{\mathcal{R}} = \mathcal{R} - 2(n-1)\mathrm{i}_{X^b}\nabla_{X_b}\mathrm{d}\lambda + (1-n)(n-2)X_c(\lambda)X^c(\lambda). \tag{6.11.5}$$

The *conformal 2-forms* C_{ab} are defined (in more than two dimensions) in terms of the curvature 2-forms and their contractions by

$$C_{ab} = R_{ab} - \frac{1}{n-2}(P_a \wedge e_b - P_b \wedge e_a) + \frac{1}{(n-2)(n-1)}\mathcal{R}e_a \wedge e_b. \tag{6.11.6}$$

These 2-forms have the important property of being invariant under conformal scalings of the metric. That is, if $\widehat{C_{ab}}$ are the conformal 2-forms of $\hat{g}$ with respect to $\{\widehat{e^a}\}$ then

$$\widehat{C_{ab}} = C_{ab}. \tag{6.11.7}$$

If the (3, 1) *conformal tensor* (or *Weyl tensor*) C is defined by

$$C = 2C^a{}_b \otimes e^b \otimes X_a \tag{6.11.8}$$

then equivalently

$$\hat{C} = C. \tag{6.11.9}$$

From their definition the conformal 2-forms C_{ab} are manifestly antisymmetric under interchange of a and b. They also satisfy (for zero torsion) analogous identities to those for the curvature 2-forms, namely

$$C_{ab} \wedge e^b = 0 \tag{6.11.10}$$

$$\mathrm{i}_{X_a}\mathrm{i}_{X_b}C_{pq} = \mathrm{i}_{X_p}\mathrm{i}_{X_q}C_{ab}. \tag{6.11.11}$$

In addition there is the identity

$$\mathrm{i}_{X^a} C_{ab} = 0. \qquad (6.11.12)$$

A manifold is *conformally flat* if its metric is conformally related to a flat one. Certainly the conformal tensor must vanish for a conformally flat space. In fact in more than three dimensions a manifold is conformally flat if and only if its conformal tensor is zero (Eisenhart 1949).

6.12 Some Curvature Relations in Low Dimensions

In two dimensions there is only one independent curvature form, which must be proportional to the volume form. We have

$$R_{ab} = \tfrac{1}{2}\mathcal{R}e_{ab}. \qquad (6.12.1)$$

Since there is only one tangent plane we write the sectional curvature simply as K. This is related to the curvature scalar by

$$K = \tfrac{1}{2}\mathcal{R}. \qquad (6.12.2)$$

The conformal 2-forms are not defined in two dimensions. However, all two-dimensional manifolds are conformally flat (Eisenhart 1949). It is often useful to exploit this by adopting coordinates in which the metric is parametrised by the scale function that relates it to a flat metric.

We can use the metric to relate the (3, 1) curvature tensor to a (4, 0) tensor, $R = 2R_{ab} \otimes e^{ab}$. Both factors in the tensor product are 2-forms, and it is often convenient to have a notation for the tensor obtained by taking the Hodge dual of either factor. We write

$$*R = 2*R_{ab} \otimes e^{ab} \qquad (6.12.3)$$

and

$$R* = 2R_{ab} \otimes *e^{ab} \qquad (6.12.4)$$

In three dimensions the dual of a 2-form is a 1-form, so

$$R* = 2R_{ab} \otimes e^c \wedge \mathrm{i}_{X_c} *e^{ab} = 2R_{ab}\mathrm{i}_{X_c} *e^{ab} \otimes e^c.$$

The first factor now involves the Einstein forms, which were given in (6.8.3). So if

$$\mathcal{G} \equiv 2G_c \otimes e^c \qquad (6.12.5)$$

we have $R* = \mathcal{G}$, or

$$R = \mathcal{G}*^{-1}. \qquad (6.12.6)$$

Now

$$\mathscr{G}*^{-1} = -G_c \mathrm{i}_{X_a}\mathrm{i}_{X_b}*^{-1}e^c \otimes e^{ab} = (-\mathscr{R}*e_c + 2*P_c)\mathrm{i}_{X_a}\mathrm{i}_{X_b}*^{-1}e^c \otimes e^{ab}$$

by (6.8.4). To simplify the first term write

$$\begin{aligned}\mathrm{i}_{X_a}\mathrm{i}_{X_b}*^{-1}e^c e_c &= \mathrm{i}_{X^c}\mathrm{i}_{X_a}\mathrm{i}_{X_b}*^{-1}1 e_c = \mathrm{i}_{X^c}(\mathrm{i}_{X_a}\mathrm{i}_{X_b}*^{-1}1 \wedge e_c) + 3\mathrm{i}_{X_a}\mathrm{i}_{X_b}*^{-1}1 \\ &= \mathrm{i}_{X^c}\{\mathrm{i}_{X_a}(\mathrm{i}_{X_b}*^{-1}1 \wedge e_c) - \mathrm{i}_{X_b}*^{-1}1 g_{ac}\} + 3*^{-1}e_{ba} \\ &= \mathrm{i}_{X^c}\mathrm{i}_{X_a}(e_c \wedge *^{-1}e_b) - \mathrm{i}_{X_a}\mathrm{i}_{X_b}*^{-1}1 + 3*^{-1}e_{ba} \\ &= g_{bc}\mathrm{i}_{X^c}\mathrm{i}_{X_a}*^{-1}1 + 2*^{-1}e_{ba} = *^{-1}e_{ba}.\end{aligned}$$

In exactly the same way we obtain

$$P_c \mathrm{i}_{X_a}\mathrm{i}_{X_b}*^{-1}e^c = \mathrm{i}_{X_b}P_c *^{-1}e_a{}^c - \mathrm{i}_{X_a}P_c *^{-1}e_b{}^c + \mathscr{R}*^{-1}e_{ba}.$$

Using the symmetry of the Ricci tensor, (6.7.17), gives

$$P_c \mathrm{i}_{X_a}\mathrm{i}_{X_b}*^{-1}e^c = *^{-1}(e_a \wedge P_b - e_b \wedge P_a + \mathscr{R}e_{ba}).$$

so we have

$$\mathscr{G}*^{-1} = 2(\tfrac{1}{2}\mathscr{R}e_{ba} + P_a \wedge e_b - P_b \wedge e_a)\otimes e^{ab}.$$

Thus (6.12.6) shows that in three dimensions

$$R_{ab} = \tfrac{1}{2}\mathscr{R}e_{ba} + P_a \wedge e_b - P_b \wedge e_a. \tag{6.12.7}$$

The first immediate consequence is that the conformal 2-forms are identically zero in three dimensions. It also follows that in three dimensions any Einstein space is necessarily of constant curvature.

Exercise 6.8

(i) Use the conformal scalings of (6.11.2)–(6.11.5) to show that if in n dimensions $Y_a \equiv \mathrm{D}P_a - [2(n-1)]^{-1}\mathrm{d}\mathscr{R} \wedge e_a$ then $\widehat{Y_a} = \exp(-\lambda) \times [Y_a + (n-2)X^b(\lambda)C_{ba}]$.

(ii) Show that $Y_a \wedge e_b - Y_b \wedge e_a = (2-n)\mathrm{D}C_{ab}$ and $Y_a \wedge e^a = 0$.

In three dimensions $C_{ab} \equiv 0$ and so in this case the tensor $Y_a \otimes e^a$ is conformally invariant. Thus the vanishing of $Y_a \otimes e^a$ is a necessary condition for conformal flatness: in fact it is also a sufficient condition (Eisenhart 1949). In three dimensions the (2, 0) tensor **SEY** $\equiv *Y_a \otimes e^a$ is conformally covariant, symmetric and traceless, by (ii).

(iii) Show that in three dimensions $\mathrm{D}Y_a = 0$.

In four dimensions there are useful identities involving the 'left and right' duals of the curvature tensor. Setting

$$R^{\pm} \equiv \tfrac{1}{2}(R \pm *^{-1}R*) \tag{6.12.8}$$

we have

$$R^- = (P_p \wedge e_q - P_q \wedge e_p - \tfrac{1}{2}\mathscr{R}e_{pq})\otimes e^{pq} \tag{6.12.9}$$

and

$$R^+ = C + \tfrac{1}{6}\mathcal{R}e_{pq} \otimes e^{pq} \tag{6.12.10}$$

where $C = 2C_{pq} \otimes e^{pq}$. These relations can be verified in exactly the same way as their three-dimensional analogues.

6.13 Killing's Equation

In §4.14 we introduced Killing vectors, these being vector fields that generate local isometries on a pseudo-Riemannian manifold. Because the pseudo-Riemannian connection is determined by the metric structure there are several useful relations between Killing vectors and this connection. Indeed, Killing vectors are often characterised by being solutions of Killing's equation, which is a differential equation for a vector field involving the pseudo-Riemannian connection.

It is convenient at this point to introduce the operator

$$A_X \equiv \mathcal{L}_X - \nabla_X \qquad \forall X \in \Gamma TM. \tag{6.13.1}$$

It immediately follows that A_X is a derivation on tensor fields that commutes with contractions, also satisfying $A_X f = 0 \ \ \forall f \in \mathcal{F}(M)$. In particular

$$A_X(g(Y, Z)) = 0 = (A_X g)(Y, Z) + g(A_X Y, Z) + g(Y, A_X Z).$$

For ∇ metric compatible $A_X g = \mathcal{L}_X g$ so the above becomes

$$g(A_X Y, Z) + g(Y, A_X Z) = -(\mathcal{L}_X g)(Y, Z).$$

Since for any vector field Y we have $A_X Y = [X, \ Y] - \nabla_X Y$, if ∇ is torsion free then $A_X Y = -\nabla_Y X$, hence

$$g(\nabla_Y X, Z) + g(\nabla_Z X, Y) = (\mathcal{L}_X g)(Y, Z). \tag{6.13.2}$$

If $\widetilde{X}$ is the 1-form related by the metric to X then it is often convenient to rewrite the above in the equivalent form

$$\mathrm{i}_Z \nabla_Y \widetilde{X} + \mathrm{i}_Y \nabla_Z \widetilde{X} = (\mathcal{L}_X g)(Y, Z). \tag{6.13.3}$$

If K is a Killing vector then (6.13.2) becomes Killing's equation:

$$g(\nabla_Y K, Z) + g(\nabla_Z K, Y) = 0 \qquad \forall Y, Z \in \Gamma TM. \tag{6.13.4}$$

The relation (6.13.3) is often useful in applications. Subsequently we shall need a related result for the 2-form $\mathrm{d}\widetilde{X}$. If V and Y are arbitrary vector fields then by (6.7.4)

$$
\begin{aligned}
\nabla_V \mathrm{d}\widetilde{Y} &= \nabla_V e^a \wedge \nabla_{X_a}\widetilde{Y} + e^a \wedge \nabla_V \nabla_{X_a}\widetilde{Y} \\
&= -\, e^a(\nabla_V X_b) e^b \wedge \nabla_{X_a}\widetilde{Y} + e^a \wedge \nabla_V \nabla_{X_a}\widetilde{Y} \\
&= -\, e^b \wedge \nabla_{\nabla_V X_b}\widetilde{Y} + e^a \wedge \nabla_V \nabla_{X_a}\widetilde{Y} \\
&= e^a \wedge (\nabla_V \nabla_{X_a} - \nabla_{\nabla_V X_a})\widetilde{Y} \\
&= e^a \wedge (\boldsymbol{R}(V, X_a) + \nabla_{X_a}\nabla_V + \nabla_{[V, X_a]} - \nabla_{\nabla_V X_a})\widetilde{Y} \\
&= e^a \wedge (\boldsymbol{R}(V, X_a) + \nabla_{X_a}\nabla_V - \nabla_{\nabla_{X_a}V})\widetilde{Y} \text{ since } \nabla \text{ is torsion-free,} \\
&= e^a \wedge \boldsymbol{R}(V, X_a)\widetilde{Y} + \mathrm{d}\nabla_V \widetilde{Y} - e^a \wedge \nabla_{\nabla_{X_a}V}\widetilde{Y}. \qquad (6.13.5)
\end{aligned}
$$

Now

$$
\begin{aligned}
\mathrm{i}_X \mathrm{d}\widetilde{Y} &= \mathrm{i}_X e^a \nabla_{X_a}\widetilde{Y} - e^a \wedge \mathrm{i}_X \nabla_{X_a}\widetilde{Y} \\
&= \nabla_X \widetilde{Y} - e^a \wedge (\mathrm{i}_X \nabla_{X_a} + \mathrm{i}_{X_a}\nabla_X)\widetilde{Y} + \nabla_X \widetilde{Y}
\end{aligned}
$$

so that

$$
\nabla_X \widetilde{Y} = \tfrac{1}{2}\mathrm{i}_X \mathrm{d}\widetilde{Y} + \tfrac{1}{2} e^a \wedge (\mathrm{i}_X \nabla_{X_a}\widetilde{Y} + \mathrm{i}_{X_a}\nabla_X \widetilde{Y}).
$$

Using (6.13.3) we have

$$
\nabla_X \widetilde{Y} = \tfrac{1}{2}\mathrm{i}_X \mathrm{d}\widetilde{Y} + \tfrac{1}{2}\mathscr{L}_Y g(X, X_a) e^a. \qquad (6.13.6)
$$

This gives

$$
\begin{aligned}
e^a \wedge \nabla_{\nabla_{X_a}V}\widetilde{Y} &= \tfrac{1}{2} e^a \wedge \mathrm{i}_{\nabla_{X_a}V}\mathrm{d}\widetilde{Y} + \tfrac{1}{2}\mathscr{L}_Y g(\nabla_{X_a}V, X_b) e^{ab} \\
&= \tfrac{1}{2} e^a \wedge (\nabla_{X_a}(\mathrm{i}_V \mathrm{d}\widetilde{Y}) - \mathrm{i}_V \nabla_{X_a}\mathrm{d}\widetilde{Y}) + \tfrac{1}{2}\mathscr{L}_Y g(\nabla_{X_a}V, X_b) e^{ab} \\
&= \tfrac{1}{2}\mathrm{d}\mathrm{i}_V \mathrm{d}\widetilde{Y} + \tfrac{1}{2}\mathrm{i}_V(e^a \wedge \nabla_{X_a}\mathrm{d}\widetilde{Y}) - \tfrac{1}{2}\nabla_V \mathrm{d}\widetilde{Y} \\
&\quad + \tfrac{1}{2}\mathscr{L}_Y g(\nabla_{X_a}V, X_b) e^{ab} \\
&= \tfrac{1}{2}\mathrm{d}\mathrm{i}_V \mathrm{d}\widetilde{Y} - \tfrac{1}{2}\nabla_V \mathrm{d}\widetilde{Y} + \tfrac{1}{2}\mathscr{L}_Y g(\nabla_{X_a}V, X_b) e^{ab} \qquad (6.13.7)
\end{aligned}
$$

since $\mathrm{d}^2 = 0$. Using (6.13.6) once again

$$
\begin{aligned}
\mathrm{d}\nabla_V \widetilde{Y} &= \tfrac{1}{2}\mathrm{d}\mathrm{i}_V \mathrm{d}\widetilde{Y} + \tfrac{1}{2}\mathrm{d}(\mathscr{L}_Y g(V, X_a) e^a) \\
&= \tfrac{1}{2}\mathrm{d}\mathrm{i}_V \mathrm{d}\widetilde{Y} + \tfrac{1}{2}\nabla_{X_b}\mathscr{L}_Y g(V, X_a) e^{ba} + \tfrac{1}{2}\mathscr{L}_Y g(\nabla_{X_b}V, X_a) e^{ba}. \qquad (6.13.8)
\end{aligned}
$$

Returning now to (6.13.5) with (6.13.7) and (6.13.8) produces

$$
\nabla_V \mathrm{d}\widetilde{Y} = 2e^a \wedge \boldsymbol{R}(V, X_a)\widetilde{Y} + \nabla_{X_b}\mathscr{L}_Y g(V, X_a) e^{ba}.
$$

This can be expressed in terms of the curvature 2-forms as

$$
\nabla_V \mathrm{d}\widetilde{Y} = 2Y^a V^b R_{ab} + \nabla_{X_b}\mathscr{L}_Y g(V, X_a) e^{ba} \qquad (6.13.9)
$$

where $Y^a = e^a(Y)$ etc. Operating on this with the interior product gives an expression with the Ricci forms:

$$
\mathrm{i}_{X^c}\nabla_{X_c}\mathrm{d}\widetilde{Y} = -2Y^a P_a + \nabla_{X_c}\mathscr{L}_Y g(X^c, X_a) e^a - \nabla_{X_b}\mathscr{L}_Y g(X^a, X_a) e^b
$$

or, by (6.9.1)

$$\delta \mathrm{d} \widetilde{Y} = 2Y^a P_a - \nabla_{X_c} \mathscr{L}_Y g(X^c, X_a) e^a + \nabla_{X_b} \mathscr{L}_Y g(X^a, X_a) e^b. \quad (6.13.10)$$

Obviously such expressions are particularly useful for vectors that generate symmetries.

Exercise 6.9

A vector field K is called a *conformal Killing vector* if $\mathscr{L}_K g = 2\lambda g$ for some function λ. Show that K satisfies

$$\text{(i) } \delta \widetilde{K} = n\lambda \quad (6.13.11)$$

$$\text{(ii) } \delta \mathrm{d} \widetilde{K} = 2K^a P_a + 2(n-1)\mathrm{d}\lambda. \quad (6.13.12)$$

Exercise 6.10

For some calculations one needs to be able to commute a Lie derivative past a covariant derivative. If

$$\mathrm{D}_X(Y) \equiv [\mathscr{L}_Y, \nabla_X] - \nabla_{[Y, X]} \quad (6.13.13)$$

show that

(i) $\mathrm{D}_X(Y)$ is a tensor derivation that commutes with contractions

(ii) $\mathrm{D}_{fX}(Y) = f\mathrm{D}_X(Y)$ for $f \in \mathscr{F}(M)$

(iii) $\mathrm{D}_X(Y)fS = f\mathrm{D}_X(Y)S$ for any tensor field S

(iv) $\mathrm{D}_X(Y)Z = \mathrm{D}_Z(Y)X$ (since ∇ is torsion-free)

If $\mathrm{D}_{X_a}(Y)X_b \equiv M_{ab}{}^c(Y)X_c$ show that

$$M_{ab}{}^p(Y) = \tfrac{1}{2} g^{cp}(\nabla_{X_a} \mathscr{L}_Y g(X_c, X_b) - \nabla_{X_c} \mathscr{L}_Y g(X_b, X_a) + \nabla_{X_b} \mathscr{L}_Y g(X_a, X_c)). \quad (6.13.14)$$

Hint: starting from $\mathrm{D}_{X_a}(Y)(g(X_b, X_c)) = 0$ follow the procedure for solving for the connection coefficients given in §6.6.

Bibliography

Eisenhart L P 1949 *Riemannian Geometry* (Princeton, NJ: Princeton University Press)

Helgason S 1978 *Differential Geometry, Lie Groups, and Symmetric Spaces* (New York: Academic)

7

Gravitation

7.1 Lorentzian Connections

As we noted in §6.2 the space $\mathbb{R}^n$ has a natural connection. This is defined such that a natural coordinate basis is parallel. We have already seen how Newtonian dynamics may be described with the natural connection on $\mathbb{R}^3$. In Chapter 5 Minkowski spacetime was modelled on $\mathbb{R}^4$, the natural coordinate basis being declared orthonormal with respect to a Lorentzian metric. Such a field of global orthonormal frames is parallel with respect to the natural $\mathbb{R}^4$ connection, and thus we may now recognise the class of inertial frames as consisting of all frames that are parallel with respect to this connection. More generally on any spacetime we may use the unique torsion-free metric-compatible connection (the *Lorentzian connection*) to evaluate the acceleration of curves. If a particle of mass μ is modelled on a unit timelike curve C then the acceleration $\nabla_{\dot{C}}\dot{C}$ may be attributed to a four-force $\mathcal{F}$: $\mathcal{F} = \nabla_{\dot{C}}(\mu\dot{C})$.

For example, if C describes a particle of electric charge q moving in a background electromagnetic field described by the 2-form F then the force is given by the Lorentz rule $\mathcal{F} = q\widetilde{i_{\dot{C}}F}$. Hence C may be determined by solving the equation

$$\nabla_{\dot{C}}(\mu\dot{C}) = q\widetilde{i_{\dot{C}}F}. \tag{7.1.1}$$

(Since the particle may radiate an electromagnetic field this equation should be coupled with the Maxwell field equations (the particle produces a source of electric current) to determine F properly.) It is instructive to compare a Minkowski four-dimensional description with our earlier Newtonian formulation. We may express F in terms of electric and magnetic fields observed by an inertial observer ∂_t,

$F = \widetilde{\boldsymbol{E}} \wedge \mathrm{d}t + B$. Similarly we express the trajectory four-velocity $\dot{C}$ in terms of the Newtonian velocity v^k, $k = 1, 2, 3$, with respect to the same inertial observer, as $\dot{C} = \gamma(\partial_t + v^k \partial_k)$, where $\gamma^{-1} = (1 - v^k v_k)^{1/2}$. It is straightforward to calculate

$$\nabla_{\dot{C}}(\mu \dot{C}) = \dot{C}(\mu\gamma)\partial_t + \dot{C}(\mu\gamma v^k)\partial_k \tag{7.1.2}$$

and

$$\widetilde{\mathrm{i}_{\dot{C}} F} = -\gamma E_j v^j \partial_t - \gamma E^j \partial_j + \gamma v^j \widetilde{\mathrm{i}_{\partial_j} B}. \tag{7.1.3}$$

We have written $\widetilde{\boldsymbol{E}} = E_j \mathrm{d}x^j$ and used $\mathrm{i}_{\dot{C}} \mathrm{d}t = \gamma$, $\widetilde{\mathrm{d}t} = -\partial_t$, $\widetilde{\mathrm{d}x^j} = \partial_j$. If we write $\widetilde{\mathrm{i}_{\partial_k} B} = -\varepsilon_k{}^{lm} B_l \partial_m$ (where ε_{klm} is totally antisymmetric k, l, $m = 1, 2, 3$ and $\varepsilon_{123} = 1$) then in an inertial chart for Minkowski spacetime (7.1.1) becomes

$$\dot{C}(\mu\gamma v_m) = -q\gamma(E_m + v^k \varepsilon_{klm} B^l)$$

$$\dot{C}(\mu\gamma) = -q\gamma E_m v^m.$$

Since $\dot{C} = C_* \partial_\tau$, $\dot{C}(t) = \mathrm{d}t/\mathrm{d}\tau = \gamma$ relates the inertial time variable t to the proper time τ at points on the curve. Similarly $\dot{C}(x^k) = \mathrm{d}x^k/\mathrm{d}\tau$ $= \gamma v^k = (\mathrm{d}t/\mathrm{d}\tau)v^k$, hence $v^k = (\mathrm{d}x^k/\mathrm{d}t)$. Setting $p^k = \mu\gamma v^k$, $\mathscr{E} = \mu\gamma$ gives the equations in the form

$$\frac{\mathrm{d}}{\mathrm{d}t}(p_m) = -q(E_m + v^k \varepsilon_{klm} B^l)$$

$$\frac{\mathrm{d}}{\mathrm{d}t}\mathscr{E} = -qE_i v^i.$$

We see that the Newtonian equations of motion are recovered for $v^k v_k \ll 1$. For many practical calculations it is, however, often easier to use (7.1.1) directly without passing to an inertial chart.

Example 7.1

Use the transformation from the inertial Minkowski coordinates (t, x, y, z) to the coordinates (ξ, η, y', z'). $t = \xi \sinh \eta$, $x = \xi \cosh \eta$, $y' = y$, $z' = z$ to express the Minkowski metric tensor in the form

$$g = -\xi^2 \mathrm{d}\eta \otimes \mathrm{d}\eta + \mathrm{d}\xi \otimes \mathrm{d}\xi + \mathrm{d}y' \otimes \mathrm{d}y' + \mathrm{d}z' \otimes \mathrm{d}z'$$

on a patch defined by $0 \leqslant \xi, \eta, y', z' < \infty$. Verify that the only non-vanishing connection components in this chart are given by $\nabla_{\partial_\xi} \partial_\eta = (1/\xi)\partial_\eta = \nabla_{\partial_\eta} \partial_\xi$ and $\nabla_{\partial_\eta} \partial\eta = \xi \partial_\xi$. Show that $\dot{C} = \mathscr{G}\partial_\eta$, $\mathscr{G} \in \mathscr{R}$, solves (7.1.1) for a constant electric field expressed in the inertial chart as $F = E_0 \mathrm{d}x \wedge \mathrm{d}t$ if $\mathscr{G} = -qE_0/m$. Hence derive the hyperbolic orbit ($\xi = \mathscr{G}^{-1}$, $\eta = \mathscr{G}\tau$, $y' = 0$, $z' = 0$) and show that this asymptotes to a light cone. Note $\mathscr{G}$ is the norm of the constant four-acceleration of the particle:

$$g(\nabla_{\dot{C}} \dot{C}, \nabla_{\dot{C}} \dot{C}) = \mathscr{G}^2.$$

7.2 Fermi–Walker Transport

If C is any geodesic of an arbitrary spacetime ($\nabla_{\dot{C}}\dot{C} = 0$) then if $g(X, \dot{C}) = 0$ at any point on the curve then X remains orthogonal to $\dot{C}$ at all points if $\nabla_{\dot{C}}X = 0$. But if the acceleration field $A_C = \nabla_{\dot{C}}\dot{C}$ along C is not zero then this property is lost. However, on any given C we may usefully define a new connection $\hat{\nabla}$ in terms of ∇ and the metric tensor field g. Acting on any vector field X restricted to C

$$\hat{\nabla}_{\dot{C}}X \equiv \nabla_{\dot{C}}X + g(\dot{C}, X)A_C - g(A_C, X)\dot{C}. \tag{7.2.1}$$

This connection is called a *Fermi–Walker* or ***F**-connection* on C. Its construction manifestly depends on the parametrised curve C itself. An immediate consequence of the definition is that

$$\dot{C}(g(X, Y)) = g(X, \hat{\nabla}_{\dot{C}}Y) + g(\hat{\nabla}_{\dot{C}}X, Y) \qquad \forall X, Y \text{ on } C \tag{7.2.2}$$

so $\hat{\nabla}$ is compatible with the metric tensor g. If C is an observer curve ($g(\dot{C}, \dot{C}) = -1$) then $g(A_C, \dot{C}) = 0$ and hence $\hat{\nabla}_{\dot{C}}\dot{C} = 0$: so a velocity vector is also $\boldsymbol{F}$-parallel. For any vector field Y on C, $\dot{C}(g(Y, \dot{C})) = g(\hat{\nabla}_{\dot{C}}Y, \dot{C})$, so if Y is $\boldsymbol{F}$-parallel ($\hat{\nabla}_{\dot{C}}Y = 0$) then the metric projection of Y on $\dot{C}$ (or the *angle* between Y and $\dot{C}$) is preserved along C. In particular a g-orthonormal frame $\{X_a\}$ at one point of C, with a timelike basis vector $X_0 = \dot{C}$, will remain orthonormal with $X_0 = \dot{C}$ at all points along C if parallel transported with respect to the Fermi–Walker connection. Such an $\boldsymbol{F}$-parallel frame is said to be *non-rotating* along C and gives one a way of determining whether any spacelike vector undergoes spatial rotation along C: spatial rotation being measured by the components with respect to the $\boldsymbol{F}$-parallel basis on C.

It is generally believed that in spacetime an $\boldsymbol{F}$-parallel spacelike vector S satisfying the orthogonality condition $g(S, \dot{C}) = 0$ along a timelike curve C models the behaviour of an ideal gyroscope (one that experiences no non-gravitational torques) on C. Three such mutually orthogonal gyroscopes ($g(S_i, S_j) = \delta_{ij}$) together with $\dot{C}$ then define a non-rotating frame along C. It is interesting to note that this concept of frame rotation is determined by the metric properties of spacetime. The relation of these properties to gravitational fields is explored in the next few sections.

7.3 The Einstein Field Equations

The theory of Newtonian gravitation provides an excellent description for a large class of natural phenomena. The gravitational interaction between macroscopic distributions of matter is defined in terms of a Newtonian force derivable most simply from a real scalar field on Newtonian spacetime. As originally formulated, no account is taken of the propagation velocity of this interaction. It is regarded as an instantaneous or static interaction. When Einstein introduced the special theory of relativity the notion of simultaneity became observer dependent. The recognition that Maxwell's equations of electromagnetism could be formulated as a set of tensor equations on a four-dimensional spacetime encouraged Einstein to reformulate all the basic laws of classical physics in terms of spacetime tensor fields.

According to Einstein the Lorentzian metric of spacetime should also be governed by partial differential equations so that the geometry itself has a dynamical status along with the fields of matter. The idea that the matter and geometry of a spacetime form a mutually sustaining dynamical system found fruition in the general theory of relativity proposed by Einstein in 1916. Despite its title this theory proposes that there is an absolute spacetime arena in which the classical events of physics take place. This needs qualifying as follows. If g is any spacetime metric tensor field satisfying Einstein's equations on a manifold M then for $\varphi : M \to \varphi M$ a diffeomorphism, $\varphi^* g$ will solve the diffeomorphic image of Einstein's equations on φM. Any such manifold isometric to M under a diffeomorphism is regarded as describing the same physical phenomena. The choice of field equations was partly inspired by the need to recover Newton's laws of gravity in the limit in which propagation effects could be neglected and partly by the aesthetic desire to maintain a tensorial description of spacetime events in which the coordinates of such events were to be relegated to the labelling conventions adopted by different observers. The field equations involve the curvature tensor of the Lorentzian connection and tensors constructed out of various matter fields describing the sources of the gravitational field. There are many ways to formulate these field equations. In the early literature one finds the tensor components of the field equations written out in some local chart from the manifold atlas. There is some virtue in writing out the local equations in full tensorial form since as we shall show this often facilitates their solution and simplifies their presentation. One should, however, note that each local solution of the coupled system of field equations may in general be extended to the whole manifold in different ways. If the global properties of the spacetime manifold are constrained then the class of solutions that can be defined globally will be similarly constrained.

Whereas in principle all the physical consequences of such a theory should follow from the Einstein equations for gravity together with the field equations for the matter tensors, an often used approximation models macroscopic 'test' particles that interact solely with gravitation by geodesic world lines.

Let us first write Einstein's equations in terms of exterior forms on some neighbourhood of the spacetime manifold M. If $\{G_c\}$ are the Einstein 3-forms associated with the Lorentzian connection, given in (6.8.3), then Einstein's equations for g are

$$\kappa G_c + \tau_c(g, \Phi) = 0 \qquad c = 0, 1, 2, 3 \tag{7.3.1}$$

where $\{\tau_c(g, \Phi)\}$ is a set of stress 3-forms determined in this co-frame by some choice of matter fields, denoted generically here by Φ, and κ is some (positive) coupling constant. (The notation indicates that τ_c depends on g and Φ rather than being contracted on these fields.) We shall supplement these equations with a set of matter field equations denoted collectively by

$$\mathscr{E}(g, \Phi) = 0. \tag{7.3.2}$$

We cannot choose the stress forms arbitrarily, since for zero torsion (6.8.6) and (6.8.7) reduce to

$$\mathrm{D}G_a = 0 \tag{7.3.3}$$

and

$$G_a \wedge e_b = G_b \wedge e_a. \tag{7.3.4}$$

The matter stress forms defined with respect to $\{e^a\}$ determine the *stress energy tensor field*

$$\mathscr{T} = *^{-1}\tau_a \otimes e^a. \tag{7.3.5}$$

Any matter model for Einstein's equations must therefore give rise to a symmetric second-rank stress tensor $\mathscr{T} = \mathscr{T}_{ab} e^a \otimes e^b$ that is divergenceless: $\nabla . \mathscr{T} = 0$. In many cases given a matter model there is a well defined procedure for generating such a stress tensor. Indeed the most economical way to summarise the whole coupled system is in terms of an action functional whose extremal equations generate the full set of field equations including the consistent stress forms. Although it is straightforward to set up a heuristic scheme for applying a variational calculus to obtain all the field equations it would take us too far afield to set up a decent formalism for this purpose. (The precise formulation of a variational scheme involving spinors requires particular care.) We shall be content in this chapter to give some examples of matter models in exterior form together with their associated stresses. Such matter models have featured prominently in many theoretical discussions of gravitational interactions with fields.

Exercise 7.1

Show, by contracting (4.8.4) and using (7.3.1), that in n dimensions Einstein's equations can be written as

$$2\kappa P_c = *^{-1}\tau_c - \frac{\mathrm{i}_{X^a} *^{-1}\tau_a}{n-2} e_c.$$

The conditions that the stress tensor be symmetric and divergenceless are required for it to be equated to the Einstein tensor of a metric-compatible torsion-free connection. In addition further 'energy' conditions are usually required to hold in order for the stress tensor to be physically reasonable. The *weak energy condition* is that $\mathcal{T}(V, V) \geqslant 0$ for all timelike V. This condition is motivated by assuming that an observer whose curve is tangent to V would interpret $\mathcal{T}(V, V)$ as an energy density. The *dominant energy condition* is similarly motivated. This can be phrased as requiring that j_V be a future-pointing non-spacelike vector for all future-pointing timelike V, where $\widetilde{j_V} = -*\tau_V$ for $\tau_V = \tau_a e^a(V)$. Alternatively one can impose conditions on the stress tensor by requiring that the corresponding (via Einstein's equations) Einstein tensor has certain properties, resulting in gravity being, in some sense, attractive. The condition on the stress tensor such that $\mathbf{Ric}(V, V) \geqslant 0$ for all timelike V is called the *strong energy condition*. Details of these energy conditions can be found in Hawking and Ellis.

7.4 Conservation Laws

In Newtonian dynamics the total energy and momentum of a system may be defined to be certain dynamical variables that remain fixed as the system evolves. Such constants of the motion have their origin in the existence of certain symmetries of the equations of motion. Similarly in the dynamics of continuous media the vanishing divergence of the Newtonian energy–momentum tensor affords a succinct description of the equations of motion, and the associated constants of motion may be obtained by integrating densities constructed from the components of such a tensor. On a curved manifold, however, caution is required in correlating conservation laws to the existence of a divergenceless stress tensor. In general it is necessary for the spacetime metric to admit some kind of symmetry in order to construct conserved quantities.

Let $\mathcal{T}$ be a symmetric (2, 0) tensor whose metric related (0, 2) tensor has components $\mathcal{T}^{ab}$ in some orthonormal frame $\{X_a\}$. For any vector field V we have $\mathcal{L}_V g(X_a, X_b) + g(\mathcal{L}_V X_a, X_b) + g(X_a, \mathcal{L}_V X_b) = 0$ since $\mathcal{L}_V[g(X_a, X_b)] = 0$ for any orthonormal frame $\{X_a\}$. Hence since $\mathcal{T}^{ab} = \mathcal{T}^{ba}$ and ∇ is torsion free:

$$\begin{aligned}
&\mathcal{L}_V g(X_a, X_b)\mathcal{T}^{ab}*1 \\
&= -2g(\mathcal{L}_V X_a, X_b)\mathcal{T}^{ab}*1 = -2g(\nabla_V X_a - \nabla_{X_a} V, X_b)\mathcal{T}^{ab}*1 \\
&= -\{g(\nabla_V X_a, X_b) + g(X_a, \nabla_V X_b)\}\mathcal{T}^{ab}*1 + 2g(\nabla_{X_a} V, X_b)\mathcal{T}^{ab}*1.
\end{aligned}$$

Now $g(\nabla_V X_a, X_b) + g(X_a, \nabla_V X_b) = 0$ since $V\{g(X_a, X_b)\} = 0$ and so

$$\tfrac{1}{2}\mathcal{L}_V g(X_a, X_b)\mathcal{T}^{ab}*1 = g(\nabla_{X_a}V, X_b)\mathcal{T}^{ab}*1$$

$$= \nabla_{X_a}\{g(V, X_b)\mathcal{T}^{ab}\}*1 - g(V, \nabla_{X_a}X_b)\mathcal{T}^{ab}*1 - g(V, X_b)\nabla_{X_a}\mathcal{T}^{ab}*1.$$

Now for any $(n-1)$-form J we may write

$$\mathrm{d}J = e^a \wedge \nabla_{X_a}J = \nabla_{X_a}(e^a \wedge J) - \nabla_{X_a}e^a \wedge J.$$

So introducing $j^a = e^a \wedge J$ we have $\mathrm{d}J = \nabla_{X_a}j^a - (\nabla_{X_a}e^a)(X_b)j^b$. Thus we have

$$\tfrac{1}{2}\mathcal{L}_V g(X_a, X_b)\mathcal{T}^{ab}*1$$

$$= \nabla_{X_a}\{g(V, X_b)\mathcal{T}^{ab}*1\} - (\nabla_{X_a}e^a)(X_b)g(V, X_c)\mathcal{T}^{bc}*1$$

$$+(\nabla_{X_a}e^a)(X_b)g(V, X_c)\mathcal{T}^{bc}*1$$

$$- g(V, \nabla_{X_a}X_b)\mathcal{T}^{ab}*1 - g(V, X_b)\nabla_{X_a}\mathcal{T}^{ab}*1$$

$$= \mathrm{d}\{V_b\mathcal{T}^{ab}*e_a\} - \{e^a(\nabla_{X_a}X_b)V_c\mathcal{T}^{bc} + g(V, \nabla_{X_a}X_b)\mathcal{T}^{ab} + V_b\nabla_{X_a}\mathcal{T}^{ab}\}*1$$

$$= \mathrm{d}\{V^b\mathcal{T}_{ab}*e^a\} - \{\nabla_{X^a}e^b(X_a)\mathcal{T}_{bc}e^c + \nabla_{X^a}e^b\mathcal{T}_{ab} + X^a(\mathcal{T}_{ab})e^b\}(V)*1.$$

We may write this in terms of the $(n-1)$-form $J_V = V^b\mathcal{T}_{ab}*e^b$ as

$$\tfrac{1}{2}(\mathcal{L}_V g)(X_a, X_b)\mathcal{T}^{ab}*1 = \mathrm{d}J_V - (\nabla\cdot\mathcal{T})(V)*1. \qquad (7.4.1)$$

From this relation we conclude that if the spacetime admits a conformal Killing vector field C, $\mathcal{L}_C g = 2\lambda g$, then

$$\lambda\mathcal{T}_a{}^a*1 = \mathrm{d}J_C - (\nabla\cdot\mathcal{T})(C)*1.$$

Hence a closed $(n-1)$-form may be constructed out of a divergenceless traceless stress tensor in a spacetime with conformal isometries. If the vector field K is Killing ($\mathcal{L}_K g = 0$) then irrespective of the trace of $\mathcal{T}$

$$\mathrm{d}J_K = 0.$$

If P_0, (P_i) are Killing vector fields on four-dimensional spacetime generating open timelike (spacelike) integral curves then the integrals of the corresponding 3-forms over a spacelike 3-chain Σ define the energy (momentum) contributed by $\mathcal{T}$ to Σ. Similarly if J_i are three Killing vector fields that generate the closed integral curves corresponding to the orbits of the rotation group SO(3) then the corresponding integrals may be taken as defining the angular momentum in Σ.

There is a useful analogy between solutions of Einstein's equations, coupled to matter, admitting symmetries and solutions to Maxwell's equations coupled to charged matter. The closed 3-forms constructed out of the stress tensor and the Killing vector are the analogues of the closed electromagnetic current 3-form. Maxwell's equations have the important property that one may define the total charge contained in a compact region by the integral of the 2-form $*F$, which is closed in any source-free region, over any closed 2-chain. (Electric charge may be defined by a de-Rham period.) Einstein's equations give rise to analogous 2-forms that are closed in source-free regions of spacetimes with symmetries. Einstein's equations imply that when the stress tensor vanishes the spacetime is Ricci flat. So if the spacetime admits a Killing vector K then, from (4.13.12) the 2-form $*\mathrm{d}\widetilde{K}$ is closed. In such spacetimes we shall refer to $*\mathrm{d}\widetilde{K}$ as a *Komar form*, the component expression having been introduced into general relativity by Komar [11].

7.5 Some Matter Fields

The Einstein–Klein–Gordon system

The massive real scalar field $\varphi \in \Gamma\Lambda_0 M$ is taken to satisfy

$$\mathrm{d}{*}\mathrm{d}\varphi = \mu^2 {*}\varphi + U'(\varphi){*}1 \qquad (7.5.1)$$

where μ is some real parameter and U is a polynomial in φ. The stress forms in the local frame $\{X_a\}$ are given by

$$\tau_a = \tfrac{1}{2}(\mathrm{i}_a \mathrm{d}\varphi \wedge {*}\mathrm{d}\varphi + \mathrm{d}\varphi \wedge \mathrm{i}_a {*}\mathrm{d}\varphi) - (\tfrac{1}{2}\mu^2\varphi^2 + U){*}e_a \qquad (7.5.2)$$

where $\mathrm{i}_a \equiv \mathrm{i}_{X_a}$. The stress associated with a constant U is sometimes attributed to a 'cosmological term'.

As we have remarked, in order to be consistently equated to the Einstein tensor, the stress forms should satisfy $\mathrm{D}\tau_a = 0$. Taking the expression in (7.5.2) gives

$$\mathrm{D}\tau_a = \tfrac{1}{2}(\mathrm{D}\mathrm{i}_{X_a}\mathrm{d}\varphi \wedge {*}\mathrm{d}\varphi + \mathrm{i}_{X_a}\mathrm{d}\varphi \wedge \mathrm{d}{*}\mathrm{d}\varphi - \mathrm{d}\varphi \wedge \mathrm{D}\mathrm{i}_{X_a}{*}\mathrm{d}\varphi) \\ - (\mu^2\varphi + U')\mathrm{d}\varphi \wedge {*}e_a. \qquad (7.5.3)$$

Now we may use (6.7.11) (for zero torsion):

$$\mathrm{D}\tau_a = \tfrac{1}{2}(\nabla_{X_a}\mathrm{d}\varphi \wedge {*}\mathrm{d}\varphi + \mathrm{i}_{X_a}\mathrm{d}\varphi \wedge \mathrm{d}{*}\mathrm{d}\varphi - \mathrm{d}\varphi \wedge \nabla_{X_a}{*}\mathrm{d}\varphi + \mathrm{d}\varphi \wedge \mathrm{i}_{X_a}\mathrm{d}{*}\mathrm{d}\varphi) \\ - \mathrm{i}_{X_a}\mathrm{d}\varphi(\mu^2\varphi + U'){*}1.$$

Since ∇ is metric-compatible $\mathrm{d}\varphi \wedge \nabla_{X_a}{*}\mathrm{d}\varphi = \mathrm{d}\varphi \wedge {*}\nabla_{X_a}\mathrm{d}\varphi = \nabla_{X_a}\mathrm{d}\varphi \wedge {*}\mathrm{d}\varphi$ and so the terms involving ∇ cancel. Since $\mathrm{d}\varphi \wedge \mathrm{i}_{X_a}\mathrm{d}{*}\mathrm{d}\varphi = -\mathrm{i}_{X_a}(\mathrm{d}\varphi \wedge \mathrm{d}{*}\mathrm{d}\varphi) + \mathrm{i}_{X_a}\mathrm{d}\varphi \wedge \mathrm{d}{*}\mathrm{d}\varphi$, and $\mathrm{d}\varphi \wedge \mathrm{d}{*}\mathrm{d}\varphi$ is a 5-form in four dimensions

$$\mathrm{D}\tau_a = X_a(\varphi)(\mathrm{d}{*}\mathrm{d}\varphi - \mu^2{*}\varphi - U'{*}1).$$

Thus whenever the field equations (7.5.1) hold $\mathrm{D}\tau_a = 0$.

Exercise 7.2
Show that $\mathcal{T}(X_0, X_0)*1 = \tau_0 \wedge e^0$ and that for (7.5.2)

$$\tau_0 \wedge e^0 = \left(\frac{1}{2}\sum_{a=0}^{3}(X_a(\varphi))^2 + \tfrac{1}{2}\mu^2\varphi^2 + U\right)*1$$

that is, for a suitable potential U the weak energy condition is satisfied.

The Einstein–Proca system

The 'massive' real 1-form field $\hat{A}$ is taken to satisfy

$$\mathrm{d}{*}\mathrm{d}\hat{A} = -\mu^2{*}\hat{A} \tag{7.5.4}$$

with μ some real non-zero constant. The associated stress forms are

$$\tau_a = \tfrac{1}{2}(\mathrm{i}_a\mathrm{d}\hat{A} \wedge {*}\mathrm{d}\hat{A} - \mathrm{i}_a{*}\mathrm{d}\hat{A} \wedge \mathrm{d}\hat{A}) + \tfrac{1}{2}\mu^2(\mathrm{i}_a\hat{A} \wedge {*}\hat{A} + \hat{A} \wedge \mathrm{i}_a{*}\hat{A}). \tag{7.5.5}$$

It may be noted that an integrability condition follows by applying $*\mathrm{d}$ to (7.5.4):

$$\delta\hat{A} = 0. \tag{7.5.6}$$

The Einstein–Maxwell system

For the electromagnetic field 2-form F we have the curved space Maxwell equations

$$\mathrm{d}{*}F = 0 \tag{7.5.7}$$

$$\mathrm{d}F = 0 \tag{7.5.8}$$

with associated stresses

$$\tau_a = \tfrac{1}{2}(\mathrm{i}_aF \wedge {*}F - \mathrm{i}_a{*}F \wedge F). \tag{7.5.9}$$

The Einstein Yang–Mills system

Let $\boldsymbol{A} = A_iT^i$ be a Lie-algebra-valued 1-form, $A_i \in \Gamma\Lambda_1M$ and $\{T^i\}$ a basis for some Lie algebra, with Lie bracket $[T^i, T^j]$. The Yang–Mills field strength is the Lie-algebra-valued 2-form $\boldsymbol{F} = \mathrm{d}\boldsymbol{A} + \frac{1}{2}[\boldsymbol{A}, \boldsymbol{A}] = F_iT^i$ where the bracket between a Lie-algebra-valued p-form $\boldsymbol{H}$ and a Lie-algebra-valued q-form $\boldsymbol{B}$ is

$$[\boldsymbol{H}, \boldsymbol{B}] = H_i \wedge B_j[T^i, T^j] = \boldsymbol{H} \wedge \boldsymbol{B} - (-1)^{pq}\boldsymbol{B} \wedge \boldsymbol{H}$$

and $\mathrm{d}\boldsymbol{A} = \mathrm{d}A_iT^i$. It is useful to define an exterior covariant derivative on the Lie-algebra-valued p-forms $\boldsymbol{H}$:

$$\mathbf{D}\boldsymbol{H} = \mathrm{d}\boldsymbol{H} + [\boldsymbol{A}, \boldsymbol{H}]. \tag{7.5.10}$$

From the definition of $\boldsymbol{F}$ we have the Bianchi identity

$$\mathbf{D}\boldsymbol{F} = 0. \tag{7.5.11}$$

The field equation analagous to (7.5.7) is

$$\mathbf{D}{*}\boldsymbol{F} = 0. \tag{7.5.12}$$

The system is coupled to Einsteinian gravity with the stress forms

$$\tau_a = \sum_i \tfrac{1}{2}(\mathrm{i}_a F_i \wedge {*}F_i - \mathrm{i}_a{*}F_i \wedge F_i). \tag{7.5.13}$$

The Einstein–Maxwell-charged scalar system

In this case an electrically charged complex scalar field Φ couples to both gravity and electromagnetism. The Maxwell equations now have electric current sources $j[g, \Phi]$:

$$\mathrm{d}{*}F = j \tag{7.5.14}$$

$$\mathrm{d}F = 0 \tag{7.5.15}$$

where the current 3-form is

$$j = \mathrm{Im}(\Phi{*}\mathscr{D}\Phi^*) \tag{7.5.16}$$

and the U(1) *exterior covariant derivative* is defined by

$$\mathscr{D}\Phi = \mathrm{d}\Phi + \mathrm{i}A\Phi$$

in terms of the 1-form A satisfying $F = \mathrm{d}A$. Under the maps $A \mapsto A - \mathrm{d}\lambda$, $\Phi \mapsto e^{\mathrm{i}\lambda}\Phi$ for λ any real function on M, $\mathscr{D}\Phi \mapsto e^{\mathrm{i}\lambda}\mathscr{D}\Phi$. All electrically charged tensors and their U(1) covariant derivatives belong to some representation of the group U(1). The Maxwell stress forms are now supplemented by

$$\tau_a[g, A, \Phi] = \tfrac{1}{2}\mathrm{Re}(\mathrm{i}_a\mathscr{D}\Phi \wedge {*}\mathscr{D}\Phi^* + \mathscr{D}\Phi \wedge \mathrm{i}_a{*}\mathscr{D}\Phi^*)$$
$$- \tfrac{1}{2}(\mu^2|\Phi|^2 + U(|\Phi|^2)){*}e_a. \tag{7.5.17}$$

The U(1) covariant field equation for Φ is

$$\mathscr{D}{*}\mathscr{D}\Phi = \mu^2{*}\Phi + U'\Phi{*}1 \tag{7.5.18}$$

with $U' = \mathrm{d}U/\mathrm{d}|\Phi|^2$.

Exercise 7.3

Show that the total stress tensor, the sum of those in (7.5.9) and (7.5.17), satisfies $\mathrm{D}\tau_a = 0$ when the coupled Maxwell–Klein–Gordon equations, (7.5.14), (7.5.16) and (7.5.18), are satisifed.

Ideal-fluid stress

In astrophysical problems one often models massive fluids on a timelike vector field. If V is a local vector field with $g(V, V) = -1$ each integral curve is considered to describe the world line of a massive fluid element. If the fluid has mass density specified by the 0-form ρ, the 3-form mass current is

$$j = \rho * \widetilde{V} \tag{7.5.19}$$

and the mass in a spacelike 3-surface Σ is $\int_\Sigma j$. If the number of particles in the fluid remains constant then $\mathrm{d}j = 0$. We examine the symmetric tensor field

$$\widetilde{\widetilde{\mathcal{T}}} = \rho V \otimes V. \tag{7.5.20}$$

Since

$$\nabla_{X_a} \widetilde{\widetilde{\mathcal{T}}} = (X_a \rho) V \otimes V + \rho \nabla_{X_a} V \otimes V + \rho V \otimes \nabla_{X_a} V$$

then

$$(\nabla_{X_a} \widetilde{\widetilde{\mathcal{T}}})(e^a, \;) = (X_a \rho) V^a V + \rho(\nabla_{X_a} V)(e^a) V + \rho V^a \nabla_{X_a} V.$$

The symmetric tensor field $\widetilde{\widetilde{\mathcal{T}}}$ has divergence

$$\nabla . \widetilde{\widetilde{\mathcal{T}}} = V(\rho) V + \rho \nabla . V V + \rho \nabla_V V$$

but $(\nabla_{X_a}(\rho V))(e^a) = V(\rho) + \rho \nabla . V$, hence

$$\begin{aligned} \nabla . \widetilde{\widetilde{\mathcal{T}}} &= \nabla .(\rho V) V + \rho \nabla_V V = -\delta(\rho \widetilde{V}) V + \rho \nabla_V V \\ &= -(*\mathrm{d}j) V + \rho \nabla_V V. \end{aligned}$$

Thus for $\widetilde{\widetilde{\mathcal{T}}}$ to be divergenceless the acceleration of V must be proportional to V. But if V is timelike with constant norm its acceleration is orthogonal to itself. So the divergence of $\widetilde{\widetilde{\mathcal{T}}}$ is zero if and only if $\mathrm{d}j = 0$ and V is a geodesic vector field, $\nabla_V V = 0$.

Electrically charged fluid stress

Suppose that each integral curve of V models the world line of an electrically charged fluid element. Let the charge density ρ_e of the fluid be $(e/m)\rho$. Thus each world line may be taken to correspond to a point particle with electric charge e and mass m. The gravitational field equations are the Maxwell–Einstein equations where the Maxwell equations have as 3-form current source

$$J_e = *(\rho_e \widetilde{V}). \tag{7.5.21}$$

The symmetric stress tensor for the system of electromagnetic fields and fluid is

$$\mathcal{T} = \mathcal{T}_{(\mathrm{M})} + \rho \widetilde{V} \otimes \widetilde{V}$$

where $\mathcal{T}_{(\mathrm{M})}$ is the Maxwell stress tensor. If $\mathrm{d}{*F} = J_e$ then from (7.5.9) $\mathrm{D}\tau_{(\mathrm{M})a} = F \wedge \mathrm{i}_{X_a} J_e$. Since $\mathcal{T}$ enjoys similar properties to the Einstein tensor $\mathcal{G}$, an argument analagous to that leading to (4.8.9) shows that $\nabla . \mathcal{T} = *^{-1} \mathrm{D}\tau_a e^a$. So for any X, $\nabla . \mathcal{T}_{(\mathrm{M})}(X) = *^{-1}(F \wedge \mathrm{i}_X J_e)$. Repeatedly using (1.4.7) with $** = -\eta$:

$$\begin{aligned} F \wedge \mathrm{i}_X J_e &= F \wedge \mathrm{i}_X {**} J_e = F \wedge *({*J_e} \wedge \widetilde{X}) \\ &= ({*J_e} \wedge \widetilde{X}) \wedge {*F} = {*F} \wedge {*J_e} \wedge \widetilde{X}. \end{aligned}$$

so that

$$\begin{aligned} *(F \wedge \mathrm{i}_X J_e) &= *({*F} \wedge {*J_e} \wedge \widetilde{X}) = \mathrm{i}_X *({*F} \wedge {*J_e}) \\ &= \mathrm{i}_X \mathrm{i}_{\widetilde{*J_e}} {**F} = -\mathrm{i}_X \mathrm{i}_{\widetilde{*J_e}} F \end{aligned}$$

and

$$\nabla . \mathcal{T}_{(\mathrm{M})}(X) = \mathrm{i}_X \mathrm{i}_{\widetilde{*J_e}} F = \mathrm{i}_{\widetilde{*J_e}} F(X).$$

Thus $\nabla . \mathcal{T}_{(\mathrm{M})} = \mathrm{i}_{\widetilde{*J_e}} F = (e\rho/m) \mathrm{i}_V F$, by (7.5.21), so

$$\widetilde{\nabla . \mathcal{T}} = \frac{e\rho}{m} \widetilde{\mathrm{i}_V F} + \rho \nabla_V V - ({*\mathrm{d}j}) V. \tag{7.5.22}$$

As we noted before, if V is of constant norm then its acceleration is orthogonal to itself, and $\widetilde{\mathrm{i}_V F}(V) = \mathrm{i}_V \mathrm{i}_V F = 0$. Thus by equating to zero the components of $\widetilde{\nabla . \mathcal{T}}$ parallel and orthogonal to V we see that $\mathcal{T}$ is divergenceless if and only if the particle number is conserved,

$$\mathrm{d}j = 0$$

and

$$\nabla_V V = -\frac{e}{m} \widetilde{\mathrm{i}_V F}. \tag{7.5.23}$$

We recognise this as the Lorentz force law equation for charged world lines.

7.6 The Reissner–Nordström Solution

In principle one can take an assumed form of metric and matter fields, parametrised by a set of functions, and compute the Einstein and stress tensors to obtain equations for the unknown functions. The resulting equations will be non-linear coupled partial differential equations. If the assumed form of solution is not appropriately parametrised then these differential equations will not admit a solution, whilst usually a very general form of trial solution merely results in intractable equations. Thus, in practice, such a 'brute force' approach is somewhat limited in obtaining physically interesting solutions to Einstein's equations: the generation of such solutions being a specialised pursuit.

The imposition of symmetries on the fields is one obvious way of restricting the number of free parameters. We here consider a static spherically symmetric metric. A metric is *stationary* if it admits a timelike Killing vector. If, in addition, this Killing vector is orthogonal to a family of spacelike hypersurfaces then the metric is called *static*. We consider a metric tensor that can be written in a local polar spacetime chart (t, r, θ, φ) as

$$g = -H_0(r)^2 \mathrm{d}t\otimes\mathrm{d}t + H_1(r)^2\mathrm{d}r\otimes\mathrm{d}r + r^2\mathrm{d}\theta\otimes\mathrm{d}\theta + r^2\sin^2\theta\mathrm{d}\varphi\otimes\mathrm{d}\varphi. \tag{7.6.1}$$

The chart is specified by $\{0 \leqslant \theta < \pi,\ 0 \leqslant \varphi < 2\pi,\ 0 < t < \infty\}$ and r is bounded to keep H_0 and H_1 real. This metric is invariant under an SO(3) group of transformations generated by the rotational Killing vectors given in (5.4.7). It is also static since $\mathcal{L}_{(\partial/\partial t)}g = 0$ and $(\partial/\partial t)$ is orthogonal to the hypersurfaces with t = constant. As we pointed out in Chapter 6 it is convenient to choose an orthonormal co-frame in which to compute the connection forms. Choosing the local co-frame:

$$\{e^0 = H_0\mathrm{d}t,\ e^1 = H_1\mathrm{d}r,\ e^2 = r\mathrm{d}\theta,\ e^3 = r\sin\theta\,\mathrm{d}\varphi\}$$

one computes the non-vanishing connection forms

$$\begin{aligned}
\omega_{01} &= -\omega_{10} = -\frac{H_0'}{H_0H_1}e^0\\
\omega_{12} &= -\omega_{21} = -\frac{1}{rH_1}e^2\\
\omega_{13} &= -\omega_{31} = -\frac{1}{rH_1}e^3\\
\omega_{23} &= -\omega_{32} = -\frac{\cot\theta}{r}e^3.
\end{aligned}$$

(The co-frames here are a special case of those used to compute the connection forms given in table 6.1.) The curvature forms now follow from the definition (6.4.12):

$$R_{23} = \frac{1}{r^2}\left(1 - \frac{1}{H_1^2}\right)e^{23}$$

$$R_{01} = \left(\frac{H_0'}{H_1}\right)'\frac{1}{H_0H_1}e^{01}$$

$$R_{31} = \frac{1}{rH_1}\left(\frac{1}{H_1}\right)' e^{13}$$

$$R_{02} = \frac{H_0'}{rH_1^2 H_0} e^{02}$$

$$R_{12} = -\frac{1}{rH_1}\left(\frac{1}{H_1}\right)' e^{12}$$

$$R_{03} = \frac{H_0'}{rH_1^2 H_0} e^{03}.$$

Taking $*1 = e^{0123}$ the Einstein forms are calculated:

$$G^0 = -2R_{12} \wedge e^3 - 2R_{23} \wedge e^1 - 2R_{31} \wedge e^2$$

$$= 2\left[\frac{2}{rH_1}\left(\frac{1}{H_1}\right)' - \frac{1}{r^2} + \frac{1}{r^2 H_1^2}\right] e^{123}$$

$$G^1 = 2\left(2\frac{H_0'}{rH_1^2 H_0} - \frac{1}{r^2} + \frac{1}{r^2 H_1^2}\right) e^{023}$$

$$G^2 = -2J(r) e^{013}$$

$$G^3 = 2J(r) e^{012}$$

where

$$J(r) = \left(\frac{H_0'}{H_1}\right)' \frac{1}{H_0 H_1} + \frac{H_0'}{rH_1^2 H_0} + \frac{1}{rH_1}\left(\frac{1}{H_1}\right)'.$$

The vaccuum equations $G^a = 0$ are now all satisfied by

$$H_0 = \frac{1}{H_1} = \left(1 + \frac{\mu}{r}\right)^{1/2}$$

for some constant μ. This solution has the property that for large r the metric looks like the metric of Minkowski spacetime.

To illustrate the effect of the electromagnetic field on the geometry of spacetime consider a spherically symmetric static Einstein–Maxwell system. In the above chart we choose a gauge in which $A = f(r)\mathrm{d}t$, ensuring that $\mathscr{L}_{K_j} F = 0$ for $F = \mathrm{d}A$ and K_j any Killing vector of the spherically symmetric static metric. The Maxwell 2-form is $F = L(r) e^1 \wedge e^0$ where $L(r) = f'/(H_0 H_1)$. Integrating the differential equations $\mathrm{d}*F = 0$ gives $Lr^2 = q$ for some constant q. From (7.5.9) the Maxwell stress forms follow simply

$$\tau^0 = \frac{q^2}{2r^4} e^{123}, \ \tau^1 = \frac{q^2}{2r^4} e^{023}, \ \tau^2 = \frac{q^2}{2r^4} e^{013}, \ \tau^3 = -\frac{q^2}{2r^4} e^{012}.$$

The presence of the stress modifies the equations above to

$$\kappa\left[\frac{2}{rH_1}\left(\frac{1}{H_1}\right)' - \frac{1}{r^2} + \frac{1}{r^2H_1^2}\right] + \frac{q^2}{4r^4} = 0$$

$$\kappa\left(2\frac{H_0'}{rH_1^2H_0} - \frac{1}{r^2} + \frac{1}{r^2H_1^2}\right) + \frac{q^2}{4r^4} = 0$$

$$\kappa J(r) - \frac{q^2}{4r^4} = 0.$$

These equations are all satisfied by

$$H_0 = \frac{1}{H_1} = \left(1 + \frac{\mu}{r} + \frac{q^2}{4\kappa r^2}\right)^{1/2}. \tag{7.6.2}$$

The electromagnetic 2-form field is $F = (q/r^2)e^1 \wedge e^0$, so we may interpret this solution as the gravitational field of a spherically symmetric static electrically charged source. It is known as the Reissner–Nordström solution.

In the above solution we have two arbitrary constants μ and q. The latter we have identified with a source of electric charge. The former may be identified with a Newtonian gravitational mass. However, classical gravitation is observed to give rise always to an attractive interaction between macroscopic masses. This feature implies that μ should be chosen to be a negative constant. The examples below are intended to convince the reader of this identification.

Exercise 7.4

Consider the geodesic motion of an uncharged test particle in a spacetime metric described by the local orthonormal co-frame

$$\{e^0 = F\mathrm{d}x^0,\ e^k = F^{-1}\mathrm{d}x^k \qquad k = 1, 2, 3\}$$

with F a function of the three spatial coordinates. Show that the geodesic

$$C : I \to M,\ \tau \longmapsto (x^0(\tau), x^k(\tau))$$

is determined by

$$\ddot{x}^0 + 2\dot{x}^0 F^{-1}\dot{C}(F) = 0$$

$$\ddot{x}^i + [(\dot{x}^0)^2F^3 + \dot{x}^j\dot{x}_jF^{-1}]\partial_iF - 2\dot{x}^iF^{-1}\dot{C}(F) = 0.$$

For $\dot{C}$ timelike choose a proper-time parametrisation to replace these with

$$g(\dot{C}, \dot{C}) = -1$$

$$\ddot{x}^i + \tfrac{1}{2}\partial_iF^2 + 2F^{-1}(\dot{x}^j\dot{x}_j\partial_i - \dot{x}^i\dot{x}^j\partial_j)F = 0.$$

If now $|\dot{x}^i| \ll 1$ and $F^2 = 1 - h$ with $h \ll 1$ then these approximate to

$$\ddot{x}^i = \tfrac{1}{2}\partial_i h.$$

By comparing with Newton's law of motion for a slowly moving particle in a Newtonian gravitational potential Φ, make the weak field identification

$$\Phi = -h/2.$$

Exercise 7.5
In the above metric (7.6.1), set $q = 0$ and make the coordinate transformation

$$r = R - \frac{\mu}{2} + \frac{\mu^2}{16R}$$

to write it in the *isotropic* form

$$g = -\left(\frac{4R + \mu}{4R - \mu}\right)^2 \mathrm{d}t\otimes\mathrm{d}t + \left(1 - \frac{\mu}{4R}\right)^4 (\mathrm{d}R\otimes\mathrm{d}R + R^2\mathrm{d}\theta\otimes\mathrm{d}\theta + R^2\sin^2\theta\mathrm{d}\varphi\otimes\mathrm{d}\varphi).$$

In a region where $\mu \ll 4R$ this is of the type considered in exercise 7.4, (change from standard R^3 polar to R^3 Cartesian coordinates.)

Recall that for a point source of Newtonian gravity due to a mass M, the potential $\Phi = -GM/r$ where G is the Newtonian gravitational coupling constant. Hence from $h = GM/r$ identify the constant in the Schwarzschild solution; $\mu = -2GM$.

Exercise 7.6
In the metric in exercise 7.4 above verify that for $h \ll 1$, $G^0 = -2(\partial_k\partial^k h)e^1 \wedge e^2 \wedge e^3$. For an ideal fluid of density ρ show that $\tau^0 = \rho e^1 \wedge e^2 \wedge e^3$ in the frame $\{X_a\}$ in which its velocity $V = X_0$. Hence use the Newtonian Poisson equation $\nabla^2\varphi = 4\pi G\rho$ to relate our κ to the Newtonian coupling G by

$$\kappa = \frac{1}{16\pi G}.$$

Exercise 7.7
Use the result of exercise 7.1 to rewrite Einstein's equations in the form

$$P_c - \tfrac{1}{2}e_c\mathcal{R} = 8\pi G *^{-1}\tau_c.$$

In the absence of the electromagnetic field ($q = 0$) the Reissner–Nordström metric reduces to the *Schwarzschild metric*. That is, we have a vacuum spacetime with metric

$$g = -\left(1 - \frac{2M}{r}\right)\mathrm{d}t\otimes\mathrm{d}t + \left(1 - \frac{2M}{r}\right)^{-1}\mathrm{d}r\otimes\mathrm{d}r + r^2(\mathrm{d}\theta\otimes\mathrm{d}\theta + \sin^2\theta\mathrm{d}\varphi\otimes\mathrm{d}\varphi) \quad (7.6.3)$$

where the coordinate r is restricted to be greater than $2M$. Some properties of this spacetime can be understood by looking at the behaviour of local light cones in this chart, where for fixed (r, t) we have a standard 2-sphere. The tangent vector $p(\partial/\partial t) + q(\partial/\partial r)$ has norm squared $(1 - 2M/r)^{-1}q^2 - (1 - 2M/r)p^2$ and is therefore timelike if

$$\left|\frac{q}{p}\right| < 1 - \frac{2M}{r}.$$

The local directions determined by all such tangent vectors lie in the local light cones attached to each point on the 2-sphere at (r, t). These light cones appear to close as the coordinate r approaches $2M$. Thus any incoming timelike or null curve will asymptote to $r = 2M$ in the (r, t) chart. On the other hand, if one calculates the scalar curvature near $r = 2M$ it appears well behaved, suggesting that the Schwarzchild coordinates may cover only part of some Lorentzian manifold. If we introduce the Eddington–Finkelstein coordinates (T, r', θ, φ) where $T = t + r + 2M\log(r - 2M)$ and $r' = r$ then it is straightforward to compute $\mathrm{d}T$ in terms of $\mathrm{d}t$ and $\mathrm{d}r$ and write the above metric in these coordinates as

$$g = -\left(1 - \frac{2M}{r'}\right)\mathrm{d}T\otimes\mathrm{d}T + \mathrm{d}T\otimes\mathrm{d}r' + \mathrm{d}r'\otimes\mathrm{d}T + r'^2(\mathrm{d}\theta\otimes\mathrm{d}\theta + \sin^2\theta\,\mathrm{d}\varphi\otimes\mathrm{d}\varphi). \quad (7.6.4)$$

The region of spacetime covered by $r \in (2M, \infty)$ $t \in (-\infty, \infty)$ is now covered by r' and T ranging over the same values. There now appears no reason to restrict r' to be less than $2M$. Thus we may regard the original coordinates as describing only part of a Lorentzian manifold, the whole of which is covered by the new coordinates with $T > 0$. Looking now in the (r', T) plane at the *forward* light cones for $r' < 2M$, in which lie the future directed timelike curves, a dramatic result is evident. No future-directed timelike (or null) curve from $r' < 2M$ ever reaches the region of spacetime with $r' > 2M$: all such curves are eventually focused to $r' = 0$. Thus there exists a *horizon* at $r' = 2M$, no causal information of any kind being received by an observer outside the horizon from points within. Furthermore, all incoming timelike curves that enter the horizon eventually (in a finite proper time) strike the line $r' = 0$ where the curvature tensor becomes unbounded. Such events do not belong to a Lorentzian manifold and prohibit any further extensions of the spacetime.

For a spherically symmetric star of mass M and radius parameter $r > M$ the Schwarzschild metric describes the unique spacetime in the vacuum exterior to the star. The spacetime inside the star will depend on its matter stresses. A star unfortunate enough to evolve to a radius parameter less than $2M$ is predicted to find all its atoms on doomed world lines and undergoes catastrophic gravitational collapse. (For an object whose Newtonian mass is n times the mass of the sun this radius is about $3n$ km.) One of the most celebrated theorems in the theory of gravitation asserts that under a number of reasonable assumptions such a phenomenon is not restricted to the idealised spherically symmetric metric discussed here. The physics of the collapse of matter to a singular state is one of the great challenges of contemporary research.

Further details of the Schwarzschild geometry can be found in, for example, Hawking and Ellis [12] and Misner, Thorne and Wheeler [13]. These books give a more complete account of the possible extensions to the exterior Schwarzschild solution.

7.7 Gravitation with Torsion

Einstein's theory of gravitation is written in terms of a metric-compatible torsion-free connection. There have been many attempts to generalise these equations. One direction is to maintain their form but to relax the requirement that the connection has zero torsion. One must then supplement them with further equations that determine the torsion tensor. They may be regarded as geometrical descriptions of interactions that depend on tensor (and spinor) fields other than the metric. One may also contemplate gravitational theories in which the metric compatibility of the connection is relaxed although such approaches have attracted little attention so far. Needless to say the adoption of a particular connection for the geometrical description of physical phenomena depends on the physics of the situation. Sometimes (as in the case of theories with supergravity) a connection with a torsion determined by a spinor field equation provides an elegant formulation of a theory. Rewriting the theory in terms of the Levi–Civita connection is always possible, but possibly at a cost of algebraic complexity.

As a simple example of a model written in terms of a metric-compatible connection with torsion, consider a self-interacting real scalar field α coupled to gravity according to the field equations [14]

$$\tfrac{1}{2}\alpha^2 G^a = -\tau^a[\alpha] + \lambda\alpha^4 {*e^a} \qquad (7.7.1)$$

$$c\,\mathrm{d}{*\mathrm{d}}\alpha^2 = 2\lambda\alpha^3 {*1} \qquad (7.7.2)$$

$$T^a = e^a \wedge \frac{d\alpha}{\alpha} \tag{7.7.3}$$

with

$$\tau^a = \tfrac{1}{2}c(i^a d\alpha \wedge *d\alpha + d\alpha \wedge i^a * d\alpha). \tag{7.7.4}$$

The non-vanishing real parameters λ and c are coupling constants. (For $\lambda = 0$ this model is equivalent to a theory of gravitation proposed by Brans and Dicke REF [15].) The equation (7.7.3) involving the torsion may be solved for the connection forms (6.6.8):

$$\omega_{ab} = \Omega_{ab} + \left(\frac{i_b d\alpha}{\alpha}\right)e_a - \left(\frac{i_a d\alpha}{\alpha}\right)e_b \tag{7.7.5}$$

in terms of the torsion-free connection forms Ω_{ab}. It is an interesting exercise to rewrite the above system of equations in terms of the Einstein forms associated with the torsion-free connection. In such a reformulation the torsional effects due to the scalar field coupling to gravity may be interpreted as an additional contribution to the stress forms. In addition c becomes replaced by $c - 6$.

Bibliography

Adler R, Bazin M and Schiffer M 1975 *Introduction to General Relativity* (New York: McGraw-Hill)

O'Niel B 1983 *Semi-Riemannian Geometry with Applications in Physics* (New York: Academic)

Thorpe J A 1975 *Proc. Symp. in Pure Mathematics* vol XXVII, p425

8

Clifford Calculus on Manifolds

The first three chapters of this book are purely algebraic. They deal with tensor, exterior and Clifford algebras of an arbitrary vector space. In the following chapters when dealing with manifolds, and applications in physics, we have assimilated the material of Chapter 1 by taking that vector space to be the cotangent space. We shall now similarly incorporate Chapter 2.

In Chapter 2 we identified the Clifford algebra with the vector space of exterior forms with the product given in (2.1.7). Hence on a pseudo-Riemannian manifold M we have the structure of a Clifford algebra on each fibre of the exterior bundle. The exterior bundle equipped with this multiplication in the fibres will be called the Clifford bundle $\mathbf{C}(M)$. The situation is that we have a vector bundle with two different rules for turning it into an algebra bundle; so we shall freely interchange the terms Clifford bundle and exterior bundle (for a pseudo-Riemannian manifold) depending on which aspect we wish to emphasise. Similarly we may sometimes refer to 'Clifford forms' to emphasise that we are thinking of the differential forms as elements of a Clifford rather than exterior algebra.

Just as one can develop an efficient exterior calculus of differential forms with the exterior derivative (and more generally the covariant exterior derivative) and Hodge dual, one can efficiently calculate using the covariant derivative ∇ and Clifford multiplication (equation (2.1.19) relating the Hodge dual to Clifford multiplication). Unlike the exterior algebra the Clifford algebra is not Z-graded. So Clifford multiplication of differential forms will naturally involve us with inhomogeneous differential forms; that is, sums of differential forms of different degrees. Certain equations involving forms of differing degrees can be conveniently expressed in terms of Clifford products.

The utility of being able to Clifford multiply differential forms really becomes apparent when we come to spinor fields (these carrying representations of the Clifford—as opposed to exterior—algebra). An inspection of many calculations involving spinors in theoretical physics reveals that often the components of a vector (or co-vector) are saturated with a set of γ-matrices that generate a Clifford algebra. (Indeed there is even a special notation for such objects!) It is conceptually, as well as notationally, simpler to work directly with the Clifford algebra of differential forms.

In this chapter we shall frequently use the notation, and results, of Chapter 2. In particular we shall juxtapose differential forms to denote their Clifford product.

8.1 Covariant Differentiation of Clifford Products

If Φ is an arbitrary inhomogeneous differential form and A an arbitrary 1-form on a pseudo-Riemannian manifold M then (2.1.7) gives

$$A\Phi = A \wedge \Phi + i_{\tilde{A}}\Phi \, .$$

If ∇ is the pseudo-Riemannian connection then $\nabla_X(i_{\tilde{A}}\Phi) = i_{\nabla_X\tilde{A}}\Phi + i_{\tilde{A}}\nabla_X\Phi$, since ∇_X commutes with contractions, and $\nabla_X\tilde{A} = \widetilde{\nabla_X A}$ since ∇ is metric compatible. Hence

$$\nabla_X(A\Phi) = \nabla_X A\Phi + A\nabla_X\Phi \tag{8.1.1}$$

and it follows that ∇_X is a derivation on Clifford products. (This does not require zero torsion.) Adding and subtracting equations (2.1.7) and (2.1.8) gives us relations that permit $A \wedge \Phi$ and $i_{\tilde{A}}\Phi$ to be expressed in terms of Clifford products:

$$A\Phi + \Phi^{\eta}A = 2A \wedge \Phi \tag{8.1.2}$$

$$A\Phi - \Phi^{\eta}A = 2i_{\tilde{A}}\Phi \, . \tag{8.1.3}$$

For $\{e^a\}$ a local orthonormal co-frame we denote $e^a \wedge e^b$ by e^{ab}. Then (8.1.3) gives

$$[e^{bc}, e^a] = 2(\eta^{ac}e^b - \eta^{ab}e^c) \tag{8.1.4}$$

where the left-hand side is a Clifford commutator and η^{ab} are the orthonormal components of the metric. So if we use the connection 1-forms to introduce the 2-form

$$\sigma_X \equiv \tfrac{1}{4}\omega_{bc}(X)e^{bc} = \tfrac{1}{4}\nabla_X e^a \wedge e_a \tag{8.1.5}$$

we can write (6.3.3) as

$$\nabla_X e^a = [\sigma_X, e^a] \, . \tag{8.1.6}$$

If we introduce an orthonormal multibasis $\{e^I\}$ for $\Gamma\Lambda M$ then, since an exterior product of mutually orthogonal 1-forms is the same as a Clifford product

$$\nabla_X e^I = [\sigma_X, e^I] . \tag{8.1.7}$$

If we expand an arbitrary differential form as $\Phi = \Phi_I e^I$ then

$$\nabla_X \Phi = (X\Phi_I)e^I + [\sigma_X, \Phi] . \tag{8.1.8}$$

If S is any invertible element of the Clifford algebra and $E^a \equiv Se^aS^{-1}$ then it follows from (8.1.6) that $\nabla_X E^a = [\Sigma_X, E^a]$ with $\Sigma_X = S\sigma_X S^{-1} + \nabla_X SS^{-1}$. If $s \in {}_{\pm}\Gamma^{\pm}$ then $\{e^{a\prime} \equiv se^as^{-1}\}$ is another orthonormal frame. If σ'_X denotes the expression in (8.1.5) computed with the connection forms in this new basis then

$$\sigma'_X = s\sigma_X s^{-1} + \nabla_X ss^{-1}. \tag{8.1.9}$$

Certainly the two sides of this expression can only differ by an element of the centre. Since σ_X is a 2-form and $s \in {}_{\pm}\Gamma^{\pm}$ then $s\sigma_X s^{-1}$ is a 2-form and we need only check that $\nabla_X ss^{-1}$ is a 2-form. For $s \in {}_{\pm}\Gamma^{\pm}$ we can write $s = x^1x^2 \ldots x^h$ where the x^i are 1-forms such that $(x^i)^2 = \pm 1$, then

$$\begin{aligned}\nabla_X ss^{-1} &= (\nabla_X x^1x^2 \ldots x^h + x^1\nabla_X x^2 \ldots x^h + \ldots \\ &\quad + x^1 \ldots x^{h-1}\nabla_X x^h)[(x^h)^{-1} \ldots (x^2)^{-1}(x^1)^{-1})] \\ &= \nabla_X x^1(x^1)^{-1} + x^1[\nabla_X x^2(x^2)^{-1}](x^1)^{-1} + \ldots \\ &\quad + x^1 \ldots x^{h-1}[\nabla_X x^h(x^h)^{-1}](x^1 \ldots x^{h-1})^{-1} .\end{aligned}$$

Since $(x^i)^2$ is a constant $\nabla_X x^i$ anticommutes with x^i and hence with $(x^i)^{-1} = x^i/(x^i)^2$. So $\nabla_X x^i(x^i)^{-1} = \frac{1}{2}(\nabla_X x^i(x^i)^{-1} - (x^i)^{-1}\nabla_X x^i) = \nabla_X x^i \wedge (x^i)^{-1}$. It follows that $\nabla_X ss^{-1}$ is a 2-form.

If $\{e^I\}$ is an orthonormal multibasis for $\Gamma\Lambda M$ then differentiating (8.1.7) expresses the curvature operator as $\boldsymbol{R}(X, Y)e^I = [\mathcal{R}_{XY}, e^I]$ for

$$\mathcal{R}_{XY} = \nabla_X\sigma_Y - \nabla_Y\sigma_X - [\sigma_X, \sigma_Y] - \sigma_{[X, Y]} . \tag{8.1.10}$$

Since the curvature operator is $\mathcal{F}$-linear then for any $\Phi \in \Gamma\Lambda M$

$$\boldsymbol{R}(X, Y)\Phi = [\mathcal{R}_{XY}, \Phi] . \tag{8.1.11}$$

It can be verified that $\mathcal{R}_{XY}$ is unchanged if $\sigma_X \longmapsto S\sigma_X S^{-1} + \nabla_X SS^{-1}$ for any invertible S. The forms $\mathcal{R}_{XY}$ are certainly related to the curvature 2-forms R_{ab}; we now establish the exact relationship. Differentiating (8.1.5) and using (8.1.7) gives $\nabla_X\sigma_Y = \frac{1}{4}X(\omega_{bc}(Y))e^{bc} + [\sigma_X, \sigma_Y]$, and hence

$$\mathcal{R}_{XY} = \tfrac{1}{4}\{X(\omega_{bc}(Y)) - Y(\omega_{bc}(X)) - \omega_{bc}([X, Y])\}e^{bc} + [\sigma_X, \sigma_Y] .$$

Referring to (4.10.3) we can simplify the first three terms: $\mathcal{R}_{XY} = \frac{1}{2}d\omega_{bc}(X, Y)e^{bc} + [\sigma_X, \sigma_Y]$. To recognise the last term we will use the following useful relation:

$$[e^{ab}, e^{cd}] = 2\eta^{bd}e^{ca} - 2\eta^{ad}e^{cb} + 2\eta^{bc}e^{ad} - 2\eta^{ac}e^{bd}. \qquad (8.1.12)$$

We can use this and the antisymmetry of the connection forms, to write

$$[\sigma_X, \sigma_Y] = \tfrac{1}{4}(\omega_{ab}(X)\omega^b{}_c(Y) - \omega_{ab}(Y)\omega^b{}_c(X))e^{ac}$$
$$= \tfrac{1}{2}(\omega_{ba} \wedge \omega^a{}_c)(X, Y)e^{bc}.$$

So we have

$$\mathcal{R}_{XY} = \tfrac{1}{2}R_{bc}(X, Y)e^{bc} = -\tfrac{1}{4}\mathrm{i}_X\mathrm{i}_Y R_{bc}e^{bc}\ . \qquad (8.1.13)$$

This can be rewritten, using the 'pairwise symmetric' Bianchi identity for zero torsion (6.7.16), as

$$\mathcal{R}_{XY} = \tfrac{1}{2}e^a(X)e^b(Y)R_{ab}\ . \qquad (8.1.14)$$

Exercise 8.1
Use (2.1.7) and (2.1.8) to show that (for zero torsion):

$$R^a{}_b e^b = P^a \qquad (8.1.15)$$

$$P_a e^a = \mathcal{R} \qquad (8.1.16)$$

$$R_{ab}e^{ba} = \mathcal{R}. \qquad (8.1.17)$$

8.2 The operator $\not{d}$

Many equations in physics can be elegantly formulated in terms of the exterior derivative d and the co-derivative δ. In Chapter 6 we showed how these operators could be expressed in terms of the pseudo-Riemannian connection. We now define an operator $\not{d}$ on $\Gamma\Lambda M$ by

$$\not{d} \equiv e^a\nabla_{X_a}\ . \qquad (8.2.1)$$

Thus from (6.7.4) and (6.9.1) we have

$$\not{d} = \mathrm{d} - \delta \qquad (8.2.2)$$

with δ defined in (5.4.2). The operator $\not{d}$ is sometimes called the Hodge de-Rham operator. Unlike d and δ separately, $\not{d}$ is not a homogeneous operator on differential forms; whereas d increases the degree of a form by one, δ decreases the degree by one. The square of $\not{d}$ is homogeneous for since d and δ are nilpotent

$$\not{d}^2 = \Delta \qquad (8.2.3)$$

where Δ is the Laplace–Beltrami operator of (5.4.5).

We can trivially rewrite the pair of Maxwell equations $\mathrm{d} * F = J$, $\mathrm{d}F = 0$ as

$$\not{d}F = j \tag{8.2.4}$$

where $j = -*^{-1}J$. As an example of manipulating Clifford expressions we now re-express the Maxwell stress tensor in terms of Clifford products and evaluate its divergence. The stress tensor is related to the stress forms by $\mathcal{T} = *^{-1}\tau_a \otimes e^a = *^{-1}\tau_a(X_b)e^b \otimes e^a$. For a four-dimensional Lorentzian spacetime $** = -\eta$, and the stress tensor components are $\mathcal{T}_{ba} = \mathrm{i}_b * \tau_a$. From (7.5.9)

$$2\tau_a = \mathrm{i}_a F \wedge *F - \mathrm{i}_a *F \wedge F \,.$$

First we use (8.1.2) to exchange the exterior products for Clifford products:

$$4\tau_a = \mathrm{i}_a F *F + *F\mathrm{i}_a F - \mathrm{i}_a *FF - F\mathrm{i}_a *F.$$

Now we use (8.1.3)

$$8\tau_a = (e_a F - Fe_a)*F + *F(e_a F - Fe_a) - (e_a *F - *Fe_a)F$$
$$- F(e_a *F - *Fe_a) \,.$$

Finally we use (2.1.19) to write the Hodge dual in terms of the volume 4-form z:

$$\tau_a = \tfrac{1}{2}Fe_a Fz \,.$$

We have used $F^\xi = -F$ since F is a 2-form and $z\Phi = \Phi^\eta z$. Once again we use (8.1.3) to obtain the stress tensor components

$$\mathcal{T}_{ba} = \tfrac{1}{4}(Fe_a Fe_b + e_b Fe_a F) \,. \tag{8.2.5}$$

When covariantly differentiating the stress tensor the derivatives of the co-frames in the above components will cancel the derivatives of the tensor basis, hence

$$(\nabla\cdot\mathcal{T})_a = \tfrac{1}{4}(\nabla_{X_c} Fe_a Fe^c + Fe_a \nabla_{X_c} Fe^c + e^c \nabla_{X_c} Fe_a F + e^c Fe_a \nabla_{X_c} F) \,.$$

We want to use the Maxwell equations (8.2.4) to simplify this, but the terms $\nabla_{X_c} F$ and e^c do not all occur in the right order to write them as $\not{d}$. The above expression is certainly a 0-form, so by applying the homogeneous projector (cf (2.1.12)) $\mathcal{S}_0$ we do nothing. Under this projector, factors in the Clifford product can be cyclically permuted (2.1.17). (We cannot, of course, then remove the projector.) So we have

$$(\nabla\cdot\mathcal{T})_a = \tfrac{1}{2}\mathcal{S}_0(\not{d}Fe_a F + \nabla_{X_c} Fe^c Fe_a) \,.$$

Since $F^\xi = -F$, then $\nabla_{X_c} Fe^c = -(\not{d}F)^\xi$. We can insert this in the above and then use $\mathcal{S}_0\Phi = \mathcal{S}_0\Phi^\xi$ to obtain $(\nabla\cdot\mathcal{T})_a = \mathcal{S}_0(\not{d}Fe_a F)$. We can now use the Maxwell equations (8.2.4):

$$\nabla \cdot \mathcal{T} = \mathcal{S}_0(Fje_a)e^a = \mathcal{S}_1(Fj)$$

using (2.1.18). Since j is a 1-form and F is a 2-form then $Fj = j \wedge F - \mathrm{i}_{\tilde{j}} F$ and so finally

$$\nabla \cdot \mathcal{T} = -\mathrm{i}_{\tilde{j}} F \ . \tag{8.2.6}$$

(We earlier obtained this result in the discussion of the electrically charged fluid stress in Chapter 7.) We have somewhat laboured the above calculation in order to illustrate some of the techniques that are useful in practice and to show how one can always interchange any exterior expression for a Clifford one and vice versa.

8.3 The Kahler Equation

In 1928 Darwin [16] was experimenting with tensor equations in order to understand the properties of electrons. He eventually made contact with Dirac's spinor wave equation (to be discussed later) but considered his method uneconomical. Apparently Landau and Ivanenko [17] had similar intentions around the same time. These were perhaps precursors of the equation introduced in 1961 by Kahler [18] for a complex inhomogeneous differential form Φ on a pseudo-Riemannian manifold:

$$\not{d}\Phi = \mu\Phi - \mathrm{i}A\Phi \ . \tag{8.3.1}$$

The term involving A describes the electromagnetic coupling to the Maxwell field $F = \mathrm{d}A$. He was apparently motivated to develop a 'calculus of infinitesimals' in which relations of the form $\mathrm{d}x^\mu \wedge \mathrm{d}x^\mu = 0$ and $\mathrm{d}x^\mu \vee \mathrm{d}x^\nu + \mathrm{d}x^\nu \vee \mathrm{d}x^\mu = 2g^{\mu\nu}$ could co-exist on a pseudo-Riemannian manifold. Kahler recovered Dirac's solution describing the wave mechanics of a relativistic electron of mass μ in a hydrogen atom when he analysed (8.3.1) in flat Minkowski spacetime.

It was a desire to find a first-order equation, such that the components satisfied the second-order Klein–Gordon equation, that motivated Dirac to formulate his celebrated equation in 1928 [19]. Because of (8.2.3), and since the Laplace–Beltrami operator is homogeneous, the p-form components $\mathcal{S}_p(\Phi)$ of an arbitrary solution to (8.3.1), in the absence of an electromagnetic field, satisfy

$$\Delta\mathcal{S}_p(\Phi) = \mu^2\mathcal{S}_p(\Phi) \ . \tag{8.3.2}$$

However, an arbitrary complex differential form on spacetime has sixteen complex components; whereas a spinor of the complexified

Clifford algebra has four complex components. Thus an arbitrary solution to (8.3.1) has more components than a solution to Dirac's equation. To understand the Kahler equation better, and its relationship to the Dirac equation, we examine the possibility of solutions lying in minimal left ideals—these carrying irreducible representations of the Clifford algebra. A set of four pairwise-orthogonal primitive idempotents may be used to project an arbitrary element of the Clifford algebra into minimal left ideals. In flat Minkowski space we can always choose inertial coordinates $\{x^a\}$ in which $e^a = \mathrm{d}x^a$, $a = 0, 1, 2, 3$ constitute an orthonormal basis. We can construct a set of globally defined primitive idempotents $\{P_i\}$ out of this parallel co-frame. The resulting idempotents will also be parallel, $\nabla_X P_i = 0 \quad \forall X \in \Gamma TM$. Thus if $\varphi_i \equiv \Phi P_i$ then φ_i is in a minimal left ideal. If Φ satisfies (8.3.1) then multiplying (8.3.1) on the right by P_i gives

$$\not{d}\varphi_i = \mu\varphi_i - \mathrm{i}A\varphi_i \qquad i = 1, 2, 3, 4 \tag{8.3.2}$$

since P_i is parallel. Thus Kahler's equation decouples into four equivalent equations for elements lying in minimal left ideals. (If Kahler's equation was written in exterior form then the coupled equations for the homogeneous p-forms would not be very transparent.)

A general solution of the Kahler equation has more degrees of freedom than a solution to the Dirac equation. This raises the question of the significance of (8.3.1) for the description of those particles in Nature (such as the electron–positron field) that are conventionally described by the Dirac equation. If one uses a spacetime 3+1 decomposition to perform a non-relativistic reduction then one obtains from (8.3.1) four copies of the Pauli–Schrödinger equation [20]. The wave mechanics of a particle described by such a system is indistinguishable from a non-relativistic description of an electron in an external electromagnetic field except in one respect: all single-particle (quantum) states have an extra fourfold degeneracy. For example, if a beam of such hypothetical particles was passed through an inhomogeneous static magnetic field (a Stern–Gerlach experiment) it would be split into two components. This is what happens with electrons on atoms in a real experiment. Furthermore, no electromagnetic field could be devised that would split the degeneracy of each beam. However, a (powerful) inhomogeneous gravitational field would in general break the degeneracy, producing four distinct beams in the field. Electrons described by the Dirac equation are not predicted to behave in this way. Although such an experiment has never been done with real electrons, our understanding of the periodic table of the elements is based on the Pauli principle for electrons with two internal states rather than four. Without a major reformulation of this principle it is difficult to reconcile our current understanding of the quantum mechanics of electrons with the four copies of the Pauli–Schrödinger equation obtained from (8.3.1). In an arbitrary curved spacetime (gravitational field) the Kahler equation will not decouple into four minimal left ideas (there will not be globally defined parallel primitives). Although the experimental significance of this is far from clear the fact that the degeneracy of the Minkowski space system can be broken would seem to lead to interpretational problems for the quantum theory.

Exercise 8.2

Define in the usual Minkowski spacetime polar chart (t, r, θ, φ) the local 1-forms

$$S_k^m = r^{1-k} \not{d}(r^k Y_k^m(\theta, \varphi)) = kY_k^m(\theta, \varphi) + r\mathrm{d}Y_k^m$$
$$k = 0, 1, 2 \ldots \qquad -k \leqslant m \leqslant k$$

in terms of standard spherical harmonics satisfying $\not{d}^2(r^k Y_k^m) = 0$. Verify that

$$\not{d} S_k^m = \frac{(1-k)}{r} \mathrm{d}r\, S_k^m$$

and that for any inhomogeneous differential form R independent of $\mathrm{d}t$:

$$\not{d}(RS_k^m) = \left(\not{d}R + R^{\eta\xi}\frac{1-k}{r}\,\mathrm{d}r\right)S_k^m .$$

Verify that a solution of Kahler's equation with a Coulomb 1-form potential $A = (e/r)\mathrm{d}t$ in this spacetime may be written

$$\Psi = \sum_{\varepsilon=\pm} \sum_{k} \sum_{m=-k}^{k} R_{km}^{\varepsilon}(r, \theta, \varphi) T^{\varepsilon}(t)$$

where $R_{km}^{\varepsilon} = \{f_k^{\varepsilon}(r) + g_k^{\varepsilon}(r)\mathrm{d}r\}S_k^m$ and $T^{\varepsilon}(t) = \exp(\mathrm{i}\omega^{\varepsilon} t)(1 + \mathrm{i}\varepsilon \mathrm{d}t)$ and for each ε, k the 0-forms f and g satisfy the ordinary differential equations:

$$f' + \frac{(1-k)}{r}f - \frac{e^2}{r}g + (\omega - \mu)g = 0$$
$$g' + \frac{(1+k)}{r}g + \frac{e^2}{r}f - (\omega + \mu)f = 0.$$

Exercise 8.3

The 1-form harmonics S_k^m may also be used to analyse Maxwell's equations $\not{d}F = 0$. First observe that the 1-forms $\alpha_{km}^{\varepsilon} = Z_k^{\varepsilon}(\lambda r)S_k^m$ obey $\not{d}^2\alpha = -\lambda^2\alpha$ and the 2-forms $\beta_{km}^{\varepsilon} = Z_k^{\varepsilon}(\lambda r)\mathrm{d}rS_k^m$ obey $\not{d}^2\beta = -\lambda^2\beta$, where Z_{-k}^{ε} label the independent Bessel solutions of the equation

$$\rho''(r) + \frac{2}{r}\rho'(r) + \left(\lambda^2 - \frac{(k^2-k)}{r^2}\right)\rho(r) = 0 \qquad \lambda \neq 0 .$$

Writing $F = E\mathrm{d}t + B$ with $\dot{E} = \mathrm{i}\omega E$ and $\dot{B} = \mathrm{i}\omega B$ write the harmonic component Maxwell equations as the complex pair:

$$\not{d}E = -\mathrm{i}\omega B$$
$$\not{d}B = -\mathrm{i}\omega E$$

and seek solutions of the form $E = \rho_k(r)S_k^m$ for some 0-forms ρ_k. Hence construct the multipole expansions:

$$E^I = \sum_{\varepsilon,k,m} \mathcal{S}_1(A^\varepsilon_{km}\Pi^\varepsilon_{km})\exp(\mathrm{i}\omega^\varepsilon t) \qquad B^I = \frac{\mathrm{i}}{\omega}\not{d}E^I$$

$$B^{II} = \sum_{\varepsilon,k,m} \mathcal{S}_2(B^\varepsilon_{km}\beta^\varepsilon_{km})\exp(\mathrm{i}\omega^\varepsilon t) \qquad E^{II} = \frac{\mathrm{i}}{\omega}\not{d}B^{II}$$

where

$$\Pi^\varepsilon_{km} = Z^\varepsilon_k(\omega r)\hat{*}1\mathrm{d}rS^m_k \qquad \omega \neq 0$$
$$\beta^\varepsilon_{km} = Z^\varepsilon_k(\omega r)\mathrm{d}rS^m_k \qquad \omega \neq 0$$

$\hat{*}1 = e^{123}$ and A^ε_{km}, B^ε_{km} are any complex constants.

Exercise 8.4

The stress tensor for the Einstein–Kahler coupled system (with $A = 0$) is

$$T = \tfrac{1}{4}\mathcal{S}_0(\Phi^{\xi\eta}e_a\nabla_{X_c}\Phi e^c e_b + \Phi^{\xi\eta}e_b\nabla_{X_c}\Phi e^c e_a)e^a \otimes e^b\ .$$

Verify that

$$\nabla\cdot T = 0.$$

Hint. Since the co-frames with contracted indices will not contribute to the divergence concentrate on the terms

$$\begin{aligned}4(\nabla\cdot T)_b = \mathcal{S}_0(&\nabla_{X_a}\Phi^{\xi\eta}e^a\nabla_{X_c}\Phi e^c e_b + \Phi^{\xi\eta}e^a\nabla_{X_a}\nabla_{X_c}\Phi e^c e_b\\ &+ \Phi^{\xi\eta}e^a\nabla_{X_c}\Phi\nabla_{X_a}e^c e_b + \Phi^{\xi\eta}e_b e_a\nabla_{X_a}\nabla_{X_c}\Phi e^c e^a\\ &+ \Phi^{\xi\eta}e_b\nabla_{X_c}\Phi\nabla_{X_a}e^c e^a).\end{aligned}$$

Note $\mathcal{S}_0(\nabla_{X_a}\Phi^{\xi\eta}e_b\nabla_{X_c}\Phi e^c e^a) = 0$ since $\mathcal{S}_0(\Psi^\xi) = \Psi$ for any Ψ. Using (8.3.1) and its iterate, $\Delta\Phi = \mu^2\Phi$, the above terms cancel with the aid of the relations

$$\mathrm{d}\Phi^\xi = -(\mathrm{d}\Phi)^{\xi\eta}$$
$$\delta\Phi^\xi = (\delta\Phi)^{\xi\eta}$$
$$\mathrm{d}\Phi^\eta = -(\mathrm{d}\Phi)^\eta$$
$$\delta\Phi^\eta = -(\delta\Phi)^\eta$$
$$\nabla_{X_a}\Phi e^a = -(\mathrm{d}\Phi + \delta\Phi)^\eta.$$

The last relation follows from (8.1.3).

8.4 The Duffin–Kemmer–Petiau Equations

After the success of the Dirac equation in describing the electron there were attempts made to find first-order equations suitable for describing integer spin particles. The Duffin–Kemmer–Petiau equations are an example [21].

The Kahler equation is not unique in being a first-order equation for an inhomogeneous differential form which iterates to the Laplace–Beltrami equation. For example, consider

$$\mathrm{d}\Phi_+ - \delta\Phi_- = \mu\Phi \tag{8.4.1}$$

where $\Phi_\pm \equiv \frac{1}{2}(1 \pm \eta)\Phi$. This corresponds to the Duffin–Kemmer–Petiau equation. Writing this in terms of Clifford products,

$$e^a \nabla_{X_a} \Phi + \nabla_{X_a} \Phi e^a = 2\mu\Phi$$

we see that the second term prevents the decoupling of the equation into minimal left ideals in Minkowski space. Since d and δ map even (odd) forms to odd (even) ones (8.4.1) is equivalent to

$$\mathrm{d}\Phi_+ = \mu\Phi_-$$

$$\delta\Phi_- = -\mu\Phi_+ \, .$$

As a consequence $\delta\Phi_+ = 0$ and $\mathrm{d}\Phi_- = 0$ so any solution to (8.4.1) will also satisfy the Kahler equation for Φ.

In the massless case (8.4.1) exhibits the generalised gauge symmetry

$$\Phi_+ \longmapsto \Phi_+ + \mathrm{d}\chi_-$$

$$\Phi_- \longmapsto \Phi_- + \delta\chi_+$$

and describes what in the physics literature are often called *antisymmetric tensor gauge fields.*

Bibliography

Chisholm J S R and Common A K (ed) 1986 *NATO ASI Series* **183**

9

Spinor Fields

In §2.5 spinors (or semi-spinors) were defined as carrying irreducible representations of the Clifford algebra. Any such irreducible representation is equivalent to that carried by a minimal left ideal of the Clifford algebra. We thus took any minimal left ideal as the space of spinors. The Clifford bundle of a pseudo-Riemannian manifold M has as fibre at p, the Clifford algebra of the cotangent space of M at p. Any minimal left ideal of this fibre algebra carries the spinor representation. If we could smoothly assign a minimal left ideal of the fibre algebra to each p in M then we would have a bundle over M with each fibre carrying an irreducible representation of the corresponding fibre of the Clifford bundle. Such a bundle of spinor spaces would be a sub-bundle of the Clifford bundle. For such a bundle to exist the topology of M would have to be severely restricted. Requiring the bundle of spinor spaces to be contained in the Clifford bundle is unduly restrictive. Therefore, rather than requiring that the spinor spaces be minimal left ideals of the Clifford algebra, we only require that they carry a representation equivalent to that carried by any minimal left ideal.

Locally any bundle of spinor spaces will be isomorphic to a sub-bundle of the Clifford bundle, with fibres being minimal left ideals of the Clifford algebra. As we shall show, if any bundle of spinor spaces exists we can always form a bundle by patching together the minimal left ideals of the Clifford algebra in such a way that locally a spinor field may be represented by a differential form lying in a minimal left ideal of the Clifford algebra.

9.1 Spinor Bundles

We assume first that the pseudo-Riemannian manifold M is even dimensional so that the real Clifford algebra is central simple. Thus

$C(T^*_pM,g) \simeq \mathcal{M}_r(\mathbb{R}) \otimes D(\mathbb{R})$, where $\mathcal{M}_r(\mathbb{R})$ is the algebra of all order-r real matrices and the real central division algebra D must be either the real numbers $\mathbb{R}$ or the quaternions H. Any minimal left ideal of $C(T^*_pM,g)$ carries the spinor representation. Thus minimal left ideals are r-dimensional right D-modules, Clifford multiplication inducing a D-linear transformation. As we noted above we are not now going to require that our spinor spaces be identified with any minimal left ideal, only that they carry an equivalent representation. Thus our spinor spaces will be right D-linear spaces such that Clifford multiplication is D-linear. Let $\mathcal{S}(M)$ be a bundle over M such that for each $p \in M$ the fibre above p is a right D-linear space carrying an irreducible representation of $C(T^*_pM,g)$. Any such bundle will be called a (real) *spinor bundle*, sections being called *spinor fields*. If any spinor bundle exists then M is called a *spin manifold*. A discussion of the topological restrictions on M in order for it to be a spin manifold are beyond the scope of this book. However, the reason that there is some restriction will become apparent later. Whereas M may have no spinor bundle, it may also have many. These can be split into equivalence classes. Two spinor bundles $\mathcal{S}(M)$ and $\mathcal{S}'(M)$ are *equivalent* if and only if there is a diffeomorphism relating them such that fibres of $\mathcal{S}(M)$ above p are mapped into fibres of $\mathcal{S}'(M)$ above p with the diffeomorphism commuting with Clifford multiplication. An equivalence class of spinor bundles constitutes a *spinor structure* for $C(M)$. (This definition of spinor structure is equivalent to the more usual one to be found in, for example, Milnor [22].)

Let us assume that M is a spin manifold with $\mathcal{S}(M)$ a spinor bundle. Fibres of the Clifford bundle are isomorphic to the algebra of D-valued matrices. If $\{e^{a(\alpha)}\}$ is a local orthonormal co-frame defined on the open neighbourhood U_α of M then an isomorphism between $C(T^*_pM,g)$ and D-valued matrices may be given at each $p \in U_\alpha$ in terms of the generators $\{e^{a(\alpha)}|_p\}$ and the constant matrices $\{\gamma^a\}$ satisfying

$$\gamma^a\gamma^b + \gamma^b\gamma^a = 2g^{ab}\mathbf{1}. \tag{9.1.1}$$

For a given choice of D-valued γ-matrices we may correlate a local orthonormal co-frame with a local basis of sections of $\mathcal{S}(M)$. On U_α there is a local basis for spinor fields $\{b_i{}^{(\alpha)}\}$ such that

$$e^{a(\alpha)}b_i^{(\alpha)} = b_j^{(\alpha)}\gamma^a_{ji}. \tag{9.1.2}$$

(Note that we juxtapose symbols to denote the Clifford action of sections of $\mathbf{C}(M)$ on sections of $\mathcal{S}(M)$.) (Thus the basis $\{b_i^{(\alpha)}\}$ transforms under Clifford multiplication just like the 'first column' of a matrix basis for the Clifford algebra.) Notice that (9.1.2) does not uniquely determine the spinor basis. If $\{b_i^{(\alpha)\prime}\}$ also satisfies (9.1.2) then $f^{(\alpha)}$ is a non-zero function on U_α such that

$$b_i^{(\alpha)\prime} = f^{(\alpha)} b_i^{(\alpha)}. \tag{9.1.3}$$

On $U_{\alpha\beta} \equiv U_\alpha \cup U_\beta$ there must be some local section of $\mathbf{C}(M)$, $s^{(\beta\alpha)}$, such that $b_i^{(\beta)} = s^{(\beta\alpha)} b_i^{(\alpha)}$. But $e^{a(\beta)} b_i^{(\beta)} = b_j^{(\beta)} \gamma_{ji}^a = s^{(\beta\alpha)} b_j^{(\alpha)} \gamma_{ji}^a = s^{(\beta\alpha)} e^{a(\alpha)} b_i^{(\alpha)}$. So $e^{a(\beta)} s^{(\beta\alpha)} b_i^{(\alpha)} = s^{(\beta\alpha)} e^{a(\alpha)} b_i^{(\alpha)}$ and

$$e^{a(\beta)} = s^{(\beta\alpha)} e^{a(\alpha)} s^{(\beta\alpha)^{-1}}. \tag{9.1.4}$$

Thus certainly $s^{(\beta\alpha)}$ is in the Clifford group Γ. If the spinor bases are changed as in (9.1.3) then $s^{(\beta\alpha)\prime} = f^{(\beta)} s^{(\beta\alpha)} f^{(\alpha)^{-1}}$. It turns out that we can, in fact, always choose the local bases in (9.1.2) such that the Clifford elements $s^{(\beta\alpha)}$ relating them on overlaps are in ${}_{\pm}\Gamma$. (It is a standard result that any Γ bundle is reducible to a ${}_{\pm}\Gamma$ bundle since $\Gamma/{}_{\pm}\Gamma \simeq \mathbb{R}^+$, see for example Kobayashi and Nomizu [23].) On triple overlaps $U_\alpha \cup U_\beta \cup U_\gamma \equiv U_{\alpha\beta\gamma}$ the Clifford elements relating spinor bases satisfy the *coherence* condition

$$s^{(\alpha\beta)} s^{(\beta\gamma)} = s^{(\alpha\gamma)}. \tag{9.1.5}$$

If M is both space and time orientable then we may choose local orthonormal co-frames related on overlaps by an element of $SO^+(p,q)$. Then if $\mathcal{S}(M)$ is a spinor bundle we may choose local spinor frames, as above, related on overlaps by an element of ${}_+\Gamma^+$. It is important to know that such local bases exist; we shall call them *standard spinor frames*. (Strictly speaking our definition of a spinor bundle is equivalent to the usual one only in the orientable case. Without orientability our definition is equivalent to what would usually be called a pinor structure.)

If M is *any* pseudo-Riemannian manifold then we can choose local orthonormal frames, related on overlaps by an orthogonal transformation, $\Lambda^{(\beta\alpha)}$ say. We can choose an $s^{(\beta\alpha)} \in {}_{\pm}\Gamma$ such that $\chi(s^{(\beta\alpha)}) = \Lambda^{(\beta\alpha)}$. On triple overlaps we must have $s^{(\alpha\beta)} s^{(\beta\gamma)} = \pm s^{(\alpha\gamma)}$. In general, we cannot choose the $\{s^{(\alpha\beta)}\}$ so as to eliminate all the minus signs in these relations. We can do this if and only if M is a spin manifold.

In the case in which $D = H$ we have required the spin bundle to have a right H-linear structure. Thus spinor fields can be multiplied by quaternions. This condition could be relaxed. We know that each spinor space is a right H-linear space, so locally any spinor bundle must have this structure. But we could consider the more general case in which spinor fields can be multiplied by sections of a non-trivial quaternion bundle, this multiplication commuting with the Clifford action. The existence of a spinor bundle without the H-linear structure is equivalent to the weaker condition of having a *generalised spinor structure* [24].

So far we have only considered bundles of real spinors for the case in which M is even dimensional. If M is odd dimensional with signature such that the Clifford algebra is reducible then the central idempotents

$\frac{1}{2}(1 \pm z)$, with z the volume n-form, decompose the Clifford algebra into simple ideals. So if M is orientable the Clifford bundle splits into two bundles of simple algebras. In this case we can define spinor bundles exactly as above and show that there are standard spinor frames related on overlaps by an element of ${}_{+}\Gamma^{+}$. When the Clifford algebra is isomorphic to the algebra of complex matrices then certainly any bundle carrying an irreducible representation of the Clifford bundle has local bases related on overlaps by elements of the Clifford group. But in this case we cannot argue that they can be chosen in ${}_{+}\Gamma^{+}$ (assuming orientability); rather they will be elements of ${}_{+}\Gamma^{+}$ multiplied by unimodular complex functions. The existence of such a bundle is equivalent to having a $Spin^C$ structure, this being a weaker condition than having a *Spin* structure. The case of the complexified Clifford bundle is like that just discussed. If we assume orientability then the existence of a bundle carrying an irreducible representation is equivalent to having a $Spin^C$ structure.

In the following we shall assume that M is a spin manifold. Unless we specifically say otherwise we shall mean by spinor bundle a bundle carrying an irreducible representation of the Clifford bundle, or its complexification, such that we have standard spinor frames related on overlaps by an element of ${}_{+}\Gamma^{+}$. For the case of odd dimensions, or the complexified case, this is a stronger requirement than that the bundle simply carry an irreducible representation of the Clifford bundle.

9.2 Inner Products on Spinor Fields

In Chapter 2 we took the space of spinors to be any minimal left ideal of the Clifford algebra, projected by some primitive idempotent P. In §2.6 we constructed spin-invariant products on the space of spinors with values in the division algebra $P\mathbf{C}(V,g)P \simeq D$. We now want to define spin-invariant products on spinor fields with values in D. Although we shall use the same notation as in §2.6 now our spinors need not lie in any minimal left ideal of the Clifford algebra, and the product will take values in D which is the 'standard' algebra isomorphic to $P\mathbf{C}(V,g)P$ for any primitive P.

If we had an inner product defined on sections of the spinor bundle then we could use this product to establish local canonical bases (orthonormal, symplectic etc.). On overlaps these canonical bases would be related by transformations in the invariance group of the product. Conversely we can use a set of local bases related on overlaps by an element of ${}_{+}\Gamma^{+}$ to define a ${}_{+}\Gamma^{+}$-invariant product on spinor fields. For the sake of definiteness we assume that the (real or complexified)

Clifford algebra is isomorphic to the algebra of all (real or complex) matrices, with the involution $\xi\eta$ similar to transposition. In this case for matrices as in (9.1.1) there is a matrix $\mathbf{C}$, symmetric or skew, such that

$$\mathbf{C}\gamma^{aT}\mathbf{C}^{-1} = -\gamma^a. \tag{9.2.1}$$

If $\{b_i^{(\alpha)}\}$ is a standard spinor frame, satisfying (9.1.2), then a bilinear product on local spinor fields is specified by defining

$$(b_i^{(\alpha)}, b_j^{(\alpha)})_{(\alpha)} = C_{ij}^{-1}. \tag{9.2.2}$$

The product has been labelled with the subscript (α) since in principle we have a different product for each U_α. We want to show that on $U_{\alpha\beta}$ the products $(\ ,\ )_{(\alpha)}$ and $(\ ,\ )_{(\beta)}$ coincide, for then we have a well defined product on spinor fields. First we show that these local products are spin invariant. For any such local product then (suppressing the (α)-labelling)

$$(b_i, e^a b_j) = (b_i, b_k \gamma^a_{kj}) = C^{-1}_{ik}\gamma^a_{kj} = (C^{-1}\gamma^a)_{ij} = -(\gamma^{aT}C^{-1})_{ij}$$

by (9.2.1), so

$$(b_i, e^a b_j) = -\gamma^{a\,T}_{ik}C^{-1}_{kj} = -\gamma^a_{kj}C^{-1}_{kj} = -\gamma^a_{ki}(b_k, b_j) = -(e^a b_i, b_j).$$

Thus for any spinor fields and $m \in \Gamma\mathbf{C}(M)$ $(\varphi, m\psi)_{(\alpha)} = (m^{\xi\eta}\varphi, \psi)_{(\alpha)}$ and hence these local products are spin invariant, having $\xi\eta$ as adjoint involution. On $U_{\alpha\beta}$ the standard spinor frames are related by $b_i^{(\beta)} = s^{(\beta\alpha)}b_i^{(\alpha)}$ for $s^{(\beta\alpha)} \in {}_+\Gamma^+$. So on $U_{\alpha\beta}$

$$\begin{aligned}(b_i^{(\beta)}, b_j^{(\beta)})_{(\alpha)} &= (s^{(\beta\alpha)}b_i^{(\alpha)}, s^{(\beta\alpha)}b_j^{(\alpha)})_{(\alpha)} = (s^{(\beta\alpha)^{\xi\eta}}s^{(\beta\alpha)}b_i^{(\alpha)}, b_j^{(\alpha)})_{(\alpha)}\\ &= (b_i^{(\alpha)}, b_j^{(\alpha)})_{(\alpha)} = (b_i^{(\beta)}, b_j^{(\beta)})_{(\beta)}.\end{aligned}$$

Thus for any local spinor fields $(\varphi, \psi)_{(\alpha)} = (\varphi, \psi)_{(\beta)}$. Hence we have a well defined product on spinor fields and so omit the neighbourhood labelling.

We demonstrated the existence of a spin-invariant product on spinor fields by constructing one using a special basis. That construction does not, in fact, specify a unique product. For given local orthonormal co-frames and γ-matrices the standard local spinor frames are not unique. If the local orthonormal co-frames are related by $\Lambda^{(\alpha\beta)}$ then the $s^{(\alpha\beta)} \in {}_+\Gamma^+$ such that $\chi(s^{(\alpha\beta)}) = \Lambda^{(\alpha\beta)}$ is determined up to a sign. So if $\{b_i^{(\alpha)\prime}\}$ is also a standard spinor frame with $b_i^{(\alpha)\prime} = f^{(\alpha)}b_i^{(\alpha)}$ for a local function $f^{(\alpha)}$ then on overlaps we must have $f^{(\beta)} = \pm f^{(\alpha)}$. So a non-zero function f is defined on M by $f|_{U_\alpha} = (\operatorname{sgn} f^{(\alpha)})f^{(\alpha)}$, with $b_i^{(\alpha)\prime} = \pm f b_i^{(\alpha)}$. So if $(b_i^{(\alpha)\prime}, b_j^{(\alpha)\prime})' = (b_i^{(\alpha)}, b_j^{(\alpha)})$, then for any spinor fields $f^2(\varphi, \psi)' = (\varphi, \psi)$. It is easily seen that the choices of orthonormal co-frames, γ-matrices and matrix $\mathbf{C}$ cannot affect the spinor product by more than a conformal scaling. Thus this prescription determines a class of conformally related spin-invariant products.

Although in the above we assumed for definiteness that $\xi\eta$ was similar to transposition in a total matrix algebra, the above construction obviously goes through similarly in general. We may analogously construct spin-invariant products with adjoint involution ξ or, for the complexified algebras, ξ^* or $\xi\eta^*$.

If, for M even dimensional, $\mathcal{I}(M)$ is a bundle of spinors carrying an irreducible representation of the complexified Clifford bundle then we may define charge conjugation on spinor fields. Once again, although we know that we can do this locally, we have to check that we can do it globally. We therefore give the definition locally using a standard spinor frame and make sure that it is consistent on overlaps. From (2.7.9) we know that there is a matrix $\mathbf{m}$ such that

$$\gamma^{a*} = \mathbf{m}^{-1}\gamma^a\mathbf{m}, \text{ with } \mathbf{m}^* = \pm\mathbf{m}^{-1}. \tag{9.2.3}$$

On U_α the operator $c(\alpha)$ is defined by

$$\psi^{c(\alpha)} = (b_i^{(\alpha)}\psi^i)^{c(\alpha)} = b_j^{(\alpha)}m_{ji}\psi^{j*}. \tag{9.2.4}$$

If $\#(\alpha)$ is the local operation on spinor fields that complex conjugates the components in the $b_i^{(\alpha)}$ basis then we use the same symbol to denote the automorphism of the complexified Clifford algebra defined by

$$(a\psi)^{\#(\alpha)} = a^{\#(\alpha)}\psi^{\#(\alpha)}. \tag{9.2.5}$$

Thus if $ab_i^{(\alpha)} = b_j^{(\alpha)}a_{ji}$ then $a^{\#(\alpha)}b_i^{(\alpha)} = b_j^{(\alpha)}a_{ji}{}^*$. (Care is needed with the notation. By $a_{ji}{}^*$ we mean the complex conjugate of the components of a, whereas $a^*{}_{ji}$ are the components of the Clifford element a^*. The difference between these is the difference between $*$ and $\#(\alpha)$.) If $m^{(\alpha)}$ is the local Clifford form such that $m^{(\alpha)}b_i^{(\alpha)} = b_j^{(\alpha)}m_{ji}$ then it follows from (9.2.3) that

$$a^{\#(\alpha)} = m^{(\alpha)^{-1}}a^*m^{(\alpha)} \tag{9.2.6}$$

so

$$\begin{aligned}(a\psi)^{c(\alpha)} &= (b_j^{(\alpha)}a_{ji}\psi^i)^{c(\alpha)} = b_k^{(\alpha)}m_{kj}a_{ji}{}^*\psi^{i^*} = m^{(\alpha)}b_j^{(\alpha)}a_{ji}{}^*\psi^{i^*} \\ &= m^{(\alpha)}a^{\#(\alpha)}b_i^{(\alpha)}\psi^{i^*} = m^{(\alpha)}a^{\#(\alpha)}m^{(\alpha)^{-1}}\psi^{c(\alpha)} = a^*\psi^{c(\alpha)}.\end{aligned}$$

If we expand ψ as $\psi = b_i^{(\beta)}\psi^i$ then $\psi^{c(\beta)} = b_j^{(\beta)}m_{ji}\psi^{i*}$, but $\psi = s^{(\beta\alpha)}b_i^{(\alpha)}\psi^i$ so

$$\psi^{c(\alpha)} = s^{(\beta\alpha)^*}b_j^{(\alpha)}m_{ji}\psi^{i^*} = b_j^{(\beta)}m_{ji}\psi^{i^*}$$

since $s^{(\beta\alpha)^*} = s^{(\beta\alpha)}$ for $s^{(\beta\alpha)} \in {}_+\Gamma^+$. Hence the operations $c(\alpha)$ and $c(\beta)$ agree on $U_{\alpha\beta}$ and we have a well defined operation of charge conjugation, denoted c. If M is odd dimensional the complexified Clifford algebra is semi-simple. In this case either $*$ or η^* is a conjugate-linear involuntary automorphism of the simple component algebras. In the latter case we can define charge conjugation using η^* instead of $*$.

In even dimensions we have spin-invariant products on the real spinor bundle with adjoint involutions ξ and $\xi\eta$. The automorphism η is inner with $a^\eta = zaz^{-1}$ for z the volume form. Using a subscript to label the product by its adjoint involution we have

$$(\psi, \varphi)_{\xi\eta} = (\psi, z\varphi)_\xi. \tag{9.2.7}$$

In the complexified case we have similarly

$$(\psi, \varphi)_{\xi^*} = (\psi^c, \varphi)_\xi \tag{9.2.8}$$

and

$$(\psi, \varphi)_{\xi\eta^*} = (\psi^c, z\varphi)_\xi\ . \tag{9.2.9}$$

For a semi-simple real Clifford algebra there is a product on the semi-spinors associated with either ξ or $\xi\eta$. When the real Clifford algebra is isomorphic to complex matrices then either ξ or $\xi\eta$ is associated with a complex bilinear product, the other being associated with a conjugate-linear product; the products being related by 'charge conjugation' defined using η. For the bundle of complex semi-spinors in odd dimensions then either ξ or $\xi\eta$ is associated with a complex bilinear product; either ξ^* or $\xi\eta^*$ being associated with a conjugate-linear one. The products are related by 'charge conjugation' defined with either $*$ or η^*.

9.3 Covariant Differentiation of Spinor Fields

In a similar way to that used to show the existence of a spin-invariant product we can define covariant differentiation of spinor fields using a standard spinor frame. We will first follow this most direct approach. We may then observe that the spinor covariant derivative has certain properties. In fact these properties completely determine this covariant derivative as we will then show. For most purposes it is sufficient to know that a unique covariant derivative having these properties exists. It is customary to use the symbol ∇ to denote covariant differentiation of spinor fields as well as of tensor fields; the meaning depending on what it acts on. We prefer to use a separate symbol S to denote covariant differentiation of spinor fields. Although we shall only really be concerned with the pseudo-Riemannian connection on M it should be apparent that the discussion here is equally applicable in the case of non-zero torsion.

If $\{e^{a(\alpha)}\}$ is a local orthonormal co-frame then, from (8.1.5) and (8.1.6), we have $\nabla_X e^{a(\alpha)} = [\sigma_X^{(\alpha)}, e^{a(\alpha)}]$ where $\sigma_X^{(\alpha)} = \frac{1}{4}\omega_{bc}^{(\alpha)}(X)e^{bc(\alpha)}$. We can use this local orthonormal co-frame to define a standard spinor frame satisfying (8.1.2). We can introduce a covariant derivative $S_X^{(\alpha)}$ of local spinor fields by defining

$$S_X^{(\alpha)} b_i^{(\alpha)} = \sigma_X^{(\alpha)} b_i^{(\alpha)}. \tag{9.3.1}$$

If $\psi^{(\alpha)}$ is an arbitrary local spinor field then $S_X^{(\alpha)}$ is defined by

$$S_X^{(\alpha)} \psi^{(\alpha)} = S_X^{(\alpha)}(b_i^{(\alpha)} \psi^i) = S_X^{(\alpha)} b_i^{(\alpha)} \psi^i + b_i^{(\alpha)} X(\psi^i). \tag{9.3.2}$$

The components ψ^i are D-valued functions and the above requires that we know how to differentiate these. Quaternionic or complex-valued functions are differentiated as ordered quadruples or pairs of real functions; that is, the algebra D has a parallel basis. Of course, we will want to show that if ψ is a local spinor field defined on $U_{\alpha\beta}$ then $S_X^{(\alpha)}\psi = S_X^{(\beta)}\psi$. We will then have a well defined covariant derivative on arbitrary sections of the spinor bundle and can drop the label (α). First we show that a consequence of the definition (9.3.1) is that the local spinor covariant derivatives obey a 'Leibnitz' rule. If A is an arbitrary 1-form on M with $\psi^{(\alpha)}$ a local spinor field then

$$\begin{aligned}
S_X^{(\alpha)}(A\psi^{(\alpha)}) &= S_X^{(\alpha)}(A_a e^a b_i^{(\alpha)} \psi^i) = S_X^{(\alpha)}(A_a b_j^{(\alpha)} \gamma^a_{ji} \psi^i) \\
&= X(A_a) b_j^{(\alpha)} \gamma^a_{ji} \psi^i + A_a \sigma_X^{(\alpha)} b_j^{(\alpha)} \gamma^a_{ji} \psi^i + A_a b_j^{(\alpha)} \gamma^a_{ji} X(\psi^i) \\
&= X(A_a) e^a \psi^{(\alpha)} + \sigma_X^{(\alpha)} A \psi^{(\alpha)} + A b_i^{(\alpha)} X(\psi^i) \\
&= \nabla_X A \psi^{(\alpha)} + A \sigma_X^{(\alpha)} \psi^{(\alpha)} + A b_i^{(\alpha)} X(\psi^i)
\end{aligned}$$

by (8.1.8), so

$$S_X^{(\alpha)}(A\psi^{(\alpha)}) = \nabla_X A \psi^{(\alpha)} + A S_X^{(\alpha)} \psi^{(\alpha)}.$$

Since this is true for all local $\psi^{(\alpha)}$ and the 1-forms generate the Clifford algebra we have

$$S_X^{(\alpha)}(a\psi^{(\alpha)}) = \nabla_X a \psi^{(\alpha)} + a S_X^{(\alpha)} \psi^{(\alpha)} \tag{9.3.3}$$

for any Clifford form a. If now ψ is any spinor field then on $U_{\alpha\beta}$ we have

$$S_X^{(\beta)} \psi = S_X^{(\beta)}(b_i^{(\beta)} \psi^i) = \sigma_X^{(\beta)} \psi + b_i^{(\beta)} X(\psi^i).$$

But on $U_{\alpha\beta}$ we have $b_i^{(\beta)} = s^{(\beta\alpha)} b_i^{(\alpha)}$ for $s^{(\beta\alpha)} \in {}_+\Gamma^+$, so

$$\begin{aligned}
S_X^{(\alpha)} \psi &= S_X^{(\alpha)}(s^{(\beta\alpha)} b_i^{(\alpha)} \psi^i) \\
&= \nabla_X s^{(\beta\alpha)} b_i^{(\alpha)} \psi^i + s^{(\beta\alpha)} \sigma_X^{(\alpha)} b_i^{(\alpha)} \psi^i + s^{(\beta\alpha)} b_i^{(\alpha)} X(\psi^i) \\
&= \nabla_X s^{(\beta\alpha)} s^{(\beta\alpha)^{-1}} \psi + s^{(\beta\alpha)} \sigma_X^{(\alpha)} s^{(\beta\alpha)^{-1}} \psi + b_i^{(\beta)} X(\psi^i).
\end{aligned}$$

Hence from (8.1.9) we see that $S_X^{(\alpha)}\psi = S_X^{(\beta)}\psi$ and we have a covariant derivative S_X defined on arbitrary spinor fields such that $S_X\psi^{(\alpha)} = S_X^{(\alpha)}\psi^{(\alpha)}$.

We have shown the existence of this spinor covariant derivative by specifying it in standard local spinor frames. These standard spinor frames were also used to introduce a spin-invariant product. Suppose that (,) is any such D-valued product with $(b_i^{(\alpha)}, b_j^{(\alpha)}) = C_{ij}^{-1}$ for some constant matrix $\mathbf{C}$. Then if ψ and φ are arbitrary spinor fields with $\psi = b_i^{(\alpha)}\psi^i$ on U_α

$$(S_X\psi, \varphi) + (\psi, S_X\varphi) \\ = (\sigma_X^{(\alpha)}\psi + b_i^{(\alpha)}X(\psi^i), \varphi) + (\psi, \sigma_X^{(\alpha)}\varphi + b_i^{(\alpha)}X(\varphi^i)).$$

Since $\sigma_X^{(\alpha)}$ is a real 2-form $\sigma_X^{(\alpha)\mathscr{I}} = -\sigma_X^{(\alpha)}$ for any $\mathscr{I}$ that is the adjoint involution of a spin-invariant product. Hence

$$\begin{aligned}(S_X\psi, \varphi) + (\psi, S_X\varphi) &= (b_i^{(\alpha)}X(\psi^i), \varphi) + (\psi, b_i^{(\alpha)}X(\varphi^i)) \\ &= (X(\psi^i))^j C_{ik}^{-1}\varphi^k + (\psi^i)^j C_{ik}^{-1}X(\varphi^k)\end{aligned}$$

where the product is D^j-linear in the first variable. Since $(X(\psi^i))^j = X((\psi^i)^j)$ and the matrix $\mathbf{C}^{-1}$ is constant

$$(S_X\psi, \varphi) + (\psi, S_X\varphi) = X(\psi, \varphi)\ . \tag{9.3.4}$$

Thus, in this sense, the spinor covariant derivative is compatible with any spin-invariant product for which the standard spinor frames are a 'canonical' basis. In particular, for the complexified case, S_X is compatible with both a complex bilinear and a Hermitian product, related as in (9.2.8). Thus the covariant derivative commutes with charge conjugation,

$$S_X\psi^c = (S_X\psi)^c. \tag{9.3.5}$$

This follows directly from (9.2.4) since the matrix m is constant.

Having defined a covariant derivative in a particular basis we have observed the properties (9.3.3), (9.3.4) and (9.3.5). We will now show how any covariant derivative satisfying these axioms is unique. Obviously S_X should map spinor fields to spinor fields. We shall require $\mathscr{F}$-linearity in X

$$S_{fX} = fS_X \tag{9.3.6}$$

the 'Leibnitz' rule

$$S_X(a\psi) = \nabla_X a\psi + aS_X\psi \qquad \forall\, a \in \Gamma\mathbf{C}(M),\ \forall\, \psi \in \Gamma\mathscr{I}(M) \tag{9.3.7}$$

and compatibility with some spin-invariant product

$$(S_X\psi, \varphi) + (\psi, S_X\varphi) = X(\psi, \varphi). \tag{9.3.8}$$

Given an S that satisfies these axioms, is it unique? Suppose that S'_X also satisfied the axioms above. Then if $L_X \equiv S'_X - S_X$ we have

$$L_X : \Gamma\mathcal{J}(M) \to \Gamma\mathcal{J}(M) \tag{9.3.9}$$

$$L_{fX} = fL_X \tag{9.3.10}$$

$$L_X(a\psi) = aL_X\psi \tag{9.3.11}$$

$$(\varphi, L_X\psi) + (L_X\varphi, \psi) = 0. \tag{9.3.12}$$

Equation (9.3.11) says that L_X commutes with Clifford multiplication, so $L_X\psi = \psi\rho_X$ for some D-valued function ρ_X. Putting this in (9.3.12) gives

$$(\varphi, \psi\rho_X) + (\varphi\rho_X, \psi) = 0.$$

If the product is D^j-linear in the first variable then, since $\rho_X \in D$.

$$(\varphi, \psi)\rho_X + \rho_X^j(\varphi, \psi) = 0. \tag{9.3.13}$$

The D-linearity in ψ ensures that $\varphi, \psi \to (\varphi, \psi)$ maps $\Gamma\mathcal{J}(M) \times \Gamma\mathcal{J}(M)$ onto D, so we can choose φ and ψ such that $(\varphi, \psi) = 1$. This shows that $\rho_X^j = -\rho_X$, and if this is substituted into (9.3.13) then we see that ρ_X must be in the centre of D. If D is one of the central algebras R or H then we must have $\rho_X = 0$. Similarly if $D \simeq \mathbb{C}$ with j the identity involution. However, for the remaining case of $D \simeq \mathbb{C}$ and j complex conjugation then ρ_X can be any imaginary function. Since the mapping $X \to \rho_X$ is required to be $\mathcal{F}$-linear (by (9.3.6)) then if S_X satisfies (9.3.6)–(9.3.8) then so does S'_X, with

$$S'_X\psi = S_X\psi + \mathrm{i}A(X)\psi \tag{9.3.14}$$

for any real 1-form A. Thus if the spinors carry an irreducible representation of a (real or complexified) Clifford algebra that is isomorphic to complex matrices then requiring compatibility with a pseudo-Hermitian spinor product leaves the freedom to add an arbitrary U(1) term to the covariant derivative. We can remove this arbitrariness by also requiring (9.3.5) to hold. This is equivalent to requiring that the covariant derivative also be compatible with a complex bilinear product. Because the different spin-invariant products are related as in (9.2.7) and (9.2.8) then the spinor covariant derivative is simultaneously compatible with all.

Exercise 9.1
Show that if S_X satisfies (9.3.5)–(9.3.8) then there are standard spinor frames such that $S_X b_i^{(\alpha)} = \sigma_X^{(\alpha)} b_i^{(\alpha)}$.

In §2.6 we used a D-valued spin-invariant product to map a spinor into the D-linear dual space. We will use the definition and notation of

(2.6.7) for spinor fields. When our spinor space was a minimal left ideal of the Clifford algebra then the D-linear dual space is naturally identified with a minimal right ideal, and for a spinor φ and dual spinor $\widetilde{\psi}$ we have $\varphi\widetilde{\psi}$ in the Clifford algebra. Although the notation of simply juxtaposing the spinors is a slight liberty when the spinor fields are not in the Clifford algebra we still have a mapping taking a spinor and a dual spinor to the Clifford algebra; $\varphi,\widetilde{\psi} \mapsto \varphi\widetilde{\psi}$ where

$$(\varphi\widetilde{\psi})\rho = \varphi(\widetilde{\psi}\rho) \equiv \varphi(\psi,\rho) \qquad \forall \rho \in \Gamma\mathcal{S}(M)\ . \tag{9.3.15}$$

If the adjoint spinor $\widetilde{\psi}$ is defined with respect to a product with which S_X is compatible then we have the useful relation

$$\nabla_X(\varphi\widetilde{\psi}) = S_X\varphi\widetilde{\psi} + \varphi\widetilde{S_X\psi}. \tag{9.3.16}$$

This follows by differentiating (9.3.15); using the Leibnitz property on the left-hand side and the metric compatibility on the right-hand side.

The curvature operator of S is defined in the obvious way,

$$\mathbf{S}(X,Y) = [S_X,S_Y] - S_{[X,Y]}. \tag{9.3.17}$$

There is always a local basis in which $S_X b_i = \sigma_X b_i$, and hence $\mathbf{S}(X,Y)b_i = \mathcal{R}_{XY}b_i$ where $\mathcal{R}_{XY}$ is defined in (8.1.10). Since the curvature operator is $\mathcal{F}$-linear then for any spinor field

$$\mathbf{S}(X,Y)\psi = \mathcal{R}_{XY}\psi. \tag{9.3.18}$$

Using (8.1.13) and (for zero torsion) (8.1.14) we can write this in terms of the curvature 2-forms giving

$$\mathbf{S}(X,Y)\psi = -\tfrac{1}{4}\mathrm{i}_X\mathrm{i}_Y R_{ab}e^{ab}\psi \tag{9.3.19}$$

or

$$\mathbf{S}(X,Y)\psi = \tfrac{1}{2}e^a(X)e^b(Y)R_{ab}\psi. \tag{9.3.20}$$

9.4 Lie Derivatives of Spinor Fields

Because the Clifford product involves the metric then unless the vector field V is Killing the Lie derivative $\mathcal{L}_V$ will not be a derivation on Clifford products. It follows immediately that there can be no 'Lie derivative' on spinor fields such that the obvious analogue of the 'Leibnitz' rule (9.3.7) holds for arbitrary vectors. Although one could call any operator a 'Lie derivative on spinor fields' the utility of such a definition depends on the consequent properties. So we can anticipate that any definition of a Lie derivative on spinor fields will really only be useful for Killing vectors. We shall notationally distinguish the Lie derivative operator on spinor fields from that on tensor fields by using the symbol $\mathscr{L}_X$.

We shall first parallel the initial treatment of the spinor covariant derivative by using a standard spinor frame. We shall show that for a Killing vector the Lie derivative of an orthonormal co-frame can be written as a Clifford commutator. Thus defining the Lie derivative of the associated spinor frame to be mulplication by the element that enters into that commutator ensures the 'Leibnitz' property. In (6.13.1) we introduced the operator $A_V \equiv \mathscr{L}_V - \nabla_V$, satisfying $A_V(f\varphi) = fA_V\varphi$ for any function f and differential form φ. Since A_V is a derivation on the exterior algebra we have

$$A_V\varphi = A_V e^a \wedge i_{X_a}\varphi \qquad \forall\, \varphi \in \Gamma\Lambda M.$$

We can use (8.1.2) and (8.1.3) to write the interior and exterior products in terms of Clifford products, producing

$$A_V\varphi = \tfrac{1}{4}[A_V e^a \wedge e_a, \varphi] + \tfrac{1}{2}i_{X_a}A_V e^a\varphi - \tfrac{1}{4}(A_V e^a\varphi^\eta e_a + e_a\varphi^\eta A_V e^a).$$

The Clifford commutator is a Clifford derivation. The 2-form $A_V e^a \wedge e_a$ can be written in terms of the exterior derivative of $\widetilde{V}$. Since A_V commutes with contractions and $A_V f = 0$ for $f \in \mathscr{F}(M)$, if $\{e^a\}$ and $\{X_a\}$ are dual bases and $A_V X_a = m_a{}^b X_b$ for some matrix $m_a{}^b$ then $A_V e^b = -m_a{}^b e^a$. Then $A_V e^a \wedge e_a = -m_b{}^a e^b \wedge e_a = m_a{}^b e_b \wedge e^a$, using the antisymmetry of the exterior product, so $A_V e^a \wedge e_a = \widetilde{A_V X_a} \wedge e^a$. Now $A_V X_a \equiv [V, X_a] - \nabla_V X_a$, so if ∇ is torsion free $A_V X_a = -\nabla_{X_a} V$, thus

$$A_V e^a \wedge e_a = e^a \wedge \widetilde{\nabla_{X_a} V} = e^a \wedge \nabla_{X_a}\widetilde{V} = \mathrm{d}\widetilde{V} \qquad \text{(by (4.7.4))}.$$

The remaining terms in the expression for A_V in general prevent it from being a Clifford derivation. If written in terms of the matrix $m_a{}^b$ then only the symmetric part enters:

$$\tfrac{1}{2}i_{X_a}A_V e^a\varphi - \tfrac{1}{4}(A_V e^a\varphi^\eta e_a + e_a\varphi^\eta A_V e^a)$$
$$= -\tfrac{1}{2}m_a{}^a\varphi + \tfrac{1}{8}(m_{ba} + m_{ab})(e^b\varphi^\eta e^a + e^a\varphi^\eta e^b)$$

using the usual index-lowering convention. Since $g(A_V X_a, X_b) = m_{ab}$ and A_V commutes with contractions

$$m_{ab} + m_{ba} = -A_V g(X_a, X_b).$$

The metric compatibility of ∇ enables us to write $A_V g = \mathscr{L}_V g$ and

$$A_V\varphi = [\tfrac{1}{4}\mathrm{d}\widetilde{V}, \varphi] + \tfrac{1}{4}\mathscr{L}_V g(X_a, X^a)\varphi$$
$$- \tfrac{1}{8}\mathscr{L}_V g(X_a, X_b)(e^b\varphi^\eta e^a + e^a\varphi^\eta e^b). \qquad (9.4.1)$$

Thus, as expected, A_V, and hence $\mathscr{L}_V$, is a Clifford derivation if and only if V is a Killing vector.

If K is a Killing vector then the above simplifies to

$$\mathscr{L}_K\varphi = \nabla_K\varphi + [\tfrac{1}{4}\mathrm{d}\widetilde{K}, \varphi]\,. \tag{9.4.2}$$

So if $\{e^a\}$ is an orthonormal co-frame we have, from (8.1.6)

$$\mathscr{L}_K e^a = [\sigma_K + \tfrac{1}{4}\mathrm{d}\widetilde{K}, e^a] \tag{9.4.3}$$

where $\sigma_K = \frac{1}{4}\mathrm{i}_K\omega_{pq}e^{pq}$. Under Lie transport along the flow of an isometry an orthonormal frame undergoes an orthogonal transformation. The Lie derivative gives the infinitesimal transformation, representing the Lie algebra of the orthogonal group on the frame. Analogous to the way in which we introduced the covariant derivative we can define the Lie derivative on the associated standard spinor frame to be given by left multiplication by the element that appears in this commutator: that is

$$\mathcal{L}_K b_i = (\sigma_K + \tfrac{1}{4}\mathrm{d}\widetilde{K})b_i.$$

If $\psi = b_i\psi^i$ and $\mathcal{L}_K\psi = b_iK(\psi^i) + \mathcal{L}_K b_i\psi^i$ then, recalling the definition of the covariant derivative, we have

$$\mathcal{L}_K\psi = S_K\psi + \tfrac{1}{4}\mathrm{d}\widetilde{K}\psi. \tag{9.4.4}$$

Such a definition can (and will) be taken for the Lie derivative on spinors with respect to an arbitrary vector, but only in the case of Killing vectors is there a clear geometrical interpretation with $\mathcal{L}$ having useful properties.

When K is a Killing vector then, like S_K, $\mathcal{L}_K$ satisfies a 'Leibnitz' property:

$$\mathcal{L}_K(a\psi) = \mathscr{L}_K a\psi + a\mathcal{L}_K\psi. \tag{9.4.5}$$

This follows from (9.4.2) and (9.3.7). If $\widetilde{\psi}$ is the spinor adjoint to ψ, with respect to any spin-invariant product, then for K Killing

$$\mathscr{L}_K(\varphi\widetilde{\psi}) = \mathcal{L}_K\varphi\widetilde{\psi} + \varphi\widetilde{\mathcal{L}_K\psi}. \tag{9.4.6}$$

If the Lie derivative is written using (9.4.2) then this follows from the analogous property of S_X, (9.3.16).

Equations (6.13.13) and (6.13.14) give the commutator of a Lie derivative with a covariant derivative. We now obtain the analogous expression for the spinor operators. This will be useful for examining the covariances of spinor equations in the next chapter. Straight from the definition we have

$$[\mathcal{L}_K, S_V] - S_{[K,V]} = \boldsymbol{S}(K,V) - \tfrac{1}{4}\nabla_V\mathrm{d}\widetilde{K}.$$

The curvature of S is given in (9.3.20), and $\nabla_V\mathrm{d}\widetilde{K}$ can be expressed as in (6.13.9) to give

$$[\mathcal{L}_K, S_V] - S_{[K,V]} = -\tfrac{1}{4}\nabla_{X_b}\mathscr{L}_K g(V, X_a)e^{ba}. \tag{9 4.7}$$

For the special case of K a conformal Killing vector with

$$\mathcal{L}_K g = 2\lambda g \qquad \lambda \in \mathcal{F}(M) \tag{9.4.8}$$

the above simplifies to

$$[\mathcal{L}_K, S_V] - S_{[K,V]} = -\tfrac{1}{2}\mathrm{d}\lambda \wedge \widetilde{V}. \tag{9.4.9}$$

We can use the commutator of the Lie derivative with a covariant derivative to evaluate the commutator of two Lie derivatives,

$$\begin{aligned}&[\mathcal{L}_X, \mathcal{L}_Y]\psi - \mathcal{L}_{[X,Y]}\psi \\ &\qquad = [\mathcal{L}_X, S_Y]\psi - S_{[X,Y]}\psi + \tfrac{1}{4}\mathcal{L}_X(\mathrm{d}\widetilde{Y}\psi) - \tfrac{1}{4}\mathrm{d}\widetilde{Y}\mathcal{L}_X\psi - \tfrac{1}{4}\mathrm{d}\widetilde{[X, Y]}\psi.\end{aligned}$$

From (9.4.1)

$$\begin{aligned}\mathcal{L}_X(\mathrm{d}\widetilde{Y}\psi) - \mathrm{d}\widetilde{Y}\mathcal{L}_X\psi &= \mathcal{L}_X\mathrm{d}\widetilde{Y}\psi - \tfrac{1}{4}\mathcal{L}_X g(X_a, X^a)\mathrm{d}\widetilde{Y}\psi \\ &\quad + \tfrac{1}{8}\mathcal{L}_X g(X_a, X_b)(e^b\mathrm{d}\widetilde{Y}e^a + e^a\mathrm{d}\widetilde{Y}e^b)\psi\end{aligned}$$

and since

$$e^b\mathrm{d}\widetilde{Y}e^a + e^a\mathrm{d}\widetilde{Y}e^b = 2g^{ab}\mathrm{d}\widetilde{Y} - 2(e^a \wedge \mathrm{i}_{X^b}\mathrm{d}\widetilde{Y} + e^b \wedge \mathrm{i}_{X^a}\mathrm{d}\widetilde{Y})$$

then

$$\begin{aligned}&\tfrac{1}{8}\mathcal{L}_X g(X_a, X_b)(e^b\mathrm{d}\widetilde{Y}e^a + e^a\mathrm{d}\widetilde{Y}e^b) \\ &\qquad = \tfrac{1}{4}\mathcal{L}_X g(X_a, X^a)\mathrm{d}\widetilde{Y} - \tfrac{1}{2}\mathcal{L}_X g(X_a, X_b)e^a \wedge \mathrm{i}_{X^b}\mathrm{d}\widetilde{Y}.\end{aligned}$$

It follows from the definition of $\widetilde{Y}$ that

$$\mathcal{L}_X\widetilde{Y} = \widetilde{\mathcal{L}_X Y} + \mathcal{L}_X g(Y, X_a)e^a.$$

Since the Lie and exterior derivatives on differential forms commute

$$\begin{aligned}\mathcal{L}_X\mathrm{d}\widetilde{Y} &= \mathrm{d}\widetilde{[X,Y]} + \mathrm{d}(\mathcal{L}_X g(Y, X_a)e^a) \\ &= \mathrm{d}\widetilde{[X,Y]} + \nabla_{X_b}\mathcal{L}_X g(Y,X_a)e^{ba} + \mathcal{L}_X g(\nabla_{X_b}Y,X_a)e^{ba}\end{aligned}$$

thus

$$\begin{aligned}&\mathcal{L}_X(\mathrm{d}\widetilde{Y}\psi) - \mathrm{d}\widetilde{Y}\mathcal{L}_X\psi - \mathrm{d}\widetilde{[X, Y]}\psi \\ &\qquad = \nabla_{X_b}\mathcal{L}_X g(Y,X_a)e^{ba}\psi + \mathcal{L}_X g(\nabla_{X_b}Y,X_a)e^{ba}\psi \\ &\qquad\quad - \tfrac{1}{2}\mathcal{L}_X g(X_a,X_b)e^a \wedge \mathrm{i}_{X^b}\mathrm{d}\widetilde{Y}.\end{aligned}$$

Returning now to the commutator of the Lie derivatives we use (9.4.7) to obtain

$$[\mathcal{L}_X, \mathcal{L}_Y] - \mathcal{L}_{[X,Y]} = \tfrac{1}{4}\mathcal{L}_X g(\nabla_{X_b}Y, X_a)e^{ba} - \tfrac{1}{8}\mathcal{L}_X g(X_a, X_b)e^a \wedge \mathrm{i}_{X^b}\mathrm{d}\widetilde{Y}.$$

The right-hand side may be simplified so as to exhibit explicitly the antisymmetry in X and Y:

$$e^a \wedge \mathrm{i}_{X^b}\mathrm{d}\widetilde{Y} = e^a \wedge \nabla_{X^b}\widetilde{Y} - \mathrm{i}_{X^b}\nabla_{X_c}\widetilde{Y}e^{ac}$$

so

$$2\mathcal{L}_X g(\nabla_{X_b} Y, X_a) e^{ba} - \mathcal{L}_X g(X_a, X_b) e^a \wedge \mathrm{i}_{X^b} \mathrm{d} \widetilde{Y}$$
$$= -\mathcal{L}_X g(X_a, X_b) \mathrm{i}_{X^b} \nabla_{X_c} \widetilde{Y} e^{ac} - \mathcal{L}_X g(X_a, X_b) \mathrm{i}_{X_c} \nabla_{X^b} \widetilde{Y} e^{ac}.$$

Use of Killing's equation, (6.13.3), produces the final result

$$[\mathcal{L}_X, \mathcal{L}_Y] - \mathcal{L}_{[X,Y]} = -\tfrac{1}{8}\mathcal{L}_X g(X_a, X_b) \mathcal{L}_Y g(X^b, X_c) e^{ac}. \quad (9.4.10)$$

If either X or Y is conformal Killing then the right-hand side vanishes.

Exercise 9.2
Show that if $\{K_i\}$ is an algebra of Killing vectors in flat space then

$$[\tfrac{1}{4}\mathrm{d}\widetilde{K}_i, \tfrac{1}{4}\mathrm{d}\widetilde{K}_j] = \tfrac{1}{4}\mathrm{d}[\widetilde{K_i, K_j}].$$

Hint: write out the commutator of two spinorial Lie derivatives in terms of the curvature of S.

9.5 Representing Spinor Fields with Differential Forms

When M is even dimensional we can take as spinor bundle any bundle carrying an irreducible representation of the real Clifford bundle $\mathbf{C}(M)$. For the special case in which M is topologically $\mathbb{R}^n$ with a flat pseudo-Riemannian metric then we have a spinor sub-bundle of the Clifford bundle. Let $\{\hat{e}^a\}$ be a global parallel orthonormal co-frame. Then for some choice of constant γ-matrices there is a global matrix basis $\{\mathbf{e}_{ij}\}$ for Clifford forms such that $\hat{e}^a = \gamma^a_{ij}\mathbf{e}_{ij}$. Elements of this matrix basis can be written as Clifford polynomials of the parallel co-frames with constant coefficients, and so are parallel. Then $\mathcal{I}(M)$ is a spinor sub-bundle of $\mathbf{C}(M)$ if the fibres of $\mathcal{I}(M)$ are the minimal left ideals spanned by $\{\mathbf{e}_{i1}\}$. Sections of $\mathcal{I}(M)$ (spinor fields) are inhomogeneous differential forms. The pseudo-Riemannian connection ∇ induces a connection on $\mathcal{I}(M)$. In fact this is easily seen to be the spinor covariant derivative, generally denoted S, for this particular spinor bundle. We can of course always choose non-parallel co-frames, say $e^a = s\hat{e}^a s^{-1}$ for $s \in {}_+\Gamma^+$, with $\nabla_X e^a = [\sigma_X, e^a]$ for $\sigma_X = \nabla_X s s^{-1}$. The corresponding standard spinor basis is $\{b_i = s\mathbf{e}_{i1}\}$ satisfying $\nabla_X b_i = \sigma_X b_i$.

If T denotes the involution of transposition in the matrix basis $\{\mathbf{e}_{ij}\}$ and C is the Clifford element such that $a^{\xi\eta} = Ca^T C^{-1}$ then a spin-invariant product on sections of $\mathcal{I}(M)$ is given by

$$(\varphi, \psi) = \mathcal{S}_0(C^{-1}\varphi^{\xi\eta}\psi). \quad (9.5.1)$$

Notice that the 0-form projector $\mathcal{S}_0$ gives a product with values in the real numbers rather than the isomorphic algebra with $\mathbf{e}_{11}$ as identity. For the special spinor bundle here this product accords with the general prescription of §9.2.

Although for this particular spinor bundle the connections S and ∇ coincide there is still a need to distinguish $\mathfrak{L}_K$ from $\mathcal{L}_K$. For K a Killing vector these are seen, using (9.4.2), to be related by

$$\mathfrak{L}_K \psi = \mathcal{L}_K \psi + \tfrac{1}{4}\psi \mathrm{d}\widetilde{K}. \tag{9.5.2}$$

The Lie derivative $\mathcal{L}_K$ does not induce an operator on the sub-bundle $\mathcal{I}(M)$: it does not preserve the minimal left ideals. The addition of the second term ensures that $\mathfrak{L}_K \psi \in \Gamma\mathcal{I}(M)$ for all $\psi \in \Gamma\mathcal{I}(M)$.

In the above we showed how in flat space we had a spinor sub-bundle of the Clifford bundle. This is a very special situation. In general a manifold can admit a spinor structure without the Clifford bundle having a spinor sub-bundle. The following exercise illustrates this point.

Exercise 9.3

(i) Let I be any minimal left ideal of $C_{2,0}(\mathbb{R})$. Show that there is a unique vector a such that $\psi a = \psi$, $\forall\, \psi \in I$. Hint: Take an orthonormal frame $\{e^1, e^2\}$ and construct a matrix basis using $P_\pm = \frac{1}{2}(1 \pm e^1)$. Then if $I_0 = C_{2,0}(\mathbb{R})P_+$ then $I = I_0 S$ for some invertible S. Expand S in the previously constructed matrix basis and explicitly construct the a such that $P_+ S a = P_+ S$.

(ii) Argue that the real Clifford bundle of a two-dimensional sphere does not contain a spinor sub-bundle of minimal left ideals (since there is no non-vanishing vector field on a sphere). The sphere does, however, admit a spinor structure.

We have emphasised that we cannot in general find a spinor sub-bundle of the Clifford bundle, and thus cannot in general identify spinor fields with certain differential forms. However, we can if we wish always do this locally. For each open neighbourhood U_a of M we can choose a local basis for the Clifford algebra $\{\mathbf{e}_{ij}^{(\alpha)} Q_k^{(\alpha)}\}$. The local matrix frame $\{\mathbf{e}_{ij}^{(\alpha)}\}$ commutes with the basis $\{Q_k^{(\alpha)}\}$ for the division algebra. On $U_{\alpha\beta}$ there is a local Clifford form $S^{(\alpha\beta)}$ such that $\mathbf{e}_{ij}^{(\beta)} = S^{(\beta\alpha)} \mathbf{e}_{ij}^{(\alpha)} (S^{(\beta\alpha)})^{-1}$ and $Q_k^{(\beta)} = S^{(\beta\alpha)} Q_k^{(\alpha)} (S^{(\beta\alpha)})^{-1}$. If $I^{(\alpha)}$ is the minimal left ideal spanned by the first column of $\mathbf{e}_{ij}^{(\alpha)}$ and D is the 'standard' division algebra with basis $\{q_k\}$ then $I^{(\alpha)}$ is a right D-module with the rule $\mathbf{e}_{i1}^{(\alpha)} q_k \equiv \mathbf{e}_{i1}^{(\alpha)} Q_k^{(\alpha)}$. If we can choose the $S^{(\beta\alpha)}$ coherently, that is $S^{(\alpha\beta)} S^{(\beta\gamma)} = S^{(\alpha\gamma)}$ on $U_{\alpha\beta\gamma}$, then we can define an equivalence relation between $I^{(\alpha)}$ and $I^{(\beta)}$ on $U_{\alpha\beta}$ to form a spinor bundle. Thus the $S^{(\alpha\beta)}$ can be chosen coherently if and only if M is a spin manifold. If this is the case then for $\psi_p^{(\alpha)} \in I_p^{(\alpha)}$ and $\varphi_q^{(\beta)} \in I_q^{(\beta)}$, $p, q \in U_{\alpha\beta}$ we define the equivalence relation by

$$\psi_p^{(\alpha)} \sim \varphi_q^{(\beta)} \qquad \text{iff } p = q \text{ and } \varphi_p^{(\beta)} = \psi_p^{(\alpha)} (S^{(\beta\alpha)})^{-1}. \tag{9.5.3}$$

The resulting equivalence classes of differential forms form a bundle. On U_α we may represent a section of this bundle by a differential form lying in the minimal left ideal $I^{(\alpha)}$, on U_β we may choose a representative form in $I^{(\beta)}$, these being related on $U_{\alpha\beta}$ by the above relation. If a is an arbitrary Clifford form and $q \in D$ then for $\varphi^{(\beta)} \sim \psi^{(\alpha)}$ we have $a\varphi^{(\beta)}q \sim a\psi^{(\alpha)}q$ so, indeed, this bundle is a spinor bundle, carrying an irreducible representation of the Clifford bundle with a D-linear structure. Although sections of this bundle are not differential forms, but rather equivalence classes of local differential forms, we may represent local sections with any differential form in the class. However, the connection ∇ does not induce a connection on this bundle (in general). The pseudo-Riemannian connection will not preserve the minimal left ideals $I^{(\alpha)}$, and we need to distinguish between it and the spinor connection S.

Although it can be convenient to represent a spinor field locally by a differential form this can never be more than a matter of taste. Given that the spinor bundle carries an irreducible representation of the Clifford bundle we can define spin-invariant products, covariant differentiation etc, and the properties of these do not depend on how we choose to represent spinor fields.

Bibliography

Geroch R P 1967 *J.Math.Phys.* **8** 782

—— 1968 *J.Math.Phys.* **9** 1739

—— 1970 *J.Math.Phys.* **11** 11

Greub W and Petry H R 1978 *Lecture Notes on Mathematics* vol 675 (Heidelberg: Springer)

Isham C 1978 Spinor fields in 4-dimensional space–times *Proc.R.Soc.* A **364** 591

Kosman Y 1971 *Annali di Matematica* **25** 317–95

Lee K K 1973 *General Relativity and Gravitation* vol 4 p 421

Penrose R and Rindler W 1984 *Spinors and Space–Time* vol 1,2 (Cambridge: Cambridge University Press)

Petry H R 1984 *Spin Structures on Lorentz Manifolds, Trieste ISAS-44/84*

Pressley A and Segal G 1987 *Loop Groups* (Oxford: Oxford University Press)

10

Spinor Field Equations

10.1 The Dirac Operator

The Dirac operator gets its name from its appearance in Dirac's wave equation for the electron. It is now usual to extrapolate the nomenclature from this spacetime setting to mean by Dirac operator any operator of the form of that occurring in Dirac's wave equation. There is no clear concensus on how far this extrapolation is to go. We shall use the terminology as follows: if S_X denotes covariant differentiation with respect to X of sections of a bundle carrying an irreducible representation of the (real or complexified) Clifford bundle then the *Dirac operator* on sections is $\not{S} \equiv e^a S_{X_a}$. The co-frame $\{e^a\}$ is dual to the arbitrary tangent frame $\{X_a\}$. Sometimes mathematicians use the terminology more liberally to mean by Dirac operator any operator of the above form where S_X is any covariant derivative on sections of a bundle carrying any representation of the Clifford bundle. We will mostly be concerned with the Dirac operator on sections of a spinor bundle with the covariant derivative S_X of §9.3.

The *Dirac equation* for a complex spinor field ψ is

$$\not{S}\psi = \mu\psi \tag{10.1.1}$$

where μ is a complex constant. The nature of the manifold may restrict the eigenvalue μ to certain real or imaginary values. In other cases we may only be interested in real or imaginary eigenvalues for physical reasons. If $S_X^{(0)}$ is the standard spinor covariant derivative of §9.3.1 and A is a U(1) connection 1-form then a U(1)-covariant spinor derivative is given by

$$S_X^{(q)}\psi = S_X^{(0)}\psi + qiA(X)\psi \tag{10.1.2}$$

where q is the 'charge' coupling constant. The original equation of Dirac involved such a U(1)-charged covariant derivative

Exercise 10.1
Show that $\mathbf{S}^{(q)}(X, Y)\psi = \mathbf{S}^{(0)}(X, Y)\psi + \mathrm{i}q\,\mathrm{i}_X \mathrm{i}_Y F\psi$ where $F = \mathrm{d}A$.

In even dimensions the spinor representation of the complexified Clifford algebra induces a reducible representation of the even subalgebra. If $\check{z}$ is proportional to the volume form with $\check{z}^2 = 1$ then a complex spinor ψ is reduced into 'Weyl' spinors $\psi^{\pm}$ carrying irreducible representations of the even subalgebra by

$$\psi^{\pm} = \tfrac{1}{2}(1 \pm \check{z})\psi. \tag{10.1.3}$$

The projectors $\frac{1}{2}(1 \pm \check{z})$ anticommute with members of the co-frame $\{e^a\}$ and are parallel. So if ψ satisfies a massless ($\mu = 0$) Dirac equation then so do the Weyl spinors $\psi^{\pm}$. Such massless equations for the Weyl spinors are known in physics as *Weyl equations*.

Spinors of the real Clifford algebras can also be subjected to the Dirac equation (10.1.1) (with μ real). For signature (p, q) satisfying $p - q = 0, 2 \bmod 8$ the real Clifford algebra is a total real matrix algebra and the spinors are known in physics as Majorana spinors. In this case the Dirac equation may be known as a Majorana–Dirac equation. (Although the eigenvalue μ in (10.1.1) can be taken to be any real constant such an equation can not be obtained from a variational principle. Without recourse to 'anticommuting' parameters a variational principle will only give a Majorana–Dirac equation with zero eigenvalue.)

As we remarked at the beginning of §9.5, for the special case of a flat parallelisable manifold the Clifford bundle contains a spinor sub-bundle of minimal left ideals. The pseudo-Riemannian connection ∇ induces the spinor covariant derivative on this sub-bundle. Thus in this case the operator $\not{d}$, restricted to sections of this spinor sub-bundle, is a Dirac operator on spinor fields.

One of Dirac's requirements for his equation for the electron was that the components of the field should satisfy a Klein–Gordon equation. As we have just noted above the operator $\not{d}$, which squares to the Laplace–Beltrami operator, induces a Dirac operator on spinor fields in flat space. So this Dirac operator squares to the Laplace-Beltrami operator, acting on differential forms in the spinor sub-bundle. More generally, the square of the Dirac operator is known as the *spinor Laplacian*. We have

$$\begin{aligned}
(\not{S})^2\psi &= e^a S_{X_a}(e^b S_{X_b}\psi) \\
&= \not{d}e^b S_{X_b}\psi + \tfrac{1}{2}(e^a e^b + e^b e^a) S_{X_a} S_{X_b}\psi + \tfrac{1}{2}(e^a e^b - e^b e^a) S_{X_a} S_{X_b}\psi \\
&= \not{d}e^a S_{X_a}\psi + S_{X_a} S_{X^a}\psi + \tfrac{1}{2}e^{ab}[S_{X_a}, S_{X_b}]\psi \\
&= \not{d}e^a S_{X_a}\psi + S_{X_a} S_{X^a}\psi + \tfrac{1}{2}e^{ab}\mathbf{S}(X_a, X_b)\psi + \tfrac{1}{2}e^{ab} S_{[X_a, X_b]}\psi.
\end{aligned}$$

Now $[X_a, X_b] = \mathrm{i}_{X_a}\mathrm{i}_{X_b}\mathrm{d}e^c X_c$, and so

$$\tfrac{1}{2}e^{ab}S_{[X_a,\,X_b]}\psi = -\mathrm{d}e^c S_{X_c}\psi.$$

gives

$$\not{S}^2\psi = \mathrm{i}_{X^a}\nabla_{X_a}e^b S_{X_b}\psi + S_{X_a}S_{X^a}\psi + \tfrac{1}{2}e^{ab}\mathbf{S}(X_a, X_b)\psi.$$

Using (9.3.20) the curvature operator of S can be written in terms of the curvature 2-forms to give

$$\tfrac{1}{2}e^{ab}\mathbf{S}(X_a, X_b)\psi = \tfrac{1}{4}R_{pq}e^{pq}\psi.$$

From (8.1.17) we have, for zero torsion, $R_{cd}e^{cd} = -\mathcal{R}$, the curvature scalar, and so

$$\not{S}^2\psi = (S_{X^a} + \mathrm{i}_{X_b}\nabla_{X^b}e^a)S_{X_a}\psi - \tfrac{1}{4}\mathcal{R}\psi. \tag{10.1.4}$$

Exercise 10.2

Analogously express the Laplace–Beltrami operator as

$$\not{d}^2\Phi = (\nabla_{X^a} + \mathrm{i}_{X_b}\nabla_{X^b}e^a)\nabla_{X_a}\Phi - \tfrac{1}{4}\mathcal{R}\Phi - \tfrac{1}{4}R_{cd}\Phi e^{cd}.$$

10.2 Covariances of the Dirac Equation and Conserved Currents

Generally we expect equations formulated on pseudo-Riemannian manifolds to have a covariance corresponding to any isometries. For example, in §5.4 we showed how the Lie derivative with respect to a Killing vector maps solutions to Maxwell's equations into new solutions. In the same way we may use the Lie derivative on spinors to obtain new solutions to the Dirac equation in spaces with isometries.

For a vector field K we have

$$\mathcal{L}_K\not{S} = (\nabla_K e^a + \tfrac{1}{4}[\mathrm{d}\widetilde{K}, e^a])S_{X_a} + e^a\mathcal{L}_K S_{X_a}.$$

If now K is a conformal Killing vector, with $\mathcal{L}_K g = 2\lambda g$, then for A any 1-form $\mathcal{L}_K A = \nabla_K A + \tfrac{1}{4}[\mathrm{d}\widetilde{K}, A] + \lambda A$. This follows from (9.4.1) and the observation that for X_p a p-form

$$e_a X_p e^a = (n - 2p)X_p^\eta \tag{10.2.1}$$

so

$$\begin{aligned}\mathcal{L}_K\not{S} &= \mathcal{L}_K e^a S_{X_a} - \lambda\not{S} + e^a\mathcal{L}_K S_{X_a}\\ &= \mathcal{L}_K e^a S_{X_a} - \lambda\not{S} + \not{S}\mathcal{L}_K + e^a S_{[K,\,X_a]} - \tfrac{1}{2}e^a(\mathrm{d}\lambda \wedge e_a)\end{aligned}$$

by (4.4.9). Since $\mathcal{L}_K(e^a(X_b)) = 0$ then $\mathcal{L}_K e^a S_{X_a} + e^a S_{[K,\,X_a]} = 0$, and

$e^a(\mathrm{d}\lambda \wedge e_a) = e^a \wedge (\mathrm{d}\lambda \wedge e_a) + \mathrm{i}^a(\mathrm{d}\lambda \wedge e_a) = X_a(\lambda)e^a - n\mathrm{d}\lambda = (1 - n)\mathrm{d}\lambda$

so $[\mathcal{L}_K, \not{S}] = -\lambda\not{S} - \tfrac{1}{2}(1 - n)\mathrm{d}\lambda$. Since $\not{S}(\lambda\psi) = \mathrm{d}\lambda\psi + \lambda\not{S}\psi$ this may be written as

$$[\mathcal{L}_K + \tfrac{1}{2}(n-1)\lambda, \not{S}] = -\lambda\not{S}. \tag{10.2.2}$$

If K is a Killing vector ($\lambda = 0$) then $\mathcal{L}_K$ commutes with the Dirac operator and if ψ satisfies the Dirac equation (10.1.1) then so does $\mathcal{L}_K\psi$. For the massless case ($\mu = 0$) we also have a covariance for K a conformal Killing vector: if $\not{S}\psi = 0$ then $\not{S}[\mathcal{L}_K + \frac{1}{2}(n-1)\lambda]\psi = 0$.

Out of any two solutions to the Dirac equation we may construct a closed $(n-1)$-form. For definiteness we take $(\, , \,)$ to be a Hermitian-symmetric product on complex spinors with $\xi\eta^*$ as adjoint involution. Then $\mathrm{Re}(\, , \,)$ is a real-valued symmetric product. If we express an $(n-1)$-form $\mathcal{J}$ as $\mathcal{J} = j_a e^a z$, with z the volume n-form, then

$$\mathrm{d}\mathcal{J} = e^b \wedge \nabla_{X_b}(j_a e^a z) = e^b \wedge (\nabla_{X_b} j_a e^a z + j_a \nabla_{X_b} e^a(X_c) e^c z).$$

Now $e^b \wedge (e^a z) = e^b \wedge \mathrm{i}_{X^a} z = -\mathrm{i}_{X^a}(e^b \wedge z) + g^{ab} z = g^{ab} z$, so

$$\mathrm{d}\mathcal{J} = (\nabla_{X^a} j_a + \mathrm{i}_{X^b} \nabla_{X_b} e^a j_a) z. \tag{10.2.3}$$

Taking

$$\mathcal{J} = \mathrm{Re}(\psi, e_a\varphi) e^a z \tag{10.2.4}$$

gives

$$\mathrm{d}\mathcal{J} = \mathrm{Re}(S_{X_a}\psi, e^a\varphi)z + \mathrm{Re}(\psi, \not{S}\varphi)z = -\mathrm{Re}(\not{S}\psi, \varphi)z + \mathrm{Re}(\psi, \not{S}\varphi)z$$

where the covariant derivative S_X is compatible with the spinor product. (This covariant derivative could contain a U(1) coupling.) Thus if $\not{S}\psi = \mu\psi$, for μ real, and similarly for φ, then $\mathrm{d}\mathcal{J} = 0$. In this way we obtain a conserved current (a closed $(n-1)$-form) from any pair of solutions to the field equations. (Had we taken a spinor product with ξ^* as adjoint involution then the form $\mathcal{J}$ would be closed for spinors satisfying the Dirac equation for an imaginary eigenvalue.) If $\widetilde{\psi}$ is the adjoint to ψ with respect to the Hermitian-symmetric product then $(\psi, e_a\varphi)e^a = \mathcal{S}_0(\widetilde{\psi} e_a \varphi)e^a = \mathcal{S}_0(\varphi\widetilde{\psi} e_a)e^a = \mathcal{S}_1(\varphi\widetilde{\psi})$. So the $(n-1)$-form in (10.2.4) can be written as

$$\mathcal{J} = \mathrm{Re}\mathcal{S}_1(\varphi\widetilde{\psi})z = *\mathrm{Re}\mathcal{S}_1(\varphi\widetilde{\psi}). \tag{10.2.5}$$

In particular, taking $\varphi = \mathrm{i}\psi$ in (10.2.4) gives the U(1) current

$$\mathcal{J} = (\psi, \mathrm{i}e_a\psi)e^a z. \tag{10.2.6}$$

This current would provide a source for the equation (such as Maxwell's equation) for any U(1) field entering into the spinor covariant derivative.

We now only consider the Dirac equation without a U(1) coupling. The presence of isometries, generated by a Killing vector K, ensures that if ψ is a solution to the field equations then so is $\mathcal{L}_K\psi$. We thus have the associated closed currents

$$\mathcal{J}_K = \text{Re}(\psi, e_a \mathcal{L}_K \psi) e^a z. \qquad (10.2.7)$$

10.3 The Dirac Equation in Spacetime

In Chapter 5 Maxwell's theory of Electromagnetism was formulated in a Lorentzian spacetime. Together with relativistic mechanics this theory provides a good description of phenomena involving the electromagnetic interactions of charged matter. However, new phenomena sometimes occur (for example, when the energies involved in the interactions exceed certain critical values) that cannot be understood in terms of this theory. For instance, a faint green beam of light continues to liberate electrons from the surface of certain metals even when its intensity is reduced. Or, a strong magnetic field can be used to create pairs of particles. Furthermore, the very stability of atomic matter is not readily comprehensible in terms of a classical theory that predicts radiation from accelerating charged particles. For these and other reasons quantum mechanics was devised. Originally it provided an explanation of non-relativistic phenomena in domains in which classical mechanics was inadequate. The many-body version of this approach (in which the behaviour of a fixed but indefinite number of particles is accommodated) gave rise to a new formalism known as field quantisation. These methods were successfully extended to Maxwell's theory, in which the role of the classical field was replaced by some operator in an infinite-dimensional projective space of photon states. Historically it soon became clear that the classification of elementary particle types in Nature was intimately connected with the dynamical equations involving the respective field operators. Fields were clasified as bosons or fermions according to the observed behaviour of the respective many-body states. This classification was correlated according to whether they carried a representation of the rotation group SO(3) or its covering group SU(2).

It was Dirac's famous equation for the electron–positron field that gave the impetus to the development of relativistic field quantisation and remains a cornerstone in the development of quantum field theory. As a single-particle theory (that is, where particle and antiparticle creation can be ignored to a first approximation) this equation gave a more accurate account of certain atomic spectra and the behaviour of electron beams in weak electromagnetic fields. Ingenious methods have since been invented to include the quantised radiation field in the theory. Some of the refined predictions of quantum electrodynamics provide examples of the most successful predictions in theoretical physics.

Although it is beyond the scope of this book to enter into the realms of the quantum field theory of electrons and positrons it may be noted that such a formalism does require as an important ingredient a basis of solutions to the Dirac equation. These are put into correspondence with a basis of states used in the construction of the quantum theory. In Minkowski space a basis of such free-particle states may be labelled by the eigenvalues of a set of Lie derivatives with respect to a set of commuting Killing vectors.

In recent years field theories on non-flat spaces have become increasingly relevant. We mention three examples. In order to study the behaviour of electrons in a superconducting toroid one must look at spinor fields on a space with a non-trivial topology. Phenomena associated with different types of boundary conditions on the electron field arise and may provide a geometrical interpretation of low-temperature electron states. Secondly, spinor fields on a dynamical string can be formulated in terms of a Dirac equation on a two-dimensional surface. Some believe that such a picture may underlie a viable model for all the basic forces in Nature. Finally we mention that in 1976 great excitement was generated by the construction of certain theories in which spin-$\frac{3}{2}$ fields were coupled to gravity in a manner that gave rise to new symmetries. Such supersymmetries were expected to ameliorate certain difficulties that arose when attempts were made to extend to gravitation the methods used to make successful quantum electrodynamical predictions. It is now thought that such effective-field theories are phenomenological remnants of a more general theory in which spinor fields in higher dimensions play a crucial role.

In any phenomenological description of spinor fields and gravitation there is one aspect that deserves comment here. Although it is possible to construct a symmetric divergenceless stress tensor for a spinor field (this is given in the next section) it does not manifestly satisfy the positive-energy conditions mentioned in Chapter 7. This is analagous to the indefinite sign of the energy of a Dirac field in flat spacetime and is a reflection of the existence of antiparticle states in that case. This is one reason why a quantum interpretation is mandatory in order to give a cogent interpretation to Dirac's theory. In an arbitrary gravitational field, however, there is no natural way to define positive- and negative-energy states and the simple interpretational scheme used to interpret the quantum field theory in a flat space evaporates. It may be of course that the energy conditions are excessively restrictive when applied to spinor fields coupled to gravity, or that in a more fundamental theory of gravitation involving many fields no relevance should be attached to the stress properties of a single field. Although the resolution of this dilemma must await a more coherent synthesis of quantum field theory and geometry it is unlikely that the formulation and properties of spinor field equations on a manifold will cease to be important.

The Dirac equation for a complex spinor (a Dirac spinor with unit charge) ψ on spacetime is

$$\not{\partial}\psi + \mathrm{i}A\psi = m\psi \tag{10.3.1}$$

where we have explicitly exhibited the U(1) interaction with the electromagnetic 1-form potential A. The real eigenvalue m will be interpreted as a mass. The spinor field provides an electromagnetic current 1-form j,

$$j = \mathcal{S}_1(\mathrm{i}\psi\widetilde{\psi}) \tag{10.3.2}$$

where $\widetilde{\psi}$ is the spinor adjoint of ψ with respect to the pseudo-Hermitian product whose adjoint involution is $\xi\eta^*$. The Maxwell 2-form $F = \mathrm{d}A$ satisfies

$$\delta F = j \tag{10.3.3}$$

with δ the co-derivative of (5.4.2).

The electromagnetic current 1-form j is future-pointing and timelike for any spinor ψ. The argument that this is so is algebraic. We first consider the charge density $\rho = (\psi, \mathrm{i}e^0\psi)$. If we took the spinor adjoint as in (2.8.13) then the positivity of ρ would follow immediately. The fact that the spinor adjoint can be cast in this form follows ultimately from the positivity of the metric on the three-dimensional spacelike subspaces. It is instructive to argue the positivity of ρ directly from properties of the various spinor products. Let $\{e^a\}$ be a local orthonormal co-frame and $\hat{z} \equiv \mathrm{i}e^{123}$ such that $\hat{z}^2 = 1$. Let u_ε be a spinor such that $\hat{z}u_\varepsilon = \varepsilon u_\varepsilon$, with $\varepsilon = \pm 1$. Then u_ε carries a semi-spinor representation of the subalgebra generated by $\{e^1, e^2, e^3\}$. Let $(\ ,\ )_{\xi *}$ be the pseudo-Hermitian product associated with $\xi *$ then

$$(u_\varepsilon, u_{\varepsilon'})_{\xi *} = (\varepsilon\hat{z}u_\varepsilon, u_{\varepsilon'})_{\xi *} = \varepsilon(u_\varepsilon, \hat{z}^{\xi *}u_{\varepsilon'})_{\xi *} = \varepsilon\varepsilon'(u_\varepsilon, u_{\varepsilon'})_{\xi *}.$$

If a four-dimensional spinor ψ is decomposed as $\psi = u_+ + u_-$ then

$$(\psi, \psi)_{\xi *} = (u_+, u_+)_{\xi *} + (u_-, u_-)_{\xi *}.$$

We know from §2.7 that $\xi *$ is the adjoint of a zero-index product on the semi-spinors of the three-dimensional subalgebra, whereas the product on four-dimensional spinors is of maximal index. Let us suppose that the product on four-dimensional spinors induces a positive-definite product on u_+ and a negative-definite product on u_-. For three-dimensional semi-spinors we have

$$(u_\varepsilon, \mathrm{i}e^0u_{\varepsilon'}) = \varepsilon(u_\varepsilon, \mathrm{i}\hat{z}^{\xi\eta *}e^0u_{\varepsilon'}) = \varepsilon(u_\varepsilon, \mathrm{i}e^0\hat{z}u_{\varepsilon'}) = \varepsilon\varepsilon'(u_\varepsilon, \mathrm{i}e^0u_{\varepsilon'}).$$

So the charge density ρ is diagonal in the three-dimensional semi-spinors:

$$\rho = (u_+, \mathrm{i}e^0u_+) + (u_-, \mathrm{i}e^0u_-).$$

Now

$$(u_\varepsilon, \mathrm{i}e^0 u_\varepsilon) = \varepsilon(u_\varepsilon, \mathrm{i}e^0\mathrm{i}e^{\mathrm{i}23} u_\varepsilon) = \varepsilon(u_\varepsilon, z u_\varepsilon).$$

The volume 4-form z relates the products associated with $\xi\eta^*$ and ξ^* so that we have $(u_\varepsilon, \mathrm{i}e^0 u_\varepsilon) = \varepsilon(u_\varepsilon, u_\varepsilon)_{\xi^*}$ and

$$\rho = (u_+, u_+)_{\xi^*} - (u_-, u_-)_{\xi^*}. \tag{10.3.4}$$

Thus ρ is positive-definite or zero since the first product is positive-definite and the second negative-definite.

To show that the charge density is positive-definite above we split the four-dimensional spinor into semi-spinors of the three-dimensional subalgebra. This argument implies that $g(\tilde{j}, V)$ is less than or equal to zero for all future pointing timelike vectors V and consequently that $\tilde{j}$ must be a forward-pointing timelike or null vector field. It is instructive to rederive this result using the often useful Fierz rearrangement technique. To this end we will this time split the spinor into two semi-spinors of the even subalgebra. Let $\psi^\pm \equiv \frac{1}{2}(1 \pm \mathrm{i}z)\psi$, that is $\mathrm{i}z\psi^\pm = \pm\psi^\pm$, then

$$(\psi^\varepsilon, a\psi^{\varepsilon'}) = \varepsilon\varepsilon'(\mathrm{i}z\psi^\varepsilon, a\mathrm{i}z\psi^{\varepsilon'}) = \varepsilon\varepsilon'(\psi^\varepsilon, zaz\psi^{\varepsilon'}) = -\,\varepsilon\varepsilon'(\psi^\varepsilon, a^\eta\psi^{\varepsilon'}).$$

So the components of j are diagonal in ψ^+ and ψ^-:

$$j^a = (\psi^+, \mathrm{i}e^a\psi^+) + (\psi^-, \mathrm{i}e^a\psi^-) \equiv j^a_+ + j^a_-. \tag{10.3.5}$$

The norm of j_ε is given by

$$-j^2_\varepsilon = (\psi^\varepsilon, e^a\psi^\varepsilon)(\psi^\varepsilon, e_a\psi^\varepsilon) = \widetilde{\psi}^\varepsilon e^a\psi^\varepsilon\widetilde{\psi}^\varepsilon e_a\psi^\varepsilon.$$

Using (10.2.1)

$$e^a\psi^\varepsilon\widetilde{\psi}^\varepsilon e_a = \sum_{p=0}^{4}(4 - 2p)(-1)^p\mathscr{S}_p(\psi^\varepsilon\widetilde{\psi}^\varepsilon).$$

Now

$$(\psi^\varepsilon\widetilde{\psi}^\varepsilon)^\eta = \mathrm{i}z\psi^\varepsilon\widetilde{\psi}^\varepsilon\mathrm{i}z = -\mathrm{i}z\psi^\varepsilon\widetilde{\mathrm{i}z\psi^\varepsilon} = -\psi^\varepsilon\widetilde{\psi}^\varepsilon$$

and so only odd p enter into the sum. We have

$$\mathscr{S}_3(\psi^\varepsilon\widetilde{\psi}^\varepsilon) = \mathscr{S}_1(\psi^\varepsilon\widetilde{\psi}^\varepsilon\mathrm{i}z)\mathrm{i}z = -\mathscr{S}_1(\psi^\varepsilon\widetilde{\mathrm{i}z\psi^\varepsilon})\mathrm{i}z = -\varepsilon\mathscr{S}_1(\psi^\varepsilon\widetilde{\psi}^\varepsilon)\mathrm{i}z$$

thus

$$\begin{aligned}-j^2_\varepsilon &= -2\widetilde{\psi}^\varepsilon\mathscr{S}_1(\psi^\varepsilon\widetilde{\psi}^\varepsilon)\psi^\varepsilon - 2\varepsilon\widetilde{\psi}^\varepsilon\mathscr{S}_1(\psi^\varepsilon\widetilde{\psi}^\varepsilon)\mathrm{i}z\psi^\varepsilon = -4\widetilde{\psi}^\varepsilon\mathscr{S}_1(\psi^\varepsilon\widetilde{\psi}^\varepsilon)\psi^\varepsilon \\ &= -4\mathscr{S}_0(\psi^\varepsilon\widetilde{\psi}^\varepsilon e_a)\widetilde{\psi}^\varepsilon e^a\psi^\varepsilon = -4(\psi^\varepsilon, e_a\psi^\varepsilon)(\psi^\varepsilon, e^a\psi^\varepsilon) = 4j^2_\varepsilon.\end{aligned}$$

So j_+ and j_- are both null and, since $\rho \geqslant 0$, future pointing. Since the sum of two future-pointing null vectors lies in or on the forward light cone the current j is future pointing, timelike or null.

Exercise 10.3

Consider the 1-form of (10.3.2) on an arbitrary even-dimensional Lorentzian manifold (not necessarily four dimensional). Show that the density ρ is always positive semidefinite but that the argument for j being timelike or null only holds in 2, 4, 6 and 10 dimensions.

We now consider the covariances of the Maxwell–Dirac equations under the isometry group of Minkowski space—the Poincaré group. We noted in §10.2 the covariance of the free ($A = 0$) Dirac equation under Lie derivatives with respect to Killing vectors. To analyse the covariances of the coupled Maxwell–Dirac system it is convenient to work with the finite diffeomorphisms rather than the Lie derivatives. This will also allow a discussion of the discrete orientation-changing transformations.

Let $\{x^a\}$ be global inertial coordinates for Minkowski space, such that $\{dx^a\}$ is a global orthonormal co-frame. We can label the diffeomorphisms forming the Lorentz isometry group by a parallel element of the Clifford group. The diffeomorphism $\pi(s) : M \to M$ is such that

$$\pi^*(s)dx^a = s dx^a s^{-1}. \tag{10.3.6}$$

If a is an arbitrary differential form then $a = a_I dx^I$ with the multi-index I labelling a parallel basis for the exterior (or Clifford) algebra. Then

$$\pi^*(s)a = (a_I \circ \pi(s)) s dx^I s^{-1} \tag{10.3.7}$$

the components of the pulled-back form being composed with the diffeomorphism whilst the change in the basis is effected by Clifford multiplication. This suggests how we can induce an action of the diffeomorphism on a spinor field. Let $\{\mathbf{b}_i\}$ be a standard parallel spinor frame associated with the co-frame $\{dx^a\}$. Then if $\psi = \psi^i \mathbf{b}_i$, we can define

$$\pi(s)\cdot\psi = (\psi^i \circ \pi(s)) s \boldsymbol{b}_i. \tag{10.3.8}$$

Since $dx^I \mathbf{b}_i = \Gamma^I_{ji} \mathbf{b}_j$ for Γ^I_{ji} constants, it follows from (10.3.7) that (10.3.8) satisfies

$$\pi(s)\cdot(a\psi) = (\pi^*(s)a)(\pi(s)\cdot\psi) \qquad \forall\, a \in \Gamma \mathbf{C}(M). \tag{10.3.9}$$

If X is an arbitrary vector field we also have

$$\pi(s)\cdot S_X\psi = S_{\pi_*^{-1}(s)X}(\pi(s)\cdot\psi). \tag{10.3.10}$$

This follows from (10.3.8) since $\nabla_X s = 0$ and $X(\psi^i) \circ \pi(s) = (\pi_*^{-1}(s)X)(\psi^i \circ \pi(s))$. Since $\pi(s)$ is an isometry, the pullback of the Clifford product of two forms is the product of the pulled-back forms. If $\{e^a\}$ and $\{X_a\}$ are dual bases then so are $\{\pi^*(s)e^a\}$ and $\{\pi_*^{-1}(s)X_a\}$, thus

$$\pi(s)\cdot\not{D} = \not{D}\cdot\pi(s). \tag{10.3.11}$$

It immediately follows that if ψ and A satisfy (10.3.1) then so do $\pi(s)\cdot\psi$ and $\pi^*(s)A$ for $\pi(s)$ any Lorentz transformation. The pullback map commutes with the exterior derivative and, in the case of an orientation-preserving isometry, with the Hodge map and hence the co-derivative δ. The pullback with an orientation-reversing isometry picks up a minus sign in moving past a Hodge dual, but since δ involves two duals (or no choice of orientation) the pullback still commutes with it. So if F and j satisfy (10.3.3) then so do $\pi^*(s)F$ and $\pi^*(s)j$. But j is a functional of the spinor field ψ—to symbolise this we will here write $j(\psi)$ for the 1-form determined by (10.3.2). Is it the case that $j(\pi(s)\cdot\psi) = \pi^*(s)j(\psi)$? Equation (10.3.2) involves the spinor adjoint with respect to a product whose invariance group does not contain the whole Clifford group, but only ${}^+\Gamma$. (This is the subgroup defined with the norm μ, so $s^{\xi\eta} = s^{-1}$ for $s \in {}^+\Gamma$.) The image under the vector representation of ${}^+\Gamma$ is the orthochronous Lorentz group. So if $T \in {}^\pm\Gamma$ is such that $\chi(T)$ is a reflection changing the time orientation, then $\pi(T)\cdot\psi$ and $\pi^*(T)A$ will not satisfy the coupled Maxwell–Dirac equations given that ψ and A do.

We know that Lorentz transformations of the cotangent space extend to inner automorphisms of the real Clifford algebra and hence, by complex linearity, to inner automorphisms of the complexified algebra. These inner automorphisms will commute with complex conjugation, and so composing them with complex conjugation gives an outer automorphism of the complexified algebra. A spin transformation on each of a pair of spinors induces an inner automorphism on the Clifford elements formed with a spinor adjoint with respect to a spin-invariant product. That is, $s\varphi\widetilde{s\psi} = \chi(s)\cdot(\varphi\widetilde{\psi})$ if (and only if) s is in the invariance group of the spinor product used to define $\widetilde{\psi}$. As we will see, if instead s is in the real subalgebra such that $(s\varphi, s\psi) = (\varphi, \psi)^*$ for a product on complex spinors then $s\varphi\widetilde{s\psi} = \chi(s)\cdot(\varphi\widetilde{\psi})^*$.

For the four-dimensional Lorentzian case that we are considering the space of complex spinors is the complexification of the real spinor space. The skew-symmetric product on real spinors with adjoint involution $\xi\eta$ is extended by complex bilinearity to a product on complex spinors, $(\ ,\ )_{\xi\eta}$. In an appropriate basis, charge conjugation simply complex conjugates the spinor components and we have

$$(\varphi^c, \psi^c)_{\xi\eta} = (\varphi, \psi)^*_{\xi\eta}. \tag{10.3.12}$$

If we now define

$$(\varphi, \psi) \equiv (\mathrm{i}\varphi^c, \psi)_{\xi\eta} \tag{10.3.13}$$

then $(\ ,\ )$ certainly has $\xi\eta*$ as adjoint involution. The factor of i ensures that the product is Hermitian symmetric:

$$(\varphi, \psi) = (\mathrm{i}\varphi^c, \psi)_{\xi\eta} = -(\mathrm{i}\varphi, \psi^c)^*_{\xi\eta} = (\psi^c, \mathrm{i}\varphi)^*_{\xi\eta} = (\mathrm{i}\psi^c, \varphi)^*_{\xi\eta} = (\psi, \varphi)^*.$$

Since charge conjugation is involutory and the complex bilinear product in (10.3.13) is skew symmetric we have

$$(\varphi^c, \psi^c) = -(\varphi, \psi)^* \tag{10.3.14}$$

If $\widetilde{\psi}$ is the adjoint of ψ with respect to (,) then for any three spinors

$$(\varphi\widetilde{\psi})^*\rho = ((\varphi\widetilde{\psi})\rho^c)^c = ((\psi, \rho^c)\varphi)^c = (\psi, \rho^c)^*\varphi^c = (\psi^{c^2}, \rho^c)^*\varphi^c$$
$$= -(\psi^c, \rho)\varphi^c$$

by (10.3.14). So $(\varphi\widetilde{\psi})^*\rho = -(\varphi^c\widetilde{\psi}^c)\rho$ and

$$(\varphi\widetilde{\psi})^* = -\varphi^c\widetilde{\psi}^c. \tag{10.3.15}$$

Consider now the element $T \in {}^{\pm}\Gamma$ with $\chi(T)$ a time-orientation-changing reflection. Then $T^{\xi\eta *} = T^{\xi\eta} = -T^{-1}$, so

$$T\varphi^c\widetilde{T\psi^c} = -T\varphi^c\widetilde{\psi}^cT^{-1} = T(\varphi\widetilde{\psi})^*T^{-1}$$

by (10.3.15). We now define

$$\mathcal{T}.\psi \equiv \pi(T).\psi^c \tag{10.3.16}$$

and then have

$$\mathcal{T}.\varphi\widetilde{\mathcal{T}.\psi} = \pi^*(T)(\varphi\widetilde{\psi})^*. \tag{10.3.17}$$

The operation $\mathcal{T}$ is known as *Wigner time reversal* on spinors. It obviously satisfies

$$\mathcal{T}.(a\psi) = \pi^*(T)a^*(\mathcal{T}.\psi). \tag{10.3.18}$$

We now examine the covariances of the Maxwell–Dirac system under this operation. If A and ψ satisfy (10.3.1) then so do $-\pi^*(T)A$ and $\mathcal{T}\cdot\psi$. It follows from (10.3.17) that $j(\mathcal{T}\cdot\psi) = -\pi^*(T)j(\psi)$ and hence $-\pi^*(T)A$ and $\mathcal{T}\cdot\psi$ also satisfy (10.3.3). (Notice that whereas $-\pi^*(T)A$ and $\pi(T)\cdot\psi$ satisfy (10.3.3) they do not satisfy (10.3.1).)

Plane-wave solutions play an important part in the physical interpretation of the free ($A = 0$) Dirac equation, and to these we now turn. If $\mathbf{b}$ is a parallel spinor then we look for a solution to (10.3.1), for $\boldsymbol{A} = 0$, of the form $\psi = \exp(\mathrm{i}f)\mathbf{b}$ for f some real function. Then $\not{\partial}\psi = \mathrm{i}\mathrm{d}f\psi$ and we require $\mathrm{i}\mathrm{d}f\psi = m\psi$. It follows that the 1-form $\mathrm{d}f$ must be timelike, with

$$(\mathrm{d}f)^2 = -m^2. \tag{10.3.19}$$

We can write the algebraic condition on $\mathbf{b}$ as

$$\tfrac{1}{2}(1 + \mathrm{i}\mathrm{d}f/m)\mathbf{b} = \mathbf{b}. \tag{10.3.20}$$

If s is a unit spacelike 1-form orthogonal to $\mathrm{d}f$ and z is the volume 4-form then $(zs)^2 = -sz^2s = s^2 = 1$ and $zs\mathrm{d}f = -z\mathrm{d}fs = \mathrm{d}fzs$. So $\frac{1}{2}(1 + zs)$ is an idempotent orthogonal to $\frac{1}{2}(1 + \mathrm{i}\mathrm{d}f/m)$ so that $\frac{1}{2}(1 + \mathrm{i}\mathrm{d}f/m)\frac{1}{2}(1 + zs)$ is primitive. With ε and σ taking the values ± 1 a complete set of pairwise orthogonal primitive idempotents is given by

$$\{P_{\varepsilon\sigma} = \tfrac{1}{2}(1 + \varepsilon \mathrm{i}\mathrm{d}f/m)\tfrac{1}{2}(1 - \sigma z s)\}. \qquad (10.3.21)$$

We can choose a basis of spinors such that each is an eigenspinor of one of these primitive idempotents. An inertial observer would use inertial coordinates $\{t, x, y, z\}$ to interpret $\mathrm{d}f(\partial/\partial t)$ as an energy and $\mathrm{d}f(\partial/\partial x)$ as a component of momentum along the x-axis.

If we assume that $\mathrm{d}f$ and s are parallel then we can choose inertial coordinates $\{t, x, y, z\}$ such that $f = mt$ and $s = \mathrm{d}x$. Then we can label plane-wave solutions by ε and σ,

$$\psi_{\varepsilon\sigma} = \exp(\mathrm{i}\varepsilon m t)\boldsymbol{b}_{\varepsilon\sigma} \qquad (10.3.22)$$

where $\boldsymbol{b}_{\varepsilon\sigma} = P_{\varepsilon\sigma}\boldsymbol{b}_{\varepsilon\sigma}$ for

$$P_{\varepsilon\sigma} = \tfrac{1}{2}(1 + \mathrm{i}\varepsilon \mathrm{d}t)\tfrac{1}{2}(1 + \sigma \mathrm{d}y\mathrm{d}z\mathrm{d}t) \qquad (10.3.23)$$

and we have chosen $z = \mathrm{d}x\mathrm{d}y\mathrm{d}z\mathrm{d}t$. If we choose some parallel $\mathbf{b}_{++}$ then we can build up the rest of the spinor basis by taking Clifford products. For example, we have $\mathrm{d}xP_{\varepsilon\sigma} = P_{-\varepsilon-\sigma}\mathrm{d}x$ and $\mathrm{d}yP_{\varepsilon\sigma} = P_{-\varepsilon\sigma}\mathrm{d}y$ and hence $\mathrm{d}x\mathrm{d}yP_{\varepsilon\sigma} = P_{\varepsilon-\sigma}\mathrm{d}x\mathrm{d}y$. So we may choose the basis as

$$\{\mathbf{b}_{++}, \mathbf{b}_{--} = \mathrm{d}x\mathbf{b}_{++}, \mathbf{b}_{-+} = \mathrm{d}y\mathbf{b}_{++}, \mathbf{b}_{+-} = \mathrm{d}x\mathrm{d}y\mathbf{b}_{++}\}. \qquad (10.3.24)$$

If (,) has $\xi\eta^*$ as adjoint then we may use the algebraic properties of this basis to work out the non-vanishing products, we have

$$\begin{aligned}(\mathbf{b}_{\varepsilon\sigma}, \mathbf{b}_{\varepsilon'\sigma'}) &= (P_{\varepsilon\sigma}\mathbf{b}_{\varepsilon\sigma}, P_{\varepsilon'\sigma'}\mathbf{b}_{\varepsilon'\sigma'}) = (\mathbf{b}_{\varepsilon\sigma}, P_{\varepsilon\sigma}{}^{\xi\eta^*}P_{\varepsilon'\sigma'}\mathbf{b}_{\varepsilon'\sigma'}) \\ &= (\mathbf{b}_{\varepsilon\sigma}, P_{-\varepsilon\sigma}P_{\varepsilon'\sigma'}\mathbf{b}_{\varepsilon'\sigma'}) = \delta_{-\varepsilon\varepsilon'}\delta_{\sigma\sigma'}(\mathbf{b}_{\varepsilon\sigma}, \mathbf{b}_{\varepsilon'\sigma'}).\end{aligned}$$

Thus the only non-zero independent products are $(\mathbf{b}_{++}, \mathbf{b}_{-+})$ and $(\mathbf{b}_{+-}, \mathbf{b}_{--})$. If we choose the basis as in (10.3.24) then these are related for

$$(\mathbf{b}_{+-}, \mathbf{b}_{--}) = (\mathrm{d}x\mathrm{d}y\mathbf{b}_{++}, \mathrm{d}x\mathbf{b}_{++}) = (\mathbf{b}_{++}, \mathrm{d}y\mathbf{b}_{++}) = (\mathbf{b}_{++}, \mathbf{b}_{-+}).$$

So by suitably scaling $\mathbf{b}_{++}$ we have

$$(\mathbf{b}_{++}, \mathbf{b}_{-+}) = (\mathbf{b}_{+-}, \mathbf{b}_{--}) = 1. \qquad (10.3.25)$$

Thus the two-dimensional subspaces with fixed ε are isotropic, whilst those with fixed σ are unitary subspaces of maximal index.

The ε-label of $\psi_{\varepsilon\sigma}$ specifies the eigenvalue of the spinor Lie derivative in the $\partial/\partial t$ direction,

$$\mathcal{L}_{\partial/\partial t}\psi_{\varepsilon\sigma} = \mathrm{i}\varepsilon m\psi_{\varepsilon\sigma}. \qquad (10.3.26)$$

Similarly σ may be used to label the eigenvalue of the Lie derivative with respect to the vector $y\partial/\partial z - z\partial/\partial y$ that generates rotations about the x-axis, we have

$$\mathcal{L}_{(y\partial/\partial z - z\partial/\partial y)}\Psi_{\varepsilon\sigma} = \tfrac{1}{4}\overline{\mathrm{d}(y\partial/\partial z - z\partial/\partial y)}\psi_{\varepsilon\sigma}$$
$$= \tfrac{1}{2}\mathrm{d}y\mathrm{d}z\psi_{\varepsilon\sigma} = \tfrac{1}{2}\mathrm{i}\mathrm{d}y\mathrm{d}z\mathrm{d}t\mathrm{i}\mathrm{d}t\psi_{\varepsilon\sigma} = \tfrac{1}{2}\mathrm{i}\varepsilon\mathrm{d}y\mathrm{d}z\mathrm{d}t\psi_{\varepsilon\sigma}$$
$$\mathcal{L}_{(y\partial/\partial z - z\partial/\partial y)}\Psi_{\varepsilon\sigma} = \tfrac{1}{2}\mathrm{i}\varepsilon\sigma\psi_{\varepsilon\sigma}. \tag{10.3.27}$$

The eigenvalues of $\pm\frac{1}{2}\mathrm{i}$ lead to the physical interpretation of an intrinsic spin of a half for the electron. More generally, the functional dependence of the components will contribute an orbital angular momentum, the eigenvalue of the Lie derivative being interpreted as the total angular momentum.

10.4 The Stress Tensor

Although we have not done so the Dirac equation can be obtained from a variational principle. This ensures the existence of a symmetric stress tensor which is divergenceless when the field equations hold. We here simply present such a tensor and explicitly demonstrate (not so simply) that its divergence is zero for solutions to the Dirac equation.

For definiteness we take (,) to be a Hermitian-symmetric spinor product with $\xi\eta^*$ as adjoint involution, then Re(,) is real valued and symmetric. Let

$$\mathcal{T}_{ab} = \mathrm{Re}(\psi, e_a S_{X_b}\psi) + \mathrm{Re}(\psi, e_b S_{X_a}\psi). \tag{10.4.1}$$

If S is compatible with the spinor product then

$$\begin{aligned} X^a(\mathcal{T}_{ab}) &= \mathrm{Re}(S_{X^a}\psi, e_a S_{X_b}\psi) + \mathrm{Re}(\psi, \nabla_{X^a}e_a S_{X_b}\psi) \\ &\quad + \mathrm{Re}(\psi, e_a S_{X^a}S_{X_b}\psi) + \mathrm{Re}(S_{X^a}\psi, e_b S_{X_a}\psi) \\ &\quad + \mathrm{Re}(\psi, \nabla_{X^a}e_b S_{X_a}\psi) + \mathrm{Re}(\psi, e_b S_{X^a}S_{X_a}\psi). \end{aligned} \tag{10.4.2}$$

Changing the order of the covariant derivatives

$$\begin{aligned} e_a S_{X^a}S_{X_b}\psi &= e_a \mathbf{S}(X^a, X_b)\psi + e_a S_{X_b}S_{X^a}\psi + e^a S_{[X_a, X_b]}\psi \\ &= e^a \mathbf{S}(X_a, X_b)\psi + S_{X_b}\not{S}\psi - \nabla_{X_b}e^a S_{X_a}\psi \\ &\quad + e^c(\nabla_{X_a}X_b - \nabla_{X_b}X_a)e^a S_{X_c}\psi \end{aligned}$$

if ∇ is torsion free. Now

$$e^c(\nabla_{X_b}X_a)e^a = -\nabla_{X_b}e^c(X_a)e^a = -\nabla_{X_b}e^c$$

and $e^c(\nabla_{X_a}X_b) = -\nabla_{X_a}e^c(X_b)$. From (9.3.20) $e^a\mathbf{S}(X_a, X_b)\psi = \frac{1}{2}e^p R_{pb}\psi$, and for zero torsion this can be written in terms of the Ricci forms, $e^a\mathbf{S}(X_a, X_b)\psi = \frac{1}{2}P_b\psi$, so

$$e_a S_{X^a} S_{X_b} \psi = e^a \mathbf{S}(X_a, X_b)\psi + S_{X_b} \$\psi - \nabla_{X_a} e^c(X_b) e^a S_{X_c} \psi. \tag{10.4.3}$$

From (10.1.4) we have

$$S_{X^a} S_{X_a} \psi = \$^2 \psi - \nabla_{X_a} e^c(X_a) S_{X_c} \psi + \tfrac{1}{4} \mathcal{R} \psi. \tag{10.4.4}$$

We can rewrite $\nabla_{X^a} e_a$ as $(\nabla_{X^a} e_a)(X_c) e^c = e^c(\nabla_{X_a} X^a) e_c = -\nabla_{X_a} e^c(X^a) e_c$, and similarly $\nabla_{X^a} e_b = \nabla_{X^a} e_b(X_c) e^c = -e_b(\nabla_{X^a} X_c) e^c = -\nabla_{X^a} e^c(X_b) e_c$, collecting terms,

$$\begin{aligned}
X^a(\mathcal{T}_{ab}) = {} & \mathrm{Re}(S_{X^a}\psi, e_a S_{X_b}\psi) + \mathrm{Re}(S_{X^a}\psi, e_b S_{X_a}\psi) \\
& - \nabla_{X_a} e^c(X^a) \mathrm{Re}(\psi, e_c S_{X_b}\psi) - \nabla_{X_a} e^c(X^a) \mathrm{Re}(\psi, e_b S_{X_c}\psi) \\
& - \nabla_{X_a} e^c(X_b) \mathrm{Re}(\psi, e^a S_{X_c}\psi) - \nabla_{X_a} e^c(X_b) \mathrm{Re}(\psi, e_c S_{X^a}\psi) \\
& + \tfrac{1}{2}\mathrm{Re}(\psi, P_b\psi) + \mathrm{Re}(\psi, S_{X_b}\$\psi) \\
& + \mathrm{Re}(\psi, e_b \$^2 \psi) + \tfrac{1}{4}\mathrm{Re}(\psi, e_b \mathcal{R}\psi).
\end{aligned} \tag{10.4.5}$$

If $\mathcal{T} = \mathcal{T}_{ab} e^a \otimes e^b$ then

$$\begin{aligned}
\nabla_{X_a} \mathcal{T}(X^a, X_b) &= X_a(\mathcal{T}^a{}_b) + \nabla_{X_a} e^c(X^a) \mathcal{T}_{cb} + \nabla_{X_a} e^c(X_b) \mathcal{T}^a{}_c \\
&= \mathrm{Re}(S_{X^a}\psi, e_b S_{X_a}\psi) + \mathrm{Re}(S_{X^a}\psi, e_a S_{X_b}\psi) \\
&\quad + \mathrm{Re}(\psi, S_{X_b}\$\psi) + \mathrm{Re}(\psi, e_b \$^2\psi) \\
&\quad + \tfrac{1}{2}\mathrm{Re}(\psi, P_b\psi) + \tfrac{1}{4}\mathrm{Re}(\psi, e_b \mathcal{R}\psi).
\end{aligned}$$

Since the spinor product is symmetric with $\xi\eta^*$ as adjoint then $\mathrm{Re}(\varphi, A\psi) = -\mathrm{Re}(\psi, A\varphi)$ for A any real 1-form and

$$\nabla_{X_a} \mathcal{T}(X^a, X_b) = -\mathrm{Re}(\$\psi, S_{X_b}\psi) + \mathrm{Re}(\psi, S_{X_b}\$\psi) + \mathrm{Re}(\psi, e_b \$^2\psi).$$

It follows that $\nabla . \mathcal{T} = 0$ if $\$\psi = m\psi$ with m real.

The above is seen to go through unaltered for real spinors with a spinor product whose adjoint involution is $\xi\eta$. Had we taken a real-valued skew-symmetric spinor product on complex spinors with ξ^* as adjoint then $\mathcal{T}$ would be divergenceless for $\$\psi = \mathrm{i}m\psi$. For real spinors and a skew-symmetric product with ξ as adjoint the stress tensor would be divergenceless if $\$\psi = 0$.

Exercise 10.4
Show that the Maxwell–Dirac stress tensor is divergenceless when the coupled equations are satisfied.

For the stress tensor of (10.4.1) the trace is given by $\mathcal{T}_a{}^a = 2\mathrm{Re}(\psi, \$\psi)$. When the Dirac equation is satisfied we have

$$\mathcal{T}_a{}^a = 2m(\psi, \psi). \tag{10.4.6}$$

Certainly for m zero the trace is zero. In general the spinor product will be pseudo-Hermitian and so for $m \neq 0$ the trace can still vanish.

We have already noted in §7.4 that we can construct a closed $(n - 1)$-form from the stress tensor and a Killing vector, namely $J_K = \mathcal{T}_{ab}K^b e^a z$ where $\mathcal{T}_{ab}$ are the components of the stress tensor $\mathcal{T}$, given by (10.4.1), which is divergenceless when the field equations $\not{S}\psi = m\psi$ are imposed. In §10.2 we obtained by inspection a closed $(n - 1)$-form $\mathcal{J}_K$ for each Killing vector K. These two forms, J_K and $\mathcal{J}_K$, in fact differ by an exact form modulo the field equations, as we now demonstrate. We are going to have to recognise the exterior derivative of an $(n - 2)$-form when we see one, so first we note that if $H = H_{ab}e^{ab}z$ then

$$\mathrm{d}H = 2\{X^b(H_{ba}) - \nabla_{X_c}e^c(X^b)H_{ba} - \nabla_{X^b}e_a(X^c)H_{bc}\}e^a z \equiv (\mathrm{d}H)_a e^a z. \tag{10.4.7}$$

A fairly tedious calculation produces

$$\begin{aligned}\mathrm{Re}(\psi, \widetilde{K}S_{X_a}\psi) = \tfrac{1}{2}\mathrm{Re}(\psi, e_a\mathrm{d}\widetilde{K}\psi) - \tfrac{1}{8}(\mathrm{d}H)_a + \mathrm{Re}(\psi, e_a S_K\psi)\\ - \tfrac{1}{2}\mathrm{Re}(\psi, e_a\widetilde{K}\not{S}\psi) + \tfrac{1}{2}\mathrm{Re}(\psi, \widetilde{K}e_a\not{S}\psi)\end{aligned} \tag{10.4.8}$$

where $H_{ba} = \mathrm{Re}(\psi, e_b\widetilde{K}e_a\,\psi) - \mathrm{Re}(\psi, e_{ba}\widetilde{K}\psi)$. As well as frequently using the defining anticommutation relation of the Clifford algebra the calculation uses the fact that since the spinor product has $\xi\eta^*$ as adjoint then $\mathrm{Re}(\psi, A\psi) = 0$ for A any real 1-form. Thus for example $\mathrm{Re}(\psi, e_a\widetilde{K}e^b\psi) = -\mathrm{Re}(\psi, e^b\widetilde{K}e_a\psi)$, as is necessary for $H_{ab} = -H_{ba}$. (Although it is tedious we recommend that the reader verify (10.4.8), as it does help develop the calculational proficiency that unfortunately is sometimes required.) It follows from (10.4.8) that

$$\begin{aligned}&\mathrm{Re}(\psi, e_a S_K\psi) + \mathrm{Re}(\psi, \widetilde{K}S_{X_a}\psi)\\ &\qquad = 2\mathrm{Re}(\psi, e_a\mathcal{L}_K\psi) - \tfrac{1}{8}(\mathrm{d}H)_a + \mathrm{Re}(\psi, (\widetilde{K}\wedge e_a)\not{S}\psi).\end{aligned}$$

If we use the field equations, $\not{S}\psi = m\psi$, then

$$\mathrm{Re}(\psi, (\widetilde{K}\wedge e_a)\not{S}\psi) = m\mathrm{Re}(\psi, (\widetilde{K}\wedge e_a)\psi) = 0$$

since a real 2-form changes sign under $\xi\eta^*$, the adjoint involution of the spinor product. Thus $J_K = 2\mathcal{J}_K$ modulo an exact form, modulo the field equations.

Exercise 10.5

Repeat the analysis with a skew product whose adjoint is ξ^* with field equations $\not{S}\psi = \mathrm{i}m\psi$.

Example 10.1 Gravitational and Neutrino Waves
Consider a spacetime in which the metric takes the form

$$g = 2(\mathrm{d}u \otimes \mathrm{d}v + \mathrm{d}v \otimes \mathrm{d}u - 2H\mathrm{d}u \otimes \mathrm{d}u + \mathrm{d}z \otimes \mathrm{d}z^* + \mathrm{d}z^* \otimes \mathrm{d}z)$$

in coordinates (u, v, x^1, x^2) with $z \equiv x^1 + \mathrm{i}x^2$ and H a real function of u, z and z^*. It is here most convenient to adopt a null basis. We choose the null co-frame $\{n^a\}$ $a = 1, 2, 3, 4$ where $n^1 = \mathrm{d}u$, $n^2 = \mathrm{d}v - H\mathrm{d}u$, $n^3 = \mathrm{d}z$, $n^4 = \mathrm{d}z^*$ and the duals are $X_1 = \partial/\partial u + H\partial/\partial v$, $X_2 = \partial/\partial v$, $X_3 = \partial/\partial z$, $X_4 = \partial/\partial z^*$. The non-vanishing components of the metric are $g_{12} = g_{21} = g_{34} = g_{43} = 2$, or $g^{12} = g^{21} = g^{34} = g^{43} = \frac{1}{2}$. Since the components of the metric are constant in this basis we can evaluate the connection forms by (6.6.8), the non-vanishing ones being

$$\omega_{31} = -\omega_{13} = 2H_{,z}n^1 \qquad \omega_{41} = -\omega_{14} = 2H_{,z^*}n^1.$$

The only non-zero Ricci form is $P_1 = 2H_{,z^*z}n^1$.

We now adapt a spinor frame to this null co-frame. Let $\mathbf{b}_1$ be a spinor such that $n^1\mathbf{b}_1 = n^3\mathbf{b}_1 = 0$. We then form the spinor frame

$$\{\mathbf{b}_1, \mathbf{b}_2 = n^2\mathbf{b}_1, \mathbf{b}_3 = n^4\mathbf{b}_1, \mathbf{b}_4 = n^2n^4\mathbf{b}_1\}.$$

(We can represent the spinor $\mathbf{b}_1$ by the differential form n^1n^3, this lying in a minimal left ideal of the complexified Clifford algebra. The other spinors $\{\mathbf{b}_i\}$ are then seen to complete the basis for the minimal left ideal.) If $\{X_a\}$ is the frame dual to $\{n^a\}$ then, for σ_X defined in (8.1.5), we have $\sigma_{X_1} = (H_{,z}n^3 + H_{,z^*}n^4)n^1$ with all other σ_{X_i}, zero. It follows that the spinors $\mathbf{b}_1$ and $\mathbf{b}_3$ are parallel. Hence if h_1 and h_2 are arbitrary complex functions of u and $\psi = h_1(u)\mathbf{b}_1 + h_2(u)\mathbf{b}_3$ then $\not{S}\psi = 0$. To obtain Einstein's equations we now need to evaluate the spinor stress tensor. If (,) is the Hermitian-symmetric spinor product with $\xi\eta^*$ as adjoint then we can use the algebraic properties of the spinor frame to evaluate the products. For example,

$$(\mathbf{b}_1, \mathbf{b}_3) = (\mathbf{b}_1, n^4\mathbf{b}_1) = (n^4\mathbf{b}_1, \mathbf{b}_1)^* = -(\mathbf{b}_1, n^3\mathbf{b}_1)^* = 0$$

since $n^3\mathbf{b}_1 = 0$. Also $\mathbf{b}_2 = n^2\mathbf{b}_1$ and so $n^1\mathbf{b}_2 = n^1n^2\mathbf{b}_1 = (1 - n^2n^1)\mathbf{b}_1 = \mathbf{b}_1$ and hence $(\mathbf{b}_1, \mathbf{b}_1) = (n^1\mathbf{b}_2, n^1\mathbf{b}_2) = -(\mathbf{b}_2, n^1n^1\mathbf{b}_2) = 0$. In this way we can show that the non-vanishing products are specified by the imaginary components $(\mathbf{b}_1, \mathbf{b}_2) = (\mathbf{b}_3, \mathbf{b}_4)$. By suitably normalising $\mathbf{b}_1$ we have

$$(\mathbf{b}_1, \mathbf{b}_2) = (\mathbf{b}_3, \mathbf{b}_4) = \mathrm{i}.$$

The only non-zero component of the stress tensor of (10.4.1) is then

$$\mathcal{T}_{11} = 4\mathrm{Re}(\mathrm{i}h_1^*h_1' + \mathrm{i}h_2^*h_2').$$

Since for zero mass the spinor stress tensor is traceless we can write the Einstein equations as $2\kappa P_c = *^{-1}\tau_c$, and so the coupled system reduces to the equation

$$\kappa H_{z^*z} = \mathrm{Re}(\mathrm{i}h_1^* h_1' + \mathrm{i}h_2^* h_2').$$

10.5 Tensor Spinors

Starting with the spinor representation of the spin group we can build up higher-dimensionsal irreducible representations by forming tensor products. That is, tensor products of the spinor space and its dual space carry representations of the spin group, this space of tensors being decomposable into irreducible representation spaces. The covariant derivative on spinor fields induces a covariant derivative on these spin tensors and one can consider various field equations. We have already noted that elements of the Clifford algebra can be identified with (1, 1) tensors on the space of spinors. Certain higher-dimensional half-integral irreducible representations of the spin group can be found by taking the tensor product of tensors on the vector space V with the spinor space of $\mathbf{C}(V, g)$. Such objects can be thought of as spinor-valued tensors.

As an example we consider a spinor-valued 1-form Ψ on spacetime. Then we can write this in any co-frame $\{e^a\}$ as

$$\Psi = \psi_a \otimes e^a \tag{10.5.1}$$

where each ψ_a is a spinor. We can think of Ψ as a mapping from vector to spinor fields:

$$\Psi(X) = \psi_a e^a(X) \qquad \forall X \in \Gamma TM\ . \tag{10.5.2}$$

Equivalently if $\{\mathbf{b}_i\}$ is any standard spinor frame with $\psi_a = \psi_a^i \mathbf{b}_i$ then we can write Ψ as

$$\Psi = \mathbf{b}_i \otimes \psi^i \tag{10.5.3}$$

with the 1-forms ψ^i given by $\psi^i = \psi_a^i e^a$. These spinors could carry irreducible representations of the complexified Clifford algebra, its even subalgebra or real subalgebra (Dirac, Weyl or Majorana spinors). Let us suppose that the ψ_a are Weyl spinors, satisfying $\mathrm{i}z\psi_a = \psi_a$. Then the ψ_a carry irreducible representations of the spin group $\mathrm{Sl}(2, C)$. A 1-form is a tensor on the space of spinors, Clifford multiplication interchanging the semi-spinor spaces (since a 1-form anticommutes with the volume 4-form). So we may regard a spinor-valued 1-form as a degree-three tensor on the spinor space. If u and v are any two Weyl spinors, lying in the same semi-spinor space as the ψ_a, then we define

$$\Psi(u, v) \equiv (u, \psi_a)e^a v. \tag{10.5.4}$$

The brackets on the left-hand side signify that Ψ is evaluated on u and v, whereas the brackets on the right-hand side are the spinor product of u and ψ_a where the product has ξ as adjoint involution. (The skew-symmetric complex bilinear product on Dirac spinors induces a non-degenerate product on each of the two spaces of Weyl spinors. If $u = \mathrm{i}zu$ then the spinor product (u, ψ_a) will only involve $\frac{1}{2}(1 + \mathrm{i}z)\psi_a$.) It turns out [9] that irreducible Sl(2, C) representations are carried by spin tensors that are totally symmetric in the covariant and contravariant arguments separately. It is therefore interesting to examine the condition on Ψ such that (10.5.4) defines a mapping symmetric in u and v. In order to do this we will need the following:

$$(u, v)w - (w, v)u = 4(u, w)v \tag{10.5.5}$$

for u, v and w any three Weyl spinors. To see this let α be another Weyl spinor and consider the expression $(u, v)(w, \alpha)$. Using $\widetilde{u}$ to denote the adjoint spinor we can write this as $\widetilde{u}\, v\, \widetilde{w}\, \alpha$. Now we can expand $v\, \widetilde{w}$ as in (2.1.18) to give

$$\begin{aligned}(u, v)(w, \alpha) &= \widetilde{u}\,\mathscr{S}_0(v\,\widetilde{w}\,e_A^{\xi})e^A\alpha = \mathscr{S}_0(\widetilde{w}\,e_A^{\xi}v)\,\widetilde{u}\,e^A\alpha \\ &= (w, e_A^{\xi}v)(u, e^A\alpha).\end{aligned}$$

Now for u and v Weyl spinors and a any Clifford form

$$(v, au) = (\mathrm{i}zv, a\mathrm{i}zu) = (v, zaz^{-1}u) = (v, a^{\eta}u)$$

so $(v, au) = 0$ for a odd. In addition

$$(v, au) = (a^{\xi}v, u) = -(u, a^{\xi}v)$$

so for $a^{\xi} = -a$ $(a^{\xi} = a)$ then (v, au) is symmetric (skew) in u and v. So

$$(u, v)(w, \alpha) - (u, w)(v, \alpha) = (w, e_A^{\xi}v)(u, e^A\alpha) - (w \leftrightarrow v)$$

and the first bracket on the right-hand side will only contain those e_A^{ξ} that are even under η and under ξ. These are the 0-forms and the 4-forms, thus

$$(u, v)(w, \alpha) - (u, w)(v, \alpha) = 2(w, v)(u, \alpha) - 2(w, zv)(u, z\alpha).$$

Since v and α satisfy $zv = -\mathrm{i}v$ and $z\alpha = -\mathrm{i}\alpha$ the terms on the right-hand side add up. We can use the skew symmetry of the product to rewrite the left-hand side, producing

$$(w, u)(v, \alpha) - (v, u)(w, \alpha) = 4(w, v)(u, \alpha).$$

Since this is true for all α and the spinor product is non-degenerate

$$(w, u)v - (v, u)w = 4(w, v)u.$$

This is just (10.5.5) with the spinors cyclically permuted. We can now use (10.5.5) and (10.5.4) to see that

$$\Psi(u, v) - \Psi(v, u) = 4(u, v)e^a \psi_a.$$

Thus the spinor-valued 1-form is an irreducible spin tensor if it is 'traceless':

$$e^a \psi_a = 0. \tag{10.5.6}$$

Exercise 10.6
Use the correspondence between 1-forms and (1, 1) spin tensors given at the end of §2.8 to label the components of a spinor-valued 1-form with one 'dotted' and two 'undotted' indices. Show that the 'tracelessness' condition is equivalent to symmetry in the two like indices.

The spinor covariant derivative S_X and the covariant derivative ∇_X can be extended by the Leibniz rule to a covariant derivative, also denoted S_X, on spinor-valued 1-forms. In the obvious way

$$S_X \Psi = S_X \psi_a \otimes e^a + \psi_a \otimes \nabla_X e^a . \tag{10.5.7}$$

(If any confusion is likely between the covariant derivative on spinor-valued 1-forms and that on spinors we can write the former as $S_X^{3/2}$.) A representation of the Clifford algebra on spinor-valued 1-forms can be defined by

$$a\Psi \equiv (a\psi_b) \otimes e^b \tag{10.5.8}$$

so that we have a Dirac-like equation

$$\not{S}\Psi = m\Psi. \tag{10.5.9}$$

The pair of equations (10.5.6) and (10.5.9) are the *Rarita–Schwinger equations* for spin 3/2 [25].

Exercise 10.7
Show that (10.5.6) and (10.5.9) imply the 'Lorenz' condition $(S_{X_a}\Psi)(X^a) = 0$.

In Minkowski space we can pick a parallel co-frame such that (10.5.9) reduces to four Dirac equations. We can then find plane-wave solutions as in §10.3. If $\{\mathbf{b}_{\varepsilon\sigma}\}$ is the spinor basis of (10.3.24) then we have Dirac solutions as in (10.3.22) with the sign of the frequency correlated with the ε labelling the basis spinors. By tensoring on four independent 1-forms to the two basis spinors with (say) $\varepsilon = +1$ we can form eight linearly independent spinor-valued 1-forms. We can choose four of these satisfying the tracelessness condition (10.5.6). The eight spinor-valued 1-forms can be chosen as eigenstates of the Lie derivatives with respect to vectors generating time translations and rotations about the x-axis. The 1-form basis can be chosen to have eigenvalues of $\{\mathrm{i}, -\mathrm{i}, 0, 0\}$

under the Lie derivative with respect to the rotation, whereas the spinor basis has eigenvalues $\{\frac{1}{2}\mathrm{i}, -\frac{1}{2}\mathrm{i}\}$. The four traceless spinor-valued 1-forms are then seen to have eigenvalues $\{\frac{3}{2}\mathrm{i}, \frac{1}{2}\mathrm{i}, -\frac{1}{2}\mathrm{i}, -\frac{3}{2}\mathrm{i}\}$. For the basis of (10.3.24)

$$\mathrm{d}x\mathbf{b}_{\varepsilon\sigma} = \mathbf{b}_{-\varepsilon-\sigma}, \ \mathrm{i}\mathrm{d}t\mathbf{b}_{\varepsilon\sigma} = \varepsilon\mathbf{b}_{\varepsilon\sigma}, \ \mathrm{d}y\mathbf{b}_{\varepsilon\sigma} = \sigma\mathbf{b}_{-\varepsilon\sigma}, \ \mathrm{d}z\mathbf{b}_{\varepsilon\sigma} = \mathrm{i}\varepsilon\mathbf{b}_{-\varepsilon\sigma} \tag{10.5.10}$$

so a basis for positive-frequency solutions to (10.5.6) and (10.5.9) is

$$\{\mathbf{b}_{++} \otimes (\mathrm{d}z - \mathrm{i}\mathrm{d}y), \mathbf{b}_{+-} \otimes (\mathrm{d}z - \mathrm{i}\mathrm{d}y) - 2\mathrm{i}\mathbf{b}_{++} \otimes \mathrm{d}x,$$
$$\mathbf{b}_{++} \otimes (\mathrm{d}z + \mathrm{i}\mathrm{d}y) - 2\mathrm{i}\mathbf{b}_{+-} \otimes \mathrm{d}x, \mathbf{b}_{+-} \otimes (\mathrm{d}z + \mathrm{i}\mathrm{d}y)\}. \tag{10.5.11}$$

These are eigenstates of $\mathscr{L}_{y\partial/\partial z - z\partial/\partial y}$, arranged in decreasing order of eigenvalues.

A spinor-valued 1-form features in the theory of supergravity [8]. This theory involves a connection with torsion. As we remarked in §9.3 the definition of the spinor covariant derivative S_X in terms of the metric-compatible connection ∇ does not rely on ∇ being torsion-free. So in this case we could still adopt (10.5.7) as the definition of a covariant derivative on spinor-valued 1-forms. The field equation for the spinor-valued 1-form in supergravity, however, is most readily expressed in terms of another connection. If $\{T^a\}$ are the torsion 2-forms of the connection ∇ then a covariant derivative on differential forms is defined by

$$\hat{\nabla}_X \equiv \nabla_X + \tfrac{1}{2}\mathrm{i}_X T^a \wedge \mathrm{i}_{X_a}. \tag{10.5.12}$$

From (6.7.4) we see that $\hat{\nabla}$ is just such that

$$e^a \wedge \hat{\nabla}_{X_a} = \mathrm{d}. \tag{10.5.13}$$

If S_X is the spinor covariant derivative associated with ∇ then a covariant derivative $\hat{S}_X$ on spinor-valued p-forms is defined by

$$\hat{S}_X\Psi \equiv S_X\psi_I \otimes e^I + \psi_I \otimes \hat{\nabla}_X e^I \tag{10.5.14}$$

where e^I is a p-form basis. For Ψ a spinor-valued p-form we may adopt the convention that for a any q-form

$$a \wedge \Psi \equiv \psi_I \otimes a \wedge e^I. \tag{10.5.15}$$

The *spinor covariant exterior derivative* D maps spinor-valued p-forms to spinor-valued $(p + 1)$-forms:

$$\mathrm{D}\Psi \equiv e^a \wedge \hat{S}_{X_a}\Psi. \tag{10.5.16}$$

If $\{\mathbf{b}_i\}$ is a standard spinor frame associated with some orthonormal co-frame then we may expand Ψ as $\Psi = \mathbf{b}_i \otimes \psi^i$ where the ψ^i are a set of p-forms. Then we can equivalently write the spinor covariant exterior derivative as

$$\mathrm{D}\Psi = \mathbf{b}_i \otimes \mathrm{d}\psi^i + \tfrac{1}{4}e^{pq}\mathbf{b}_i \otimes \omega_{pq} \wedge \psi^i. \qquad (10.5.17)$$

The Hodge dual of a spinor-valued p-form is defined in the obvious way, in analogy to (10.5.15). If N is a Clifford-valued q-form, $N = n^A \otimes e_A$ for n^A arbitrary Clifford forms and e_A a basis for q-forms then we choose to define

$$N\Psi = n^A\psi_I \otimes e_A \wedge e^I. \qquad (10.5.18)$$

Having adopted these conventions we consider the equation

$$e{*}\mathrm{D}\Psi = 0 \qquad (10.5.19)$$

for a spinor-valued 1-form Ψ where $e \equiv e^a \otimes e_a$. This equation is one of the field equations occurring in the theory of supergravity. Although it is usually known as the Rarita–Schwinger equation this equation is not obtained by simply putting m to zero in equations (10.5.6) and (10.5.9). The relationship between these equations is contained in the following exercise.

Exercise 10.8
(i) Show that if Ψ is a spinor-valued 1-form then

$$*(e{*}\mathrm{D}\Psi) = S_{X_a}(e_c\psi^c) \otimes e^a - \not{S}\Psi.$$

Hint: you will need $*(e^c \wedge *e^{ab}) = g^{bc}e^a - g^{ac}e^b$.
(ii) Show that if φ is a spinor field then

$$e{*}\mathrm{D}^2\varphi = e^b\mathbf{S}(X_b, X_a)\varphi \otimes *e^a.$$

Hence show that if the Ricci and torsion forms are zero (10.5.19) has the 'gauge' symmetry $\Psi \mapsto \Psi + \mathrm{D}\varphi$.

Exercise 10.9
Consider the following equation for a spinor ψ on spacetime:

$$S_X\psi - \tfrac{1}{2}\widetilde{X}\not{S}\psi = 0 \qquad \forall X \in \Gamma TM.$$

Note that this is equivalent to equating to zero a 'traceless' spinor-valued 1-form made from the covariant derivatives of ψ. Since $\widetilde{X}$ and $\not{S}$ both anticommute with the volume 4-form this equation decouples into two equations for Weyl spinors.

(i) If K is a conformal Killing vector with $\mathcal{L}_K g = 2\lambda g$ show that if ψ satisfies the above equation then so does $\mathcal{L}_K\psi - \tfrac{1}{2}\lambda\psi$. This can be shown in the same way as for the analogous (but different!) result for the massless Dirac equation.

(ii) By differentiating the equation obtain the integrability condition

$$R_{ba}\psi - \tfrac{1}{2}(e_a S_{X_b} - e_b S_{X_a})\not{S}\psi = 0.$$

Clifford multiply to obtain the contracted conditions

$$P_a\psi + \tfrac{1}{2}e_a\not{S}^2\psi + S_{X_a}\not{S}\psi = 0$$

and

$$\mathcal{R}\psi + 3\not{S}^2\psi = 0.$$

Hence obtain the integrability condition

$$C_{ab}\psi = 0.$$

(Note that $P_a \wedge e_b - P_b \wedge e_a = e_aP_b - e_bP_a$ for zero torsion.)

(iii) If $\psi = u + \mathrm{d}f\,v$, for some function f and parallel Weyl spinors u and v, show that ψ solves the above equation if $\nabla_X \mathrm{d}f = \widetilde{X}$. Hence show that this equation has a 'twistor' [9] solution with $f = \frac{1}{2}\eta_{ab}x^a x^b$, where $\{x^a\}$ are inertial coordinates for Minkowski space.

Exercise 10.10
When is a spinor a twistor?

10.6 The Lichnerowicz Theorem

We anticipated in §10.1 that the eigenvalues of the Dirac operator will depend on the properties of the manifold. Whereas the spacetime Dirac equation involves a real 'mass' eigenvalue we will see below that the Dirac operator on a compact Riemannian manifold has only imaginary eigenvalues. The Lichnerowicz theorem [26], as we will now demonstrate, shows that if the curvature scalar is positive semidefinite then there are no zero eigenvalues.

Let M be a compact Riemannian manifold. From §2.6 we know that ξ^* is the adjoint of a zero index Hermitian-symmetric product on Dirac spinors, $(\ ,\ )$. By integrating over M we introduce another Hermitian product

$$\langle \psi, \varphi \rangle \equiv \int_M (\psi, \varphi) z$$

where z is the volume n-form of M. The Dirac operator is anti-self-adjoint with respect to this product. To see this we need to recognise an exact form when we see one. To this end we write an $(n-1)$-form J as $J = j_a e^a z$ and, for ∇ torsion free, $\mathrm{d}J = (\nabla_{X^a} j_a + i_{X^b}\nabla_{X_b} e^a j_a)z$ by (10.2.3). Since $\langle\ ,\ \rangle$ has ξ^* as adjoint involution with $e^{a\xi^*} = e^a$,

$$\begin{aligned}\langle \varphi, \not{S}\psi \rangle &= \langle e^a\varphi, S_{X_a}\psi \rangle \\ &= \int_M \{\nabla_{X_a}(e^a\varphi, \psi) - \nabla_{X_a}e^a(X_b)(e^b\varphi, \psi) - (\not{S}\varphi, \psi)\}z.\end{aligned}$$

Now $\nabla_{X_a} e^a(X_b) = -e^a(\nabla_{X_a} X_b) = -\mathrm{i}_{X_a} \nabla_{X^a} e_b$, and so we may recognise an exact form in the integrand. By Stokes's theorem the integral of an exact form over a compact manifold is zero, thus

$$\langle \varphi, \not{\$}\psi \rangle = -\langle \not{\$}\varphi, \psi \rangle. \tag{10.6.1}$$

Since it is anti-self-adjoint with respect to a Hermitian product the Dirac operator on a compact Riemannian manifold has imaginary eigenvalues. As a special case of the above we have

$$\langle \not{\$}\psi, \not{\$}\psi \rangle = -\langle \not{\$}^2\psi, \psi \rangle.$$

Since $\langle\ ,\ \rangle$ is a (zero-index) Hermitian product the left-hand side is positive-semidefinite. Thus $\not{\$}^2\psi = 0 \Leftrightarrow \not{\$}\psi = 0$. Using (10.1.4) to expand the spinor Laplacian gives

$$\langle \not{\$}\psi, \not{\$}\psi \rangle = -\langle (S_{X^a} + \mathrm{i}_{X^b}\nabla_{X_b} e^a) S_{X_a}\psi, \psi \rangle + \tfrac{1}{4}\langle \mathcal{R}\psi, \psi \rangle.$$

Since

$$\begin{aligned}\langle (S_{X_a} + \mathrm{i}_{X^b} S_{X_b} e^a) S_{X_a}\psi, \psi \rangle &= \int_M \{\nabla_{X^a}(S_{X_a}\psi, \psi) - (S_{X^a}\psi, S_{X_a}\psi) \\ &\quad + \mathrm{i}_{X^b}\nabla_{X_b} e^a (S_{X_a}\psi, \psi)\} z \\ &= -\langle S_{X^a}\psi, S_{X_a}\psi \rangle\end{aligned}$$

we have

$$\langle \not{\$}\psi, \not{\$}\psi \rangle = \langle S_{X^a}\psi, S_{X_a}\psi \rangle + \tfrac{1}{4}\langle \psi, \mathcal{R}\psi \rangle. \tag{10.6.2}$$

If $\mathcal{R} \geqslant 0$ then all three terms are positive-semidefinite. If $\mathcal{R} > 0$ then there are no zero eigenvalues of the Dirac operator: if $\mathcal{R} = 0$ then $\not{\$}\psi = 0 \Leftrightarrow S_X\psi = 0\ \forall X$.

When $\mathcal{R}$ is constant, such as for the standard metric on a sphere, then we obtain a lower bound for the eigenvalues of the Dirac operator. If $\not{\$}\psi = \mathrm{i}m\psi$, with m real, then

$$(m^2 - \tfrac{1}{4}\mathcal{R})\ \langle \psi, \psi \rangle = \langle S_{X^a}\psi, S_{X_a}\psi \rangle$$

and so

$$m^2 > \tfrac{1}{4}\mathcal{R}.$$

The above arguments can be repeated with real spinors. From table 2.15 we see that the involution ξ of the real Clifford algebra is the adjoint involution of a zero-index product; the product being either R-symmetric, C^*-symmetric or $\bar{H}$-symmetric.

10.7 Killing Spinors

Because of the importance of a knowledge of the geodesics on a manifold an interesting problem in general relativity is the determination of first integrals associated with the geodesic equations. Such integrals may be identified with constants of the motion along geodesic curves. Killing symmetries play an important role in the search for such integrals. It was in this context that the notion of a Killing spinor naturally emerged [27]. Since then the same notion has been rediscovered in the context of finding classical solutions to matter field equations in background geometries [28]. In particular, Killing spinors arise in the study of the residual supersymmetries exhibited by certain solutions to supergravity models. As we shall see the existence of such spinor fields imposes interesting constraints on the geometry of a manifold.

A spinor field on some n-dimensional spin manifold M which, for some complex constant λ, satisfies

$$S_X\psi = \lambda \widetilde{X}\psi \tag{10.7.1}$$

for all vector fields X, is said to be a *Killing spinor*. The name arises from the fact that such spinor fields can be used to construct conformal Killing vectors. An immediate consequence of (10.7.1) is that a Killing spinor is an eigenspinor of the Dirac operator, $\not{S}\psi = n\lambda\psi$. We have already noted in the section above that on a compact Riemannian manifold, λ must be pure imaginary. Excluding the case in which the signature of the metric on M is (p, q) with p even and q odd then there is an Hermitian symmetric product on complex spinor (or semi-spinor) fields with $\xi*$ as adjoint involution. Let $\widetilde{\psi}$ be the adjoint spinor with respect to this product. Then a real 1-form $\widetilde{K}$ is given by

$$\widetilde{K} = \mathscr{S}_1(\psi\widetilde{\psi}). \tag{10.7.2}$$

We can expand this in a basis $\{e^a\}$ as

$$\widetilde{K} = \mathscr{S}_0(\psi\widetilde{\psi}e_a)e^a = \mathscr{S}_0(\widetilde{\psi}e_a\psi)e^a = (\psi, e_a\psi)e^a$$

so

$$\widetilde{K}^* = (\psi, e_a\psi)^*e^a = (e_a\psi, \psi)e^a = (\psi, e_a\psi)e^a$$

and $\widetilde{K}$ is indeed real. By differentiating (10.7.2)

$$\begin{aligned}
\nabla_X\widetilde{K} &= \mathscr{S}_1(S_X\psi\widetilde{\psi} + \psi\widetilde{S_X\psi}) = \mathscr{S}_1(\lambda\widetilde{X}\psi\widetilde{\psi} + \psi\widetilde{\lambda\widetilde{X}\psi}) \\
&= ((\psi, e_a\lambda\widetilde{X}\psi) + (\lambda\widetilde{X}\psi, e_a\psi))e^a \\
&= ((\psi, \lambda e_a\widetilde{X}\psi) + (\psi, \lambda^*\widetilde{X}e_a\psi))e^a \\
&= 2\mathrm{Re}(\lambda)(\psi, \psi)\widetilde{X} + 2\mathrm{i}\,\mathrm{Im}(\lambda)(\psi, (e_a \wedge \widetilde{X})\psi)e^a
\end{aligned}$$

so

$$(\nabla_X\widetilde{K})(Y) + (\nabla_Y\widetilde{K})(X) = 4\mathrm{Re}(\lambda)(\psi, \psi)g(X, Y).$$

Using Killing's equation, (6.13.3), we have

$$\mathscr{L}_K g = 4\mathrm{Re}(\lambda)(\psi, \psi)g. \tag{10.7.3}$$

If we took a Hermitian-symmetric product $\langle\ ,\ \rangle$ with $\xi\eta^*$ as adjoint (the signature does not have p odd and q even) then if $\hat{\psi}$ is the adjoint with respect to this product then

$$\widetilde{K} = \mathscr{S}_1(\mathrm{i}\psi\hat{\psi}) \tag{10.7.4}$$

is a real 1-form. This satisfies

$$\mathscr{L}_K g = -4\mathrm{Im}(\lambda)(\psi, \psi)g. \tag{10.7.5}$$

Exercise 10.11
Show that if ψ is a Killing spinor and K some Killing vector field then $\mathscr{L}_K\psi$ is also a Killing spinor with the same eigenvalue λ.

The existence of Killing spinors on a Riemannian (as opposed to a pseudo-Riemannian) manifold necessitates interesting integrability conditions. We first note that the set of first-order differential equations for the components of ψ given by (10.7.1) implies that if the spinor vanishes at some point $p \in M$ then it must vanish at all points that are arcwise connected to p [29, 30]. By differentiating (10.7.1) we may obtain an integrability condition involving the curvature. A straightforward calculation, using the zero torsion of ∇, gives

$$S(X, Y)\psi = -\lambda^2[\widetilde{X}, \widetilde{Y}]\psi \qquad \forall X, Y \in \Gamma TM.$$

This can be written in terms of the curvature 2-forms, using (10.3.20), as

$$\tfrac{1}{2}e^a(X)e^b(Y)R_{ab}\psi = -\lambda^2 e^a(X)e^b(Y)[e_a, e_b]\psi$$

or

$$R_{ab}\psi = -4\lambda^2 e_{ab}\psi. \tag{10.7.6}$$

Clifford multiplying by e^a produces the Ricci forms on the left-hand side:

$$P_b\psi = -4\lambda^2(n-1)e_b\psi.$$

Now if A is a real 1-form such that $A\psi = 0$ then certainly $A^2\psi = 0$. But $A^2 = g(A, A)$ and so for a positive-definite metric we must have $A = 0$ for ψ non-zero. Thus the above integrability condition is that

$$P_b = -4\lambda^2(n-1)e_b \tag{10.7.7}$$

and the manifold must be an Einstein space with curvature scalar given by

$$\mathscr{R} = -4n(n-1)\lambda^2. \tag{10.7.8}$$

So λ must be either real or pure imaginary. We can use (10.7.7) and (10.7.8) to rewrite (10.7.6) in terms of the conformal 2-forms. Substituting (10.7.7) and (10.7.8) into the definition (6.11.6) gives $C_{ab} = R_{ab} + 4\lambda^2 e_{ab}$ and hence (10.7.6) becomes

$$C_{ab}\psi = 0. \tag{10.7.9}$$

To go further we must make another assumption about M. A Riemannian manifold is *locally symmetric* if its curvature tensor is parallel. If M is locally symmetric then the conformal tensor is parallel and the conformal 2-forms satisfy

$$\nabla_X C_{ab} = C_{cb}\omega^c{}_a(X) + C_{ac}\omega^c{}_b(X) \qquad \forall\, X \in \Gamma TM. \tag{10.7.10}$$

Differentiating (10.7.9) and using (10.7.1) and (10.7.10) gives

$$\{C_{pb}\omega^p{}_a(X_c) + C_{ap}\omega^p{}_b(X_c)\}\psi + \lambda C_{ab}e_c\psi = 0.$$

The first two terms vanish by (10.7.9), and so for $\lambda \neq 0$ we have $C_{ab}e_c\psi = 0$. From (10.7.9) we have $e_c C_{ab}\psi = 0$ and so subtracting these gives $\mathrm{i}_c C_{ab}\psi = 0$ and hence

$$C_{ab} = 0. \tag{10.7.11}$$

Together (10.7.7), (10.7.8) and (10.7.11) show that $R_{ab} = -4\lambda^2 e_{ab}$, that is, M has a constant sectional curvature of $-4\lambda^2$. Hence the only locally symmetric Riemannian manifolds such that (10.7.1) has a solution for $\lambda \neq 0$ are the standard sphere, in which case λ is imaginary, or a hyperbolic space with λ real, or a quotient of these spaces by a discrete group.

10.8 Parallel Spinors

A spinor field ψ is parallel if

$$S_X\psi = 0 \qquad \forall\, X \in \Gamma TM. \tag{10.8.1}$$

Thus a parallel spinor is a special case ($\lambda = 0$) of a Killing spinor. Not surprisingly M must be tightly constrained if it is to admit a parallel spinor. A discussion of parallel spinors necessitates a brief mention of Kahler manifolds. A tensor field $J \in \Gamma T^1_1 M$ is an *almost complex structure* on M if

$$J^2X \equiv J(J(X)) = -X \qquad \forall\, X \in \Gamma TM. \tag{10.8.2}$$

A Riemannian manifold (M, g) with an almost complex structure J that is an isometry,

$$g(JX, JY) = g(X, Y) \qquad \forall\, X, Y \in \Gamma TM \tag{10.8.3}$$

and is parallel

$$\nabla_X J = 0 \qquad \forall\, X \in \Gamma TM \tag{10.8.4}$$

is called a *Kahler manifold*. A theorem due to Hitchin [31] states that a compact even-dimensional Riemannian spin manifold admitting a parallel spinor is a Kahler manifold. For the special case of four dimensions a direct proof requiring orientability, but not compactness, can be found in [29]. It is possible to prove rather easily a result about parallel pure spinors on even-dimensional Riemannian manifolds.

An even-dimensional Riemannian spin manifold admitting a parallel (complex) pure spinor is a Ricci-flat Kahler manifold. (10.8.5)

The Ricci flatness is just a special case of (10.7.7). Pure spinors were introduced in Chapter 3. Recall from there that pure spinors are Weyl spinors (they carry a semi-spinor representation of the complexified even subalgebra). At each point p of M a non-vanishing pure spinor ψ_p determines a maximal isotropic subspace $\mathscr{I}_p^+$ of the complexified cotangent space by

$$x\psi_p = 0 \qquad \text{for } x \in T^*_pM^{\mathbb{C}} \qquad \text{iff } x \in \mathscr{I}_p^+. \tag{10.8.6}$$

We have $T^*_pM^{\mathbb{C}} = \mathscr{I}_p^+ \otimes \mathscr{I}_p^-$ where $x^* \in \mathscr{I}_p^-$ if and only if $x \in \mathscr{I}_p^+$. So a non-vanishing pure spinor field assigns a maximal isotropic subspace to the complexified cotangent space of every point. Let $\mathscr{I}^+$ and $\mathscr{I}^-$ be the spaces of complex differential 1-forms such that $x \in \mathscr{I}^+$ if and only if $x|_p \in \mathscr{I}_p^+$. Given the subspaces $\mathscr{I}^+$ and $\mathscr{I}^-$, determined by the pure spinor, we can define an almost complex structure J by

$$\begin{aligned} Jx &= \mathrm{i}x \qquad && \forall x \in \mathscr{I}^+ \\ Jy &= -\mathrm{i}y \qquad && \forall y \in \mathscr{I}^-. \end{aligned} \tag{10.8.7}$$

(Note that we here think of J as an endomorphism of the cotangent (rather than the tangent) space.) Since it has eigenvalues $\pm\mathrm{i}$ then J is certainly an almost complex structure, and since complex conjugation interchanges $\mathscr{I}^+$ and $\mathscr{I}^-$ it is a real tensor field. Since J preserves the isotropic subspaces $\mathscr{I}^+$ and $\mathscr{I}^-$, then to check that J is an isometry we need only consider the metric evaluated on an element of $\mathscr{I}^+$ and of $\mathscr{I}^-$. Let $x \in \mathscr{I}^+$ and $y \in \mathscr{I}^-$ then $g(Jx, Jy) = g(\mathrm{i}x, -\mathrm{i}y) = g(x, y)$ and so J satisfies (10.8.3). Since ψ is parallel then the subspace $\mathscr{I}^+$ (and hence $\mathscr{I}^-$) is preserved under covariant differentiation. For if $x\psi = 0$ and ψ is parallel then $\nabla_X x\psi = 0$ and hence $\nabla_X x \in \mathscr{I}^+\ \forall x \in \mathscr{I}^+$, $\forall X \in \Gamma TM$. Since covariant differentiation commutes with complex conjugation then it also preserves $\mathscr{I}^-$. Now if $x \in \mathscr{I}^+$ we have $Jx = \mathrm{i}x$ and hence $(\nabla_X J)x + J(\nabla_X x) = \mathrm{i}\nabla_X x$. Since $\nabla_X x \in \mathscr{I}^+$ we have $\nabla_X Jx = 0$ and $\nabla_X Jx^* = 0$, hence $\nabla_X J = 0$. Thus we have established (10.8.5).

We can use the metric to construct a 2-form out of an almost complex structure satisfying (10.8.3). If $J = J_a{}^b e^a \otimes X_b$ then the usual index-lowering rule gives $J_{ab} = g(JX_a, X_b)$. If J satisfies (10.8.3) then

$$g(JX_a, X_b) = -g(JX_a, J^2X_b) = -g(X_a, JX_b) = -g(JX_b, X_a)$$

and $J_{ab} = -J_{ba}$. The 2-form

$$\Omega \equiv \tfrac{1}{2}J_{ab}e^{ab} \tag{10.8.8}$$

is called the *Kahler 2-form*. If x is any 1-form then

$$Jx = -\mathrm{i}_{\tilde{x}}\Omega = \mathcal{S}_1(\Omega x). \tag{10.8.9}$$

We showed above that an even-dimensional Riemannian manifold admitting a parallel pure spinor is a Kahler manifold. In this case the Kahler 2-form can be constructed out of the spinor. If $\widetilde{\psi}$ denotes the adjoint spinor with respect to the Hermitian spinor product whose adjoint involution is ξ^* then a real 2-form F is given by

$$F = \mathcal{S}_2(\mathrm{i}\psi\widetilde{\psi}). \tag{10.8.10}$$

For any 1-form x

$$\mathcal{S}_1(Fx) = \mathcal{S}_1\{\mathrm{i}\psi\widetilde{\psi}x - \mathcal{S}_0(\mathrm{i}\psi\widetilde{\psi})x\} = \mathcal{S}_0(\mathrm{i}\psi\widetilde{\psi}xe_a)e^a - \mathcal{S}_0(\mathrm{i}\psi\widetilde{\psi})x$$

and

$$\begin{aligned}\mathcal{S}_0(\mathrm{i}\psi\widetilde{\psi}xe_a) &= g(x, e_a)\mathcal{S}_0(\mathrm{i}\psi\widetilde{\psi}) + \tfrac{1}{2}\mathcal{S}_0(\mathrm{i}\psi\widetilde{\psi}(xe_a - e_ax)) \\ \tfrac{1}{2}\mathcal{S}_0(\mathrm{i}\psi\widetilde{\psi}xe_a) &= g(x, e_a)\mathcal{S}_0(\mathrm{i}\psi\widetilde{\psi}) - \tfrac{1}{2}\mathcal{S}_0(\mathrm{i}\psi\widetilde{\psi}e_ax) \\ &= g(x, e_a)\mathcal{S}_0(\mathrm{i}\psi\widetilde{\psi}) - \tfrac{1}{2}\mathcal{S}_0(\mathrm{i}x\psi\widetilde{\psi}e_a).\end{aligned}$$

If now ψ is a pure spinor and $x \in \mathcal{J}^+$, as determined by (10.8.6), then the last term in the above vanishes. Thus for $x \in \mathcal{J}^+$

$$\mathcal{S}_1(Fx) = \mathcal{S}_0(\mathrm{i}\psi\widetilde{\psi})x = \mathrm{i}(\psi, \psi)x.$$

Since $(\psi, \psi) > 0$ for $\psi \neq 0$ the Kahler 2-form Ω related to the almost complex structure J of (10.8.7) is given by

$$\Omega = \frac{\mathcal{S}_2(\mathrm{i}\psi\widetilde{\psi})}{(\psi, \psi)}. \tag{10.8.11}$$

By only considering parallel pure spinors we have been able to use a basically algebraic argument to see directly that M must be a Kahler manifold. If M is even dimensional and orientable, with $\dim M \leqslant 6$ then if M admits a parallel spinor then it admits a parallel pure spinor. If M is orientable with ψ parallel then the Weyl spinors $\frac{1}{2}(1 \pm \check{z})\psi$ are also parallel where $\check{z}$ is proportional to the volume form on M such that $\check{z}^2 = 1$. But for $\dim M \leqslant 6$ all Weyl spinors are pure and hence M is a Kahler manifold. Notice that we need to assume orientability but not compactness.

In the above we have studied some of the conditions that are necessary for the existence of parallel pure spinor fields. The existence of compact Ricci flat manifolds was first demonstrated by Yau [32] following a famous conjecture by Calabi. When the very stringent necessary conditions for a parallel spinor are met one can sometimes appeal to the powerful Atiyah–Singer index theorem [33] to show that a parallel spinor does in fact exist. This theorem relates the differing numbers of 'left- and right-handed' Weyl solutions of the massless Dirac equation on a compact Riemannian manifold to a topological invariant. By the Lichnerowicz theorem we know that for a Ricci-flat compact Riemannian manifold the only such solutions are parallel spinors. Thus if the topological invariant is such that the difference between the number of left- and right-handed solutions is non-zero then there must exist parallel spinors.

Exercise 10.12
Show that the almost complex structure on a Kahler manifold can be used to define a sub-bundle of minimal left ideals of the complexified Clifford bundle. Hence a Kahler manifold is a Spin^C manifold. Show that the Riemannian connection induces a connection on this sub-bundle, and hence the Kahler equation can be restricted to a minimal left ideal.

The importance of spinor fields in classical differential geometry has rarely been doubted. That they play an important role in many theories in physics is an act of faith shared by many physicists. In recent times a great deal of theoretical physics and differential geometry has become closely intertwined. The properties of Killing spinors are an example where both disciplines have gained mutual benefit from this interaction. In this book we have attempted to bring the amalgam of ideas that constitute Clifford algebras, differential geometry and the theory of spinors into a form that we hope will stimulate some readers to pursue such a synthesis further.

Appendix A

Algebra

In this appendix we have collected those algebraic results that we have referred to in the book. Thus the account here is very much tailored to our specific needs rather than giving a balanced view of the subject. The first few pages mostly define terminology that we have used. Although this is fairly standard the various 'morphisms' are used by different authors in slightly different ways, and there are some alternative terms that we have not listed. The section on algebras is much more dense, leading up to a proof of the structure theorem for simple algebras. Although the average reader will probably not want to plough through this exposition he will need to know the final result, and how it may be used to construct, for example, explicit representations of γ-matrices. The approach we have adopted is the historical one; more modern treatments prove the structure theorems for a wider class of rings than algebras over fields. We found useful the classic books of Albert (1961) [1] and Dickson (1960) [2], and the more modern book by Kochendorffer (1972) [3]. There are, of course, an abundance of books in which this material can be found, to suit all tastes.

A *group*, G, consists of a set with a binary operation, or law of composition, that satisfies four axioms. Usually multiplicative notation is used to denote this group operation, the juxtapositioning of elements denoting their composition. In view of this notation we shall often refer to the law of composition as a product. The axioms are as follows.

(i) For every a, $b \in G$ there is a unique $c \in G$ such that $ab = c$.
(ii) The product is *associative*, $(ab)c = a(bc)$.
(iii) There exists an *identity* (or unit element), denoted 1, such that
$$a1 = 1a = a \qquad \forall\, a \in G.$$
(iv) Every element a has an *inverse* a^{-1}, $aa^{-1} = a^{-1}a = 1$.

When a group consists of a finite number of elements then this number is called the *order* of the group. In general the group product is not *commutative*, $ab \neq ba$. The set of elements that commute with all other elements is called the *centre*. A group for which the product of any two elements is commutative is called *Abelian*. Often additive notation is used to denote the law of composition in an Abelian group, in which case the identity is written as 0. A subset H, of a group G, which forms a group under the product of G is called a *subgroup*. Thus H is a subgroup if and only if $uv \in H \; \forall u, v \in H$, $u^{-1} \in H \; \forall u \in H$ and $1 \in H$. For example, the centre is a subgroup. We may form a subgroup H from any subset S of a group G by taking the set of all products that can be formed from elements of S and their inverses; this group is said to be *generated* by S. A subgroup enables a group to be decomposed into equivalence classes. If we have an equivalence relation on a set such that a is equivalent to b then we write $a \sim b$. Equivalence relations satisfy $a \sim a$, $a \sim b$ for $b \sim a$, and if $a \sim b$ and $b \sim c$ then $a \sim c$. The set of all elements equivalent to an element a constitute the equivalence class of a, $[a]$. Any element of $[a]$, such as a, is called a *representative* of the class. The equivalence classes of distinct elements are either identical or non-intersecting. If H is a subgroup of G then an equivalence relation on G is defined by $a \sim b$ if $b = ah$ for some $h \in H$. The equivalence class of a is called the *left coset* of G, relative to H, generated by a. In an obvious way we define right cosets. For a special type of subgroup the cosets inherit a group structure. A subgroup H is called *normal* (or *invariant*) if $ghg^{-1} \in H \; \forall g \in G, \forall h \in H$. The notation $H \Leftarrow G$ denotes that H is a normal subgroup of G. It follows that the left and right cosets relative to a normal subgroup are equal. These cosets form a group under the product defined by $[a][b] = [ab]$. Since $[a] = [ah]$ for $h \in H$ this definition only makes sense if H is normal. This group of cosets is called the *quotient* of G modulo H, denoted G/H. We give an example. The set of integers (positive and negative) forms an Abelian group under addition, denoted $\mathbb{Z}$. Any integer n generates a subgroup H. Thus H consists of the set $\{0, \pm n, \pm 2n, \pm 3n, \ldots\}$. Any subgroup of an Abelian group is normal and so we can form the quotient, $\mathbb{Z}_n = \mathbb{Z}/H$. If m is any integer then $m = qn + r$, where $0 \leqslant r < n$, and so every element of $\mathbb{Z}$ is equivalent to a positive integer less than n. The class of the sum of two such integers is represented by their sum modulo a multiple of n. For example, $\mathbb{Z}_2$ has two elements, $[0]$ and $[1]$, and $[1] + [1] = [2] = [0]$. (The notation $\mathbb{Z}_n$ will be used to denote any group isomorphic to these quotients. For example, the set $\{1, -1\}$ forms a group under multiplication, isomorphic to $\mathbb{Z}_2$.) Roughly speaking a homomorphism is a mapping between groups that preserves the structure. Let φ be a mapping from G to G', then φ is a *homomorphism* if $\varphi(ab) = \varphi(a)\varphi(b)$. The product on the left-hand side is that of G whilst the product on the right-hand side is that of G'. If every element of G' is the image of some element of G under φ, then φ

is called *surjective* (or onto). If no two elements of G get mapped into the same element then φ is called *injective* (or one-to-one). A mapping that is both injective and surjective is called *bijective*. Groups that are related by a bijective homomorphism are called *isomorphic*, and we write $G' \simeq G$. In general a homorphism φ is not injective, and the set of elements in G mapped onto the identity of G' is called the *kernel* of φ ($\ker \varphi$). The kernel of φ is a normal subgroup of G, and we have

$$\varphi(G) \simeq G/\ker \varphi. \tag{A1}$$

(This is known as the first isomorphism theorem.)

This may be proved by introducing a map Φ,

$$\begin{aligned} \Phi : G/\ker \varphi &\longrightarrow \varphi(G) \\ [a] &\longmapsto \Phi([a]) = \varphi(a). \end{aligned}$$

The proof consists of showing that not only does such a definition make sense, but Φ is a bijection. The following is usually known as the second isomorphism theorem. If $N \Leftarrow G$ and $A \Leftarrow G$ such that $N \Leftarrow A \Leftarrow G$ then

$$\frac{G/N}{A/N} \simeq G/A. \tag{A2}$$

The conditions on the subgroups are just such as are required for this to make sense. The equivalence class of a in G given by N is written $[a]_N$; $[a]_A$ being similarly defined. The proof of (A2) is established by introducing a map φ,

$$\begin{aligned} \varphi : G/N &\longrightarrow G/A \\ [a]_N &\longmapsto \varphi\,([a]_N) = [a]_A. \end{aligned}$$

Not only is such a map well defined but it is a surjective homomorphism with kernel A/N. Then (A2) follows from (A1).

If H and K are two groups then there is a natural way in which the Cartesian product of these sets can be given a group structure. The Cartesian product set consists of ordered pairs of an element of H and an element of K. If (h_1, k_1) and (h_2, k_2) are two such pairs then we may define their product by $(h_1, k_1)(h_2, k_2) = (h_1h_2, k_1k_2)$. If G denotes the group formed by such pairs then G is the *direct product* of H and K, written $G = H \times K$. An isomorphism from a group to itself is called an *automorphism*. If φ and ψ are automorphisms of G then their product may be defined by $(\varphi\psi)(a) = \varphi(\psi(a))$. Under this product the set of all automorphisms of G forms a group, $\operatorname{Aut} G$. If t is any element of G then we have a τ in the automorphism group given by $\tau(a) = tat^{-1}$. Such an automorphism is called an *inner automorphism*.

Any automorphism that is not inner is called an *outer automorphism*. The ordered pairs consisting of an element of a group and an element of a group of automorphisms can be given a group structure other than that of direct product. If Ω is a subgroup of Aut G then for $\omega_1, \omega_2 \in \Omega$, $a_1, a_2 \in G$ we define $(a_1, \omega_1)(a_2, \omega_2) = (a_1\omega_1(a_2), \omega_1\omega_2)$. With such a product we have $(a, \omega)^{-1} = (\omega^{-1}(a^{-1}), \omega^{-1})$. The ordered pairs under this product form the *semidirect product* of G and Ω, K say, written $K = G \odot \Omega$.

A *ring* has two binary operations, addition, denoted +, and multiplication, denoted by juxtaposing elements. Under addition a ring forms an Abelian group, the additive identity being called the zero element. Multiplication is associative (unless specifically stated otherwise) and distributive over addition,

$$(a + b)c = ac + bc \qquad c(a + b) = ca + cb.$$

A *commutative ring* is one in which multiplication is commutative. The set of elements that commute with all other elements under multiplication is called the *centre*. A ring need have no *identity* (or *unit element*), denoted 1, by which is meant a unit element under multiplication. For a ring with unit element an element a is called *regular* (or *invertible*) if it has a multiplicative inverse a^{-1}, that is $aa^{-1} = a^{-1}a = 1$. A ring in which every non-zero element is regular is called a *division ring*. We have already noted that the integers, $\mathbb{Z}$, form an Abelian group under addition; with multiplication they form a ring. Similarly with multiplication being defined modulo n the group $\mathbb{Z}_n$ forms a ring.

A *field* is a commutative division ring. (Sometimes a non-commutative division ring is called a skew field.) Familiar examples of fields are the rational numbers Q, the real numbers $\mathbb{R}$ and the complex numbers $\mathbb{C}$. For p a prime number then an example of a field with a finite number of elements is $\mathbb{Z}_p$. A field F is said to be of *characteristic* p if there is a prime number p such that

$$\underbrace{a + a + a \ldots + a}_{p \text{ terms.}} = 0 \qquad \forall a \in F.$$

In this case F contains $\mathbb{Z}_p$ as a subfield. If there is no such p then F is said to be of *characteristic zero*, and in this case it contains the rational numbers as a subfield. We shall really only be concerned with the zero characteristic fields $\mathbb{R}$ and $\mathbb{C}$. The complex numbers have the property of being algebraically closed, which results in the property that we shall observe of enabling any complex number to be written as a square. The real numbers do not have this property, no negative number being a square of a real number.

A *vector space* over a field F, V, is a set (of vectors) with an operation of addition and a rule of scalar multiplication, which assigns a vector to the product of a vector with an element of the field. (In this context elements of the field are called scalars.) Under addition the vectors form an Abelian group, with multiplication by scalars satisfying the following:

(i) $(\lambda\mu)x = \lambda(\mu x)$
(ii) $(\lambda + \mu)x = \lambda x + \mu x$
$\lambda(x + y) = \lambda x + \lambda y \qquad \forall \lambda, \mu \in F, x, y \in V.$
(iii) $1x = x$, where 1 is the unit element of F.

If $\{x_i\}$ is a set of vectors such that $x = \Sigma_i \lambda^i x_i$ for $\lambda^i \in F$ then x is said to be a *linear combination* of the x_i. A set of vectors is called *linearly dependent* if any one vector can be written as a linear combination of the others. Conversely the set $\{x_i\}$ is *linearly independent* if $\Sigma_i \lambda^i x_i = 0$ implies that all λ^i are zero. A set of vectors $\{x_i\}$ is said to *span* V (or *generate* V) if any element of V can be written as a linear combination of the x_i. A linearly independent spanning set is called a *basis*, or *linear frame*. Every vector space admits a basis, and when the vector space is spanned by a finite set any basis contains the same number of vectors, called the *dimension* of the vector space V, denoted $\dim V$. Any vector can be written as a linear combination of the basis vectors, the uniquely determined scalar coefficients being termed the *components* of the vector with respect to that basis. If $\{e_i\}$ and $\{f_i\}$ are distinct bases then the elements of one basis can be written as linear combinations of the other basis vectors,

$$e_i = \sum_{j=1}^{n} A_i{}^j f_j$$

$$f_j = \sum_{i=1}^{n} B_j{}^i e_i.$$

Substituting either expression into the other gives

$$\sum_{i=1}^{n} B_j{}^i A_i{}^k = \delta_j{}^k$$

$$\sum_{k=1}^{n} A_i{}^k B_k{}^j = \delta_i{}^j$$

where the Krönecker $\delta_i{}^j$ takes the value zero unless $i = j$ when its value is one. Thus the coefficients relating the change of basis can be displayed as a non-singular $n \times n$ matrix, with entries in F. Such non-singular matrices form a group under matrix multiplication, the general linear group over F, $\mathrm{Gl}(n, F)$. In the above expressions we have chosen to position certain indices as superscripts, others as subscripts. It is often convenient to adopt the *Einstein summation convention* in which summation is implied over any repeated index, occurring once as a superscript and once as a subscript. Thus in the above expressions we would simply omit the summation sign when using the summation convention. We shall frequently use this convention without further comment. When it is not clear from the context whether a sum is implied or not we shall explicitly state, for example, no sum.

A subset U of a vector space V such that all linear combinations of vectors from U lie in U is called a *vector subspace*. The zero element and V itself are obviously vector subspaces, any other subspace being termed non-trivial. If S is any subset from V then all linear combinations of vectors from S form a vector subspace which is said to be generated, or spanned, by S. The dimension of the subspace generated by S is called the *rank* of the set. If U and W are subspaces of V then so is the intersection of these sets, $U \cap W$. This intersection is not empty since all subspaces contain the zero element: thus should we speak of non-intersecting subspaces we really mean subspaces that only intersect in the zero element. The *sum* of U and W, $U + W$, consists of vectors of the form $x = u + w$, $u \in U$ $w \in W$. In general, such a decomposition of x into elements of U and W is not unique. It is, however, when $U \cap W = 0$. In this case the sum is said to be *direct*, written $U \oplus W$. (Later we shall reserve this notation for the direct sum of algebras, all vector space sums being direct unless stated otherwise.) For any subspace U there is a subspace W such that $V = U \oplus W$; W being called the *complement* of U in V. Obviously $\dim V = \dim U + \dim W$. Any subspace U is a normal subgroup under addition. The quotient group V/U can be given a linear structure by defining $\lambda[x] = [\lambda x]$, where the bracket denotes the equivalence class of x, with $x \sim y$ if $x = y + u$ for some $u \in U$. With this structure V/U is called the *linear quotient space* of V modulo U. (In view of the additive notation the obsolescent term difference space might seem more appropriate.)

A *linear map* between two vector spaces over the same field is a group homomorphism that commutes with scalar multiplication. That is, φ is a linear map from V to W if

$$\varphi(\lambda x + \mu y) = \lambda\varphi(x) + \mu\varphi(y) \qquad \forall x, y \in V, \lambda, \mu \in F.$$

It follows that every linear map sends the zero element of V to that in W. A linear map may be completely determined by specifying its effect on some basis for V. The terms *injective*, *surjective* and *bijective* naturally apply to linear maps. A bijective linear map is called a *vector space isomorphism*. The *kernel* of a *linear map* is the kernel of the group homomorphism, and is readily seen to be a linear subspace. In an obvious way we can define addition of linear maps and multiplication by scalars such that the linear maps from V to W form a vector space, $\mathscr{L}(V, W)$. Since any such linear map may be specified by a $\dim V \times \dim W$ matrix we have $\dim \mathscr{L}(V, W) = \dim V \dim W$. A linear map from V to V will be called a *linear transformation*, or *endomorphism*, and we will also write End V for $\mathscr{L}(V, V)$. Such linear transformations can be multiplied by composing maps, $(\varphi\psi)x = \varphi(\psi(x))$. With such a product End V has the structure of an algebra, about which more will be said later. Under mulitplication the non-singular linear transformations form a group, the *automorphism group* of V, Aut V. Of special importance is the vector space of linear mappings from the vector space V to the field F, known as the *dual space*, V^*. When V is finite dimensional then $\dim V^* = \dim V$. For each basis $\{e_i\}$ of V we may establish a *natural dual basis* $\{e^{*i}\}$ of V^* such that $e^{*i}(e_j) = \delta^i{}_j$ $\forall i, j$. (Note the conventional positioning of indices.) If arbitrary elements b and B in V and V^* respectively are expanded in dual bases as

$$b = b^i e_i \qquad B = B_i e^{*i} \qquad \text{(summation convention)}$$

then $B(b) = B_i b^i$. In particular, $e^{*i}(x) = x^i$ expresses the components of x in terms of the corresponding natural dual basis action on x. Elements of V^* are sometimes called co-vectors to distinguish them from elements of V, although for V finite dimensional this terminology is reciprocal since there exists a natural way to regard V as the dual to V^*.

A vector space V is graded by an Abelian group G if V is expressible as a direct sum of subspaces that are labelled by elements of G. More precisely, V is a *G-graded vector space* if $\{V_i\}$ is a set of non-intersecting subspaces such that $V = \Sigma_i V_i$ and k injectively assigns an element $k(i)$ of G to each V_i. G is called the *group of degrees*. Elements of V_i are called *homogeneous* of degree $k(i)$, denoted

$$\deg x = k(i) \qquad \forall x \in V_i.$$

Since the zero vector lies in every subspace it is homogeneous of every degree. Paticularly when $G = \mathbb{Z}$ we will label the subspaces with elements of G. When the only element that is homogeneous of negative degree is the zero element we have a positive gradation. If we omit mention of the group G we shall mean by graded vector space a $\mathbb{Z}$-graded space with positive gradation. A *G-graded subspace* of a G-graded space V admits a direct sum decomposition in terms of subspaces contained in the homogeneous subspaces of V. If V and W are G-graded spaces with homogeneous subspaces $\{V_i\}$ and $\{W_j\}$ then a linear map φ is called *homogeneous of degree k* if there is an element $k \in G$ such that $\varphi(V_i) \subset W_{i+k}$ $\forall i \in G$. It follows that the kernel of a homogeneous map is a graded subspace of V, whilst the image is a graded subspace of W. If U is a G-graded subspace of a G-graded V then the linear quotient V/U inherits a natural G-gradation, the equivalence classes being assigned the degree of a homogeneous representative.

A bilinear mapping on V is a mapping on pairs of vectors which is linear in each argument separately. By *bilinear form* we mean a bilinear mapping on V with values in the field F. We shall also refer to such a mapping as a *metric*. Although this use of the word is not standard we adopt it due to its prevalent use in this sense for the applications we are interested in. (Such a metric will not in general satisfy the criteria for a distance function used to define a metric space!) A metric g is *symmetric* if $g(x, y) = g(y, x)\ \forall x, y \in V$ and *non-degenerate* if $g(x, y) = 0\ \forall y$ implies that $x = 0$. We shall be primarily concerned with the case of $F = \mathbb{R}$ with g symmetric and non-degenerate, and we now restrict ourselves to this situation. In this case g is said to be *positive-definite* if $g(x, x) > 0$ for all non-zero x. It is often convenient to choose a *g-orthonormal basis*, $\{e_i\}$, in which $g(e_i, e_j) = \eta_{ij}$ where $\eta_{ij} = \pm 1$ if $i = j$ or zero otherwise. The pattern of signs is known as the *signature* of g, and may be denoted (p, q) where there are p plus signs and q minus signs. The automorphism group or *invariance group* of a space with a metric is the subgroup of the group of non-singular linear transformations consisting of elements m such that $g(m(x), m(y)) = g(x, y)\ \forall x, y \in V$. For a real-valued symmetric non-degenerate g of signature (p, q) the invariance group is called the *orthogonal group*, $O(p, q)$. Such a space will also more simply be called an orthogonal space. In particular, then, orthonormal bases are related by orthogonal transformations. The metric g can be used to associate with every element $x \in V$ an element $\tilde{x} \in V^*$ by the rule that

$$\tilde{x}(y) = g(x, y) \qquad \forall y \in V.$$

We shall refer to such an $\tilde{x}$ as the *metric dual* or adjoint of x (with respect to g). If the components of g in the basis $\{e_i\}$ are given by $g_{ij} = g(e_i, e_j)$ and $\tilde{x}$ is expressed in the dual basis as $\tilde{x} = \tilde{x}_i e^{*i}$ then $\tilde{x}_j y^j = g_{ij} x^i y^j$. Since this must hold for all y^j it implies that $\tilde{x}_j = g_{ij} x^i$. Frequently a *lowering convention* is adopted for indices in which $x_j \equiv g_{ij} x^i$, such that if $x = x^i e_i$ then $\tilde{x} = x_i e^{*i}$. The metric $g: V \times V \to \mathbb{R}$ naturally induces a metric $g^*: V^* \times V^* \to \mathbb{R}$ by the rule

$$g^*(\tilde{x}, \tilde{y}) = g(x, y) \qquad \forall x, y \in V.$$

If the components of g^* in the basis $\{e^{*i}\}$ are the numbers $g^{*ij} = g^*(e^{*i}, e^{*j})$ then $g_{ij} g^{*jk} = \delta^k{}_i$. Thus the components of g^* form the inverse of the matrix of components of g. The map $\tilde{}$ from V to V^* is invertible and we denote its inverse by $\underset{\sim}{}$. Thus if $B \in V^*$ with $B = B_i e^{*i}$ then $\underset{\sim}{B} = B^i e_i$ where the index has been raised with the components of the metric, $B^i \equiv g^{*ij} B_j$. For typographical reasons we shall use the same symbol to denote the 'lowering map' $\tilde{}$ and its inverse the 'raising map' $\underset{\sim}{}$, there being little scope for confusion so long as we state in which space the elements lie.

As well as real vector spaces we shall be interested in vector spaces over the complex field. In various ways the same Abelian group can be endowed with both an $\mathbb{R}$-linear structure and a $\mathbb{C}$-linear structure. When speaking of the dimension of such a vector space it is important to distinguish between the two linear structures, and when there is possibility for confusion we use $\dim_{\mathbb{R}}$ and $\dim_{\mathbb{C}}$ to denote the dimensions associated with the different linear structures. Similarly we speak of $\mathbb{R}$-linear and $\mathbb{C}$-linear transformations when there is possibility of confusion. If V is a real vector space then an endomorphism J such that $J^2 = -I$, where I is the identity map, is called a *complex structure* on V. Such a J can only exist if V is of even dimension. A complex structure can be used to define multiplication of elements in V by complex numbers. For $\lambda + i\mu \in \mathbb{C}$, $\lambda, \mu \in \mathbb{R}$, we define

$$(\lambda + i\mu)x = \lambda x + \mu Jx \qquad \forall\, x \in V.$$

Such a $\mathbb{C}$-linear structure turns V into a complex vector space $\check{V}$, the *complex vector space associated* with V (and J). We clearly have $\dim_{\mathbb{C}} \check{V} = \frac{1}{2} \dim_{\mathbb{R}} V$.

There is another way in which a complex vector space can be fabricated out of a real vector space V. The ordered pairs of elements of V, $V \times V$ are given a real vector space structure by defining

$$(x_1, y_1) + (x_2, y_2) = (x_1 + x_2, y_1 + y_2)$$

$$\lambda(x, y) = (\lambda x, \lambda y) \qquad \lambda \in \mathbb{R}.$$

With this structure the ordered pairs form the *external direct* sum of V with itself, $V \oplus V$. This direct sum space has a natural complex structure, $J:(x, y) \to (-y, x)$. The complex vector space associated with this complex structure is called the *complexification* of V, $V^{\mathbb{C}}$. Thus $V^{\mathbb{C}} \equiv (V \oplus V)$, and $\dim_{\mathbb{C}} V^{\mathbb{C}} = \dim_{\mathbb{R}} V$. An element of $V^{\mathbb{C}}$ is an ordered pair of elements from V. But since $(x, y) = (x, 0) + i(y, 0)$ we shall write $x + iy$ instead of (x, y). If then $\lambda + i\mu \in \mathbb{C}$ this gives, as one would expect,

$$(\lambda + i\mu)(x + iy) = \lambda x - \mu y + i(\lambda y + \mu x).$$

If now we start with a complex vector space E then we automatically have an *associated real vector space*, $E^{\mathbb{R}}$, since $\mathbb{R}$ is a subfield of $\mathbb{C}$. This real vector space comes equipped with a natural complex structure, multiplication by i in E. With this complex structure $E = (E^{\mathbb{R}})^{\mathbb{C}}$.

A group homomorphism φ between complex vector spaces is called *conjugate linear* if $\varphi(\lambda x) = \lambda^* \varphi(x)$ for $\lambda \in \mathbb{C}$ and λ^* denoting the complex conjugate. In particular, if φ is an $\mathbb{R}$-linear map on a real V that has complex structure J such that, $\varphi J = -J\varphi$ then φ is a conjugate linear map on $\check{V}$.

As we have remarked, the non-singular linear transformations on a vector space V form a group under multiplication, $\operatorname{Aut} V$. If G is an arbitrary group then a *representation* of G is a homomorphism of G into $\operatorname{Aut} V$, for some V. The vector space V is said to carry the representation. The dimension of V is called the dimension of the representation. If this homomorphism is one-to-one then the representation is called *faithful*. If the image of G under the representation leaves no non-trivial subspaces of V invariant then the representation is called *irreducible*. If V may be decomposed into subspaces that are preserved under a representation of G then that representation is *reducible*, as it induces homomorphisms of G into the automorphism groups of these subspaces. If V and W carry representations φ and ρ respectively then these are termed *equivalent* if there is an isomorphism S, mapping V to W, such that the following diagram commutes for all $g \in G$, $x \in V$:

$$\begin{array}{ccc} x & \xrightarrow{\varphi(g)} & \varphi(g)x \\ S \downarrow & & \downarrow S \\ Sx & \xrightarrow{\rho(g)} & \rho(g)Sx \end{array} \qquad \text{i.e. } S\varphi(g)S^{-1} = \rho(g).$$

An *algebra* over the field F, $\mathscr{A}(F)$, consists of a vector space over F together with an algebra product, called multiplication, which satisfies

$$a(\lambda b + \mu c) = \lambda ab + \mu ac \qquad \forall\, a, b, c \in \mathscr{A}, \forall\, \lambda, \mu \in F$$

and similarly for multiplication on the right. We shall call the dimension of the vector space the *dimension* of the algebra. The algebra is associative if its product satisifes $a(bc) = (ab)c$. Thus equivalently an associative algebra $\mathscr{A}(F)$ is a ring $\mathscr{A}$ that is a vector space for which $\alpha(ab) = a(\alpha b) = (\alpha a)b\ \forall\, a, b \in \mathscr{A}, \forall\, \alpha \in F$. We may therefore apply the terminology defined for rings to algebras. A *division algebra* being, for example, a division ring that is an algebra. An algebra with a unit element that spans the centre is called *central*. When the vector space is graded by an Abelian group G and the algebra product satisfies

$$\deg(ab) = \deg a + \deg b$$

then we have a *G-graded algebra*. Unless we further specify we shall mean by algebra $\mathscr{A}$ a finite-dimensional associative algebra over F, some arbitrary field; although in this book we shall only be concerned with the real or complex field.

If the underlying vector space of an algebra $\mathscr{A}$ is the direct sum of two subspaces $\mathscr{B}$, $\mathscr{C}$ then we will write $\mathscr{A} = \mathscr{B} + \mathscr{C}$. These subspaces need not be *subalgebras*, by which we mean a vector subspace that is closed under the algebra product. The centre is an example of a subalgebra.

For subspaces $\mathcal{B}$, $\mathcal{C}$ we define the product $\mathcal{B}\mathcal{C}$ to be the vector space spanned by all products of the bases for $\mathcal{B}$ and $\mathcal{C}$. If $\mathcal{G}$ is some subspace such that $\mathcal{A} = \mathcal{G}\mathcal{G} \ldots \mathcal{G}$ then $\mathcal{G}$ is said to *generate* $\mathcal{A}$. A basis for $\mathcal{G}$ will be termed a set of generators for $\mathcal{A}$. In general the dimension of $\mathcal{B}\mathcal{C}$ will be less that the product of those of $\mathcal{B}$ and $\mathcal{C}$. In fact we have

If $\{c_i\}$ $i = 1, \ldots, s$ is a basis for $\mathcal{C}$ then $\dim \mathcal{B}\mathcal{C} = \dim \mathcal{B} \dim \mathcal{C}$ iff $\Sigma_{i=1}^{s} d_i c_i = 0$ for $d_i \in \mathcal{B}$ implies all d_i are zero. (A3)

For if $\{b_j\}$ $j = 1, \ldots, r$ is a basis for $\mathcal{B}$ then $\mathcal{B}\mathcal{C}$ is spanned by the set of all products $b_j c_i$. So $\dim \mathcal{B}\mathcal{C} = rs$ if and only if these are all linearly independent, that is, if

$$\sum_{\substack{i=1,\ldots,s \\ j=1,\ldots,r}} \lambda_{ij} b_j c_i = 0$$

for $\lambda_{ij} \in F$ implies all $\lambda_{ij} = 0$.

For $d_i = \Sigma_{j=1}^{r} \lambda_{ij} b_j$ this is just the statement of the result. As a special case we have, for some non-zero $a \in \mathcal{A}$, $a\mathcal{A} = \mathcal{A}$ if and only if there is no non-zero b such that $ab = 0$. The above result enables us to make the following simple observation, to which we will later refer.

If there is an element b such that $ab = 1$ then b is the unique inverse of a. (A4)

It is obvious that if a had an inverse then it would be unique. Given $ab = 1$ we have $ab\mathcal{A} = \mathcal{A}$. But $ab\mathcal{A} \subset a\mathcal{A}$ so we must have $a\mathcal{A} = \mathcal{A}$, that is, from (A3), there is no non-zero d such that $ad = 0$. Suppose there were a c such that $bac \neq c$, that is $bac - c = d$ where $d \neq 0$. This implies that $abac - ac = ad$. If, however, $ab = 1$ then the left-hand side is zero, whereas the right-hand side cannot be, so $ab = 1$ gives $bac = c \; \forall \; c$, that is, $ba = 1$.

The structure of an arbitrary algebra may be understood in terms of certain building blocks of smaller algebras together with the rules for assembling them. One such way in which an algebra can be expressed in terms of others is as a direct sum. An algebra $\mathcal{A}$ is the *direct sum* of algebras $\mathcal{B}$ and $\mathcal{C}$, $\mathcal{A} = \mathcal{B} \oplus \mathcal{C}$, if we have a vector space direct sum and $\mathcal{B}\mathcal{C} = \mathcal{C}\mathcal{B} = 0$. This is obviously extended to sums of several algebras. An algebra that can be written as a direct sum of subalgebras is called reducible and the subalgebras are termed *components*. Reducible algebras contain invariant subalgebras, or ideals. A two-sided ideal, or simply an *ideal*, is a subspace I such that $\mathcal{A}\mathrm{I}\mathcal{A} \subset \mathrm{I}$. Obviously ideals are subalgebras. Thus the components of a reducible algebra are ideals.

Suppose $\mathcal{A} = \mathcal{B} + \mathcal{C}$, then we define an equivalence relation in $\mathcal{A}$ by $a \sim b$ if $a = b + c$ where $c \in \mathcal{C}$. We denote the equivalance class of a

by $[a]$. The elements of $\mathscr{A}$ form an Abelian group under the operation of addition; this group may be quotiented by defining

$$[a] + [b] = [a + b].$$

The equivalence classes are made into a vector space by defining

$$\lambda[a] = [\lambda a] \qquad \text{for } \lambda \text{ in } F.$$

The obvious way to try and make the equivalence classes into an algebra is by defining

$$[a][b] = [ab].$$

If, however, c, $d \in \mathscr{C}$ then

$$[a][b] = [a + c][b + d]$$

and so for consistency we would need

$$[ab] = [ab + ad + cb + cd]$$

that is, $(ad + cb + cd) \in \mathscr{C}$. This will be true for all a, $b \in \mathscr{A}$ and c, $d \in \mathscr{C}$ if, and only if, $\mathscr{C}$ is an ideal. When this is the case then what we have described is the *quotient algebra* of $\mathscr{A}$ modulo $\mathscr{C}$, denoted $\mathscr{A}/\mathscr{C}$. If $\mathscr{A}$ is a G-graded algebra with an ideal I which is a G-graded subspace then I wiil in fact be a G-graded algebra. As a vector space $\mathscr{A}/\mathrm{I}$ inherits a natural G-gradation such that, if a is homogeneous, $\deg[a] = \deg a$. This makes $\mathscr{A}/\mathrm{I}$ a G-graded algebra since

$$\begin{aligned} \deg\{[a][b]\} &= \deg[ab] \\ &= \deg ab \\ &= \deg a + \deg b \\ &= \deg[a] + \deg[b]. \end{aligned}$$

An *algebra homomorphism* is a linear transformation from an algebra $\mathscr{A}$ to an algebra $\mathscr{B}$ such that the multiplicative structure is preserved. That is, if φ is a linear transformation from $\mathscr{A}$ onto $\mathscr{B}$ then φ is an algebra homomorphism if $\varphi(ab) = \varphi(a)\varphi(b)$. When the linear transformation is a vector space isomorphism then we have an *algebra isomorphism*, two isomorphic algebras also being called *equivalent*, denoted $\mathscr{A} \simeq \mathscr{B}$. An isomorphism from an algebra to itself is called an *automorphism*. If an algebra has a unit element then for any invertible s the mapping $a \mapsto sas^{-1}$ defines an automorphism, called an *inner automorphism*. An automorphism is readily seen to map the centre onto itself. If the automorphism is inner then individual elements of the centre are left invariant. The *kernel* of a homomorphism is the kernel of the linear transformation. If a is in the kernel of a homomorphism φ,

$\varphi(a) = 0$, then $\varphi(bac) = \varphi(b)\varphi(a)\varphi(c) = 0$ for all b, c and so the kernel is an ideal. In the same way as the analogous result for groups is proved, we may show that

$$\varphi(\mathscr{A}) = \mathscr{A}/\ker\varphi. \tag{A5}$$

A different correspondence between algebras may be defined as follows. If u is a vector space isomorphism between $\mathscr{A}$ and $\mathscr{A}'$

$$u:\mathscr{A} \longrightarrow \mathscr{A}'$$

$$a \longmapsto a^u$$

such that $(ab)^u = b^u a^u$ then $\mathscr{A}$ and $\mathscr{A}'$ are termed *opposite algebras* and we shall use $\mathscr{A}^{\mathrm{op}}$ to denote the opposite to $\mathscr{A}$. In general $\mathscr{A}^{\mathrm{op}} \neq \mathscr{A}$. For the case when the opposite algebra is isomorphic to $\mathscr{A}$ then $\mathscr{A}'$ may be replaced with $\mathscr{A}$ in the definition above and we then speak of the mapping as an *anti-automorphism*. An anti-automorphism of particular interest is that which squares to the identity. We shall call this involutory anti-automorphism simply an involution.

If $\mathscr{B}$ and $\mathscr{C}$ are algebras of dimension m and n then we have already described how to form a new algebra of dimension $m + n$, namely the direct sum. We now describe how an algebra of dimension mn may be formed, the tensor product. If $\mathscr{A}$, $\mathscr{B}$, $\mathscr{C}$ are algebras over F with dimensions mn, m and n respectively such that $\mathscr{B}$ has a basis $\{b_i\}$ $i = 1, \ldots, m$ with multiplication table

$$b_i b_j = \sum_k B_{ijk} b_k$$

$\mathscr{C}$ has a basis $\{c_p\}$ $p = 1, \ldots, n$ with multiplication

$$c_p c_q = \sum_r C_{pqr} c_r$$

then $\mathscr{A}$ is the *tensor product* of $\mathscr{B}$ and $\mathscr{C}$, $\mathscr{A} = \mathscr{B} \otimes \mathscr{C}$, if it admits a basis $\{a_{ip}\}$ $i = 1, \ldots, m$; $p = 1, \ldots, n$ with multiplication given by

$$a_{ip} a_{jq} = \sum_{k,r} B_{ijk} C_{pqr} a_{kr}.$$

This criterion for $\mathscr{A}$ to be the tensor product of $\mathscr{B}$ and $\mathscr{C}$ involves particular bases for $\mathscr{B}$ and $\mathscr{C}$, thus there is now an onus to show that it is in fact independent of the bases chosen. If we have bases as defined above then we can define a bilinear map

$$\otimes:\mathscr{B} \times \mathscr{C} \longrightarrow \mathscr{A}$$

$$b_i, c_p \longmapsto b_i \otimes c_p = a_{ip}.$$

If now $\{b_i'\}$, $\{c_p'\}$ are any bases for $\mathscr{B}$, $\mathscr{C}$ then the bilinearity ensures that the set of $\{b'_i \otimes c'_p\}$ are linearly independent, and hence a basis for $\mathscr{A}$. Further, if

$$b'_i b'_j = \sum_k B'_{ijk} b'_k$$

and

$$c'_p c'_q = \sum_r C'_{pqr} c'_r$$

then we have

$$(b'_i \otimes c'_p)(b'_j \otimes c'_q) = \sum_{k,r} B'_{ijk} C'_{pqr} b'_k \otimes c'_r.$$

So indeed the definition of the tensor product is independent of the bases for $\mathscr{B}$ and $\mathscr{C}$. It should be stressed that the definition we have given for the tensor product algebra defines it only up to equivalence. This will be convenient later when we shall make use of the observation that if $\mathscr{B}$ and $\mathscr{C}$ are mutually commuting subalgebras of $\mathscr{A}$ with $\dim \mathscr{A} = \dim \mathscr{B} \dim \mathscr{C}$ then $\mathscr{A} = \mathscr{B} \otimes \mathscr{C}$. In the particular case that $\mathscr{A}$ has a unit element it will also be the unit element of $\mathscr{B}$ and $\mathscr{C}$.

A familiar example of an n^2-dimensional algebra is provided by the set of all $n \times n$ matrices (matrices of order n) with elements in F. The abstract algebra isomorphic to this will be termed a *total matrix algebra*, denoted $\mathscr{M}_n(F)$. Where no confusion is likely we will simply refer to such an algebra as a matrix algebra, and shall not exhibit the underlying field, writing $\mathscr{M}_n$. A basis for matrices of order n is obviously provided by all the elements with a unit in the ith row and jth column and zeroes elsewhere. We formalise this by defining an *ordinary matrix basis* to be

$$\{\mathbf{e}_{ij}\} \qquad i, j = 1, \ldots, n$$

$$\mathbf{e}_{ij}\mathbf{e}_{kl} = 0 \qquad j \neq k$$

$$\mathbf{e}_{ij}\mathbf{e}_{jk} = \mathbf{e}_{ik}.$$

The identity is the sum of the diagonal elements, that is $\mathbf{I} = \mathbf{e}_{11} + \ldots$ $+21\mathrm{e}_{nn}$. Matrix algebras have the following simple but important property.

$$\mathscr{M}_m(F) \otimes \mathscr{M}_n(F) = \mathscr{M}_{mn}(F). \tag{A6}$$

The proof will consist of spotting how to label the basis. If $\{\mathbf{e}_{ij}\}$ and $\{\mathbf{f}_{pq}\}$ are ordinary bases for $\mathscr{M}_m$ and $\mathscr{M}_n$ then if we set

$$\mathbf{e}_{ij} \otimes \mathbf{f}_{pq} = \mathbf{E}_{IJ}$$

where $I = (i-1)n + p$, $J = (j-1)n + q$ a basis for $\mathscr{M}_m \otimes \mathscr{M}_n$ is $\{\mathbf{E}_{IJ}\}$ $I, J = 1, \ldots, mn$. If $K = (k-1)n + r$, $L = (l-1)n + s$ then

$$\begin{aligned}\mathbf{E}_{IJ}\mathbf{E}_{KL} &= (\mathbf{e}_{ij}\otimes\mathbf{f}_{pq})(\mathbf{e}_{kl}\otimes\mathbf{f}_{rs})\\ &= \delta_{jk}\delta_{qr}\mathbf{e}_{il}\otimes\mathbf{f}_{ps}\\ &= \delta_{jk}\delta_{qr}\mathbf{E}_{IL}\\ &= \delta_{(j-1)n+q,\,(k-1)n+r}\mathbf{E}_{IL}\\ &= \delta_{JK}\mathbf{E}_{IL}.\end{aligned}$$

One reason for the importance of matrix algebras is that any associative algebra can be imbedded in a total matrix algebra. If V is a vector space then the set of all linear transformations from V to V forms an algebra, the endomorphism algebra, End V. If $M \in \text{End}\, V$ and $a \in V$ then we will usually write the transform of a by M as Ma, with no brackets. The product of linear transformations M and N will be defined by $(MN)a = M(Na)$. Occasionally it will be convenient to write the effect of a linear transformation as $M{:}a \to a^M$. In this case we will use the convention that $a^{MN} = (a^M)^N$. Normally this latter notation will be reserved for involutions. Addition of linear transformations is defined in the obvious way, and it is clear that End V is a total matrix algebra. A *representation* of an algebra $\mathscr{A}$ is a homomorphism into End V, for some V. Representations of algebras are termed *faithful, irreducible* or *equivalent* using the obvious analogue to the case of group representations. If $\mathscr{A}$ is an algebra then it is certainly a vector space and thus the algebra End $\mathscr{A}$ is associated with it. We may put elements of $\mathscr{A}$ into correspondence with certain elements of End $\mathscr{A}$ as follows. For $a \in \mathscr{A}$, $\mathrm{L}(a) \in \text{End}\,\mathscr{A}$ is defined by

$$\mathrm{L}(a)d = ad \qquad \forall\, d \in \mathscr{A}.$$

It follows that L is a linear map from $\mathscr{A}$ into End $\mathscr{A}$ such that $\mathrm{L}(a)\mathrm{L}(b) = \mathrm{L}(ab)$. Thus L is a homomorphism, called the *regular representation*. If $\mathscr{A}$ has a unit element then the regular representation is faithful. For if $\mathrm{L}(a) = \mathrm{L}(b)$ then $\mathrm{L}(a-b)d = 0$ for all d, and taking $d = 1$ gives $a = b$. So, in this case, the set $\{\mathrm{L}(a)\}$ for all $a \in \mathscr{A}$ forms an algebra, $\mathrm{L}(\mathscr{A})$, equivalent to $\mathscr{A}$. In an obvious fashion we define the mapping R such that $\mathrm{R}(a)d = da$. Then $\mathrm{R}(a)\mathrm{R}(b) = \mathrm{R}(ba)$ and so for an algebra with unit element $\mathrm{R}(\mathscr{A}) \simeq \mathscr{A}^{\mathrm{op}}$. Thus $\mathrm{L}(\mathscr{A})$ and $\mathrm{R}(\mathscr{A})$ are subalgebras of End $\mathscr{A}$ which are also mutually commuting, for

$$\begin{aligned}\mathrm{L}(a)\mathrm{R}(b)d &= \mathrm{L}(a)(db) = adb\\ &= \mathrm{R}(b)\mathrm{L}(a)d\end{aligned}$$

since $\mathscr{A}$ is associative. What is more, if $\mathscr{A}$ has a unit element and $S \in \text{End}\,\mathscr{A}$ commutes with all elements of $\mathrm{L}(\mathscr{A})$ then S must be in $\mathrm{R}(\mathscr{A})$.

For

$$(S\mathrm{L}(a))1 = Sa$$

and

$$(\mathrm{L}(a)S)1 = a(S1)$$

so if S commutes with $\mathrm{L}(a)$

$$Sa = a(S1)$$

that is

$$Sa = \mathrm{R}(S1)a \qquad \forall a.$$

If, then, $\mathscr{A}$ is an n-dimensional algebra with identity then End $\mathscr{A}$ is an n^2-dimensional algebra with $\mathrm{L}(\mathscr{A})$ and $\mathrm{R}(\mathscr{A})$ as n-dimensional commuting subalgebras. If the dimension of $\mathrm{L}(\mathscr{A})\mathrm{R}(\mathscr{A})$ were n^2 then End $\mathscr{A}$ would be the tensor product of $\mathrm{L}(\mathscr{A})$ and $\mathrm{R}(\mathscr{A})$. Although in general this will not be the case it is in the following situation.

If $\mathscr{D}$ is a central division algebra then $\mathrm{L}(\mathscr{D})\otimes\mathrm{R}(\mathscr{D}) = \mathrm{End}\,\mathscr{D}$. (A7)

If $\mathscr{D}$ is n-dimensional we need to show that $\dim\{\mathrm{L}(\mathscr{D})\mathrm{R}(\mathscr{D})\} = n^2$. The proof will require the following

Lemma

$$\mathrm{L}(\mathscr{D})\mathrm{R}(\mathscr{D}) = \mathrm{L}(\mathscr{D})u_1 + \mathrm{L}(\mathscr{D})u_2 + \ldots + \mathrm{L}(\mathscr{D})u_s$$

where $u_1, \ldots, u_s$ are in $\mathrm{R}(\mathscr{D})$ and the sums are direct vector space sums.

Since the identities of $\mathrm{L}(\mathscr{D})$ and $\mathrm{R}(\mathscr{D})$ coincide we have $\mathrm{R}(\mathscr{D}) \subset \mathrm{L}(\mathscr{D})\mathrm{R}(\mathscr{D})$. We pick a non-zero element of $\mathrm{R}(\mathscr{D})$, u_1 say, and form $\mathrm{L}(\mathscr{D})u_1$. Then either $\mathrm{L}(\mathscr{D})u_1 = \mathrm{R}(\mathscr{D})$ or we can pick a u_2 in $\mathrm{R}(\mathscr{D})$ that is not in $\mathrm{L}(\mathscr{D})u_1$, giving $\mathrm{L}(\mathscr{D})u_2 \cap \mathrm{L}(\mathscr{D})u_1 = 0$. For if $\alpha u_1 = \beta u_2$ where $\alpha, \beta \in \mathrm{L}(\mathscr{D})$ then for $\beta \neq 0$ $u_2 = (\beta^{-1}\alpha)u_1$, which contradicts $u_2 \notin \mathrm{L}(\mathscr{D})u_1$. Proceeding in this manner completes the proof of the lemma.

In the manner of the lemma we write

$$\mathrm{L}(\mathscr{D})\mathrm{R}(\mathscr{D}) = \mathrm{L}(\mathscr{D})u_1 + \ldots + \mathrm{L}(\mathscr{D})u_s.$$

Since for u regular $\dim\{\mathrm{L}(\mathscr{D})u\} = n$ we have $\dim(\mathrm{L}(\mathscr{D})\mathrm{R}(\mathscr{D})) = ns$, with $s \leqslant n$. If $s < n$ then we may extend the set $\{u_1, u_2, \ldots, u_s\}$ to a basis for $\mathrm{R}(\mathscr{D})$ by choosing $u_{s+1}, \ldots, u_n$. Since, as we stated in the lemma, $\mathrm{R}(\mathscr{D}) \subset \mathrm{L}(\mathscr{D})\mathrm{R}(\mathscr{D})$

$$u_{s+1} = \sum_i \alpha_i u_i \qquad \text{with } \alpha_i \in \mathrm{L}(\mathscr{D}).$$

However, since $L(\mathscr{D})$ and $R(\mathscr{D})$ are commuting subalgebras $[u_i, \beta] = 0$ $\forall \beta \in L(\mathscr{D})$ where the bracket denotes the commutator. In particular $[u_{s+1}, \beta] = 0$ giving

$$\sum_i [\alpha_i, \beta] u_i = 0.$$

Since the sum is direct, in the vector space sense, we must have

$$[\alpha_i, \beta] = 0 \qquad \forall \beta \in L(\mathscr{D}) \; i = 1, \ldots, s.$$

That is, the α_i are in the centre of $L(\mathscr{D})$. But $L(\mathscr{D}) \simeq \mathscr{D}$ which is central, so the α_i must all be multiples of the identity by the base field F. The expansion of u_{s+1} as a sum of the first s u_i then contradicts their F-linear independence and so we must have $s = n$ and the proof is complete.

There is another reason for the prominent role played by matrix algebras. The structure of an important class of algebras may be given in terms of matrix algebras and division algebras. More generally the recalcitrant (or interesting) parts of an algebra may be collected together into a certain ideal such that the structure of the quotient modulo this ideal is given in terms of matrix and division algebras. The existence of this ideal will now be established.

A non-zero element of an algebra is called *nilpotent* if some finite power of it vanishes. The smallest such power is called the *index* of that element. An algebra is called nilpotent of index v if v is the smallest integer such that all products of v terms vanish. We have already encountered the concept of a two-sided ideal; single-sided ideals are defined as follows. A *left ideal* of an algebra $\mathscr{A}$ is a subspace $\mathscr{L}$ such that $\mathscr{A}\mathscr{L} \subset \mathscr{L}$. Right ideals are defined in the obvious way. It follows that single-sided ideals are subalgebras, and so we may talk of nilpotent single-sided ideals.

The sum of two nilpotent left ideals is a nilpotent left ideal. (A8)

Let $\mathscr{B}$ and $\mathscr{C}$ be nilpotent left ideals of index α and β respectively. Then $\mathscr{B} + \mathscr{C}$ is certainly a left ideal. If any element of $\mathscr{B} + \mathscr{C}$ is raised to the power k then it will be a linear combination of terms of the form $a = a_1 a_2 \ldots a_k$ where the a_i are in $\mathscr{B}$ or $\mathscr{C}$. Suppose that p terms in this product are in $\mathscr{B}$, and that j is the largest integer such that $a_j \in \mathscr{B}$. Then if $a_{j-1} \in \mathscr{C}$ we set $a_{j-1} a_j = a'_j$, where $a'_j \in \mathscr{B}$, since $\mathscr{B}$ is a left ideal. Proceeding in this manner we can write $a = b_1 \ldots b_p r$ with the b_i in $\mathscr{B}$ and r in $\mathscr{C}$. Similarly, we have $a = c_1 \ldots c_q s$ where the c_i are in $\mathscr{C}$ and s is in $\mathscr{B}$. Here $p + q = k$. So if $k = \alpha + \beta - 1$ then $p < \alpha$ gives $q \geqslant \beta$, whereas $q < \beta$ gives $p \geqslant \alpha$ and so we must have $a = 0$. That is, $\mathscr{B} + \mathscr{C}$ is nilpotent with index no greater than one less than the sum of those of $\mathscr{B}$ and $\mathscr{C}$. Obviously this result is just as valid for right ideals.

If $\mathscr{L}$ is a nilpotent left ideal then $\mathscr{I} = \mathscr{L} + \mathscr{L}\mathscr{A}$ is a nilpotent two-sided ideal. (A9)

Firstly note that $\mathscr{I}$ is indeed an ideal; for $\mathscr{A}\mathscr{L} \subset \mathscr{L}$ since $\mathscr{L}$ is a left ideal and so $\mathscr{A}\mathscr{L}\mathscr{A} \subset \mathscr{L}\mathscr{A}$, thus $\mathscr{I}$ is a left ideal. Similarly $\mathscr{A}\mathscr{A} \subset \mathscr{A}$ and so $\mathscr{I}\mathscr{A} \subset \mathscr{L}\mathscr{A} \subset \mathscr{I}$, making $\mathscr{I}$ a right ideal. As well as being a left ideal $\mathscr{L}\mathscr{A}$ is nilpotent. For if $x \in \mathscr{L}\mathscr{A}$, $x = la$ for $l \in \mathscr{L}$, $a \in A$ and $x^k = ll'^{k-1}a$ where $l' = al$, is in $\mathscr{L}$. So $\mathscr{L}\mathscr{A}$ is nilpotent with index less than or equal to that of $\mathscr{L}$. Since $\mathscr{I}$ is the sum of two nilpotent left ideals, the previous theorem shows that $\mathscr{I}$ is nilpotent.

These two results have been established for the purpose of proving the following.

Every nilpotent left, right and two-sided ideal is contained in a unique maximal nilpotent ideal, the *radical*. (A10)

Let $\mathscr{N}$ be a nilpotent ideal of largest dimension. If $\mathscr{N}'$ is any nilpotent ideal then, by (A8), $\mathscr{N} + \mathscr{N}'$ is a nilpotent left ideal, and similarly it is a nilpotent right ideal and so an ideal. But $\mathscr{N}$ is of maximal dimension so we must have $\mathscr{N}' \subset \mathscr{N}$. If now $\mathscr{L}$ is a nilpotent left ideal then the above result and (A9) combine to give $\mathscr{L} \subset (\mathscr{L} + \mathscr{L}\mathscr{A}) \subset \mathscr{N}$. Similarly for right ideals.

Before making the anticipated good use of the existence of the nilpotent radical it is necessary to establish some properties of other important elements of an algebra, the idempotents. A non-zero P is *idempotent* if $P^2 = P$. An obvious example of an idempotent is the identity of a division algebra.

The identity is the only idempotent in a division algebra. (A11)

Suppose $P^2 = P$ and P is not zero. Then P is invertible and $P^{-1}P^2 = P^{-1}P$, that is $P = 1$. Of course, a unit element is a very special example of an idempotent. A more general example is provided by the diagonal elements of an ordinary matrix basis. A large class of algebras have an idempotent.

Every non-nilpotent algebra contains an idempotent. (A12)

Obviously a nilpotent algebra cannot contain an idempotent. We will show that if an algebra does not contain an idempotent then in fact it must be nilpotent. Suppose that $\mathscr{A}$ contains an a such that $\mathscr{A}a^k = \mathscr{A}a^{k-1}$ for some power k. Then if $\mathscr{B} = \mathscr{A}a^{k-1}$, $\mathscr{B}$ is a left ideal of $\mathscr{A}$, and hence an algebra, satisfying $\mathscr{B}a = \mathscr{B}$ and hence $\mathscr{B}b = \mathscr{B}$ where $b \in \mathscr{B}$ is given by $b = a^k$. So there must be some $P \in \mathscr{B}$ such that $Pb = b$, giving $(P^2 - P)b = 0$. But $\mathscr{B}b = \mathscr{B}$ means that there is no non-zero x with $xb = 0$ and so $\mathscr{B}$, and thus $\mathscr{A}$, contains an idempotent. So if $\mathscr{A}$ does not contain an idempotent we must have

$\dim(\mathcal{A}a^k) < \dim(\mathcal{A}a^{k-1})$ for all powers k of all $a \in \mathcal{A}$. The finite dimensionality of $\mathcal{A}$ means that there must be a finite α such that $\mathcal{A}a^\alpha = 0$; in particular $a^{\alpha+1} = 0$. Since this is true for all a, $\mathcal{A}$ is nilpotent.

The existence of an idempotent enables an algebra to be written as a direct vector space sum of subalgebras. Let P be an idempotent in $\mathcal{A}$, then $\mathcal{L}(P)$ is defined to be the left ideal consisting of $a \in \mathcal{A}$ such that $aP = 0$. Similarly the right ideal $\mathcal{R}(P)$ is defined to consist of all $a \in \mathcal{A}$ such that $Pa = 0$, and we define $\mathcal{I}(P) = \mathcal{L}(P) \cap \mathcal{R}(P)$. The following theorem gives the two-sided *Peirce decomposition* of $\mathcal{A}$.

If P is idempotent in $\mathcal{A}$ then

$$\mathcal{A} = P\mathcal{A}P + P\mathcal{L}(P) + \mathcal{R}(P)P + \mathcal{I}(P). \tag{A13}$$

All the terms in this sum are algebras. $P\mathcal{A}(P)$ consists of all $a \in \mathcal{A}$ such that $Pa = aP = a$; $P\mathcal{L}(P)$ consists of all $a \in \mathcal{A}$ with $Pa = a$, $aP = 0$; $\mathcal{R}(P)P$ consists of all $a \in \mathcal{A}$ with $Pa = 0$, $aP = a$ and if $a \in \mathcal{I}(P)$ $Pa = aP = 0$. So obviously these algebras are non-intersecting and what we need to show is that they span $\mathcal{A}$. To see this we write

$$a = PaP + P(a - aP) + (a - Pa)P + (a - Pa - aP + PaP)$$

where each term in the sum lies in one of the subalgebras contained in the Peirce decomposition.

Elements of $\mathcal{I}(P)$ are said to be (algebraically) *orthogonal* to P. An idempotent is called *principal* if there is no idempotent orthogonal to it. We can now go one step further from (A12) with

Every non-nilpotent algebra contains a principal idempotent. (A14)

If $\mathcal{A}$ is non-nilpotent then it certainly contains an idempotent. If u is a non-principal idempotent then there exists an idempotent v such that $uv = vu = 0$. That is, $v \in \mathcal{I}(u)$. If $P = u + v$ then P is idempotent with $Pu = uP = u$ and $Pv = vP = v$. So if $xP = 0$ then $xu = xP = 0$, and if $Px = 0$ then $ux = uPx = 0$, that is, $\mathcal{I}(P) \subset \mathcal{I}(u)$. In fact $\mathcal{I}(P)$ must be strictly contained in $\mathcal{I}(u)$ for $v \in \mathcal{I}(u)$ but not in $\mathcal{I}(P)$. If P is not principal then we set $P' = P + w$ where $w \in \mathcal{I}(P)$. Since $\mathcal{I}(P) \subset \mathcal{I}(u)$ if this process is continued it will eventually produce a principal idempotent since $\mathcal{I}(u)$ is finite dimensional.

Of fundamental importance are the primitive idempotents. An idempotent is *primitive* if it can not be written as a sum of two orthogonal idempotents. The following could have formed an alternative definition of a primitive idempotent.

P is the only idempotent of $P\mathcal{A}P$ iff P is primitive. (A15)

If P were not primitive then $P = u + v$ with $uv = vu = 0$. So

$Pu = uP = u$ and $Pv = vP = v$ and thus both u and v are in $P\mathcal{A}P$. Conversely, if u is an idempotent in $P\mathcal{A}P$ then $P - u$ is idempotent since P is the identity in $P\mathcal{A}P$. Further, $u(P - u) = (P - u)u = 0$ and so $P = (P - u) + u$, the sum of two orthogonal idempotents. The nomenclature is explained by the following.

Every non-primitive idempotent is the sum of a set of pairwise orthogonal primitive idempotents. (A16)

If P is not primitive then $P = u + v$, where u and v are orthogonal idempotents. Suppose that v is not primitive, then $v = w + x$ with w and x orthogonal. Now $vw = wv = w$ and $vx = xv = x$ and so $uw = uvw = 0$, $wu = wvu = 0$. Similarly $ux = xu = 0$ and so $\{u, w, x\}$ are pairwise orthogonal idempotents. If we continue in this way then the process must terminate due to the finiteness of $\mathcal{A}$ and we will arrive at a set of pairwise orthogonal primitives.

Attention will now be focused on algebras whose radical is zero. It will transpire that we can completely determine the structure of all such algebras. An algebra whose radical is zero is called *semi-simple*. The first consequence of the definition is

A semi-simple algebra has a unit element. (A17)

If $\mathcal{A}$ is semi-simple then it is not nilpotent and so, by (A14), contains a principal idempotent P say. The Peirce decomposition of (A13) then gives

$$\mathcal{A} = P\mathcal{A}P + \mathcal{B}$$

where $\mathcal{B} = P\mathcal{L}(P) + \mathcal{R}(P)P + \mathcal{I}(P)$. $\mathcal{B}$ is spanned by $\mathcal{L}(P)$ and $\mathcal{R}(P)$. We shall show that these single-sided ideals are nilpotent and hence contained in the radical, which is zero by hypothesis. This will give $\mathcal{A} = P\mathcal{A}P$; but P is the identity in $P\mathcal{A}P$, and hence of $\mathcal{A}$. If P is principal then $\mathcal{I}(P)$, which contains all elements orthogonal to P, can contain no idempotent and thus must be nilpotent. Since $\mathcal{R}(P)$ and $\mathcal{L}(P)$ consist of all elements annihilated by left and right multiplication by P respectively $\mathcal{R}(P)\mathcal{L}(P) \subset \mathcal{I}(P)$. So if $l \in \mathcal{L}(P)$ and $r \in \mathcal{R}(P)$ then $(rl)^{\alpha} = 0$ where α is the index of $\mathcal{I}(P)$. Since $(lr)^{\alpha+1} = l(rl)^{\alpha}r = 0$ then the ideal $\mathcal{L}(P)\mathcal{R}(P)$ is nilpotent of index less that or equal to $\alpha + 1$. Now

$$\begin{aligned}\mathcal{L}(P)\mathcal{A} &= \mathcal{L}(P)\{P\mathcal{A}P + P\mathcal{L}(P) + \mathcal{R}(P)P + \mathcal{I}(P)\} \\ &= \mathcal{L}(P)\mathcal{R}(P)P + \mathcal{L}(P)\mathcal{I}(P).\end{aligned}$$

Since $\mathcal{R}(P)$ is a right ideal $\mathcal{R}(P)P \subset \mathcal{R}(P)$ and since $\mathcal{I}(P) = \mathcal{L}(P) \cap \mathcal{R}(P)$ obviously $\mathcal{I}(P) \subset \mathcal{R}(P)$ and so $\mathcal{L}(P)\mathcal{A} \subset \mathcal{L}(P)\mathcal{R}(P)$. In particular, $\mathcal{L}(P)\mathcal{L}(P) \subset \mathcal{L}(P)\mathcal{R}(P)$. Since $\mathcal{L}(P)\mathcal{R}(P)$ is nilpotent of index $\leqslant \alpha + 1$ if $x \in \mathcal{L}(P)$ $(x^2)^{\alpha+1} = 0$, and so $\mathcal{L}(P)$ is a nilpotent left ideal, contained in the radical. In exactly the same way we show that $\mathcal{R}(P)$ is nilpotent and the proof follows.

The primitive idempotents in a semi-simple algebra have the following important property.

If P is an idempotent in a semi-simple $\mathcal{A}$ then $P\mathcal{A}P$ is a division algebra iff P is primitive. (A18)

Suppose that $P\mathcal{A}P$ is a division algebra. Then P is the identity which, by (A11), is the only idempotent in $P\mathcal{A}P$. (A15) then ensures that P is primitive. To prove the converse we shall use the following

Lemma

If P is an idempotent of a semi-simple $\mathcal{A}$ then $P\mathcal{A}P$ is semi-simple.

For suppose that $y \in \mathcal{N}_0$, the radical of $P\mathcal{A}P$. Then $\mathcal{A}y$ is a left ideal of $\mathcal{A}$. Since P is the identity in $P\mathcal{A}P$

$$\begin{aligned}(ay)^{\alpha+1} &= ayP(aPy)^{\alpha} \\ &= ay(PaPy)^{\alpha}.\end{aligned}$$

But $PaP \in P\mathcal{A}P$ and so $PaPy \subset \mathcal{N}_0$. Thus if α is the index of $\mathcal{N}_0$, $\mathcal{A}y$ is nilpotent of index $\leqslant \alpha + 1$. Since $\mathcal{A}$ is semi-simple $\mathcal{A}y = 0$ which, since $\mathcal{A}$ has a unit, gives $y = 0$ and $P\mathcal{A}P$ is semi-simple.

Suppose now that P is primitive then $P\mathcal{A}P$ is semi-simple, by the above lemma, with unity P. If a is any non-zero element of $P\mathcal{A}P$ then $P\mathcal{A}Pa$ is a non-zero left ideal of $P\mathcal{A}P$; further it is not nilpotent since $P\mathcal{A}P$ is semi-simple. (A12) ensures that $P\mathcal{A}Pa$ contains an idempotent, but any idempotent in $P\mathcal{A}Pa$ is certainly idempotent in $P\mathcal{A}P$ for which P is the only idempotent since P is primitive ((A15)). That is, $P \in P\mathcal{A}Pa$ say $P = ba$ for $b \in P\mathcal{A}P$. Since P is the identity in $P\mathcal{A}P$ this says that every non-zero a has a left inverse, and hence an inverse by (A4).

The semi-simple algebras are not quite as 'simple' as the simple ones. An algebra that is not a one-dimensional nilpotent algebra is called *simple* if the only ideals are the zero ideal and the algebra itself. Simple algebras are certainly semi-simple. To see this we need only check that simple algebras cannot be nilpotent. Suppose that $\mathcal{N}$ is a nilpotent algebra, then $\mathcal{N}\mathcal{N}$ is an ideal strictly contained in $\mathcal{N}$. If this is not the zero ideal then $\mathcal{N}$ cannot be simple. If $\mathcal{N}\mathcal{N}$ is zero but the dimension of $\mathcal{N}$ is greater than one then any linear subspace of one less dimension is a non-zero ideal of $\mathcal{N}$. The only exceptional case of a one-dimensional nilpotent algebra has to be excluded by the caveat in the definition. The study of semi-simple algebras may be reduced to the study of simple ones by the following.

An algebra is semi-simple iff it is simple or a direct sum of simple components. (A19)

A direct sum of simple algebras is obviously semi-simple since the only ideals are smaller sums of simple algebras which are not nilpotent. To go the other way we shall use two lemmas.

Lemma 1
If $\mathscr{A}$ has an ideal with a unit element then $\mathscr{A}$ is reducible.

Let $\mathscr{B}$ be an ideal of $\mathscr{A}$ and $1_{\mathscr{B}}$ be the unit in $\mathscr{B}$. The Peirce decomposition of $\mathscr{A}$ is

$$\mathscr{A} = 1_{\mathscr{B}}\mathscr{A}1_{\mathscr{B}} + 1_{\mathscr{B}}\mathscr{L}(1_{\mathscr{B}}) + \mathscr{R}(1_{\mathscr{B}})1_{\mathscr{B}} + \mathscr{I}(1_{\mathscr{B}}).$$

If $\mathscr{S}(1_{\mathscr{B}}) = 1_{\mathscr{B}}\mathscr{A}1_{\mathscr{B}} + 1_{\mathscr{B}}\mathscr{L}(1_{\mathscr{B}}) + \mathscr{R}(1_{\mathscr{B}})1_{\mathscr{B}}$ then $\mathscr{S}(1_{\mathscr{B}}) \subset \mathscr{B}$ since $1_{\mathscr{B}} \in \mathscr{B}$ which is a two-sided ideal of $\mathscr{A}$. So if $b \in \mathscr{B}$ we have $b = b_1 + b_2$ with $b_1 \in \mathscr{S}(1_{\mathscr{B}})$ and $b_2 \in \mathscr{I}(1_{\mathscr{B}})$. Then $b1_{\mathscr{B}} = b_1$ since b_2 is orthogonal to $1_{\mathscr{B}}$, but $b1_{\mathscr{B}} = b$ so in fact we must have $\mathscr{S}(1_{\mathscr{B}}) = \mathscr{B}$. Since $\mathscr{I}(1_{\mathscr{B}})$ is orthogonal to $1_{\mathscr{B}}$ it is orthogonal to $\mathscr{B}$ and so $\mathscr{A} = \mathscr{B} \oplus \mathscr{I}(1_{\mathscr{B}})$.

Lemma 2
A non-zero ideal of a semi-simple algebra is semi-simple.

Suppose that $\mathscr{B}$ is an ideal in a semi-simple $\mathscr{A}$, and that $\mathscr{N}$ is the radical of $\mathscr{B}$. Then $\mathscr{B}\mathscr{N}\mathscr{B} \subset \mathscr{N}$ since $\mathscr{N}$ is an ideal of $\mathscr{B}$ and $\mathscr{A}\mathscr{B} \subset \mathscr{B}$ since $\mathscr{B}$ is an ideal of $\mathscr{A}$. So $\mathscr{A}(\mathscr{B}\mathscr{N}\mathscr{B})\mathscr{A} \subset \mathscr{B}\mathscr{N}\mathscr{B}$ which is thus an ideal in $\mathscr{A}$; further it is nilpotent since it is contained in the radical of $\mathscr{B}$. Since $\mathscr{A}$ is semi-simple $\mathscr{B}\mathscr{N}\mathscr{B} = 0$. Now $(\mathscr{A}\mathscr{N}\mathscr{A})^3 \subset (\mathscr{A}\mathscr{N}\mathscr{A})\mathscr{N}(\mathscr{A}\mathscr{N}\mathscr{A})$ and $\mathscr{A}\mathscr{N}\mathscr{A} \subset \mathscr{B}$ so $(\mathscr{A}\mathscr{N}\mathscr{A})^3 \subset \mathscr{B}\mathscr{N}\mathscr{B}$, which we have shown is zero. That is, $\mathscr{A}\mathscr{N}\mathscr{A}$ is a nilpotent ideal in a semi-simple $\mathscr{A}$ so $\mathscr{A}\mathscr{N}\mathscr{A} = 0$. Since $\mathscr{A}$ has a unit element this gives $\mathscr{N} = 0$ and $\mathscr{B}$ is semi-simple.

We may now return to the proof of the theorem. If $\mathscr{A}$ is semi-simple but not simple then it has a non-zero ideal which, by Lemma 2, is semi-simple and hence has a unit. Lemma 1 then ensures that $\mathscr{A}$ is reducible. The components are certainly ideals and so semi-simple, and we may proceed to reduce them. If $\mathscr{A}$ is finite then we must arrive at an expression of $\mathscr{A}$ as a direct sum of irreducible components. The components are ideals, hence semi-simple, and irreducible hence simple.

The reduction of a semi-simple algebra to simple components is unique up to an ordering of the components. (A20)

Let $\mathscr{A} = \mathscr{B}_1 \oplus \ldots \oplus \mathscr{B}_r$ with the $\mathscr{B}_i$ simple. The identity of $\mathscr{A}$ can be written as a sum of the identities in the $\mathscr{B}_i$, $1 = e_i \oplus \ldots \oplus e_r$. Suppose $\mathscr{A} = \mathscr{C}_1 \oplus \ldots \oplus \mathscr{C}_s$ then $\mathscr{C}_k = \mathscr{C}_k e_1 + \mathscr{C}_k e_2 + \ldots \mathscr{C}_k e_r \forall k = 1, \ldots, s$. If $\mathscr{C}_{ki} = \mathscr{C}_k e_i$ then $\mathscr{C}_{ki} \subset \mathscr{A}e_i = \mathscr{B}_i$ and the above sum must be direct:

$$\mathscr{A}\mathscr{C}_k\mathscr{A} = \sum_{i=1}^{r} \mathscr{A}\mathscr{C}_{ki}\mathscr{A} \qquad \text{where } \mathscr{A}\mathscr{C}_{ki}\mathscr{A} \subset \mathscr{B}_i$$

so $\mathscr{C}_k$ is an ideal if and only if all the $\mathscr{C}_{ki}$ are ideals of $\mathscr{B}_i$. But the $\mathscr{B}_i$ are simple, so $\mathscr{C}_{ki} = \mathscr{B}_i$ or $\mathscr{C}_{ki} = 0$. If the $\mathscr{C}_k$ are irreducible then for a given k not more than one $\mathscr{C}_{ki}$ can be non-zero and it follows that the $\mathscr{C}_k$ are just the $\mathscr{B}_i$ up to a possible relabelling.

The above two theorems determine the structure of semi-simple algebras in terms of simple ones. Before turning to the classification of these we consider representations of semi-simple algebras. Again the representation theory will reduce to that of simple algebras and so we consider this case first.

All irreducible representations of a simple algebra are equivalent. (A21)

If $\mathscr{I}$ is any minimal left ideal of a simple $\mathscr{A}$ then we will show that any irreducible representation of $\mathscr{A}$ is equivalent to the representation on $\mathscr{I}$ induced by the regular representation.

Let ρ be some irreducible representation of $\mathscr{A}$ that maps $\mathscr{A}$ into End V, where V has no invariant subspaces under multiplication by $\rho(\mathscr{A})$. We first note that any minimal left ideal of End V, the pth column say, carries an equivalent representation to that carried by V. For if V is displayed as a 'column vector', with a basis $\{b_k\}$ consisting of zeroes except for a one in the kth row, then a basis for End V, $\{\mathbf{e}_{ij}\}$, is formed by the arrays whose only non-zero element is a one in the intersection of the ith row and the jth column. Elementary rules of matrix multiplication then give $\mathbf{e}_{ij}\boldsymbol{b}_k = \delta_{jk}\boldsymbol{b}_i$. A basis for the pth column is $\{\mathbf{e}_{kp}\}$ where k ranges over the order of the matrices, and $\mathbf{e}_{ij}\mathbf{e}_{kp} = \delta_{jk}\mathbf{e}_{ip}$. So the pth column, for any p carries a representation equivalent to that carried by V.

We introduce a linear transformation S that maps the minimal left ideal, $\mathscr{I}$, of $\mathscr{A}$ into the pth column of End V:

$$S\mathscr{I} = \rho(\mathscr{I})\mathbf{e}_{pp}.$$

Since $\mathscr{I}$ carries an irreducible representation of $\mathscr{A}$ then $\rho(\mathscr{I})$ carries an irreducible representation of $\rho(\mathscr{A})$ and so $\rho(\mathscr{I})\mathbf{e}_{pp}$ certainly transforms irreducibly under $\rho(\mathscr{A})$. But this is a subspace of the pth column which transforms irreducibly, so either S is a vector space isomorphism or $\rho(\mathscr{I})\mathbf{e}_{pp} = 0$. There must be some p for which this is non-zero, for otherwise we would have $\rho(\mathscr{I}) = 0$, which cannot be since $\mathscr{A}$ is simple. So at least for some choice of p, S is a vector space isomorphism between the minimal left ideal $\mathscr{I}$ and the pth column of End V. If $f \in \mathscr{I}$ then the following diagram shows the equivalence of the representation carried by the pth column (and hence V) and that carried by $\mathscr{I}$:

$$\begin{array}{ccc} & L(a) & \\ f & \longrightarrow & af \\ S \downarrow & & \downarrow S \\ & \rho(a) & \\ \rho(f)\mathbf{e}_{pp} & \longrightarrow & \rho(a)\rho(f)\mathbf{e}_{pp} = \rho(af)\mathbf{e}_{pp}. \end{array}$$

Thus any irreducible representation of a simple algebra is equivalent to that induced on any minimal left ideal by the regular representation.

We are now in a position to consider representations of semi-simple algebras.

Irreducible representations of a semi-simple algebra are equivalent if and only if their kernels are the same. (A22)

Equivalent representations must certainly have the same kernel, so what we need to show is that irreducible representations of a semi-simple algebra with the same kernel are in fact equivalent. A semi-simple algebra is the direct sum of simple ones, and so a representation can be irreducible only if the kernel contains all but one of the simple component algebras. Thus irreducible representations with the same kernel are irreducible representations of the same simple component algebra, and are thus equivalent by the preceeding result.

We now return to the classification of algebras by studying the simple ones. The main result is given below.

An algebra $\mathscr{A}$ is simple iff $\mathscr{A} = \mathscr{D} \otimes \mathscr{M}$ where $\mathscr{D}$ is a division algebra and $\mathscr{M}$ a total matrix algebra. (A23)

First we do the easy bit and assume $\mathscr{A} = \mathscr{D} \otimes \mathscr{M}$. Then $\mathscr{A}$ has an identity. Let b be a non-zero element of an ideal $\mathscr{I}$, then $b = \Sigma_{i,j} b_{ij} \mathbf{e}_{ij}$ with at least one (b_{pq} say) non-vanishing coefficient in $\mathscr{D}$. But

$$b_{pq} = \sum_i \mathbf{e}_{ip} b \mathbf{e}_{qi}$$

and so

$$\sum_i b_{pq}{}^{-1} \mathbf{e}_{ip} b \mathbf{e}_{qi} = 1.$$

That is $1 \subset \mathscr{A} b \mathscr{A} \subset \mathscr{I}$, giving $\mathscr{A} \subset \mathscr{I}$ and thus $\mathscr{A}$ is simple.

If now $\mathscr{A}$ is simple it has a unit element $1 = \Sigma_{i=1}^{n} P_i$ where the $\{P_i\}$ are pairwise orthogonal primitive idempotents. If $\mathscr{A}_{ij} = P_i \mathscr{A} P_j$ then the $\mathscr{A}_{ij}$ are certainly subspaces, and are in fact algebras since they are closed under multiplication. Multiplying two different algebras gives

$$\begin{aligned} \mathscr{A}_{ij} \mathscr{A}_{pk} &= \mathscr{A}_{ij} P_j P_p \mathscr{A}_{pk} = \mathscr{A}_{ij} \mathscr{A}_{jk} \delta_{jp} \\ &= P_i \mathscr{A} P_j \mathscr{A} P_k \delta_{jp}. \end{aligned}$$

Now $\mathscr{A}P_j\mathscr{A}$ is a two-sided ideal, which is not zero since it contains P_j, and so the simplicity of $\mathscr{A}$ gives $\mathscr{A}P_j\mathscr{A} = \mathscr{A}$ and hence

$$\begin{aligned} \mathscr{A}_{ij}\mathscr{A}_{pk} &= P_i\mathscr{A}P_k\delta_{jp} \\ &= \mathscr{A}_{ik}\delta_{jp}. \end{aligned}$$

In particular $\mathscr{A}_{11} = \mathscr{A}_{1j}\mathscr{A}_{j1}$ for any j. Since $P_1 \in \mathscr{A}_{11}$ there must be elements $\mathbf{e}_{j1}$, $\mathbf{e}_{1j}$ in $\mathscr{A}_{j1}$ and $\mathscr{A}_{1j}$, respectively, such that $\mathbf{e}_{1j}\mathbf{e}_{j1} = P_1$. If we now define $\mathbf{e}_{ij} \in \mathscr{A}_{ij}$ by $\mathbf{e}_{ij} = \mathbf{e}_{i1}\mathbf{e}_{1j}$ then

$$\begin{aligned} P_k\mathbf{e}_{ij} &= \mathbf{e}_{ij}\delta_{ik} \\ \mathbf{e}_{ij}P_k &= \mathbf{e}_{ij}\delta_{jk}. \end{aligned}$$

This gives

$$\begin{aligned} \mathbf{e}_{ij}\mathbf{e}_{pq} &= \mathbf{e}_{ij}P_jP_p\mathbf{e}_{pq} \\ &= \mathbf{e}_{ij}\mathbf{e}_{jq}\delta_{jp} \\ &= \mathbf{e}_{i1}\mathbf{e}_{1j}\mathbf{e}_{j1}\mathbf{e}_{1q}\delta_{jp} \\ &= \mathbf{e}_{i1}P_1\mathbf{e}_{1q}\delta_{jp} \\ &= \mathbf{e}_{i1}\mathbf{e}_{1q}\delta_{jp} \\ &= \mathbf{e}_{iq}\delta_{jp}. \end{aligned}$$

In particular the $\mathbf{e}_{ii}$ are idempotent. But $\mathbf{e}_{ii} \subset P_i\mathscr{A}P_i$ with P_i primitive, so $P_i\mathscr{A}P_i$ contains only one idempotent, namely P_i, so we must have $\mathbf{e}_{ii} = P_i$. So the $\mathbf{e}_{ij}$ span a total matrix algebra $\mathscr{M}$ whose identity is

$$\sum_i \mathbf{e}_{ii} = \sum_i P_i = 1$$

the identity of $\mathscr{A}$.

Since for each k P_k is primitive, $P_k\mathscr{A}P_k$ is a division algebra with P_k as identity. Each $\mathscr{A}_{kk}$ is an isomorphic copy of $\mathscr{A}_{11}$, say. For if $\alpha^{(1)} \in \mathscr{A}_{11}$ we define $\alpha^{(k)} \in \mathscr{A}_{kk}$ by $\alpha^{(k)} = \mathbf{e}_{k1}\alpha^{(1)}\mathbf{e}_{1k}$. Then for $\alpha^{(1)}$, $\beta^{(1)} \in \mathscr{A}_{11}$

$$\begin{aligned} (\alpha^{(1)}\beta^{(1)})^{(k)} &= \mathbf{e}_{k1}\alpha^{(1)}\beta^{(1)}\mathbf{e}_{1k} \\ &= \mathbf{e}_{k1}\alpha^{(1)}P_1\beta^{(1)}\mathbf{e}_{1k} \end{aligned}$$

since P_1 is the identity in $\mathscr{A}_{11}$

$$= \mathbf{e}_{k1}\alpha^{(1)}\mathbf{e}_{1k}\mathbf{e}_{k1}\beta^{(1)}\mathbf{e}_{1k}$$

since the $\mathbf{e}_{ij}$ are a matrix basis and so $(\alpha^{(1)}\beta^{(1)})^{(k)} = \alpha^{(k)}\beta^{(k)}$. This mapping from $\mathscr{A}_{11}$ to $\mathscr{A}_{kk}$ is obviously invertible and so indeed we have an isomorphism. By taking the direct sum of all elements in $\mathscr{A}_{11}$ with their isomorphic images in all $\mathscr{A}_{kk}$ we obtain another copy of $\mathscr{A}_{11}$, $\mathscr{D}$ say. That is, if $\alpha^{(1)} \in \mathscr{A}_{11}$ we define $\alpha \in \mathscr{D}$ to be

$$\alpha = \alpha^{(1)} \oplus \alpha^{(2)} \ldots \oplus \alpha^{(n)}.$$

It is straightforward to see that $\mathscr{D} \simeq \mathscr{A}_{11}$; further, elements of $\mathscr{D}$ commute with all the elements of $\mathscr{M}$. For if $\alpha \in \mathscr{D}$

$$\alpha \mathbf{e}_{ij} = \sum_k \alpha^{(k)} \mathbf{e}_{ij} = \alpha^{(i)} \mathbf{e}_{ij} = \mathbf{e}_{i1} \alpha^{(1)} \mathbf{e}_{1i} \mathbf{e}_{ij}$$

$$= \mathbf{e}_{i1} \alpha^{(1)} \mathbf{e}_{1j} = \mathbf{e}_{ij} \mathbf{e}_{j1} \alpha^{(1)} \mathbf{e}_{1j} = \mathbf{e}_{ij} \alpha^{(j)} = \sum_k \mathbf{e}_{ij} \alpha^{(k)}$$

$$= \mathbf{e}_{ij} \alpha.$$

For every $a \in \mathscr{A}$ set $a_{ij}{}^{(k)} = \mathbf{e}_{ki} a \mathbf{e}_{jk}$. Then

$$a_{ij}{}^{(k)} = \mathbf{e}_{k1} \mathbf{e}_{1i} a \mathbf{e}_{j1} \mathbf{e}_{1k} = \mathbf{e}_{k1} a_{ij}{}^{(1)} \mathbf{e}_{1k}$$

so if $a_{ij} = \Sigma_k a_{ij}{}^{(k)}$ then $a_{ij} \in \mathscr{D}$. Further

$$\sum_{i,j} a_{ij} \mathbf{e}_{ij} = \sum_{i,j,k} a_{ij}{}^{(k)} \mathbf{e}_{ij} = \sum_{i,j} a_{ij}{}^{(i)} \mathbf{e}_{ij}$$

$$= \sum_{i,j} \mathbf{e}_{ii} a \mathbf{e}_{ji} \mathbf{e}_{ij} = \sum_{i,j} \mathbf{e}_{ii} a \mathbf{e}_{jj} = \sum_{i,j} P_i a P_j = a.$$

Since this is true for every $a \in \mathscr{A}$ we have $\mathscr{A} = \mathscr{D} \otimes \mathscr{M}$ where $\mathscr{D}$ and $\mathscr{M}$ are as constructed in the proof.

The expression of a simple $\mathscr{A}$ as $\mathscr{A} = \mathscr{D} \otimes \mathscr{M}$ cannot be unique. For if $\mathbf{e}_{ij}$ is a matrix basis then so is $\mathbf{e}'_{ij} = s \mathbf{e}_{ij} s^{-1}$ where s is any regular element of $\mathscr{A}$. Then $a = \Sigma_{i,j} a'_{ij} \mathbf{e}'_{ij}$ with

$$a'_{ij} = \sum_k \mathbf{e}'_{ki} a \mathbf{e}'_{jk} = \sum_k s \mathbf{e}_{ki} s^{-1} a s \mathbf{e}_{jk} s^{-1} = s(s^{-1} a s)_{ij} s^{-1}$$

that is, $a'_{ij} \in s \mathscr{D} s^{-1}$. It turns out though that the choice of $\mathscr{D}$ and $\mathscr{M}$ is unique up to an inner automorphism like this. Note that if $\mathscr{A} = \mathscr{D} \otimes \mathscr{M} = \mathscr{D}' \otimes \mathscr{M}$ then we must have $\mathscr{D}' = \mathscr{D}$. For if $\alpha \in \mathscr{D}'$ we can write $\alpha = \Sigma_{i,j} \alpha_{ij} \mathbf{e}_{ij}$ with the $\alpha_{ij} \in \mathscr{D}$, and if α is to commute with $\mathscr{M}$ then $\alpha = \alpha_{11}(\mathbf{e}_{11} + \mathbf{e}_{22} + \ldots + \mathbf{e}_{nn}) = \alpha_{11}$ since the identities in $\mathscr{A}$ and $\mathscr{M}$ coincide. So if $\mathscr{A} = \mathscr{D} \otimes \mathscr{M} = \mathscr{D}' \otimes \mathscr{M}'$ where $\mathscr{M}' = s \mathscr{M} s^{-1}$ then we certainly have $\mathscr{D}' = s \mathscr{D} s^{-1}$.

If $\mathscr{A}$ is simple such that $\mathscr{A} = \mathscr{D} \otimes \mathscr{M}$ and $\mathscr{A} = \mathscr{D}' \otimes \mathscr{M}'$ then there is an $s \in \mathscr{A}$ such that $\mathscr{M}' = s \mathscr{M} s^{-1}$, $\mathscr{D}' = s \mathscr{D} s^{-1}$. (A24)

In view of the above comments it is sufficient to prove that $\mathscr{M}' = s \mathscr{M} s^{-1}$. Let $\{\mathbf{e}_{ij}\}$ $i, j = 1, \ldots, n$ be a basis for $\mathscr{M}$ and $\{\mathbf{e}'_{pq}\}$ $p, q = 1, \ldots, m$ be a basis for $\mathscr{M}'$. Without loss of generality we assume $m \geqslant n$. We write

$$\mathbf{e}'_{11} = \sum_{i,j=1}^{n} c_{ij} \mathbf{e}_{ij} \qquad c_{ij} \in \mathscr{D} \tag{i}$$

with at least one (c_{pq} say) of the c_{ij} not zero. If we set

$$a = c_{pq}^{-1}\mathbf{e}_{1p}\mathbf{e}'_{11} \tag{ii}$$

and

$$b = \mathbf{e}'_{11}\mathbf{e}_{q1} \tag{iii}$$

then $a \in \mathbf{e}_{11}\mathscr{A}\mathbf{e}'_{11}$, $b \in \mathbf{e}'_{11}\mathscr{A}\mathbf{e}_{11}$ with

$$\begin{aligned} ab &= c_{pq}^{-1}\mathbf{e}_{1p}\mathbf{e}'_{11}\mathbf{e}_{q1} \\ &= c_{pq}^{-1}\mathbf{e}_{1p}\sum_{i,j=1}^{n} c_{ij}\mathbf{e}_{ij}\mathbf{e}_{q1} \qquad \text{by (i)} \\ &= c_{pq}^{-1}c_{pq}\mathbf{e}_{11} = \mathbf{e}_{11} \end{aligned}$$

that is

$$ab = \mathbf{e}_{11}. \tag{iv}$$

Also

$$\begin{aligned} (ba)^2 &= b(ab)a \\ &= b\mathbf{e}_{11}a \qquad \text{by (iv)} \\ &= ba \end{aligned}$$

so ba is an idempotent in $\mathbf{e}'_{11}\mathscr{A}\mathbf{e}'_{11} = \mathscr{D}'\mathbf{e}'_{11}$; further it is not zero since $a(ba)b = (ab)^2 = \mathbf{e}_{11}$, by (iv). But the identity is the only idempotent in $\mathscr{D}'$ so we must have

$$ba = \mathbf{e}'_{11}. \tag{v}$$

If we now introduce

$$h = \sum_{i=1}^{n}\mathbf{e}_{i1}a\mathbf{e}'_{1i} \tag{vi}$$

and

$$g = \sum_{j=1}^{n}\mathbf{e}'_{j1}b\mathbf{e}_{1j} \tag{vii}$$

then

$$\begin{aligned} hg &= \sum_{i,j=1}^{n}\mathbf{e}_{i1}a\mathbf{e}'_{1i}\mathbf{e}'_{j1}b\mathbf{e}_{1j} = \sum_{i=1}^{n}\mathbf{e}_{i1}a\mathbf{e}'_{11}b\mathbf{e}_{1i} \\ &= \sum_{i=1}^{n}\mathbf{e}_{i1}ab\mathbf{e}_{1i} \qquad \text{since } a \in \mathscr{A}\mathbf{e}'_{11} \\ &= \sum_{i=1}^{n}\mathbf{e}_{i1}\mathbf{e}_{11}\mathbf{e}_{1i} \qquad \text{by (iv)} \\ &= \sum_{i=1}^{n}\mathbf{e}_{ii} \end{aligned}$$

that is

$$hg = 1. \tag{viii}$$

So h must be the inverse of g ((A4)) and $gh = 1$. But

$$\begin{aligned} gh &= \sum_{i,j=1}^{n} \mathbf{e}'_{j1}b\mathbf{e}_{1j}\mathbf{e}_{i1}a\mathbf{e}'_{1i} = \sum_{i=1}^{n} \mathbf{e}'_{i1}b\mathbf{e}_{11}a\mathbf{e}'_{1i} \\ &= \sum_{i=1}^{n} \mathbf{e}'_{i1}ba\mathbf{e}'_{1i} = \sum_{i=1}^{n} \mathbf{e}'_{ii} \qquad \text{by (v).} \end{aligned}$$

Since $\Sigma_{i=1}^{m}\mathbf{e}'_{ii} = 1$ we must have $m = n$, and hence $\mathcal{M}' \sim \mathcal{M}$. In fact

$$\begin{aligned} g\mathbf{e}_{ij}g^{-1} &= \sum_{p,q=1}^{n} \mathbf{e}'_{p1}b\mathbf{e}_{1p}\mathbf{e}_{ij}\mathbf{e}_{q1}a\mathbf{e}'_{1q} \\ &= \mathbf{e}'_{i1}b\mathbf{e}_{11}a\mathbf{e}'_{1j} = \mathbf{e}'_{ij} \qquad \text{by (v).} \end{aligned}$$

This completes the proof.

A consequence of this theorem is the following which we will frequently use.

If P is an idempotent in a simple $\mathcal{A}$ then $P = \Sigma_{i=1}^{r} P_i$ where the P_i are pairwise orthogonal primitives, and the uniquely determined r is called the rank of P. Two idempotents in $\mathcal{A}$ are similar iff they have the same rank. (A25)

Any idempotent can certainly be written as a sum of pairwise orthogonal primitives, this is (A16). To go further we shall use

Lemma
If P is idempotent in a simple $\mathcal{A}$ then $P\mathcal{A}P$ is simple.

Let $\mathcal{B}$ be a non-zero ideal in $P\mathcal{A}P$. Since $\mathcal{B}$ is an ideal in $P\mathcal{A}P$ the left-hand side is contained in $\mathcal{B}$. But $\mathcal{A}\mathcal{B}\mathcal{A}$ is an ideal in the simple $\mathcal{A}$, and so the right-hand side gives $P\mathcal{A}P$. Thus $\mathcal{B} = P\mathcal{A}P$.

If $\mathcal{A}$ is simple then $P\mathcal{A}P$ is simple with identity P. If $P = \Sigma_{i=1}^{r} P_i$ with the P_i primitive then $P\mathcal{A}P$ can be written as a tensor product of some division algebra and a total matrix algebra with the P_i as diagonal elements. The order of the matrices will then be r, which was shown in (A24) to be uniquely determined. It was also shown in (A24) that all matrix bases are similar, and so as a corollary all primitives are similar. If $\{P_i\}$ are pairwise orthogonal primitives then so are $\{sP_is^{-1}\}$, thus similarity preserves the rank of an idempotent. To see that having the same rank is sufficient for idempotents to be similar note that if

$$P = \sum_{i=1}^{r} P_i = \sum_{i=1}^{r} Q_i$$

with $\{P_i\}$ and $\{Q_i\}$ being different sets of pairwise orthogonal primitives then we can choose matrix bases with either the $\{P_i\}$ or the $\{Q_i\}$

as diagonals, and (A24) then ensures the existence of an $s:Q_i = sP_is^{-1}$ $\forall i$.

The theorem above applies to simple algebras. However the first part may be seen to apply to the semi-simple case. For if P is an element of a semi-simple $\mathscr{A}$ then $P = Q_1 \oplus Q_2 \oplus \ldots \oplus Q_s$ where the Q_i are in the simple components. P is idempotent if and only if all the Q_i are idempotent. By the above theorem each Q_i will have a unique rank and so the rank of an idempotent in a semi-simple algebra is uniquely determined. As a special case a primitive in a semi-simple algebra must be primitive in one of the simple components. Thus, of course, not all primitives, and hence all idempotents of the same rank, will be similar in a semi-simple algebra.

A subset of all simple algebras is provided by the central simple ones; that is those simple algebras whose centre is generated by the identity. For these algebras we have the following important result.

Every automorphism of a central simple algebra is an inner automorphism. (A26)

If $\mathscr{A}$ is central simple then $\mathscr{A} = \mathscr{D} \otimes \mathscr{M}_n$ where $\mathscr{D}$ is a central division algebra, and $\mathscr{A}^{\text{op}} = \mathscr{D}^{\text{op}} \otimes \mathscr{M}_n{}^{\text{op}}$. The existence of the involution of transposition on matrices shows that $\mathscr{M}_n{}^{\text{op}} \simeq \mathscr{M}_n$, and so $\mathscr{A} \otimes \mathscr{A}^{\text{op}} \simeq \mathscr{D} \otimes \mathscr{D}^{\text{op}} \otimes \mathscr{M}_{n^2}$, by (A6). We are now in a position, at last, to make use of (A7), giving $\mathscr{A} \otimes \mathscr{A}^{\text{op}} \simeq \text{End}\,\mathscr{D} \otimes \mathscr{M}_{n^2}$, that is $\mathscr{A} \otimes \mathscr{A}^{\text{op}} \simeq \mathscr{M}_m$ where m is the dimension of $\mathscr{A}$, and we have again used (A6). We extend any automorphism, t, on $\mathscr{A}$ to one on $\mathscr{A} \otimes \mathscr{A}^{\text{op}}$, T, by defining $(ab)^T = a^tb$ $\forall a \in \mathscr{A},\ b \in \mathscr{A}^{\text{op}}$. In the 'uniqueness theorem', (A24), we essentially proved that all automorphisms of a total matrix algebra are inner. Thus for every $x \in \mathscr{A} \otimes \mathscr{A}^{\text{op}}$, $x^T = sxs^{-1}$ where $s \in \mathscr{A} \otimes \mathscr{A}^{\text{op}}$, that is $a^t = sas^{-1}$ for $a \in \mathscr{A}$ and $b = sbs^{-1}$ for $b \in \mathscr{A}^{\text{op}}$. Thus s must commute with every element of $\mathscr{A}^{\text{op}}$. Since $\mathscr{A}^{\text{op}}$ is central simple s must be in $\mathscr{A}$, and so t is inner.

So far we have assumed that all algebras are over some field, F, which has not warranted much attention; indeed we have usually simply referred to an algebra as $\mathscr{A}$ rather than as $\mathscr{A}$ over F. In a moment we shall assume a restriction on the choice of F. The situation for the simple algebras is also such that we may regard a simple algebra over F as an algebra over certain other fields. If $\mathscr{A}$ over F is simple then the centre $\mathscr{C}$ is a commutative division algebra, that is, a field. In an obvious way $\mathscr{A}$ is an algebra over $\mathscr{C}$, making $\mathscr{A}$ over $\mathscr{C}$ central simple. In the following section we will examine involutions of a simple algebra $\mathscr{A}$ over F where F is assumed not to be of characteristic two. (As stated in the introduction for the purposes of this book F can be taken to be one of the zero characteristic fields $\mathbb{R}$ or $\mathbb{C}$.)

If $\mathscr{A}$ over F has an involution T then the set of T-symmetric quantities forms a subspace $\mathcal{S}_T$. That is, $a \in \mathcal{S}_T$ if and only if $a^T = a$. Similarly we define $\mathcal{T}_T$ to be the set of T-skew quantities, and then we have $\mathcal{A} = \mathcal{S}_T + \mathcal{T}_T$. For if $a \in \mathcal{A}$, $a = \frac{1}{2}(a + a^T) + \frac{1}{2}(a - a^T)$. The sum is direct since if $a = a^T$ and $a = -a^T$ then $a + a = 0$ which (for characteristic not two) gives $a = 0$. What is more, if the centre contains a T-skew q then $\mathcal{A} = \mathcal{S}_T + q\mathcal{S}_T$. If q is a non-zero element of the centre (of a simple algebra) then it has an inverse which is also T-skew. If $a \in \mathcal{T}_T$ then $a = qq^{-1}a$, and $(q^{-1}a)^T = a^T q^{-1T} = aq^{-1} = q^{-1}a$. The T-symmetric quantities in the centre will form a subfield of $\mathcal{C}$, $\mathcal{E}$ say. We will refer to an involution as being an involution over $\mathcal{E}$, say, when $\mathcal{E}$ is the subfield of the centre $\mathcal{C}$ left invariant by the involution.

If $\mathcal{A}$ over F is simple with J and T involutions over $\mathcal{E}$ then TJ is an automorphism of $\mathcal{A}$ over $\mathcal{C}$. (A27)

If T and J are involutions then TJ is certainly an automorphism of $\mathcal{A}$ over F. What we need to show is that it leaves elements in the centre invariant. The involutions T and J induce automorphisms of the centre, $\mathcal{C}$. An element of $\mathcal{C}$ is T-symmetric if and only if it is J-symmetric. This is, in fact, sufficient to show that T and J induce the same automorphism on $\mathcal{C}$. Let q be a non-zero J-skew element of $\mathcal{C}$ then qq^T is manifestly T-symmetric, and hence J-symmetric. But $(qq^T)^J = -qq^{TJ}$, which since q is invertible, gives $q^{TJ} = -q^T$. So $(q + q^T)^J = -(q + q^T)$. But $q + q^T$ is manifestly T-symmetric, and thus J-symmetric. Since any element that is both J-symmetric and J-skew must be zero we have $q^T = -q$. We have shown then that any J-skew element of $\mathcal{C}$ is also T-skew. But any element of $\mathcal{C}$ can be written as a sum of J-symmetric and J-skew parts and thus T and J coincide on $\mathcal{C}$. Since T and J are involutions TJ must leave all elements of $\mathcal{C}$ invariant.

The observation that if $\mathcal{A}$ over F is simple then $\mathcal{A}$ over $\mathcal{C}$ is central simple gives (A26) a wider range of applicability than might at first sight be supposed. In particular, it enables us to prove the following.

If $\mathcal{A}$ over F is simple and T is an involution over $\mathcal{E}$ then $J{:}a \mapsto a^J$ is an involution over $\mathcal{E}$ iff there exists an s with $s = \pm s^T$ such that $a^J = sa^Ts^{-1}$. (A28)

First the easy bit. If $a^J = sa^Ts^{-1}$ then $J{:}a \mapsto a^J$ is an anti-automorphism. Furthermore $a^{JJ} = s(sa^Ts^{-1})^Ts^{-1} = s(s^T)^{-1}as^Ts^{-1}$, so if $s^T = \pm s$, J is an involution. Inner automorphisms leave all elements of the centre invariant. So if T is an involution over $\mathcal{E}$ then so is J.

Conversely let J be an involution over $\mathcal{E}$, then JT is an automorphism over $\mathcal{C}$ ((A27)). (A26) then ensures the existence of a g such that

$$a^{JT} = g^{-1}ag$$

$$a^J = (g^{-1}ag)^T$$
$$= g^T a^T (g^T)^{-1}.$$

Since J is an involution

$$a = a^{JJ} = g^T(g^T a^T (g^T)^{-1})^T (g^T)^{-1}$$
$$= g^T g^{-1} a g (g^T)^{-1}.$$

Since this is true for all a we must have $g^T g^{-1} = \lambda \in \mathscr{C}$. If $\lambda = -1$ then there is nothing left to do, if not then set $s = g + g^T = g(1 + \lambda)$ and s will have the desired property. Obviously the choice of such an s is determined only up to multiplication by an element of the centre.

A familiar example of an involution is provided by transposition of matrices. In some ordinary matrix basis we define T such that $\mathbf{e}_{ij}^{\mathrm{T}} = \mathbf{e}_{ji}$. For some other basis $\{\mathbf{e}'_{ij}\}$ we define J by $\mathbf{e}'^{J}_{ij} = \mathbf{e}'_{ji}$. T and J are examples of what we shall call equivalent involutions. Two involutions, V and J, will be called *equivalent* if there is some automorphism S such that $a^J = a^{SVS^{-1}} \equiv ((a^S)^V)^{S^{-1}}$. If an inner S relates equivalent involutions J and V, related to some 'standard' involution T by

$$a^V = v a^T v^{-1}$$
$$a^J = j a^T j^{-1},$$

then $j = \lambda s v s^T$ for some $\lambda \in \mathscr{C}$.

In classifying the structure of algebras we showed first the existence of the radical. Semi-simple algebras were then defined to have zero radical. It was possible to determine the structure of a semi-simple algebra completely in terms of simple ones, whose structure was in turn given as a tensor product of a division algebra and a total matrix algebra. Most of the structure theorems for associative algebras were first given by J H M Wedderburn, and we shall refer to the expression of a simple $\mathscr{A}$ such as $\mathscr{A} = \mathscr{D} \otimes \mathscr{M}$ as the Wedderburn decomposition of $\mathscr{A}$. It is all very well to be able to determine the structure of algebras whose radical is zero, but it would be rather limiting if it told us nothing about algebras with a radical. However, this is not the case. The most important result on the structure of algebras is known as Wedderburn's principal structure theorem. It states that (subject to certain caveats relating to the underlying field) any algebra is the vector space sum of its radical and the semi-simple algebra obtained from the quotient modulo the radical. We shall not need this result and so will not give the proof. This may be found in (for example) Albert [1], Kochendorffer [3] or, for the case of zero characteristic field, in Dickson [2]. As was stated in the introduction to this Appendix we will really only be concerned in this book with algebras over the real field. For this case one can go further in determining the structure of all semi-simple algebras. The Wedderburn structure theorem reduces the classification of simple algebras over the reals to the classification of real division algebras. This had already been done by Frobenius in 1878. He showed that the only associative real division algebras are $\mathbb{R}$, $\mathbb{C}$ and H; the reals themselves, the algebra of complex numbers and the quaternion algebra. A proof may be found in Dickson [2] or Kochendorffer [3]. In view of this we now give a brief discussion of these algebras.

Let $\mathscr{A}$ be a one-dimensional algebra over $\mathbb{R}$. Then a basis is provided by u where $u^2 = \lambda u$. If $\lambda = 0$ then $\mathscr{A}$ is nilpotent of index two. If $\lambda \neq 0$ then it is invertible and if $P = \lambda^{-1}u$, P is an idempotent. For any $a \in \mathscr{A}$ we have $a = \mu P$, $\mu \in \mathbb{R}$ and $\mathrm{I}: a \mapsto \mu$ clearly establishes an isomorphism between $\mathscr{A}$ and $\mathbb{R}$.

The real algebra $\mathbb{C}(\mathbb{R})$ is a two-dimensional algebra generated by i where $\mathrm{i}^2 = -1$. This real commutative algebra is not central. It has the well known involution of complex conjugation $*: \mathrm{i} \mapsto -\mathrm{i}$.

The real quaternion algebra $\mathrm{H}(\mathbb{R})$ has a basis $\{1, \mathrm{i}, \mathrm{j}, \mathrm{k}\}$ whose multiplication table is given in table A1. The algebra is generated by the subspace spanned by $\{\mathrm{i}, \mathrm{j}\}$, say. (We note here that the other four-dimensional real simple algebra $\mathscr{M}_2(\mathbb{R})$ is generated by $\{\alpha, \beta\}$ where $\alpha^2 = 1$, $\beta^2 = -1$ and $\alpha\beta = -\beta\alpha$. For example, $\alpha = \mathbf{e}_{12} + \mathbf{e}_{21}$, $\beta = \mathbf{e}_{12} - \mathbf{e}_{21}$.) The quaternions are not commutative but the algebra is central. In the given basis, $\{\mathrm{i}, \mathrm{j}, \mathrm{k}\}$ span the subspace of *vector quaternions*, whilst the identity spans the *scalar quaternions*. The involution of *quaternion conjugation*, $q \mapsto \bar{q}$, is defined to change the sign of the vector part of every quaternion. Then $q\bar{q}$ is self-conjugate and hence in the centre. By inspection $q\bar{q}$ is seen to be strictly positive for non-zero q, say $q\bar{q} = \lambda^2$. Then $q^{-1} = \lambda^{-2}\bar{q}$ and indeed H is a division algebra. Suppose that T is some other involution, then (A28) ensures that $q^T = \mathrm{t}\bar{q}\mathrm{t}^{-1}$ where $\bar{\mathrm{t}} = \pm\mathrm{t}$. Since the only self-conjugate quaternions are in the centre, to get an involution distinct from conjugation we must have $\bar{\mathrm{t}} = -\mathrm{t}$. In particular we define $\hat{q} = \mathrm{k}\bar{q}\mathrm{k}^{-1}$ where k is one of the 'standard' basis vectors. This involution will be called a *reversion* since it leaves the generators $\{\mathrm{i}, \mathrm{j}\}$ invariant, but of course reverses their order in products. By taking any vector quaternion t we have an involution given by $q^T = \mathrm{t}\bar{q}\mathrm{t}^{-1}$. However, all such involutions are equivalent to reversion. Without loss of generality we can choose the defining t to satisfy $\mathrm{t}^2 = -1$. Then if t and k are linearly independent they generate H. To see this all we need to check is that the commutator $[\mathrm{t}, \mathrm{k}]$, which is certainly a vector quaternion since it is anticonjugate, is not a linear combination of t and k. But t and k both anticommute with $[\mathrm{t}, \mathrm{k}]$, which thus cannot be a linear combination of them. Since $\{\mathrm{k}, \mathrm{t}\}$ generate H we may define an automorphism, G, by $\mathrm{t}^G = \mathrm{k}$, $\mathrm{k}^G = \mathrm{t}$. This automorphism must be inner since H is a central division algebra and hence central simple. That is, $\mathrm{t} = g\mathrm{k}g^{-1}$ for some g, and $g^{-1} = \lambda^{-2}\bar{g}$ for some

$\lambda \in \mathbb{R}$. So if $s = \lambda^{-1}g$ then $t = sk\bar{s}$, which is the criterion for T to be equivalent to reversion.

Table A1 The quaternion algebra

	1	i	j	k
1	1	i	j	k
i	i	−1	k	−j
j	j	−k	−1	i
k	k	j	−i	−1

Just as it is important to know that any positive real number can be written as a square of a positive number, and that any complex number can be written as a square, it will prove important to know that any reversion symmetric quaternion can be written as a square of a reversion symmetric quaternion. As we have remarked $q\bar{q}$ is a positive real number and so we may introduce a norm defined by $|q|^2 = q\bar{q}$. Reversion is related to conjugation by $\bar{q} = k^{-1}\hat{q}k$, and for any q we have $q^{-1} = \bar{q}/|q|^2$, so if $y = \hat{y}$ then

$$y^{-1} = \frac{k^{-1}yk}{|y|^2} \tag{i}$$

Writing $1 + q$ as $1 + q = q^{-1}q + q = (1 + q^{-1})q$ gives $q = (1 + q^{-1})^{-1}(1 + q)$, for any q. In particular, if y_0 is a unit-norm reversion symmetric quaternion then

$$\begin{aligned} y_0 &= (1 + y_0{}^{-1})^{-1}(1 + y_0) \\ &= \frac{k^{-1}(1 + k^{-1}y_0k)k(1 + y_0)}{|1 + y_0^{-1}|^2} \qquad \text{by (i)} \\ &= \left(\frac{1 + y_0}{|1 + y_0{}^{-1}|}\right)^2 \qquad \text{since } k^2 = -1. \end{aligned} \tag{ii}$$

For any q we have $|k^{-1}qk|^2 = k^{-1}qk\bar{k}\bar{q}\bar{k}^{-1} = k^{-1}q\bar{q}k = q\bar{q} = |q|^2$, so from (i)

$$|1 + y_0^{-1}|^2 = |k^{-1}(1 + y_0)k|^2 = |1 + y_0|^2.$$

Thus (ii) gives $y_0 = x^2$, for the reversion symmetric x given by

$$x = \frac{1 + y_0}{|1 + y_0|}.$$

Then for a reversion symmetric y of arbitrary norm we can write $y = |y|y_0 = \{|y|^{1/2}x\}^2$, since any positive real number has a real square root.

It will be useful to be able to identify the tensor products of these division algebras. Obviously $\mathbb{R}\otimes\mathbb{R} \simeq \mathbb{R}$, $\mathbb{R}\otimes\mathbb{C} \simeq \mathbb{C}$ and $\mathbb{R}\otimes\mathrm{H} \simeq \mathrm{H}$.

The algebra $\mathbb{C}\otimes\mathbb{C}$ has a basis $\{1, \mathrm{i}, \mathrm{j}, \mathrm{ij}\}$ where i and j commute and $\mathrm{i}^2 = \mathrm{j}^2 = -1$. So if $P = \frac{1}{2}(1 + \mathrm{ij})$ and $Q = \frac{1}{2}(1 - \mathrm{ij})$ then P and Q are orthogonal idempotents such that $1 = P + Q$. The algebra $P(\mathbb{C}\otimes\mathbb{C})P$ has P as identity, and since P is in the centre of $\mathbb{C}\otimes\mathbb{C}$ we have $P(\mathbb{C}\otimes\mathbb{C})P = (\mathbb{C}\otimes\mathbb{C})P$, which is a two-sided ideal. Similarly for $(\mathbb{C}\otimes\mathbb{C})Q$. Since P and Q are orthogonal

$$\mathbb{C}\otimes\mathbb{C} = (\mathbb{C}\otimes\mathbb{C})P\oplus(\mathbb{C}\otimes\mathbb{C})Q.$$

We may choose $\{P, \mathrm{i}P\}$ as basis for $(\mathbb{C}\otimes\mathbb{C})P$ and so have $(\mathbb{C}\otimes\mathbb{C})P \simeq \mathbb{C}$. Similarly for the other ideal giving

$$\mathbb{C}\otimes\mathbb{C} \simeq \mathbb{C}\oplus\mathbb{C}. \tag{A29}$$

The algebra $\mathbb{C}\otimes H$ has a basis $\{1, z, \mathrm{i}, \mathrm{j}, \mathrm{k}, z\mathrm{i}, z\mathrm{j}, z\mathrm{k}\}$ where $\{1, z\}$ is a basis for the complex subalgebra that commutes with the quaternion subalgebra spanned by $\{1, \mathrm{i}, \mathrm{j}, \mathrm{k}\}$. $\mathbb{C}\otimes\mathrm{H}$ may be generated by $\{z, \mathrm{i}, \mathrm{j}\}$. The subset $\{1, z\}$ spans the centre which is thus isomorphic to $\mathbb{C}$. If $\mathbf{e}_{11} = \frac{1}{2}(1 + z\mathrm{i})$ and $\mathbf{e}_{22} = \frac{1}{2}(1 - z\mathrm{i})$ then $\mathbf{e}_{11}$, $\mathbf{e}_{22}$ are orthogonal idempotents with $1 = \mathbf{e}_{11} + \mathbf{e}_{22}$. If we choose $\mathbf{e}_{21} = \mathbf{j}\mathbf{e}_{11} = \mathbf{e}_{22}\mathrm{j}$ and $\mathbf{e}_{12} = -\mathbf{j}\mathbf{e}_{22} = -\mathbf{e}_{11}\mathrm{j}$ then the $\mathbf{e}_{ij}$ form an ordinary basis for $\mathcal{M}_2(\mathbb{R})$, so

$$\mathbb{C}(\mathbb{R})\otimes\mathrm{H}(\mathbb{R}) \simeq \mathbb{C}(\mathbb{R})\otimes\mathcal{M}_2(\mathbb{R}). \tag{A30}$$

We do not have to do any work to determine the structure of $\mathrm{H}\otimes\mathrm{H}$. The quaternion algebra is a central division algebra and, since it has the involution of conjugation, $\mathrm{H} \simeq \mathrm{H}^{\mathrm{op}}$. So from Theorem 4 we have

$$\mathrm{H}(\mathbb{R})\otimes\mathrm{H}(\mathbb{R}) \simeq \mathcal{M}_4(\mathbb{R}). \tag{A31}$$

Having completed our review of associative algebras we turn now to a generalisation of the concept of a vector space in which the field is replaced with a ring, or associative algebra, with unit element. A *right R-module M*, over the ring R is an additive Abelian group with a map from

$$M \times R \longrightarrow M : (x, q) \longmapsto xq$$

such that

$$x(q_1 q_2) = (xq_1)q_2 \tag{i}$$

$$x(q_1 + q_2) = xq_1 + xq_2 \tag{ii}$$

$$(x + y)q = xq + yq$$

$$x1 = x \tag{iii}$$

where 1 is the identity in R.

The writing of the element from R on the right-hand side is of significance in (i) when R is non-commutative; in this case the above are obviously altered to give a *left R-module*. The notion of a linear map may readily be extended to apply to left (or right) R-modules. If I is a minimal left ideal in an algebra with unity, $\mathcal{A}$, then I is an example of a left $\mathcal{A}$-module. If $\mathcal{A}$ is simple with $\mathcal{A} = \mathcal{D}\otimes\mathcal{M}$ then I is also a right $\mathcal{D}$-module, for multiplication on the right by $\mathcal{D}$ will preserve the I. In this case I is simultaneously a left $\mathcal{A}$-module and a right $\mathcal{D}$-module, with the $\mathcal{A}$ action being right $\mathcal{D}$-linear, and the $\mathcal{D}$ action being left $\mathcal{A}$-linear. Thus for simple algebras we are lead to consider right H-modules. Although the concept of linear independence extends to modules, in general an R-module need have no basis. However, H-modules do have bases, the number of basis vectors determining the quaternionic dimension, $\dim_H$. Thus, for example, if I is a minimal left ideal in $\mathcal{A} = \mathrm{H}\otimes\mathcal{M}_r$ then $\dim_H \mathrm{I} = r$, whereas $\dim_{\mathbb{R}} \mathrm{I} = 4r$.

Bibliography

Albert A 1941 *Introduction to Algebraic Theories* (Chicago: Chicago University Press)

—— 1961 *Structure of Algebras* (*Am. Math. Soc. Coll.* Publ. vol 24)

Greub W 1978 *Multilinear Algebra* 2nd edn (Berlin: Springer)

Appendix B

Vector Calculus on $\mathbb{R}^3$

As an illustration of the methods of differential calculus it is useful to make contact with the elementary vector calculus of Euclidean 3-space. Such a space regarded as a manifold has the special property of admitting a class of global charts. We might call one such a chart a Cartesian chart since the coordinate maps $\{x^i\}$ $i = 1, 2, 3$ yield the familiar Cartesian coordinates $x^i(p)$ for $p \in \mathbb{R}^3$. In such a global chart the Euclidean metric tensor is expressed as

$$g = \sum_{i=1}^{3} \mathrm{d}x^i \otimes \mathrm{d}x^i \qquad x^i(p) \in \mathbb{R}.$$

The orthonormal frames $\{X_i\} = \{\partial/\partial x^1, \partial/\partial x^2, \partial/\partial x^3\}$ and co-frames $\{e^i\} = \{\mathrm{d}x^1, \mathrm{d}x^2, \mathrm{d}x^3\}$ are in this case naturally dual to each other. Observe also that $\widetilde{\mathrm{d}x^i} = \partial/\partial x^i$. For some problems other non-global charts are useful. The familiar 'spherical polar' chart with coordinate functions (r, θ, φ) has co-domain

$$0 < r(p) < \infty$$

$$0 < \varphi(p) \leqslant 2\pi$$

$$0 < \theta(p) < \pi.$$

The polar chart is related to the Cartesian chart on the overlap by the transformation of coordinates

$$r = [(x^1)^2 + (x^2)^2 + (x^3)^2]^{1/2}$$

$$\theta = \sin^{-1} \frac{[(x^1)^2 + (x^2)^2]^{1/2}}{[(x^1)^2 + (x^2)^2 + (x^3)^2]^{1/2}}$$

$$\varphi = \cos^{-1} \frac{x^3}{[(x^1)^2 + (x^2)^2 + (x^3)^2]^{1/2}}.$$

If we tried to cover the whole surface $r = \text{constant}$ $(\neq 0)$, with a single coordinate chart there would arise an ambiguity in assigning coordinates to the poles of the sphere. Such ambiguities can give rise to 'singularities' in subsequent calculations, these pathologies reflecting only an improper use of coordinates. In a polar chart we may write

$$g = \sum_{i=1}^{3}\left(\frac{\partial x^i}{\partial r}\mathrm{d}r + \frac{\partial x^i}{\partial \theta}\mathrm{d}\theta + \frac{\partial x^i}{\partial \varphi}\mathrm{d}\varphi\right)\otimes\left(\frac{\partial x^i}{\partial r}\mathrm{d}r + \frac{\partial x^i}{\partial \theta}\mathrm{d}\theta + \frac{\partial x^i}{\partial \varphi}\mathrm{d}\varphi\right)$$

or, since

$$x^1 = r\sin\theta\cos\varphi$$

$$x^2 = r\sin\theta\sin\varphi$$

$$x^3 = r\cos\theta$$

$$g = \mathrm{d}r\otimes\mathrm{d}r + r^2\mathrm{d}\theta\otimes\mathrm{d}\theta + r^2\sin^2\theta\mathrm{d}\varphi\otimes\mathrm{d}\varphi.$$

Similarly

$$\begin{aligned} g^* &= \sum_{i=1}^{3}(\partial_{x_i}\otimes\partial_{x_i}) \\ &= \sum_{i=1}^{3}[(\partial r/\partial x^i)\partial_r + (\partial\theta/\partial x^i)\partial_\theta + (\partial\varphi/\partial x^i)\partial_\varphi]\otimes[(\partial r/\partial x^i)\partial_r \\ &\quad + (\partial\theta/\partial x^i)\partial_\theta + (\partial\varphi/\partial x^i)\partial_\varphi] \\ &= \frac{\partial}{\partial r}\otimes\frac{\partial}{\partial r} + \frac{1}{r^2}\frac{\partial}{\partial\theta}\otimes\frac{\partial}{\partial\theta} + \frac{1}{r^2\sin^2\theta}\frac{\partial}{\partial\varphi}\otimes\frac{\partial}{\partial\varphi}. \end{aligned}$$

Hence an orthonormal co-frame in this chart is $\{E^i\} = \{\mathrm{d}r, r\mathrm{d}\theta, r\sin\theta\,\mathrm{d}\varphi\}$ with dual (orthonormal) frame

$$\{Y_i\} = \left\{\frac{\partial}{\partial r}, \frac{1}{r}\frac{\partial}{\partial\theta}, \frac{1}{r\sin\theta}\frac{\partial}{\partial\varphi}\right\}.$$

The metric duals of $\mathrm{d}r$, $\mathrm{d}\theta$, $\mathrm{d}\varphi$ are the local vector fields

$$\widetilde{\mathrm{d}r} = \frac{\partial}{\partial r}, \quad \widetilde{\mathrm{d}\theta} = \frac{1}{r^2}\frac{\partial}{\partial\theta}, \quad \widetilde{\mathrm{d}\varphi} = \frac{1}{r^2\sin^2\theta}\frac{\partial}{\partial\varphi}.$$

(Observe that points p with $r(p) = 0$, $\theta(p) = 0$ are outside our working chart.)

On the overlap U of a Cartesian chart and our polar chart, for $f\in\mathcal{F}(U)$ we may write

$$\mathrm{d}f = (\partial f/\partial x^i)\mathrm{d}x^i = (\partial f/\partial r)\mathrm{d}r + (\partial f/\partial\theta)\mathrm{d}\theta + (\partial f/\partial\varphi)\mathrm{d}\varphi.$$

The metric dual of $\mathrm{d}f$ is called the *gradient* of f, sometimes written $\operatorname{grad} f$. On U

$$\begin{aligned} \operatorname{grad} f = \widetilde{\mathrm{d}f} &= (\partial f/\partial x^i)\partial/\partial x^i = (\partial f/\partial r)\widetilde{\mathrm{d}r} + (\partial f/\partial\theta)\widetilde{\mathrm{d}\theta} + (\partial f/\partial\varphi)\widetilde{\mathrm{d}\varphi} \\ &= \left(\frac{\partial f}{\partial r}\right)\frac{\partial}{\partial r} + \frac{1}{r^2}\left(\frac{\partial f}{\partial\theta}\right)\frac{\partial}{\partial\theta} + \frac{1}{r^2\sin^2\theta}\left(\frac{\partial f}{\partial\varphi}\right)\frac{\partial}{\partial\varphi}. \end{aligned}$$

In terms of the orthonormal basis $\{Y_i\}$

$$\operatorname{grad} f = \left(\frac{\partial f}{\partial r}\right)Y_1 + \frac{1}{r}\left(\frac{\partial f}{\partial \theta}\right)Y_2 + \frac{1}{r\sin\theta}\left(\frac{\partial f}{\partial \varphi}\right)Y_3.$$

If Z is a vector field on U we may write

$$Z = \xi^i \partial/\partial x^i = \xi^r \partial/\partial r + \xi^\theta \partial/\partial\theta + \xi^\varphi \partial/\partial\varphi$$

where ξ^i, ξ^r, ξ^θ, $\xi^\varphi \in F(U)$. The 'rate of change of f' in the direction specified by the vector Z, or the directional derivative of f in the direction Z, is defined as $Z(f)$. In terms of the vector field $\operatorname{grad} f$

$$Z(f) \equiv \mathrm{d}f(Z) = g(Z, \widetilde{\mathrm{d}f}) = g(Z, \operatorname{grad} f).$$

In three-dimensional Euclidean space it is customary to use a dot notation for the metric evaluated on two vectors, namely $g(X, Y) \equiv X \cdot Y$. This casts the expression for the directional derivative into the form

$$Z(f) = \operatorname{grad} f \cdot Z.$$

Let us explicitly compute the $*$ map associated with the Euclidean metric. If $\{E^i\}$ is any orthonormal co-frame with respect to this g then, with $*1 = E^1 \wedge E^2 \wedge E^3$, we find

$$*E^1 = E^2 \wedge E^3,\ *E^2 = E^3 \wedge E^1,\ *E^3 = E^1 \wedge E^2$$

$$*(E^1 \wedge E^2) = E^3,\ *(E^2 \wedge E^3) = E^1,\ *(E^3 \wedge E^1) = E^2$$

$$*(E^1 \wedge E^2 \wedge E^3) = 1.$$

Consequently, in this case, $** = 1$ on all forms. The $*$ map for Euclidean $\mathbb{R}^3$ establishes a relation between 2-forms and 1-forms. The metric dual, $\sim$, maps 1-forms to vector fields. Thus there is a correspondence given by the Euclidean metric tensor between 2-forms and vector fields on $\mathbb{R}^3$. Given two vector fields in any g-orthonormal frame, $X = \xi^i Y_i$, $Y = \zeta^j Y_j$, we have

$$\widetilde{X} \wedge \widetilde{Y} = (\xi^1\zeta^2 - \xi^2\zeta^1)\widetilde{Y}_1 \wedge \widetilde{Y}_2 + (\xi^2\zeta^3 - \xi^3\zeta^2)\widetilde{Y}_2 \wedge \widetilde{Y}_3 + (\xi^3\zeta^1 - \xi^1\zeta^3)\widetilde{Y}_3 \wedge \widetilde{Y}_1.$$

But since $\{\widetilde{Y}_j\}$ is an orthonormal co-frame

$$*(\widetilde{X} \wedge \widetilde{Y}) = (\xi^1\zeta^2 - \xi^2\zeta^1)\widetilde{Y}_3 + (\xi^2\zeta^3 - \xi^3\zeta^2)\widetilde{Y}_1 + (\xi^3\zeta^1 - \xi^1\zeta^3)\widetilde{Y}_2 \quad \in \Gamma T^*\mathbb{R}^3.$$

Hence the orthonormal components of the vector field $\widetilde{*(\tilde{X} \wedge \tilde{Y})}$ correspond to the components of the cross or vector product of two vectors with orthonormal components (ξ^i), (ζ^i) respectively. Such a correspondence also enables us to make contact with the operation curl.

For a vector field V on $U \in \mathbb{R}^3$ we define

$$\operatorname{curl} V = \widetilde{*\mathrm{d}\widetilde{V}}.$$

For example, in a Cartesian chart with $V = V^j \partial_j$:

$$\widetilde{V} = V^j \mathrm{d}x^j$$

$$\mathrm{d}\widetilde{V} = (\partial_1 V^2 - \partial_2 V^1)\mathrm{d}x^1 \wedge \mathrm{d}x^2 + (\partial_2 V^3 - \partial_3 V^2)\mathrm{d}x^2 \wedge \mathrm{d}x^3 + (\partial_3 V^1 - \partial_1 V^3)\mathrm{d}x^3 \wedge \mathrm{d}x^1$$

$$*\mathrm{d}\widetilde{V} = (\partial_1 V^2 - \partial_2 V^1)\mathrm{d}x^3 + (\partial_2 V^3 - \partial_3 V^2)\mathrm{d}x^1 + (\partial_3 V^1 - \partial_1 V^3)\mathrm{d}x^2.$$

Thus, indeed, the orthonormal components of $*\mathrm{d}\widetilde{V}$ have the expected form for the components of the curl of the vector field with orthonormal components (V^1, V^2, V^3). If we work in the polar chart with

$$\begin{aligned} V &= V^r\partial_r + V^\theta\partial_\theta + V^\varphi\partial_\varphi \\ &= V^1 Y_1 + V^2 Y_2 + V^3 Y_3 \end{aligned}$$

where $V^1 = V^r$, $V^2 = rV^\theta$, $V^3 = r\sin\theta V^\varphi$, then

$$\begin{aligned} \widetilde{V} &= V^1 E^1 + V^2 E^2 + V^3 E^3 \\ &= V^r \mathrm{d}r + r^2 V^\theta \mathrm{d}\theta + r^2 \sin^2\theta V^\varphi \mathrm{d}\varphi \end{aligned}$$

where $E^1 = \mathrm{d}r$, $E^2 = r\mathrm{d}\theta$, $E^3 = r\sin\theta\mathrm{d}\varphi$. Hence

$$\begin{aligned} \mathrm{d}\widetilde{V} &= \partial_\theta V^r \mathrm{d}\theta \wedge \mathrm{d}r + \partial_\varphi V^r \mathrm{d}\varphi \wedge \mathrm{d}r + \partial_r(r^2 V^\theta)\mathrm{d}r \wedge \mathrm{d}\theta \\ &\quad + \partial_\varphi(r^2 V^\theta)\mathrm{d}\varphi \wedge \mathrm{d}\theta + \partial_r(r^2 \sin^2\theta\, V^\varphi)\mathrm{d}r \wedge \mathrm{d}\varphi \\ &\quad + \partial_\theta(r^2\sin^2\theta\, V^\varphi)\mathrm{d}\theta \wedge \mathrm{d}\varphi \\ &= [\partial_r(r^2 V^\theta) - \partial_\theta V^r]\frac{1}{r}E^1 \wedge E^2 + [\partial_\theta(r^2\sin^2\theta\, V^\varphi) \\ &\quad - \partial_\varphi(r^2 V^\theta)]\frac{1}{r^2\sin\theta}E^2 \wedge E^3 + [\partial_\varphi V^r \\ &\quad - \partial_r(r^2\sin^2\theta\, V^\varphi)]\frac{1}{r\sin\theta}E^3 \wedge E^1 \end{aligned}$$

so

$$\begin{aligned} *\mathrm{d}\widetilde{V} &= [\partial_r(r^2 V^\theta) - \partial_\theta V^r](1/r)E^3 + [\partial_\varphi V^r \\ &\quad - \partial_r(r^2\sin^2\theta\, V^\varphi)]1/(r\sin\theta)E^2 + [\partial_\theta(r^2\sin^2\theta\, V^\varphi) \\ &\quad - \partial_\varphi(r^2 V^\theta)]1/(r^2\sin\theta)E^1. \end{aligned}$$

The orthonormal components of $*\mathrm{d}\widetilde{V}$ once again provide the classical component expression of the curl of V, here in polar coordinates.

The maps $*$ and $\sim$ also give a correspondence between vector fields and 0-forms on $\mathbb{R}^3$. The 0-form div V associated with a vector field V is defined by

$$(\text{div}\, V) = *\mathrm{d}*\widetilde{V}.$$

In a Cartesian chart

$$*\widetilde{V} = V^1\mathrm{d}x^2 \wedge \mathrm{d}x^3 + V^2\mathrm{d}x^3 \wedge \mathrm{d}x^1 + V^3\mathrm{d}x^1 \wedge \mathrm{d}x^2$$

$$\mathrm{d}*\widetilde{V} = (\partial_1 V^1 + \partial_2 V^2 + \partial_3 V^3)\mathrm{d}x^1 \wedge \mathrm{d}x^2 \wedge \mathrm{d}x^3.$$

But in this case $*1 = \mathrm{d}x^1 \wedge \mathrm{d}x^2 \wedge \mathrm{d}x^3$ so

$$*\mathrm{d}*\widetilde{V} = \partial_1 V^1 + \partial_2 V^2 + \partial_3 V^3.$$

Exercise B1

Compute div V in the polar chart above.

Thus the operations of grad, curl and div in $\mathbb{R}^3$ are seen to correspond to the application of the exterior derivative d to 0, 1 and 2 forms respectively followed by the metric correspondence relating such forms to their metric duals. It is a worthwhile exercise to verify the vector analysis identities

$$\text{grad}\,(fh) = (\text{grad}\, f)h + f(\text{grad}\, h)$$

$$\text{curl}\,(f\boldsymbol{v}) = (\text{grad}\, f) \times \boldsymbol{v} + f(\text{curl}\, \boldsymbol{v})$$

$$\text{div}\,(f\boldsymbol{v}) = g(\text{grad}\, f, \boldsymbol{v}) + f\, \text{div}\, \boldsymbol{v}$$

$$\text{div}\,(\boldsymbol{v} \times \boldsymbol{u}) = g(\boldsymbol{v}, \text{curl}\, \boldsymbol{u}).$$

by associating differential forms of the appropriate degree with the functions f, h and vectors $\boldsymbol{u}$, $\boldsymbol{v}$. These relations all follow from the properties of the Hodge map, the Leibnitz rule for d and its nilpotency, $\mathrm{d}^2 = 0$.

By composing the operator $*\mathrm{d}$ with itself one obtains a higher-order differential operator on forms. If $f \in \mathcal{F}(\mathbb{R}^3)$ then in a Cartesian chart

$$*\mathrm{d}f = \partial_1 f\, \mathrm{d}x^2 \wedge \mathrm{d}x^3 + \partial_2 f\, \mathrm{d}x^3 \wedge \mathrm{d}x^1 + \partial_3 f\, \mathrm{d}x^1 \wedge \mathrm{d}x^2$$

$$*\mathrm{d}*\mathrm{d}f = (\partial_1^2 + \partial_2^2 + \partial_3^2)f$$

this being the Laplacian operator on the function f. The Hodge map affords us an efficent way to calculate the Laplacian in any chart. The trick is to express forms in a coordinate (or natural) coframe prior to the action of d thus exploiting $\mathrm{d}^2 = 0$ for each natural basis form, but to revert to the orthonormal co-frame prior to taking a Hodge dual. For example, in any polar chart

$$\mathrm{d}f = \partial_r f\, \mathrm{d}r + \partial_\theta f\, \mathrm{d}\theta + \partial_\varphi f\, \mathrm{d}\varphi$$

$$= \partial_r f E^1 + (1/r)\partial_\theta f E^2 + 1/(r \sin\theta)\partial_\varphi f E^3$$

$$*\mathrm{d}f = \partial_r f E^2 \wedge E^3 + (1/r)\partial_\theta f E^3 \wedge E^1 + 1/(r\sin\theta)\partial_\varphi f E^1 \wedge E^2.$$

Or, reverting to a natural basis,

$$*\mathrm{d}f = \partial_r f r^2 \sin\theta \,\mathrm{d}\theta \wedge \mathrm{d}\varphi + \sin\theta \partial_\theta f \,\mathrm{d}\varphi \wedge \mathrm{d}r + (1/\sin\theta)\partial_\varphi f \,\mathrm{d}r \wedge \mathrm{d}\theta.$$

Now apply d taking notice of the fact that $\mathrm{d}\theta \wedge \mathrm{d}\theta = 0$ etc:

$$\mathrm{d}*\mathrm{d}f = \left(\partial_r(r^2 \sin\theta \partial_r f) + \partial_\theta(\sin\theta \partial_\theta f) + \frac{1}{\sin\theta}(\partial_\varphi^2 f)\right)\mathrm{d}r \wedge \mathrm{d}\theta \wedge \mathrm{d}\varphi.$$

But $*(\mathrm{d}r \wedge \mathrm{d}\theta \wedge \mathrm{d}\varphi) = 1/(r^2\sin\theta)*(E^1 \wedge E^2 \wedge E^3) = 1/r^2\sin\theta$. Thus finally

$$*\mathrm{d}*\mathrm{d}f = \frac{1}{r^2}\partial_r(r^2\partial_r f) + \frac{1}{r^2\sin\theta}\partial_\theta(\sin\theta\,\partial_\theta f) + \frac{1}{r^2\sin^2\theta}\partial_\varphi^2 f.$$

The notion of a Laplacian can be generalised to an operator on p-forms, in which case it is usually called more generally the Laplace–Beltrami operator. If $\alpha \in \Gamma\Lambda_p(U)$ then $\Delta\alpha \in \Gamma\Lambda_p(U)$ is defined in Euclidean 3-space by

$$\Delta\alpha = (-1)^{p+1}(\mathrm{d}*\mathrm{d}* - *\mathrm{d}*\mathrm{d})\alpha$$

which reduces to the above Laplacian on 0-forms. The components of the Laplace–Beltrami operator on a 1-form give the 'vector Laplacian'.

Many physical theories are formulated in terms of tensor fields satisfying field equations. Such field equations often arise as the result of setting to zero certain forms constructed out of d and $*$ and other differential forms. For instance, the static Newtonian gravitational field in Euclidean 3-space devoid of matter is described in terms of a real function Φ on $\mathbb{R}^3$ subject to the equation $\mathrm{d}*\mathrm{d}\Phi = 0$ or, after applying $*$

$$\Delta\Phi = 0.$$

Solutions to this equation define a vector field $X = \widetilde{\mathrm{d}\Phi}$ called the Newtonian gravitational field. The integral curves of X describe lines of gravitational force. A massive (test) particle experiences 'Newtonian acceleration' in the direction determined by X. To describe in more detail the interaction of this field with massive particles requires a formulation of Newton's laws of motion. Surprisingly we must wait until Chapter 6 before the notion of particle acceleration is defined. Suffice to say here that a massive particle is endowed with a parameter m, its inertial mass, such that it experiences the Newtonian gravitational 'force' $m\widetilde{\mathrm{d}\Phi}$. A smooth distribution of matter can generate a Newtonian gravitational field. If the distribution is specified by the mass density 0-form $\rho \in \mathcal{F}(\mathbb{R}^3)$, it acts as a source of Newtonian gravity according to Poisson's equation:

$$\mathrm{d}*\mathrm{d}\Phi = \rho*1.$$

(NB Both sides of this equation $\in \Gamma\Lambda_3(\mathbb{R}^3)$.)

Exercise B2

Obtain in the $\mathbb{R}^3$ cylindrical polar chart with coordinates (r, φ, z) and orthonormal co-frames $e^1 = \mathrm{d}r$, $e^2 = r\mathrm{d}\varphi$, $e^3 = \mathrm{d}z$ the component equation for the Newtonian potential Φ,

$$(1/r)\partial_r(r\partial_r\Phi) + (1/r^2)\partial_\varphi^2\Phi + \partial_z^2\Phi = \rho.$$

References

[1] Albert A 1961 *Structure of Algebras* (*Am. Math. Soc. Coll. Publ.* vol 24

[2] Dickson L 1960 *Linear Algebras (Cambridge Tracts)* (Cambridge: Cambridge University Press)

[3] Kochendorffer R 1981 *Introduction to Algebra* (Groningen: Wolters-Noordhoff)

[4] Jauch J M and Rohrlich F 1959 *The Theory of Photons and Electrons* (New York: Addison-Wesley)

[5] Cartan E 1966 *The Theory of Spinors* (Cambridge, MA: MIT Press)

[6] Chevalley C 1954 *The Algebraic Theory of Spinors* (New York: Columbia University Press)

[7] Budinich P and Dabrowski L 1985 *Math. Phys.* **10** L7
Budinich P and Trautman A 1986 *Lett. Math. Phys.* **11** 315

[8] van Nieuwenhuizen P 1983 An introduction to simple supergravity and the Kaluza–Klein program, in *Relativity and Topology II (Les Houches) 1983* (Amsterdam: North-Holland) pp 825–932

[9] Penrose R and Rindler W 1986 *Spinors and Space-time* vol 2 (Cambridge: Cambridge University Press)

[10] Adams J 1981 in *Superspace and Supergravity* ed S W Hawking and M Rocek (Cambridge: Cambridge University Press)

[11] Komar A 1959 *Phys. Rev.* **113** 934

[12] Hawking S and Ellis G 1973 *The Large Scale Structure of Space–Time* (Cambridge: Cambridge University Press)

[13] Misner C, Thorne K and Wheeler A 1973 *Gravitation* (San Francisco: W H Freeman)

[14] Dereli T and Tucker R W 1982 *Phys. Lett.* **110**B 206

[15] Brans C and Dicke R H 1961 *Phys. Rev.* **124** 925
Dicke R H 1962 *Phys. Rev.* **125** 2163

[16] Darwin C G 1928 *Proc. R. Soc.* **118** 654

[17] Ivenko D and Obukhov Y 1985 *Ann. Phys., Lpz* **42** 59

[18] Kahler E 1962 *Rend. Mat.* **21** 425

[19] Dirac P A M 1928 *Proc. R. Soc.* **117** 610, **118** 341

[20] Benn I M and Tucker R W 1983 Fermions without spinors *Commun. Math. Phys.* **89** 341

[21] Duffin R J 1938 *Phys. Rev.* **54** 1114
Kemmer N 1939 *Proc. R. Soc.* A **173** 91

[22] Milnor J W 1963 *Enseignement Math.* **9** 198

[23] Kobayashi S and Nomizu K 1963 *Principles of Differential Geometry* (New York: Interscience)

[24] Benn I M and Tucker R W 1986 in *Geometry and Spinors, Trieste Conf. 1986, Representing Spinors with Differential Forms*

[25] Rarita W and Schwinger J 1941 *Phys. Rev.* **60** 61

[26] Lichnerowicz A 1964 *Bull. Soc. Math. France* **92** 11

[27] Hughston L P, Penrose R, Sommers P and Walker M 1972 *Commun. Math. Phys.* **27** 303–8

[28] Duff M J, Nilsson B and Pope C N 1986 *Phys. Rep.* **130** 1–142
Nilsson B 1986 *Class. Quantum Grav.* **3** 141–5

[29] Cahen M, Gott A, Lemaire L and Spindel P 1986 Killing spinors, in *Geometry and Physics, Trieste Conf. 1986*

[30] Lichnerowicz A 1986 Killing spinors according to O Hijazi, and applications, in *Geometry and Physics, Trieste Conf. 1986*

[31] Hitchin N 1974 *Adv. Math.* **14** 1–55

[32] Yau T 1978 *Commun. Pure Appl. Math.* **31** 339–411

[33] Shanahan P *The Atiyah-Singer Index Theorem. An Introduction* (*Springer Lecture Notes in Mathematics* vol 638)

Index

Abelian, 308
Acceleration, 203
Adjoint involutions, 67, 71
Algebra, 307, 316
Almost complex structure, 303
Alt, alternating map, 5
Angular momentum, 197
Anti-automorphism (algebra), 319
Anticommuting spinors, 103
Antisymmetric, 4
  tensor gauge fields, 260
Atiyah–Singer index, 306
Atlas, 130
Automorphism, 3
  group, 119, 308, 313
Autoparallel, 202

Basis (vector space), 311
Bianchi's first identity, 213
Bianchi's second identity, 213
Bijective, 309, 312
Bilinear covariants, 93
Bilinear form, 314
Bispinor, 100
Boost, 186
  orbit, 187
Boundary, 125, 168
Brans–Dicke theory, 250

Calabi–Yau, 306
Central algebra, 316
Centre (ring), 310
Centre, 308
Chain rule, 135
Characteristic
  field, 310
  zero, 310
Charge
  conjugate spinor, 96
  conjugation (Dirac spinor), 287
  conjugation (of spinor fields), 266
  electric, 190
Charged scalar field, 241
Chart transformations, 131
Chiral spinor, 97
C^k map, 129
Christoffel symbols, 222
Clifford
  2-forms, 252, 253
  algebra, 23
  algebra, (complexified), 60, 80
  commutator, 50, 107
  group, 42
  group (Lie algebra of), 51
  product (relation to exterior product), 24
  sub-bundles, 276, 306
  subgroups, 46, 71
Clock, 183
Closed forms, 188
Closed sets, 125
Co-derivative, 189
Coherence (on overlaps), 263
Co-homologous, 188
Commutative, 308
  ring, 310

Commutator
of Lie and covariant derivative, 231
of Lie and spinor covariant derivative, 273
Complete vector field, 158
Complex
conjugation, 41, 81, 95
structure, 116, 315
structure (on spinor space), 59
vector space, 315
Complexification, 315
Complexified Clifford algebra, 60, 80
Components (vector), 311
Conformal
2-forms, 226
group, 192
isometry, 191
Killing vector, 231
symmetry (of Maxwell's equations), 192
tensor, 226
Conformally
flat, 227
related, 225
Conjugate
linear map, 315
space, 44
Connection
1-forms, 200, 207
components, 200
Conservation laws, 237
Conserved currents (Dirac equation), 280
Constant curvature, 225
Continuous
function, 125
map, 129
Contracted Bianchi identities, 219
Contraction map (on tensors), 17
Contragradient, 17
degree, 16
Contravariant, 141
degree, 16
Coordinate
basis, 143
chart, 130
Coset, 308
Cotangent bundle, 147
Coulomb solution, 190, 193
Covariances of Dirac equation, 280
Covariant derivative, 200, 206
of spinor fields, 267
of tensor spinors, 296
of tensors, 199
Covariant degree, 16
Covariant differentiation (Clifford forms), 252
Covariant differential, 207
Covariant exterior derivative, 216
Cross product, 344
Curl, 345
Curvature, 199
constant, 225
forms, 209
operator, 209
operator (of spinor), 271, 279
operator as Clifford commutator, 253
scalar, 219
tensor, 208
Curve, 134

Decomposable, 3, 8
Degree, 2, 313
of tensor, 2, 16
Degree, (s) group of, 313
Derivation, 4, 127
Diffeomorphism, 129, 132
Differentiable
manifold, 129
map, 129
structure, 131
Differential form, 146
Dimension, 311, 316
Dirac
adjoint spinor, 92
equation, 278, 282
matrices, (see gamma matrix)
operator, 278
spinors, 92, 104
stress tensor, 290
Direct product (group), 309
Direct sum, 3
algebra, 317
vector space, 312

Directional derivative, 138, 344
Divergence, 221, 346
  of Maxwell stress tensor, 256
Division
  algebra, 316
  ring, 310
Dominant energy condition, 237
Dual space, 313
Duality rotation, 116
Duffin–Kemmer–Petiau equations, 260

Eddington–Finkelstein coordinates, 248
Einstein
  $(n-1)$-forms, 220
  field equations, 234, 236, 247
  –Maxwell system, 240
  space, 222
  summation convention, 311
  tensor, 220
  –Yang–Mills system, 240
  –Kahler stress tensor, 259
Electric charge, 190
Electrically charged fluids, 242
Electromagnetic radiation, 185
Electron, 181
Endomorphism, 313
Energy, 186
Energy conditions on the stress tensor, 236
Equivalent
  involutions, 337
  representation, 316, 321
Eta (η)
  on Clifford algebra, 23
  on exterior algebra, 7
Euclidean
  manifolds, 174
  vector space, 123
Even subalgebra, 39, 80
Exact, 188
Exponential map, 203
Exterior
  algebra (as quotient of tensor algebra), 5
  derivative, 154
  p-form, 5
  product, 5
External direct sum, 315
  bundle, 151

f-related vector fields, 143
Faces, 167
Faithful representation, 316, 321
Falling freely, 177
Fermi–Walker or $\boldsymbol{F}$-connection, 234
Fibre, 145
Field, 310
  algebraically closed, 310
  characteristic of, 310
Fierz rearrangement, 98, 285
First structure equation, 208
'Flag' (null flag), 116
Flux, 195
Frame, 311

Galilean
  group, 177
  -relativistic, 176
Gamma (γ) matrix, 37, 86
Gauge invariance of electromagnetism, 188
General linear group, 311
Generalised spinor structure, 263
Generators, 308
  algebra, 317
  of a subgroup, 308
  of a vector subspace, 312
Geodesics, 203
Germ, 137
Graded
  algebra, 316
  subspace, 313
  vector space, 313
Gradient, 344
Gravitation with torsion, 249
Gravitational mass, 206
Gravitational waves (with neutrinos), 293
Group, 307
  representation, 316
Gyroscopes, 234

H-module, 60
Harmonic, 190
Hausdorff, 126

Hermitian, 41, 63, 84, 87, 90, 269, 300
  conjugate, 92
Hodge de Rham operator, 254
Hodge map, 13, 15, 173, 180
  and Clifford products, 28
Homeomorphism, 126
Homogeneous
  elements of a graded vector space, 313
  linear map, 313
Homogenous, 2
Homologous, 191
Homomorphism
  algebra, 318
  group, 308
Horizon, 248

Ideal, 10, 23, 317
  fluid, 242
  observer, 183
Ideal of an algebra, 317
Ideal, single sided, 323
Idempotent, 324
Identity, 307
  ring, 310
Imbedded (submanifold), 133
Imbedding, 133
Immersion, 133
Index
  of inner product, 66, 76, 85
  of nilpotent element, 323
Inequivalent involutions, 68
Inertial
  chart, 184
  mass, 347
  reference systems
Infeld, 99
Injective, 309, 312
  tangent map, 133
Inner, outer, 309
Inner automorphism
  algebra, 318
  group, 309
Inner products (on spinor fields), 264
Instantaneous, 185
Integral curve, 157
Integration, 167
Interior derivative, 4
  on Clifford algebra, 23
  on exterior forms, 9
Interior multiplication, 11
Intrinsic spin, 290
Invariance group, 314
Invariant subgroup, 308
Invertible element (ring), 310
Involutions, 4, 336
Involutary anti-automorphism (*see also* ξ), 4
Involution
  classification of involutions in the real Clifford algebras, 78
  equivalence of, 337
  inequivalent involutions of real algebras, 68
  on tensor product of algebras, 72
Irreducible representation, 316, 321
Isometry, 173
Isomorphism
  algebra, 318
  group, 309
Isotropic
  coordinates, 247
  subspace, 106

Jacobi identity, 142
Jacobian, 128

Kahler
  2-form, 305
  equation, 256
  manifold, 304
Kernel, 309, 312, 318
Killing
  currents, 196
  spinor, 300
  vector, 174
Killing's equation, 229
Klein–Gordon field, 239
Komar form, 239

Laplace–Beltrami operator, 189, 254
Laplacian operator on spinors, 279
Left and right duals, 229
Left coset, 308
Left ideal, 323

Left R-module, 341–2
Length (of a curve), 183
Levi–Civita antisymmetric symbol, 15
Lichnerowicz theorem, 299
Lie algebra of Clifford group, 51
Lie-algebra-valued p-forms, 240
Lie bracket
Lie derivative
  on spinors, 271
  on tensors, 161
Light-cone, 181
Linear
  connection, 200
  dependence, 311
  frame, 311
  map, 312
  quotient space, 312
  space of linear maps, 313
  transformation, 313
Local frame, 172
Locally symmetric space, 303
Lorentz force law, 243
Lorentzian
  Clifford algebra, 85, 113
  connection, 232
  manifold, 172
Lorenz gauge, 190
Lowering convention, 314

Majorana conjugate spinor, 95
Majorana spinor, 96, 104, 115
Majorana–Weyl spinor, 97, 104
Mass–energy, 185
Maximal
  integral curve, 158
  isotropic subspace, 107
Maxwell stress (Clifford form), 255
Maxwell stress tensor, 194, 197
Maxwell's equations, 178, 181, 188
  Clifford form, 255
Metric, 314
  compatible, 214
  compatible connection forms, 215
  dual, 14, 314
  on p-forms, 14, 27
  tensor field, 171
  topology, 126
Minimal left ideal, 55
Minkowski spacetime, 181–2
Mixed tensor, 16
Module, 340
Momentum, 186
Multi-index, 9, 27
Multilinear, 2, 16
Multipole, 191

n-form, 10
Natural
  basis, 143
  dual basis, 313
  local basis,
Neighbourhood, 124
Neutrino waves (with gravity), 293
Newtonian
  acceleration, 205, 206
  angle, 186
  gravitational coupling, 247
  length, 186
  potential, 206
  velocity, 185
Nilpotent, 323
Norm homomorphism, on Clifford group, 46
Non-associative algebra, 119
Non-degenerate metric, 314
Non-nilpotent algebra, 324
Non-rotating frame, 234
Normal
  coordinates, 203
  neighbourhood, 203
  subgroup, 308

Odd dimensions, 89, 92
  of a group, 309
  of a linear space, 313
  of an algebra, 318
One-parameter diffeomorphism, 156
Open set, 125
Opposite algebra, 4, 319
Or a ring, 310
Orbital angular momentum, 290
Order, 2, 307
Ordinary matrix algebra, 320
Orientation, 14, 132

Oriented
  r-chain, 168
  r-cube, 167
Orthochronous transformations, 47
Orthogonal
  group, 42, 314
  idempotent, 325
Orthonormal basis, 314
Outer automorphism (group), 310

p-form, 5
Parallel, 201
  along a curve, 201
  spinor, 303
  transport map, 202
  vector field, 202
Parallelism, 199
Parametrise, 171
  curve, 134
Parity-preserving orthogonal transformations, 47, 49
Period (of an automorphism), 119
Photons, 185
Physical dimensions, 178
Pierce decomposition, 325
Pin groups, 46
  example of Pin(3, 1), 53
Pinor structure, 263
Plane-wave basis (for Dirac equation), 288
Poincaré, 178
  group, 182
Polarities, 179
Potential, 188
Primitive idempotent, 325
Principal idempotent, 325
Proca field, 240
Product manifold, 145
Projection operator, 26
Proper time, 183
  parametrisation, 183
Pseudo-Riemannian, 172
  connection, 221
Pullback, 133
  map, on functions, 133
  on forms, 148
Pure spinors, 106, 108

Quantum theory, 282
Quotient algebra, 10, 25
Quaternion
  conjugation, 65, 73, 338
  reversion, 339
Quaternions, 338
Quotient
  algebra, 318
  group, 308

R-module, 340
Racah time reversal, 94
Radical, 324
Raising and lowering conventions, 19
Rank, 2, 312
  of an idempotent, 334
  of tangent map, 133
Rank-two spinor, 103
Rarita–Schwinger equations, 296
Reducible
  algebra, 317
  representation, 119, 316, 321
Reflections, 43
Regular element (ring), 310
Regular representation (algebra), 321
Reissner–Nordström solution, 243
Representation
  equivalent, reducible, faithful, 316
  of an algebra, 321
  of a group, 316
Representative, 308
  spinor, 108
Representing spinors, 275
Reversion (quaternions), 338
Ricci
  1-forms, 210
  tensor, 210
Riemannian, 172
Ring, 310
Rotational isometry, 174

Scalar field, 239
Schwarzschild metric, 247
Second structure equation, 209
Section, 146
  of a tangent bundle, 146
Sectional curvature, 223

Semi-direct product, 51
  group, 310
Semi-orientation, 48
Semi-simple (algebra), 326
Semi-spinor representation, 55
Semi-spinors, 97
Signature, 314
Simple (algebra), 327
Smooth manifold, 131
Spacetime, 181
Span, 311
Spatial direction, 185
Special orthogonal group, 45
Spherical harmonics, 258
Spin^C
  manifold, 306
  structure, 264
Spin groups, 46
  example of spin(3, 1), 53
Spin-invariant products, 62
Spin manifold, 262
Spinor
  bundle, 261
  covariant exterior derivative, 297
  field, 262
  frame, 263, 293
  Laplacian, 279
  representation, 55
  structure, 262
Spinors, 54
Standard spinor frames, 263
Star map (*see* Hodge map)
Static metric, 244
Stationary, 184
  metric, 244
  observer, 184
Stokes's theorem, 169
Stress energy tensor, 236
Stress tensor
  Dirac, 290
  fluids, 242
  Kahler, 259
  Klein–Gordon, 239
  Maxwell, 194
  Proca, 240
  Yang–Mills, 240
Strong energy condition, 237
Structure
  constants, 174
  equations, first, 208
  equations, second, 209
  functions, 215, 280
Subalgebra, 316
Subgroup, 308
Submanifold, 133
Sum (vector space), 312
Summation convention, 311
Supergravity, 249, 296
Supersymmetry, 283, 301
Surjective, 309, 312
Symmetric metric, 314
Symmetrisation, 4
$\mathscr{S}_p$ (*see* projection operators)

Tangent, 136
  bundle, 143
  map, 138
  plane, 223
  space, 127, 137
  vector, 136, 142
Tensor, 2
  algebra, 2
  algebra (mixed), 16
  field, 150
  product (of algebras), 319
  spinors, 294
  the group of all, 309
Time reversal (on spinors), 49
Topological
  manifold, 124, 127
  space, 124
  subspace, 125
Topology, 125
Torque, 197
Torsion 2-forms, 208
Torsion tensor, 208
Total matrix algebra, 320
Trace
  in Clifford algebra, 91
  of a tensor, 18
  theorems, 91
Translational isometry, 174
Translations, 182
Triality, 106, 117, 120

Twisted vector representation, 45
Twistor, 299
  equation, 298
Two-component formalism, 99

$U(1)$
  covariant derivative, 241
    of spinor, 270
  exterior covariant derivative, 241
Unit element (ring), 310
Units, 181

Valence, 103
van der Waerden formalism, 99
Vector analysis in Euclidean 3-space, 342
Vector
  field, 141
  representation, 42
    twisted, 45
  space, 310
  subspace, 312
Volumn form, 14

Weak energy condition, 237
Wedderburn (structure theorems), 337
Weyl equations, 279
Weyl
  spinor, 97, 100, 108
  tensor, 226
Wigner time reversal, 94, 288
Witt basis, 107
Witt index, 66
World line, 183

Xi (ξ)
  the involutory anti-automorphism, 4
  the involution on exterior algebras, 8
  the involution on Clifford algebras, 23

Yang–Mills field, 240

$\mathbb{Z}$(mod 2), 2, 22, 47